DATE DUE

~~AP 17 '04~~			

CHILDBIRTH

Changing Ideas and Practices in Britain and America 1600 to the Present

Series Editor

PHILIP K. WILSON
Truman State University

Assistant Editors

ANN DALLY
Wellcome Institute for the History of Medicine (London)

CHARLES R. KING
Medical College of Ohio

A GARLAND SERIES

Series Contents

VOLUME

1

MIDWIFERY THEORY AND PRACTICE

Edited with introductions by

PHILIP K. WILSON

Truman State University

GARLAND PUBLISHING, INC.
New York & London
1996

Library of Congress Cataloging-in-Publication Data

Childbirth : changing ideas and practices in Britain and America 1600 to the present / edited with introductions by Philip K. Wilson.

p. cm.

Includes bibliographical references.

Contents: v. 1. Midwifery theory and practice — v. 2. The medicalization of obstetrics: personnel, practice, and instruments — v. 3. Methods and folklore — v. 4. Reproductive science, genetics, and birth control — v. 5. Diseases of pregnancy and childbirth.

ISBN 0-8153-2230-5 (v. 1 : alk. paper). — ISBN 0–8153–2231–3 (v. 2 : alk. paper). — ISBN 0–8153–2232–1 (v. 3 : alk. paper). — ISBN 0–8153–2233–X (v. 4. : alk. paper). — ISBN 0–8153-2234–8 (v. 5 : alk. paper)

1. Childbirth—United States—History. 2. Childbirth—Great Britain—History. 3. Obstetrics—United States—History. 4. Obstetrics—Great Britain—History. I. Wilson, Philip K., 1961–.

[DNLM: 1. Obstetrics—trends. 2. Midwifery—trends. 3. Pregnancy Complications. 4. Reproduction Techniques. 5. Genetic Counseling. 6. Contraception. WQ 100C5356 1996]

RG518.U5C47 1995
618.4'0973—dc20
DNLM/DLC
for Library of Congress

96–794
CIP

Printed on acid-free, 250-year-life paper
Manufactured in the United States of America

Contents

List of Illustrations

Series Introduction

Since "most women are interested in the process of giving birth and all men have been born," it would appear, claimed Johns Hopkins University obstetrician Alan Guttmacher, that the topic of childbirth would above all other topics have "universal appeal." Birth is also one of the most individual moments in each of our lives, but although we all share the experience of being delivered, the processes of delivery have been diverse. The social gathering around the childbed common in earlier times has, for many, been replaced by a more isolated hospital bed. Maternal fears of the pain and peril of procreation have, or so prevalent historiography would have us believe, intensified with the intervention of male midwives and obstetricians bringing along new "tools" of the trade. Markedly divergent beliefs about assisting in labor have created polarized factions of attendants. Some have followed wisdom similar to what Britain's Percivall Willughby first espoused in 1640:

> Let midwives observe the ways and proceedings of nature for the production of her fruit on trees, or the ripening of walnuts or almonds, from their first knotting to the opening of the husks and falling of the nut . . . . These signatures may teach midwives patience, and persuade them to let nature alone perform her work.

Opposing factions adhered to claims similar to that of the early nineteenth-century Philadelphia midwifery professor, Thomas Denman, that belief in:

> labour, being a natural act, . . . not requiring the interference of art for either its promotion or its accomplishment . . . has, from its influence, retarded, more perhaps than any other circumstance, the progress of improvement in this most important branch of medical science.

Other comparisons among midwifery writings suggest that although expectant women may no longer avoid the same "longings and cravings" of pregnancy as did their eighteenth-century forebears, contemporary concern about exposing pregnant women and their fetuses to nicotine, alcohol, known teratogenic agents, and unwarranted stress evokes similar warnings. Indeed, as the works included in this collection illustrate, many similar concerns have been shared by expectant mothers and their labor attendants for centuries.

Although there is a substantial literature on childbirth, it typically lacks the full medical, historical, and social contextualization that these volumes provide to readers. This series attempts to fill the gap in many institutions' libraries by bringing together key articles illuminating a number of issues from different perspectives that have long concerned the expectant mother and the attendants of her delivery regarding the health of the newborn infant. Primary and secondary sources have been culled from British and American publications that focus on childbirth practices over the past three hundred years. Some represent "classic" works within the medical literature that have contributed towards a more complete understanding of pertinent topics. The series draws from historical, sociological, anthropological, and feminist literature in an attempt to present a wider range of scholarly perspectives on various issues surrounding childbirth.

Childbirth: Changing Ideas and Practices is intended to provide readers with key primary sources and exemplary historiographical approaches through which they can more fully appreciate a variety of themes in British and American childbirth, midwifery, and obstetrics. For example, general historical texts commonly claim that childbed (puerperal) fever, a disease that has claimed hundreds of thousands of maternal lives, provoked much fear throughout most of British and American history. In addition to supplying readers with historians' interpretations, *Childbirth: Changing Ideas and Practices* also provides discussions of the causes and consequences of particular cases of childbed fever taken directly from the medical literature of the nineteenth and twentieth centuries, thereby enabling a better understanding of how problematic this disorder initially was to several key individuals who, after first increasing its incidence, ultimately devised specific methods of its prevention.

The articles in this series are designed to serve as a resource for students and teachers in fields including history, women's studies, human biology, sociology, and anthropology. They will also meet the socio-historical educational needs of pre-medical and nursing students and aid pre-professional, allied health, and midwifery instructors in their lesson preparations. Beyond the content of many collections on the history of childbirth, readers frequently need access to the primary sources in order to develop their own interpretive accounts. This five-volume series expands the readily accessible knowledge base as it represents both actual experiences and socio-historical interpretations on select developments within the history of British and American childbirth, midwifery, and obstetrics.

Given the vast and expanding literature on childbirth, it is virtually impossible anymore for any single source to provide a complete coverage of such a broad topic. Selecting precisely what articles to include has been, at times, a painstaking process. We have purposefully excluded works on abortion as many of these articles have recently been reprinted elsewhere. Additionally, we have only touched upon midwifery/obstetrical education, the legal issues surrounding childbirth, marriage, sex, and the family, and genetic engineering since numerous contemporary works in print thoroughly discuss these themes. Seminal articles that are currently available in other edited collections as well as general review articles were, with a few exceptions, not considered for reprinting in this series. There are several areas, including eclampsia, the development and role of the placenta, pregnancy tests throughout history, and Native American childbirth practices, for which suitable articles are wanting. Related topics such as gynecology and gynecological diseases, pediatrics, neonatology, postnatal care and teratology, though of considerable concern to many pregnant women and health care providers, appear beyond the scope of our focus and the interest of our generalist readers. Space did not allow for me to cover childbirth from the viewpoint of what have historically been considered alternative or complementary healing professions such as herbalism, homeopathy, or osteopathy, even though thousands of healthy children have been delivered by practitioners in these professions. The exorbitant permission charges that some journals charge for reprinting their articles has prohibited us from including many important articles. Finally, we have opted not to reprint biographical articles as the typical lengthy accounts of individuals would have precluded addressing more general relevant issues.

Series Acknowledgments

I am grateful to the many individuals who offered their assistance, suggestions, and support throughout the gestation of this project. Foremost, I wish to thank my co-editors, Dr. Charles R. King and Dr. Ann Dally, both highly valued "team players" in what truly became an international collaborative creation. Their medical expertise and historiographical suggestions strengthened the content of this series. Laura Runge, my undergraduate research student and Ronald E. McNair Post-Baccalaureate Achievement Program Scholar, provided exemplary editorial assistance throughout the growth of this project. In addition, she introduced Melissa Blagg-Holcomb to our team, a truly exceptional undergraduate scholar, without whom this project would not have been completed in such a timely manner. Melissa's professional interest in nurse midwifery expanded the scope of the literature we reviewed. Our research would have been impossible without the assistance of many librarians, archivists, and other members of the research staff. In particular, I wish to thank Lyndsay Lardner (The Wellcome Institute, London), Susan Case (Clendening Medical History Library, Kansas City), and Janice Wilson (Hawaii Medical Library, Honolulu; Sterling Medical Library, New Haven, and Pickler Memorial Library, Kirksville) for their exemplary library assistance. The unfailing efforts of Sheila Swafford, in Pickler Library's Reference Department, to secure necessary material are deeply appreciated. The editors also wish to thank Jane Carver, Prof. Mark Davis, Prof. Robbie Davis-Floyd, Nancy Dellapenna, Clare Dunne, Prof. Paul Finkelman, Andy Foley, Dr. Denis Gibbs, Ferenc Gyorgyey, Gwendolyn Habel, Jack Holcomb, Charlene Jagger, Maggie Jones, Carol Lockhart, Barb Magers, Andrew Melvyn, Jean Sidwell, Prof. John Harley Warner, Prof. Dorothy C. Wertz and the staffs of the Library of the Royal Society of Medicine (London), the National Library of Medicine (Bethesda), Pickler Memorial Library (Kirksville) and Rider Drug and Camera (Kirksville) for their assistance in preparing certain parts of this series. Leo Balk of Garland Publishing, Inc., proved to be a stable sounding board during the conception stage of *Childbirth*, a role that Carole Puccino has deftly carried on throughout the later progressions of this work. I also wish to thank my colleagues at the University of Hawaii-Manoa, Yale University, and Northeast Missouri State University (soon to be Truman State University) for their support and critical commentary on this project. Northeast Missouri State University provided a Summer Faculty Research Grant which allowed for the timely completion of this project. Finally, I remain indebted to my wife, Janice, for providing astute critique, able reference library assistance, and continual support and encouragement.

Philip K. Wilson

Introduction

Prior to the Early-Modern period, midwifery was principally a female-centered event. Parturient women were, in most Western cultures, attended by "neighborhood women" and female relatives who delivered children as part of their family, social and medical responsibilities for the community. The paucity of written records or manuals of practice before 1545 suggests that much of the early "art" of midwifery was typically passed along through oral tradition.

Byrth of Mankynde; or, The Woman's Booke, first published in 1540, was the first work in the English vernacular devoted to midwifery. This work, as discussed in the following article by Ballantyne, ran through thirteen editions, the last in 1654. It was claimed that gentlewomen, visiting women in confinement (during labor) carried "this booke in theyr handes, and causynge suche part of it as doth chiefly concerne the same pourpose, to be red before the mydewife, and the rest of the wemen then beying present."[1]

Additional midwifery writings appeared in the English vernacular before 1700 by authors including C.R. Both, Jane Sharp, Eduuard Poeton, William Sermon, Percivall Willughby, James McMath, John Pechey, and Robert Barret. The most successful midwifery "primer" in English, however, was the anonymous *Aristotle's Master-Piece*. This work was designed, as Roy Porter has analyzed, to offer an "integrated and coherent" introduction to reproduction.[2] The subject of childbirth becomes more apparent in the mid-eighteenth century expanded edition of the work, entitled *Aristotle's Compleat Midwife*. Although these writings were not derived from the Greek philosopher, they certainly gained authority from the use of his name.

Midwifery, as Samuel Johnson defined it in his monumental eighteenth-century *Dictionary of the English Language,* was the assistance given at childbirth or the trade of a midwife. He described a midwife as "a woman who assists women in childbirth." From the historical tradition of women that assisted in parturition, midwifery would appear to have been a learned trade and perhaps, in later years, a professional calling. For much of the present century, histories of midwifery, particularly those written by obstetricians, have actually been histories of obstetrics, suggesting that obstetrics and midwifery are synonymous callings. Physician writers who have distinguished midwifery and obstetrics, with a few notable exceptions, tended to represent them as a dichotomy where obstetrics is good, clean, scientific, and lifesaving, whereas midwifery is bad, dirty, superstitious, and life-threatening.[3] This depiction is especially prominent in the literature of the "midwife debate." This debate appeared in several incarnations, beginning in the late eighteenth century, and continuing into the twentieth century, when midwives were legislated out of existence in the United States and brought under regulation by the medical profession in Britain.[4]

Exceptions to this portrayal of midwifery are found in writings of the late eighteenth and early nineteenth centuries, when "obstetrics," then called "man-midwifery," was challenged by the more traditional women midwives. Additionally, distinctions between midwifery and obstetrics continue in contemporary literature, where over the last two and a half decades the obstetricians' authority over childbearing has been challenged both by consumers and scholars.

The impetus for this recent challenge arose largely from consumers. When "natural childbirth" techniques within the physician-controlled hospital setting failed to meet some women's expectations for a humane and satisfying birth experience, they started looking for alternatives to the medicalized obstetrical approach. They rediscovered an insight of previous generations: midwifery provided a safe and satisfying birth experience. These consumers tended to be educated women, and they started challenging obstetrics in writing, with books like *The Birth Book* (1972), *Spiritual Midwifery* (1975), and *Immaculate Deception* (1975). In opposition to professionalized obstetrics, these authors presented statistical accounts supporting their perceptions and experiences that midwifery was at least as "safe" as obstetrics.

These scholars, and those like Doris Haire, Lester Dessez Hazell, Charlotte and Fred Ward, David and Lee Stewart, and Barbara Katz Rothman, have created a new demand for the services of midwives.[5] Presently, in addition to the handful of surviving "granny" midwives in the deep South, the United States has both direct-entry lay midwives and certified nurse-midwives (CNMs). Traditionally, certified nurse-midwives have cared for populations underserved by obstetricians since the Kentucky Committee for Mothers and Babies (later renamed the Frontier Nursing Service) was formed in 1925.[6] More recently, middle and upper class women in urban settings from Connecticut to Illinois and California have also been actively seeking the service of CNMs.

Defining midwifery in conformity with the medical and legal terms of obstetrics remains problematic. Midwives, according to the practical definitions developed in the course of the "midwife debates," serve healthy women during normal pregnancies and births. Generally, however, obstetricians, not midwives or consumers, have set the limits on the range of "normal" and defined the risk categories which, in turn, dictate the legal parameters of midwifery practice. In addition, midwives have faced changing expectations from mothers. As Irvine Loudon described, women in today's society expect to undergo childbirth "in perfect health after a perfect and satisfying delivery of a perfect baby."[7] When these expectations are not met, it is not unknown for resolution to occur in the courtroom.

In the United States, the established scope of midwifery practice varies from state to state. In states where midwifery is legal, it may be restricted entirely within the shadow of obstetrics: midwifery being permitted only with the supervision of a physician, or with pre-arranged physician back-up, or under the restrictions of a nurses' association that, itself, has been organized under the medical model of normal childbirth.

In the last few decades, social science writers have also started re-writing the history of midwifery. Barbara Ehrenreich, Dierdre English, Judy Barrett Litoff, Laurel Thatcher Ulrich, Judith Walzer Leavitt and others have started writing about midwifery "from below," expressing views of parturients and midwives.[8] Midwives Jean Towler and Joan Bramall have created a historical account, in part to overcome the "medical take-over" of childbirth and ensure that midwives, "both individually and collectively . . . do not lose the position won by the midwife pioneers of the past."[9]

In a community where midwifery is underground, and where medical authorities at times use rhetoric contradictory to research results, how do people—nurses, general practitioners, students, consumers—develop an informed opinion about the role of the contemporary midwife in birthing? This volume provides an in-depth starting point to examine the historical, social, and medical contexts within which midwifery has existed during the past four centuries in Britain and the United States. As Carolyn Leonard Carson recently argued, "Additional studies of midwives themselves are essential in order to understand the roles they played in their communities and within families."[10] The editors hope this collection of articles will incite further investigation into understanding the cultural and intellectual conceptions of midwifery.

This volume gathers key primary sources and exemplary historiographical approaches through which the readers can more fully appreciate the variety of midwifery theories and practices in Britain and America. The first section of the volume, devoted to British midwifery, begins with an examination of *Byrth of Mankynde; or, The Woman's Booke*, the first major work on midwifery in the English vernacular. Remaining works in this section further explore women's attitudes towards childbirth and midwives from the sixteenth through the eighteenth centuries. The second section of this volume covers the history of American midwifery from "Obstetrick Art" customs of the colonial period through parturition practices in the late twentieth century. Topics discussed include the entry of science into the birthing room, the portrayal of midwives as working class women, and regional differences in the practice of midwifery.

Philip K. Wilson

Notes

1. Audrey Eccles, *Obstetrics and Gynaecology in Tudor and Stuart England* (Kent, Ohio: The Kent State University Press, 1982), 12.
2. Roy Porter, "'The Secrets of Generation Display'd': *Aristotle's Master-piece* in Eighteenth-Century England," *Eighteenth-Century Life 9* (May 1985):5.
3. Writings typical of this genre include John Hobson Aveling, *English Midwives. Their History and Prospects* (1872; reprint edition, London: Hugh K. Elliott, 1967); Palmer Findle, *The Story of Motherhood* (New York: Liveright Publishing Corporation, 1937); Irving S. Cutter and Henry R. Veits, *A Short History of Midwifery* (Philadelphia: W.B. Saunders & Co., 1933) ; and Harvey Graham [Isaac Harvey Flack], *Eternal Eve: The History of Gynecology and Obstetrics* (Altrincham: William Heinemann, 1950).
4. Jean Donnison described the British struggles between midwives and obstetricians in *Midwives and Medical Men. A History of Inter-Professional Rivalries and Women's Rights* (New York: Schocken Books, 1977), whereas the American controversies are discussed in Judy Barrett Litoff, *The American Midwife Debate. A Sourcebook on Its Modern Origins* (Westport, Conn.: Greenwood Press, 1986).
5. Doris Hare, *The Cultural Warping of Childbirth* (Hillside, N. J.: International Childbirth Education Association, 1972); Lester Dessez

Hazell, *Birth Goes Home* (Seattle, Wash.: Catalyst Publishing Co., 1974); Charlotte and Fred Ward, The *Home Birth Book* (Washington, D.C.: INSCAPE Publishers, 1976); David and Lee Stewart, eds., *Safe Alternatives in Childbirth* (Chapel Hill, N. C.: NAPSAC, 1976); and Barbara Katz Rothman, "Woman's Body, Woman's Mind: In Which a Sensible Woman Persuades Her Doctor , Her Family, and Her Friends to Help Her Give Birth at Home," *Ms.* 5 (Dec. 1976): 25–32.

6. Mary Breckinridge, *Wide Neighborhoods: A Story of the Frontier Nursing Service* (New York: Harper and Bros., 1952).
7. Irvine S. L. Loudon, "Childbirth." In W.F. Bynum and Roy Porter, eds., *Companion Encyclopedia of the History of Medicine* (London: Routledge, 1994), vol. 2, 1069.
8. Barbara Ehrenreich and Dierdre English, *Witches, Midwives and Nurses: A History of Women Healers* (Old Westbury, N. Y.: Feminist Press, 1973); Judy Barrett Litoff, *American Midwives, 1860 to the Present* (Westport, Conn.: Greenwood Press, 1978); Laurel Thatcher Ulrich, *A Midwife's Tale. The Life of Martha Ballard, Based on Her Diary, 1785–1812* (New York: Alfred A. Knopf, 1990); and Judith Walzer Leavitt *Brought to Bed: Childbearing in America, 1750 to 1950* (New York: Oxford University Press, 1986).
9. Jean Towler and Joan Bramall, *Midwives in History and Society* (London: Croom Helm, 1986), ix,x.
10. Carolyn Leonard Carson, "Midwifery." In Peter E. Stearns, ed., *Encyclopedia of Social History* (New York: Garland Publishing, Inc., 1994), 490.

Further Reading

Apple, Rima D., ed. *Women, Health, and Medicine in America: A Historical Handbook*. New York: Garland Publishing, Inc., 1990.

Arms, Suzanne. *Immaculate Deception: A New Look at Women and Childbirth in America*. Boston: Houghton-Mifflin, 1975.

Aveling, J. H. *English Midwives: Their History and Prospects,* 1872; repr. London: Hugh K. Eliot, 1967.

Berman, Paul. "The Practice of Obstetrics in Rural America, 1800–1860." *Journal of the History of Medicine and Allied Sciences* 50 (1995):175–193.

Bogdan, Janet Carlisle. "Care or Cure? Childbirth Practices in Nineteenth-Century America." *Feminist Studies* 4 (1978): 92–99.

Borst, Charlotte G. "The Training and Practice of Midwives: A Wisconsin Study." *Bulletin of the History of Medicine* 62 (1988): 606–27.

Borst, Charlotte G. "Wisconsin's Midwives as Working Women: Immigrant Midwives and the Limits of a Traditional Occupation, 1870–1920." *Journal of American Ethnic History* 8 (Spring 1989): 24–59.

Breckinridge, Mary. *Wide Neighborhoods: A Story of the Frontier Nursing Service*. New York: Harper and Bros., 1952.

Buss, Fran Leeper. *La Partera: Story of a Midwife*. Ann Arbor: University of Michigan Press, 1980.

Campbell, Marie. *Folks Do Get Born*. New York: Garland Publishing, Inc., 1985. Originally published: New York: Reinhart and Co., 1946.

DeClercq, Eugene. "The Transformation of American Midwifery: 1975–1988." *American Journal of Public Health* 92, no. 5(1992): 690–94.

Donegan, Jane B. *Women and Men Midwives: Medicine, Morality, and Misogyny in Early America*. Westport, Conn.: Greenwood Press, 1978.

Donnison, Jean. *Midwives and Medical Men: A History of Inter-Professional Rivalries and Women's Rights*. New York: Schocken Books, 1977.

Dougherty, Molly C. "Southern Midwifery and Organized Health Care: Systems in Conflict." *Medical Anthropology* 6 (1982): 113–26.

Dwork, Deborah. "Born in Urban America: 1830–1860." *Clio Medica* 12 (1977): 227–53.

Dye, Nancy Schrom. "The History of Childbirth in America." *Signs: Journal of Women in Culture and Society* 6 (1980): 97–108.

Dye, Nancy Schrom. "Modern Obstetrics and the Working Class Woman: The New York Dispensary, 1890–1920." *Journal of Social History* 20 (1987): 549–64.

Eakins, Pamela S., ed. *The American Way of Birth*. Philadelphia: Temple University Press, 1986.

Ferguson, James H. "Mississippi Midwives." *Journal of the History of Medicine* 5 (1950): 85–95.

Gaskin, Ina May. *Spiritual Midwifery*. Summertown, Tenn.: The Book Publishing Company, 1977. 3rd edition, 1990.

Leap, Nicky, and Billie Hunter. *The Midwife's Tale: An Oral History from Handywoman to Professional Midwife*. London: Scarlet Press, 1993.

Leavitt, Judith Walzer. *Brought to Bed: A History of Childbirth in America, 1750–1950*. New York: Oxford University Press, 1986.

Lewis, J.S. *In the Family Way: Childbearing in the British Aristoc-*

racy, 1760–1860. New Brunswick, N.J.: Rutgers, 1986.

Litoff, Judy Barrett. *The American Midwife Debate*. Westport Conn.: Greenwood Press, 1986.

Marland, Hilary. *The Art of Midwifery: Early Modern Midwives in Europe*. London and New York: Routledge, 1993.

Mongeau, Beatrice, Harvey L. Smith, and Ann C. Maney. "The 'Granny' Midwife: Changing Roles and Functions of a Folk Practitioner." *American Journal of Sociology* 66 (1961): 497–505.

Mulligan, Joan E. "Pregnant Americans, 1918–1947: Some Public Policy Notes on Rural and Military Wives." *Women and Health* 5, no. 4 (Winter 1980): 23–38.

Nightingale, Florence. *Introductory Notes on Lying-In Institutions. Together with A Proposal for Organising an Institution for Training Midwives and Midwifery Nurses.* London: Longman, Green, and Co., 1871.

Noall, Claire. "Mormon Midwives." *Utah Historical Quarterly* 10 (1924): 84–144.

O'Dowd, Michael J., and Elliot E. Philipp. *The History of Obstetrics and Gynecology*. New York: Parthenon Publishing Group, 1994.

Osgood, K., D. L. Hochstasser, and K. W. Deuschle. "Lay Midwifery in Southern Appalachia: The Case of a Mountain Country in Eastern Kentucky." *Archives of Environmental Health* 12 (1966): 759–70.

Peterson, Karen J. "Technology as a Last Resort: The Work of Lay Midwives." *Social Problems* 30 (1983): 284–97.

Robinson, Sharon. "A Historical Development of Midwifery in the Black Community, 1600–1940." *Journal of Nurse Midwifery* 26 (1981): 28–33.

Rothman, Barbara Katz, ed. *Encyclopedia of Childbearing: Critical Perspectives*. Phoenix, Ariz.: The Oryx Press, 1993.

Scholton, Catherine. "'On the Importance of Obstetrick Art': Changing Customs of Childbirth in America, 1760–1825." *William and Mary Quarterly* 34 (1977): 426–45.

Seligman, Stanley A. "The Royal Maternity Charity: The First Hundred Years." *Medical History* 24 (1980): 403–418.

Sharp, Jane. *The Midwives Book*. New York and London: Garland Publishing, Inc., 1985. Originally published: London: S. Miller, 1671.

Sullivan, Deborah A. and Rose Weitz. *Labor Pains: Modern Midwives and Home Birth*. New Haven, Conn.: Yale University Press, 1988.

Towler, Jean and Joan Bramall. *Midwives in History and Society*. London and Dover, N.H.: Croom Helm, 1986.

Ulrich, Laurel Thatcher. *A Midwife's Tale: The Life of Martha Ballard,*

Based on her Diary, 1785–1812. New York: Random House, 1990.

Wertz, Richard W. and Dorothy C. Wertz. *Lying-In: A History of Childbirth in America*. New York: Macmillan Publishing Co., 1977.

Wilson, Adrian. *The Making of Man-Midwifery: Childbirth in England, 1660–1770*. London: UCL Press, 1995.

British Midwifery

The Journal of Obstetrics and Gynæcology of the British Empire

VOL. X. | OCTOBER, 1906. | NO. 4.

The "Byrth of Mankynde."*

(ITS AUTHOR AND EDITIONS.)

By J. W. BALLANTYNE, M.D., F.R.C.P. (Edin.), F.R.S. (Edin.), *Lecturer on Midwifery and Gynæcology in Surgeons' Hall and in the Medical College for Women, Edinburgh; Physician to the Royal Maternity Hospital, Edinburgh; and President of the Edinburgh Obstetrical Society.*

THE *Byrth of Mankynde* is perhaps the most interesting work on Midwifery in the English language. It must be admitted that an earlier manuscript work, written in English, and dealing with the same subject, has been discovered; it must also be allowed that it is in great part a translation of Rhodion's (or Rösslin's) *De Partu Hominis;* yet the *Byrth of Mankynde* retains its place of honour at the head and source of English obstetric literature, and has profoundly influenced the practice of the art of midwifery in these Islands for more than three centuries. The perusal of its pages, so beautifully printed in the old black letter, gives the reader a graphic word picture of the *vie intime* of the sixteenth century obtainable from no other work; the study of its history and of the identity of its translator or editor supplies problems of considerable difficulty, and consequently of considerable interest to the medical bibliographer; and the fact that its illustrations are believed to be among the first, if not the very first, examples of copper-plate engraving which were produced in England, gives the book an importance from the standpoint of the annals of typography which can hardly be exaggerated. Published first in the year 1540, and dedicated to "the most gracious and in all goodnesse most excellent vertuous Lady Quene Katheryne," the work lived on till 1676, new editions con-

* Communicated to the Edinburgh Obstetrical Society on June 13th, 1906.

stantly appearing with but few alterations, until finally it had to give place to more modern books with a much more transitory existence. Seeing the light in the days when Henry VIII. was King, it survived the stormy times of Edward VI. and Queen Mary, and was a strong and healthy work in the Elizabethan age of literary and national grandeur; the Tudors went and the Stuarts came, and the book reached the summit of its life's arch; it was more than a hundred years old when the Civil War broke out and a new edition appeared in the days of the Commonwealth five years after Charles I. was executed; it was now a very old book, but it lived to see the Restoration and had vitality enough to blossom out into a new and last edition in 1676. The study of so remarkable a work cannot be without interest, and I purpose in the following pages to touch upon some of the many problems and considerations which cluster round it.

The First English Book on Midwifery.

The book's claim to be regarded as the first work in English on the subject of midwifery is the initial problem which calls for consideration.

Dr. J. H. Aveling, in his "Account of the Earliest English Work on Midwifery and the Diseases of Women" (*Obstet. Journ. Gr. Brit. and Ireland,* ii., 73, 1875), gives the following interesting information: "In the British Museum, among the Sloan Manuscripts, is one of the fifteenth century, vellum in quarto, beautifully written and illuminated, two hundred and thirty-four pages of which are devoted to "Midwifery and the Diseases of Women." From a note at the beginning we learn that it belonged to Richard Ferris, who was Master of the Barbers and Surgeons Company in 1563, and Sergeant Surgeon to Queen Elizabeth, and that at his death he bequeathed it to John Felde, who was to pay the executors for it "xxxxviii s. iiii d." Who the author of this interesting MS. was is not known. It is certain, however, that he was well acquainted with the medical writings of Rogerius of Parma, who lived some time after Albucasis, and whose book was first printed in 1490. The arrangement of the chapters is almost exactly similar, and in some instances the text is a verbatim translation from Rogerius."

Dr. Aveling gives extracts from this work relating to menstruation, and to the sixteen "unkind" ways in which the child may come forth. He reproduces also some of the drawings. The midwife's chief duty seems to have been to push the presenting part up again if it did not come down rightly, and to give various drugs. In cases of retained placenta, "the mydwif should anoynt her hondes and with hir nayles pullen owte the secondine if she mowe."

There can be no doubt that this work by an unknown author is considerably older than the *Byrth of Mankynde*, and Dr. Aveling deserves the thanks of the profession for having unearthed the manuscript and for having given so interesting a description of it. There is, however, no evidence that it was ever printed, and it could, therefore, never have served the same purpose as did the *Byrth of Mankynde* in the education and for the guidance of the midwives of the sixteenth century. Whilst, therefore, we cannot claim for the *Byrth* priority as the earliest work on Midwifery in English, we are certainly justified in regarding it as the first printed English book on that subject.

The Authorship.

The question of the authorship of the *Byrth of Mankynde* is not so easy of solution as might be thought from the perusal of the title page of the 1545 edition or of any of the numerous editions which followed it. The book is uniformly ascribed to Thomas Raynalde, Physician: the spelling may differ, and we find Raynold, Raynald, Reynald, and Reignalds, but obviously the same person is meant; and the designation of the alleged author may also differ, for we find Phisition, Physitian, and Dr. in phisick, but obviously a medical man is meant. When, however, we come to consult the first edition of the *Byrth*, that of 1540, we discover that the authorship or rather the translatorship is claimed by one Richard Jonas, and that the work was "imprynted at London by T.R." As will be seen immediately, "T.R." were almost certainly the initials of Thomas Raynalde, a well-known printer of the sixteenth century. Herein, then, lies the problem of the authorship. It is a triple problem, and may be best stated in the form of three questions: (1) Who was Thomas Raynalde the printer? (2) Who was Thomas Raynalde the physician, and what relationship, if any, did he bear to Thomas Raynalde the printer? and (3) Who was Richard Jonas, and what relationship did he bear to the Raynaldes?

When we wish to know what is to be known with certainty about anyone who has made his mark upon his times and upon the history of the British Isles, we turn, as a matter of course, to the great storehouse of all such information, the *Dictionary of National Biography*. In this case, however, we find even that trustworthy authority giving an uncertain sound; for the article on Thomas Raynalde, to be found in vol. xlvii., contains statements which are very considerably modified in the volume of *Errata*. Nothing could more surely prove the existence of a mystery than the striking diversity between the accounts given in the original article and in the later volume of *Errata*. The following paragraphs will serve to make this plain.

According to the *Dictionary of National Biography* (vol. xlvii., p. 347, 1896), there were two Thomas Raynaldes living at the same time, the one a physician and the other a printer in London. To the former is ascribed the translation of Rösslin's *De Partu Hominis* which bears the name "*The Byrth of Mankynde,*" and was first printed in 1540. In the volume of *Errata*, however, which was published in 1904, this statement is altered, and we are told that the original writer of the translation was one Richard Jonas, and that Thomas Raynalde's work consisted in editing the second edition of the *Byrth* which appeared in 1545. Having got thus far in my investigation, it seemed to me to be essential that I should go to the sources from which W. A. Shaw, the writer of the article in the *Dictionary of National Biography* had got his information. I, therefore, consulted Ames's *Typographical Antiquities*, both in the edition to which Shaw refers (that augmented by W. Herbert) and in that considerably enlarged by Thomas Frognall Dibdin, as well as in the original work as it left Ames's hands in 1749. It was fortunate that I did so, otherwise I should have been almost as much in the dark as before. The following statements are based upon what I found in Ames's work and its later issues, as well as upon what Dr. Aveling has to tell in his *English Midwives* (p. 11, 1872). Ames's information, it must be remembered, relates to Raynalde the printer, and it is only by accident, as it were, that it contains references to Raynalde the physician.

The following is what Ames had to say about Thomas Raynalde, the printer. I quote from the *Typographical Antiquities*, London, 1749, p. 219. "Thomas Raynalde dwelt in St. Andrew's Parish, in the Wardrop, and kept shop in St. Paul's Churchyard. It may be queried whether he was not the noted physician, who set forth the woman's book or birth of mankind, so often printed." Then follow the descriptions of the 1540 and 1545 editions, which I transcribe from Ames's book. "1540. The birthe of mankind, otherwise called the woman's boke. The first translation into English, with many small copper cuts, is dedicated to Queen Catherine, wife to King Henry VIII., and the first rowling press cuts I have seen in English books. To this edition the letters T.R. only. Quarto."

"1545. The byrth of mankynde, otherwyse named the woman's booke, newly set furth, corrected and augmented, whose contentes ye maye rede in the table of the booke, and more playnly in the prologue. By Thomas Raynold, phisition. This edition has Thomas Ray, which, if not the name of another printer, is a contraction from Raynalds. Quarto."

These statements from Ames's work seem to have been the source of most of the comments upon the *Byrth of Mankynde* and upon its supposed author during a long series of years. In a later edition of Ames's *Typographical Antiquities*, that edited and augmented by William Herbert, the same facts are stated (vol. i., p. 581, 1785); and this is the edition to which the writer of the life of Raynalde in the *Dictionary of National Biography* tells us that he had access. We thus see how it was that the idea that Raynalde the printer and Raynalde the physician were the same person arose, and we also note how the edition of 1540 came to be ascribed to Raynalde ("to this edition the letters T. R. only").

In another volume of this same edition of Ames's *Typographical Antiquities*, however, a piece of information is found which throws doubt upon the identification of Thomas Raynalde the printer with Thomas Raynalde the physician. In the third volume, published in London in 1790, and on page 1570, is the description of another book ascribed to Thomas Raynalde. It was published in 1551, and is thus described by Ames's editor (Herbert). "Dr. Thomas Raynalde, his declaration of the vertues of a lately invented oyle, called for the worthiness thereof, the Oyle Imperiall, with the manner howe the same is to be used against innumerable diseases. Printed at Venice by J. O. Griphius. Octavo." A copy of this book is in the British Museum, and is thus catalogued: "A compendious Declaration of the .... Vertues of a Certain lateli invented Oile, called .... Oile Imperial with the maner how the same is to be used, etc. J. Gryphius: Venetiis, 1551. 8°." Now the existence of this second medical book by Dr. Raynalde, its publication in Venice by a Venetian printer, and its dedication by Raynalde 'to his singular friend, Francis Mery, merchant, of the city of London,' seem almost of necessity to lead to the conclusion that Thomas Raynalde of the *Byrth of Mankynde* was not Thomas Ray or Raynalde the printer. But, further, when we come to consult a still later edition of Ames's *Typographical Antiquities*, that edited and enlarged by Thomas Frognall Dibdin and published in 1810, we get more light upon the question of the authorship of the *Byrth of Mankynde* as well as some very interesting facts regarding the 1540 edition of that work.

In Dibdin's edition of Ames's *Antiquities* (vol. iii., p. 563, 1816), the following notes appear: "Thomas Raynald. Of this ingenious printer we are unluckily without any authentic sources of information, except that he dwelt at first in St. Andrew's Parish, in the Waredrop, and in 1549 kept a shop at the Signe of the Star in St. Paul's churchyard. It may be queried (says Ames) whether he was not the noted physician, who set forth the woman's book, or birth of mankind, so often printed? But there seems to be very slight grounds for this quaere, as it is nowhere apparent, but upon the face of a colophon, that Raynald the printer was a physician." Dibdin then goes on to describe very fully the 1540 edition of the *Byrth of Mankynde*, from the examination of a perfect copy in "the richly stored library of my friend Mr. Herbert"; and I here give his description, which, in view of the extreme rarity of this edition, is exceedingly valuable. "*The Byrth of Mankynde.* 1540. Quarto. 'The byrth of mankynde newly translated out of Laten into Englysshe. In the which is entreated of all suche thynges the which chaunce to women in theyr labor, and all suche infyrmitees whiche happen unto the Infantes after they be delyuered. And also at the latter ende or in the thyrde or last boke is entreated of the conception of mankynde, and howe manye wayes it may be letted or furtheryd, with diuers other fruytefull thynges, as doth appere in the table before the booke. *Cum priv. regal. ad impr. sol.*' On the back of the title-page is a sort of religious 'admonicion to the reader,' which is followed by a dedication of six pages 'Unto the most gracious and in all goodnesse most excellent vertuous Lady Quene Katheryne wyfe and most derely belouyd spouse unto the most myghty sapient Christen prynce Kynge henry the VIII. *Richard Ionas* wyssheth perpetuall ioye and felicyte.'" Dibdin concludes his description by stating that "at the end, we read this colophon: 'Imprynted at London by T. R. Anno Domini MCCCCCXL.'"

Here, then, we have an indication of the authorship of the earliest known edition of the *Byrth of Mankynde*: the author, or rather the translator of Rösslin's *De Partu Hominis* (for a translation is what the 1540 edition really was), was a person called Richard Jonas; and the T.R. of the colophon referred to Tho. Ray the printer. But when we come now to ask who Richard Jonas was, we are almost brought to a stop. If we turn, however, to the Prologue which precedes the first book of the later editions of the *Byrth*, we find a reference which must surely apply to Jonas. I quote from the 1552 edition (but I believe the 1545 is identical). "Wherfore now to come to our purpose, ye shal understand that about a thre or foure yeres past, a certayne studious and diligent clarke, at the request and desyre of diuers honest and sad matrones, beynge of hys acquayntaunce, dyd translate out of Latin into English a great part of thys boke, entiteling it according to the Latine inscription (de partu hominis, that is to saye: of the byrth of mankynd) which we nowe do name (The womans boke) (for so moch as the most parte or well neare all therein entreated of, doth concerne and touche onely women). In whych hys translation he varied, or declyned nothing at al from the steppes of his Latine auctor. Obseruynge more fydelytye in translatynge, then choyse or dyscretion at that tyme in admyttynge and alowyng many thynges in the same Boke, greately neadyng admonityon and wary aduyse or counsell to the readers, whyche otherwyse myghte sometymes use that for a helpe, thee whyche shoulde turne to a hynderaunce, wherefore I reuoluyng and earnestly reuysinge from top to too ye sayd boke, and here withal considering the manifold utilite and profyte, whych thereby moughts ensue to all women (as touching that purpose) yf it were more narowly loked ouer, and wyth a strayghter iudgement more exactely euery thing therein pondred and tryed, thoughte my labour and paynes shuld not be euyl employed, ne unthankefully accepted and receaued of al honest, discrete and sage wemen. If I after good and dylygent perusing thereof dyd correcte and amende soche fautes in it, as semed worthy of the same, and to aduyse the readers what thynges were good or intollerable to be used, which were daungerous, and which were utterly to be eschued. The which thinge I haue not only so done, but ouer this haue ther unto adioyned and annexed dyuers more experimented and more familier medicines."

In these lines of the prologue of the 1545 and subsequent editions we are told all that we know of Richard Jonas, for I do not think it can be doubted the "studious and diligent clarke" referred to was Richard Jonas. This, at any rate is the view of the matter taken by the authorities at the British Museum and by Dr. Aveling. In the *British Museum Catalogue* of printed books, under the name *Jonas* (Richard) is placed *The Byrth of Mankynde*, 1540; and in another volume of the same catalogue the same book appears under Roesslin, showing that the work was recognized as a translation of the latter author's *De Partu Hominis,* and that the translation was ascribed to Richard Jonas. Unfortunately the British Museum copy is imperfect, for it lacks the plates which are described in the text. To this edition I shall again refer, in the meantime I simply repeat that it seems to be undoubted that *it was the first edition of the Byrth of Mankynde, that it consisted of a translation of Rhodion's or Rösslin's "De Partu Hominis," that the translator was Richard Jonas, and that it differed in several respects from the later editions.*

Although nothing more is known with certainty about Richard Jonas, I may touch upon some circumstances which may serve to throw a little light upon his personality. By the end of 1540 Henry VIII. had led five wives to the altar. In 1531 Catharine of Aragon was put away, although the marriage was not annulled as void until May 1533, four months after the King had privately married Anne Boleyn. In 1536 Anne was executed, and on the day after the execution Henry married Jane Seymour; she died in childbirth. In 1540 Henry was for a few months the husband of Anne of Cleves. The Lutheran princess was soon divorced and Katherine Howard put in her place; but she was beheaded in 1542 (February), and another Katherine (Parr) became Henry's wife in 1543. The writer of the life of Thomas Raynalde in the *Dictionary of National Biography* states that it was to Queen Catherine Parr that the *Byrth of Mankynde* was dedicated; but, since the work was published in 1540, it must have been Queen Katherine Howard who was indicated in the dedication. It is impossible now to determine the circumstances of the publication of this edition; but some guesses may be hazarded. In the first place, it is significant, I think, that Queen Jane Seymour should have died in childbirth, for this death would draw attention very sharply to the perils of parturition, and would lead midwives and surgeons to inquire whether any more trustworthy guides to practice existed in the English or in foreign languages than were commonly known or used. Queen Anne (of Cleves) was a sister-in-law of the Elector Frederick of Saxony, and had other family ties with the Lutheran Princes of Germany. Justus Jonas, of Wittenberg, was a distinguished disciple of Luther, and it is just possible that to him was related the London Richard Jonas (who may even have come over in the train of Queen Anne) the translator of the *Byrth of Mankynde*. Rhodion's book was well known in Germany at this time. Further, it may have been intended that the work should be dedicated to Queen Anne; her early dismissal, and her succession by Queen Katherine Howard may account for its dedication to the latter.

I am now in a position to proceed to the discussion of who T.R. of the 1540 edition was, and of the relationship he bore to Tho. Ray and to Thomas Raynalde the printer, and Thomas Raynalde the physician. In the first place I think there can be no doubt that T.R. of the 1540 edition (Jonas's translation) was the printer whose name appears as Tho. Ray in the colophons of the 1545 and 1552 editions. In a later work printed by Thomas Raynalde his name is indicated by the initials T.R., I refer to the *Processionale* published in 1555; and it seems too artificial a suggestion to suppose that two printers with names having the same initials should have printed the first and second editions of the *Byrth of Mankynde.* But it may be said that these reasons do not disprove the possibility that Tho. Ray and Thomas Raynalde the printer were different people. Neither they do; but there is no evidence whatever that any printer of the name of Tho. Ray existed; if there had been any such evidence either Ames himself or one of his two distinguished editors would almost with certainty have found it out. It may be concluded, therefore, *that T.R. of the 1540 edition and Tho. Ray of the 1545 and 1552 editions were one and the same person, and that he was Thomas Raynalde the printer in St. Paul's Churchyard, London.*

Ames's query whether Thomas Raynalde the printer was not the noted physician who set forth the *Woman's Book* or the *Byrth of Mankynde* now calls for consideration. Raynalde, the printer, sent out from his press in London several works between the years 1540 and 1555; among these were the two editions of the *Byrth of Mankynde* (1540 and 1545), a *Treatise against the Masse* (1548), a *Plaister for a Galled Horse* (1548), the *Boke of Barthra Priest* (1549), *Certaine Psalmes* (by Sir Thomas Wyatt, 1549), and the folio *Byble* (1549). No works seem to have come from his press after 1555, when the *Processionale* made its appearance. He flourished, therefore, between the years 1540 and 1555. As it happens these were also the years during which we have any information regarding Thomas Raynalde, the physician. His name appears upon the title page of the 1545 and 1552 editions of the *Byrth of Mankynde* as well as upon a second book, published at Venice in 1551 and already referred to (*vide antea*). From internal evidence in these two works we learn further that he was at Venice (in 1551), in Padua, and in Paris. His work on *Oyle Imperiall,* is described in Ames's *Typographical Antiquities* (Herbert's edition, iii., p. 1570, 1790) among those books "printed abroad by Englishmen, and other printers for them," and the epistle dedicatory is dated from Venice (March 1, 1551) to the author's 'singular friend, Francis Mery, merchant, of the city of London,' and in it there is an acknowledgement of the author's indebtedness to this merchant who had purchased from him a large quantity of the oil. A copy of this work is in the British Museum, catalogued under Thomas Raynalde's name; the printer's name is J. Gryphius. In the 1560 edition of the *Byrth of Mankynde* appears for the first time the ninth figure in the Anatomical plates, that showing the uterus cut open; to this is annexed (on folio li.) the following description: "And the nynth figure sheweth the Matrix cut forth of the body, being of that bygnesse as it was sene taken foorth of a woman at the laste Anothomye, which I dyd se[1] at the uniuersitie of Padua in Italy." In the 1552 edition there is a reference to medical practice, which seems to throw some light upon the doings of the writer or editor; it is as follows:—"Yet agayne to confyrme myne opinion, that the Termes do onely issue oute of the vaute of the matrix: ye shall understande, that at sundry tymes I haue had two dyuers wemen in cure thone in London and thother in Paris: of which thone by a fal from her horse, the other by a violent thrust and squat on the buttockes upon the harde stones in the strete, beynge both neare the tyme of theyr purgation, chaunced also both of them into one maner of disease (that is to say) immediately here upon, theyr Termes dyd issue." The writer continues his description of these two cases, and draws some deductions from them regarding menstruation (*vide* folio xliiii. of this edition).

These facts are, I think, sufficient to show that it is very improbable that Thomas Raynalde the printer, and Thomas Raynalde the physician, were one and the same person; for, if they were, then we must imagine the printer travelling widely on the Continent (Venice, Padua, Paris), having a book printed for him at Venice by another printer called Gryphius, while all the time books were issuing from his own press in St. Paul's Churchyard in London. To my mind it is much more likely that they were individuals bearing the same name. It is possible that they were cousins or uncle and nephew. At any rate we can easily imagine the printer asking the physician to revise and add to Jonas's work, the first book which had come from his press and one which had evidently sold well since a second issue was needed at the end of five years; we can also easily understand that the physician would be led to incorporate his personal experience in the second edition and add new chapters to the work in order to make it more widely useful. To Thomas Raynalde's pen must, I think, be ascribed the greater part of the Prologue to the Women Readers and most of the first book of the 1545 and subsequent editions; the remaining parts are translations from Rösslin's *De Partu Hominis* made doubtless by the "studious and diligent clarke," whom we must regard as Richard Jonas. Of course it is just possible that Thomas Raynalde may have been the maker of the translation of 1540, and that he took the *nom de plume* of Richard Jonas to save himself from the possible obloquy which might attach itself to the physician who concerned himself (in those times) with

[1] This sentence is suspiciously similar to that in which Vesalius describes the same figure in his work, published in 1543.

the subject of midwifery. There are other possible solutions of the problem of authorship; but it seems to me to be *probable that the 1540 edition was the work of Richard Jonas and was a simple translation of Rösslin's "De Partu Hominis"* (most likely the 1538 edition published at Paris "apud Joannem Foucher"), *that the second edition (1545) was revised and considerably added to by Dr. Thomas Raynalde, and that he also added to and edited the third and probably the fourth editions* (1552 *and* 1560). I think, therefore, it is permissible to call the work Raynalde's *Byrth of Mankynde* rather than Jonas's Byrth of Mankynde or Jonas's English Translation of Rösslin's *De Partu Hominis,* on account of the Prologue and the new material in Book I., and for the reason that for more than three hundred years it has been so named. To the first edition, however, it is more correct to give the name Jonas's Translation of Rösslin's *De Partu Hominis,* and to catalogue it either under Jonas or Rösslin. Further consideration will be given to this subject when the peculiarities of the various editions have been described.

I may here conveniently summarize the conclusions at which I have arrived regarding the authorship of the *Byrth of Mankynde.* (1) The 1540 edition of the book was a translation of Rösslin's or Rhodion's *De Partu Hominis* made by Richard Jonas, the "studious and diligent clarke" referred to by the editor of the 1545 edition. (2) It was printed by T.R.; and these were the initials of Tho. Ray or Thomas Raynalde, the printer, who had a shop in St. Paul's Churchyard, and from whose press various works issued between the years 1540 and 1555. (3) The second edition, which appeared in 1545, was revised "from top to toe" by Thomas Raynalde, the physician, and was considerably added to by him, the additions consisting of the prologue and most of Book I., as well as of short paragraphs in the other books. (4) Thomas Raynalde, the physician, seems to have practised in London and Paris and to have visited Venice and Padua; he may have been a relative of his namesake the printer. (5) He may have edited the third edition (1552) and the fourth (1560); but subsequent editions were almost identical, and were printed by different printers (Richard Jugge, R. Watkins, T. Adams), and probably Thomas Raynalde, the physician had no hand in them.

The Editions of the "Byrth of Mankynde."

(a) *The 1540 Edition.*

I have already referred to (p. 301) and quoted from Dibdin's description of the 1540 edition, given in his edition of Ames's *Typographical Antiquities* (vol. iii., p. 563, 1816). His description is of great value, for it was made from a perfect copy in Herbert's library. There is a copy at present in the British Museum; but it is imperfect, consisting of 84 numbered leaves instead of 86 (in Herbert's copy), and it lacks the plates. The title-page differs widely in its wording from that of later editions: the book is named "The byrth of mankynde, newly translated out of Laten into Englysshe," etc., and the alternative title "the womans booke," does not appear; there is no author's or translator's name on it; and there is no allusion to the prologue but only to "the table before the booke." The "sort of religious admonicion to the reader" which appears upon the back of the title page is evidently peculiar to this edition, and is probably replaced by what I have termed the Aristarchus preface of all later editions. The dedication (of six pages) to "the most gracious and in all goodnesse most excellent vertuous Lady Quene Katheryne" is also peculiar to this edition, and has been referred to already.

In addition to what has been stated already about the 1540 edition of the *Byrth of Mankynde* I may give the following particulars (Ames's *Typographical Antiquities*, vol. iii., p. 563, 1816). "The book contains 86 numbered leaves, and is printed on stout paper, in a fine large Gothic type. Between folio xxiv. and v., there are four copper-plate engravings containing seventeen illustrations of children in utero. These are very curious, as being among the rarest early specimens of this kind of engraving [the first ever seen by Ames] published in this country." These plates call for further consideration. I quote agan from Ames's *Typographical Antiquities*, as edited by Dibdin (vol. i., pp. 24, 25, 26, 1810). "While the foreign presses were putting forth the most beautiful books, and the gravers of Albert Dürer and Aldegraver were exhibiting equally beautiful specimens in the sister art, we were, I fear, generally obliged to content ourselves with the importation of worn and battered blocks, or with foreign artists discarded from want of capacity. This may account for the tardy appearance of the first Copper-Plate impressions in this country, executed by means of a roller—which are supposed to be the frontispiece to Galenus De Temperamentis printed at Cambridge in 1521, ... and the cuts in Raynalde's *Birth of Mankind*, 1540. . . . . Although Strutt says: 'it is by no means certain that these (latter) plates were engraved in England or the work of English artists,' yet, when the rudeness of the figures and the absence of any similar style of engraving preceding the date of publication of both works are considered, we may be disposed to conclude that they are the earliest known specimens of impressions from Copper Plates in this country." It is a misfortune that the copy of the 1540 edition in the British Museum lacks these plates, as we are unable to judge of their characters. The reproduction of one of them given in Ames' work (vol. i., p. 25, 1810) unfortunately is "printed from a wooden block for the convenience of press work," and probably gives little idea of the original, although the editor states that "it will be found to exhibit a very faithful character of the original."

I have seen the so-called 1540 edition of Raynalde's *Byrth of Mankynde*, which is in the Library of the Obstetrical Society of London. Even a cursory examination of the work makes it very doubtful whether it is what is claimed for it in the Catalogue, and a more careful inspection makes it practically certain that it is really the 1634 edition. The title page is in manuscript, and so forms no indication of the real age of the work; but the type is only partly black letter, the marginalia are in ordinary type, the Aristarchus preface is in English not in Latin, the pagination is by pages not by folios, and the spelling is that of the seventeenth and not of the sixteenth century. There are several other features in which the copy resembles the later rather than the earlier editions, and I do not doubt but that it is really the 1634 edition. The absence of the title-page and of the last two or three pages at the end, and their replacement by transcriptions make it impossible to be quite sure (of course the *transcriptions* may have been made from other editions). It is certainly not the 1626 edition.

There is another copy of the *Byrth of Mankynde* which carries the date 1540 on the title page; this is in the Library of the University of Glasgow. The date, however, is written in, there being no printed intimation of it. This copy is, therefore, probably the undated edition of 1564 (q.v.). It came from the Collection of James Hotson, of Airdrie, and was presented to the University by Dr. Thomas Reid, the oculist.

This edition of 1540, it will be remembered, is not in any sense a Raynalde one except in so far that it was printed by Thomas Raynalde, the printer. It is a translation into English of Rösslin's *De Partu Hominis*, made by Richard Jonas; and it ought to be catalogued either as Jonas's *Byrth of Mankynde* or as Rösslin's *De Partu Hominis* in English.

I have carefully collated the contents of this first edition of the Byrth with those of later editions. The 1540 edition has its contents divided into three instead of four books; it lacks the Prologue to the Women Readers which appears in later editions; but it has the dedication to Quene Katheryne; it lacks also the Plates of Anatomical

Figures which later editions have, although it possesses the Birth Figures (indeed, all the editions have the latter). The collation of the chapters is as follows.[1]

1540 Edition.	1552 Edition.
First Book.	First Book.
Chapter I.	Chapter XIV.
	Second Book.
Chapter II.	Chapter I.
Chapter III.	Chapter II.
Chapter IIII.	Chapter III.
Chapter V.	Chapter V. (error).
Chapter VI.	Chapter V.
Chapter VII.	Chapter VI.
Chapter VIII.	Chapter VII.
Chapter IX.	Chapter IX. (error).
	Chapter X.
	Third Book.
Chapter X.	Chapter I.
	Chapter II.
Second Book (not in chapters)	Chapter III.
Third Book	Fourth Book.
Chapter I.	Chapter I.
Chapter II.	Chapter II.
Chapter III.	Chapter III.
Chapter IIII.	Chapter IIII.
Chapter V.	Chapter V.

Briefly stated, the first and second books of the 1540 edition correspond to the second and third books of the later editions; and the third book of the former edition to the fourth book of the later editions. The first chapter of the first book of the 1540 edition, however, corresponds to the fourteenth of the first book of the later editions; and the tenth chapter of the second book of the later editions has no corresponding chapter in the 1540 edition. This collation of the chapters is of value, further, because it explains some of the irregularities in the numbering of the chapters in the 1552 and 1560 editions; the printer has evidently forgotten sometimes that the two series of chapters did not exactly coincide, and has attempted

[1] I have used the 1552 edition for purposes of collation, but it very closely resembles the 1545.

occasionally to make them do so. The greater part, then, of the first book of the 1545 and subsequent editions is not represented in the 1540 one, and is, along with the prologue, to be ascribed to Raynalde.

The collation of the 1540 edition with Rösslin's *De Partu Hominis* (the Paris edition of 1538) shows that the first and second books of the former correspond to the twelve chapters into which the latter is divided; the third book of the former is not represented in Rösslin's work. The Byrth Fygures also correspond in the two works, but are not arranged in the same order.

(b) *The 1545 Edition.*

The Royal College of Physicians of London has a fine copy of the 1545 edition of the *Byrth of Mankynde;* through the courtesy of the College I have been able to examine this book. The title page has a woodcut border, but the degree of ornamentation and illumination is small as compared with what some of the later editions show. The date (Anno. M.D. XLV.) is rendered slightly uncertain, for the Library Stamp of the College covers it and makes it appear as if there were another figure between the D and the X; but I think it is correctly given in the Catalogue as 1545, and that the figure is only a full stop and a blur. The date is not repeated at the end of the work, so that it cannot thus be checked. The setting of the title is as follows:—

The byrth of
mankynde, otherwyse named
the womans booke, Newly
set furth, corrected and
augmented, whose co-
tentes ye maye rede
in the table of the
booke and most
playnly in the
prologue

By Thomas Raynold
Phisition

Anno M. D. XLV

At the end of the work is the word "Finis," followed by "Imprynted at London by Tho. Ray." "Imprynted" is preceded by the paragraph mark (¶), which is very commonly employed in this edition although it does not appear upon the title page. The book is printed entirely in black letter including the marginalia; and there are woodcut capitals. There are 148 leaves or folia, but the numbering of them is irregular and seems to be erroneous[1]; there are four preliminary leaves, and the prologue occupies 18 leaves; so that, in all, there are 170 leaves or folia (about 340 pages of a modern book). Inserted at folio 46 are two plates representing dissections of the anterior abdominal wall; at folio 63 are two leaves, printed on both sides, (*i.e.* four plates) of copper plates, consisting of the "Birthe Figures," viz. "the Woman's Stoule," and 17 figures showing the presentations of the child in utero. The anatomical plates are from Vesalius, the "Birthe Figures" from Rösslin. On folio 108, at the end of the second book, is the following interesting passage which I here quote. "And as touching the aboue Trociskes and emplasters described here in this place, ye shall fynde them alwayes readye made at the signe of the thre Dooues in Boucklars bery[2] in London: the name of the good man of ye house is wyllam Normewyll. In whose shoppe I haue caused the sayde thinges faithfully to be made, for because that I am certayne that he is one of the moostt fydell and faithfullest Apothecaries in London: And suche as wyll not spare for any cost to acquire and obtayne of the best and moost singular symples and droges in there kynde, that may be gotten for monye." This passage throws an interesting sidelight upon the life of the time, and on the relations of medical men and apothecaries, as well as upon the pharmacy of the middle of the Sixteenth century. It may also be compared with the corresponding passages in some of the other editions, to which reference will be made under these editions.

It is to be noted that the preface, commencing "Quamvis possit Aristarchus," is in Latin in this edition as it is also in that of 1552 (q.v.).

(c) *The 1552 Edition.*

The edition of 1552 is a small quarto of 149 folia or double pages, not including twenty folia devoted to the table of contents and prologue. The title page, reduced very slightly in size, is reproduced in Fig. 1. It is noteworthy that in the ornamental border on the right side is the figure of a midwife (or of a pregnant woman) with the forefinger of each hand extended as if for a vaginal examination. The date (MDLii) is clearly set forth, and the name of the author is

[1] After signature J, follows Hh8; and then Hhh6; it contains Y10 in eights; A, only 4.

[2] Bucklersbury was in Cheapside, and was noted for its druggists' and grocers' shops. There is an allusion to it in Shakespeare's *Merry Wives of Windsor* (Act. iii., sc. iii., 19): "Many of these lisping hawthornbuds, that come like women in men's apparel, and smell like Bucklersbury in simple time."

Fig. I.

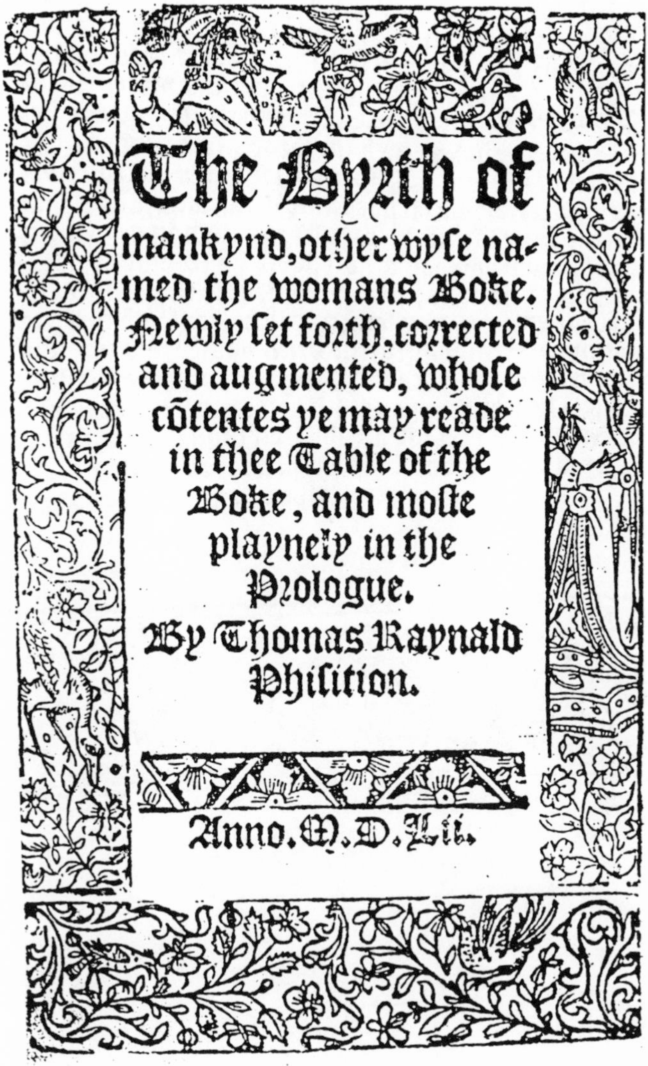

The Byrth of
mankynd, otherwyse na-
med the womans Boke.
Newly set forth, corrected
and augmented, whose
cõtentes ye may reade
in thee Table of the
Boke, and moste
playnely in the
Prologue.
By Thomas Raynald
Phisition.

Anno. M. D. Lii.

Title Page of the 1552 Edition.

Fig. II.

Quamuis possit Aristarchus quispiam cu-
rans nonnullam, diligentiamq; in hoc o-
pere desiderare: nullus tamē tam iniquus
erit, qui inficias iuerit operę precium hic
aliquod factum esse: quum & multo locu-
pletius, auctius, emendatiusq; quam vel
Latine, vel Anglice, prius habebatur, iam prodeat. Et
nisi plurimis, hisq; necessariis curis familiaribus di-
stracti (interimdum excudebatur) fuissemus, alius pla
ne, factus foret hic liber. Haud ignari, non pauca hic
congeri parum admodum necessaria, que emitti potu-
issent absq; vllo lectoris dispendio. Rursum alia desi-
derari que adscita, non leuiter profuissent ad presens
negotium, decreuimus igitur breui opusculum hoc sub
incudem reuocare, & exactiore lima, omnia (pro viribus
nostris) nitidius expolire: Interim precantes vt
non minore successu ab obstetricibus, reliquis
& matronis omnibus amplectatur, quam
est affectu scriptum.

Latin Preface to the 1552 Edition.

given as Thomas Raynald, Phisition. On the reverse side of the title page is the preface in Latin. In the editions of 1560, 1564 (?), 1565, 1598, 1626, 1634, 1654, and I presume all the others, the preface appears in English. The Latin is reproduced in Fig. 2, and I give here the translation as it appears in the edition of 1560: "Albeit some Aristarchus[1] may perhappes finde some lacke of faithfulnesse and diligence in this work: yet there is none so froward to deny but that there is some fruit and profyte to bee founde therein, seyng that it commeth nowe abrode much more enlarged and encreased, and more diligently corrected then it was before, eyther in the Latine or in the Englyshe. And where before in the other printes, there lacked matter necessary to the openyng and declaration of the figures pertaynyng to the inner partes: it is nowe so plainely set foorth, that the simplest Mydwyfe which can reade, maye both understand for her better instruction, and also other women that haue nede of her help, the more commoditie. Wherefore my desyre is, that it maye be receaued and practysed of Mydwyues and all other Matrones, with no lesse successe, then it is with good wyll and desyre wrytten to profite and to do good to other."

Then follow the table of contents and "A Prologe to the women Readers." Frequently throughout the table of contents and the prologue there occurs the old sign for paragraph (something like a reversed D, thus ₵); it is seen also throughout the whole book, but it does not appear on the title page as it does on that of the editions of 1560 (*vide* Fig. 3), and 1564 (?).

The first book extends from folio 1 to folio 56 (*i.e.* 112 pages); it consists of fifteen chapters and of "the declaration by letters of the fygures followyng." Eleven figures in all are described. From the description given, the first and second figures represented dissections of a man's body showing the peritoneum and the parts below it. At the end of the descriptive notes is the following remark: "Here ye shal be aduertysed that although these two fyrst fygures be made principally for ye man, yet may they serue as wel to expresse the woman; for the man and woman differ in nothyng but in the pryuie partes." In the 1560 edition they have therefore disappeared (reducing the number of figures to nine); further, the declaration in that edition reads as follows: "The declaration by letters of the fygures folowing, wherein be set forth to the eye, euery parte in woman mentioned in thys boke before: Which in the former Printinges hath ben corrupted, but nowe truely set forth." The

[1] Aristarchus of Samothrace was a great critic; so the word came to be employed with the meaning of a severely critical person.

eighth, ninth, tenth and eleventh figures of the 1552 edition are the fifth, sixth, seventh, and eighth of the 1560 and subsequent editions. The ninth figure of the 1560 edition ("the matrix . . . . taken foorth of a woman at the laste Anothomye, which I dyd se at the vniuersitie of Padua in Italy") does not appear in the present (1552) edition nor in the 1545; perhaps it took the place of the fifth and seventh figures of the 1552 edition. All these anatomical plates are to be found in Vesalius' (*De Humani Corporis Fabrica*, 1543).

The second book extends from folio 57 to folio 112; it consists of nine chapters, numbered I., II., III., V., V., VI., VII., IX., and X. In the second part of the third chapter is the description of most of the Birth Figures, viz., the first to the sixteenth (folio 71 to 74). The seventeenth and last of the Birth Figures is referred to on folio 60; it exhibits a two-headed monster. At the end of the second book is the interesting paragraph about the Trochiskes, which reads as follows: "And as touching the aboue Trochiskes and emplasters described here in this place, ye shal fynd them alwayes ready made at the signe of the thre Doues in Boucklers bery in London: the name of the good man of that house is Wylliam Normeuyl. In whose shoppe I haue caused the sayd thinges faithfully to be made, for because that I am certayne that he is one of the moost fydell and faythfullest Apothecaries in London. And soch as wyl not spare for any coste to acquyre and obtayne of the best, and moost singular simples and droges in theyr kynde, that maye be gotten for moneye."

The third book consists of three chapters and extends from folio 113 to folio 136. It deals largely with children's diseases. The fourth book has six chapters, extends from folio 137 to folio 149, and deals with sterility and certain cosmetics or "bellyfying receptes."

The only copy of the 1552 edition which I have had an opportunity of examining is that in the Library of the Royal College of Physicians of Edinburgh. It lacks the illustrations, but is otherwise perfect; and it was given to the College in 1846 by Dr. W. Beilby. Here and there throughout the volume are written comments or remarks, and some of these are interesting. They are all very old, as the caligraphy proves, *vide* the written title at the top of the title page. On the margin of folio 36 is written "The birth of man by Tho: Reignalds Dr in phisick," and on folio 18 is a somewhat similar inscription. Again on folio 86 is written "Joshua Jones his book 1688," and on the opposite page is "Richard Jones."[1] These were

[1] The occurrence of this name (Richard Jones) suggests the speculation whether the translator of the first edition may not have been a Richard Jones and not a Richard Jonas, and perhaps an ancestor of this Richard!

Fig. III.

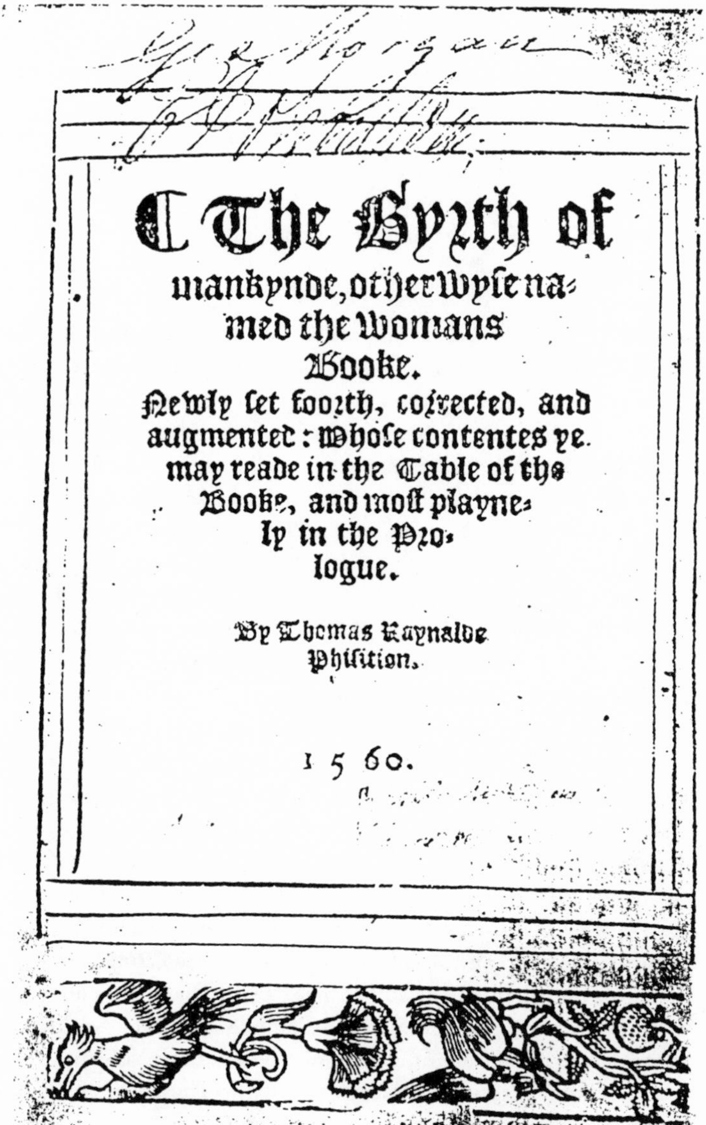

¶ The Byrth of mankynde, otherwyse named the Womans Booke.
Newly set foorth, corrected, and augmented: Whose contentes ye may reade in the Table of the Booke, and most playnely in the Prologue.

By Thomas Raynalde Phisition.

1560.

Title Page of the 1560 Edition.

doubtless former owners of this copy of Raynalde. On folio 60 occurs the marginal MS. note "The birth of man easier y^n ye birth of wom." There are some other notes mostly of the nature of comments on the text.

At the end of the work is the word FINIS in Gothic capitals followed by "Imprinted at London by Tho. Ray." The colophon, however, is not supplied with any date. I have already discussed the relationship of Tho. Ray to Thomas Raynalde.

(d) *The 1560 Edition.*

The edition of 1560 has a comparatively plain title page, that of my copy is reproduced in Fig. 3. It is noteworthy that even the title in this edition is prefixed by the old paragraph sign (¶). There is a good deal of carelessness visible in the headings of the leaves; thus: "the first booke" is put for "the seconde booke" on leaves 58, 59, 60, 61, 74, 75, 76, 77 and 92, whilst "the seconde booke" appears instead of "the thyrde booke" on leaves 102 and 108. The pagination, or, rather, foliation, does not include the "Prologue to the women readers" nor the table of contents; the Prologue covers nineteen as against thirty-three pages in the 1552 edition; the first book covers fifty-one leaves (102 pages) as against fifty-six leaves (112 pages) in the 1552 edition; in all, the four books occupy 131 leaves (262 pages) as against 149 (299 pages) in the 1552 edition. My copy has two former owners' names on the title page, viz., Geo. Morgan and E. Johnston; it lacks the plates of the anatomical relations of the genital organs, but has the "Birth Fygures" including the "Womans Stoole" and the seventeen figures of the presentations; the latter, consisting of four plates, are placed immediately after the table of contents, and not in the third chapter of the second book where they are described in the text; and it is incomplete at the end, some leaves of the third and the whole of the fourth book being absent. The seventeenth of the birth figures, that of the double monster "such as of late was sene in to the dominion of Werdenberg" is referred to in the second chapter of the second book (leaf 54). There are no written notes or comments in my copy except the statement on the margin of the first leaf of the second book: "Wrote by Thos. Raynalde M.D. A.Domini ——1560" (*cf.* the account of the 1552 edition, the copy of the Royal College of Physicians of Edinburgh). The ninth figure of the anatomical plates makes its first appearance in this edition. It "sheweth the Matrix cut forth of the body . . . of a woman at the laste Anothomye, which I dyd se at the uniuersitie of Padua in Italy." It, also, is taken from Vesalius (*De Humani Corporis Fabrica*, 27th

Plate in 5th book, 1543). The description suggests that Raynalde himself saw the post mortem, but a comparison of his words with those of Vesalius throws a doubt upon this point.

There is a perfect copy of the 1560 edition in the University of Aberdeen. It formed part of the Bute gift to that University. No printer's name occurs on this edition, either on the title page or in the colophon; indeed, there is simply the word "Finis" on the last page. All the illustrations are present in the Aberdeen copy; the nine anatomical figures (from Vesalius) appear at leaves 44, 46, 49 and 51, and two of the plates are folded; the Birth Fygures and the Woman's Stoole are inserted opposite leaf 63. On the flyleaf opposite the title page is written "Rebekah Cox her Book," and below it is "Edward Challis, His Book, paid 1d at ye second hand 27th March 1669."

The illustrations of the 1560 edition (which are also those of all the later issues) are reproduced in Figs. 4 to 11.

(*e*) *The 1565 Edition.*

The edition of 1565 was a fine one, and a good copy of it is in the Library of the University of Edinburgh. It is a small quarto of 131 folia or double pages, not including 14 folia devoted to the title page, table of contents and prologue. The title page is ornate: the right border contains the figure of a woman and the left that of a man; there is a lion at each side of the base; at the top is a scroll with a monogram which seems to contain the letters R, E, G, and G. The arrangement and spelling of the words of the title are as follows:—

The birth of
mankynde, other-
wyse named the
womans booke.

Newely set foorth, corrected and
augmented. Whose con-
tentes ye may reade in
the Table of the
booke, and
most
playnely in the
prologue
By Thomas Raynalde
Phisition

Fig. IV.

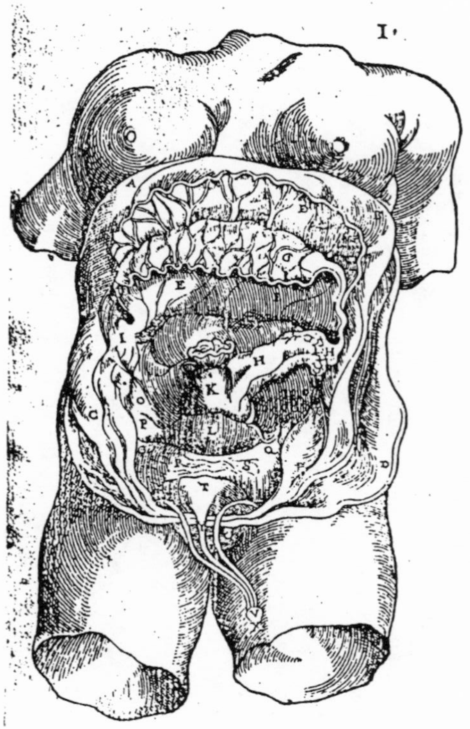

Fig. V.

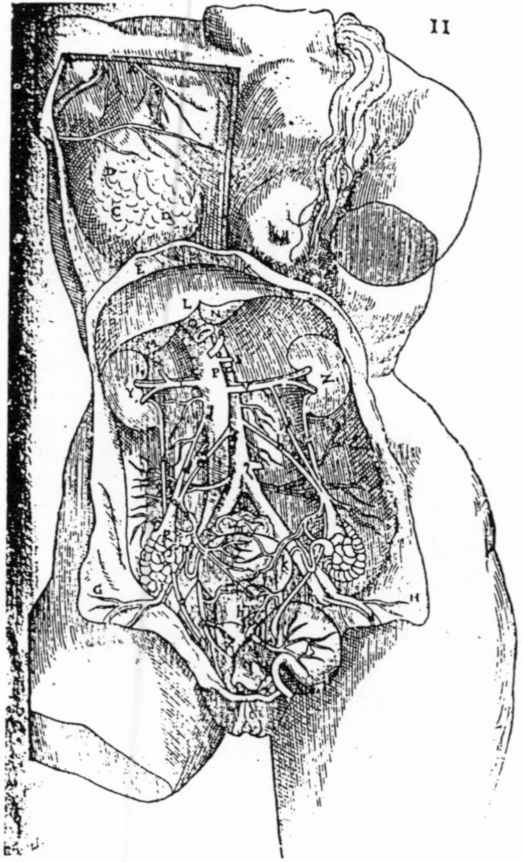

Anatomical Figures in the "Byrth of Mankynde."

Fig. VI.

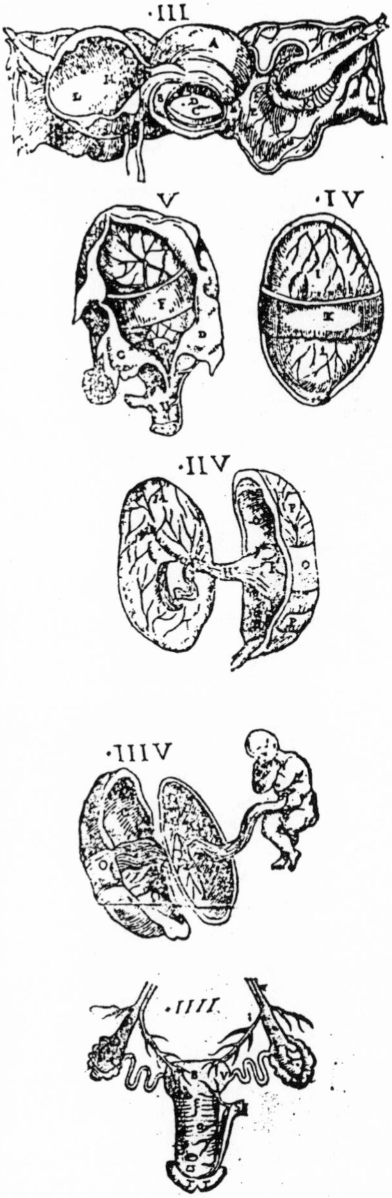

Fig. VII.

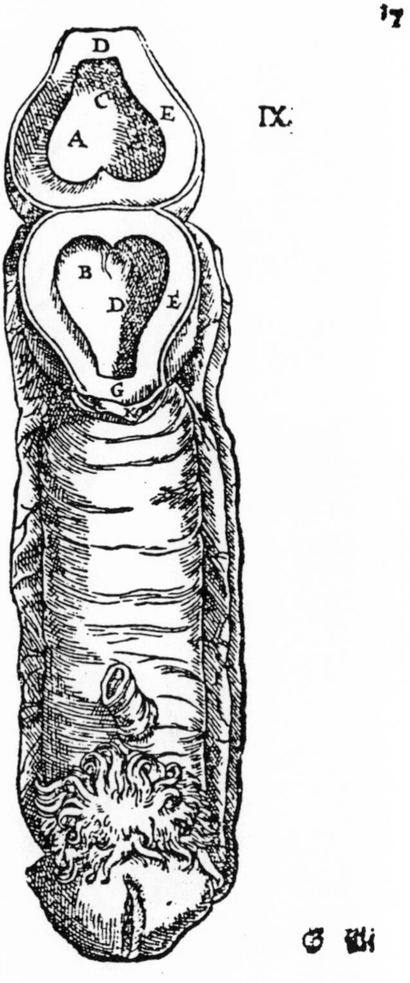

Anatomical Figures in the Byrth of Mankynde.

Fig. VIII.

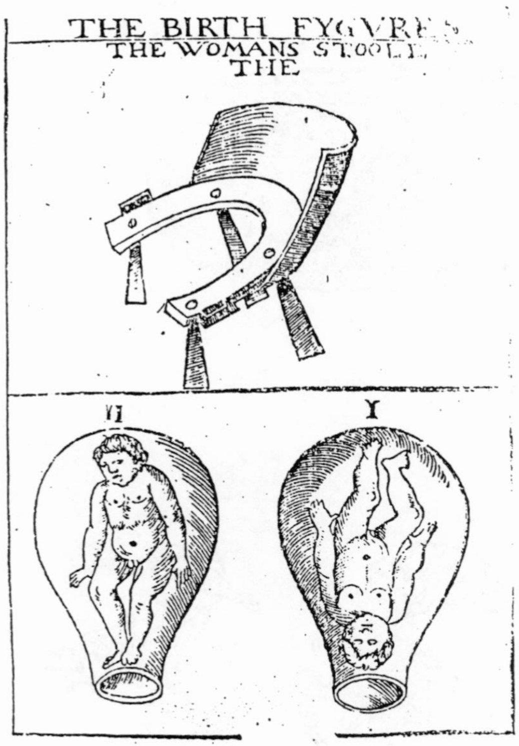

Fig. IX.

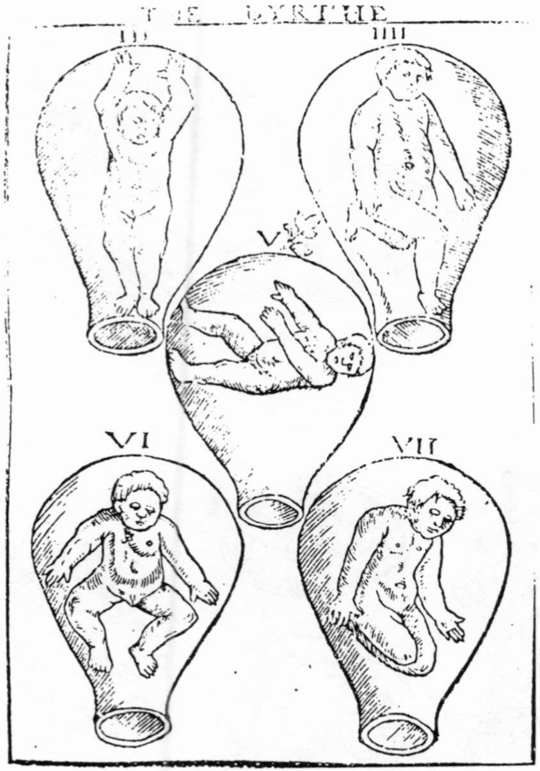

Birth Figures in the "Byrth of Mankynde."

Fig. X.

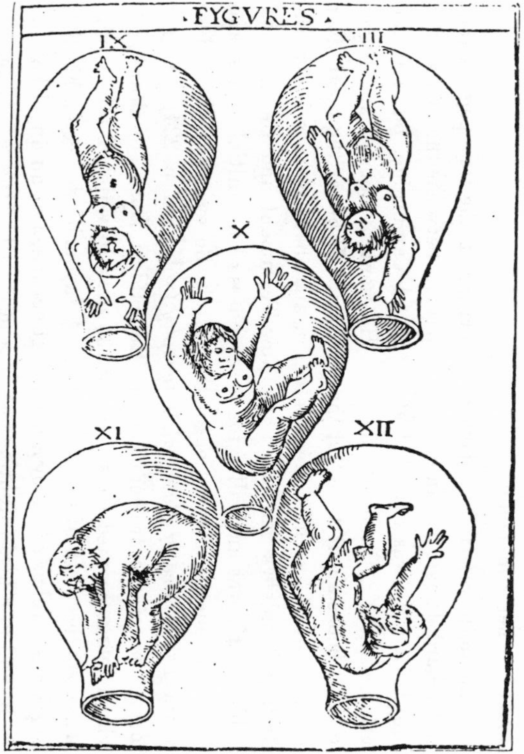

Fig. XI.

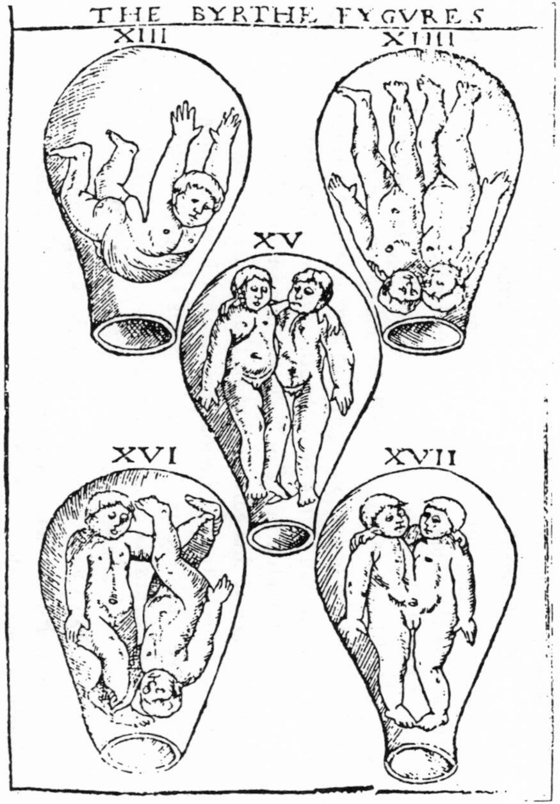

Birth Figures in the "Byrth of Mankynde."

The table of contents has some fine initial letters (F, O and I), and the initial A of the Preface is also fine and resembles that of the 1598 edition. The type is black letter throughout, but of varying size. The Edinburgh University copy contains no plates, but the description of the figures is found as usual at the end of the first book and is the same as that in the 1560 edition, containing, as it does, "The declaration of the Karacters of the nynth figure of women." The ninth figure, it will be remembered, was "the matrix cut foorth of the bodye, beyng of that bygnesse as it was seene taken foorth of a woman at the laste Anothomye, which I did see at the vniuersitie of Padua in Italie." There is no indication of the place of publication of this edition; but in the British Museum Catalogue of Printed Books the suggestion is made that it was London. At the end of the work occurs the word "Finis" with "1565" below it. Below that, again, is a fine ornamental emblem with the following sentences incorporated in it: "Cogita mori;" "Pro lege, rege, et grege;"[1] "Love kepythe lawe, obeyeth the kynge, and is good to the common welthe." The paragraph at the end of the second book closely resembles that in the 1560 edition and widely differs from the corresponding passages in the 1552 and 1545 editions. It runs as follows: "And as touchyng the aboue Trochiskes and emplasters described in this place, ye shall fynde them alwayes redye made in Boucklers berie in London."

The motto referred to above as occurring on the title page is that of Richard Jugge, a printer, who flourished between the years 1531 and 1577. He was educated at Eton and Cambridge, but left the latter without a University degree. He went to London and began to print books about 1548 "at the sign of the Bible at the north door of St. Paul's Church." He printed, in all, about seventy books, including Tyndale's version of the New Testament and the Bishops' Bible. "He was unrivalled for the richness of his initial letters, and for the handsome disposition of the text." So writes H. R. Tedder in the *Dictionary of National Biography* (vol. xxx., p. 224); and my inspection of the two editions of Raynalde's *Byrth of Mankynde* which Jugge printed (this one and the next to which I refer) fully bears out the truth of the statement. A favourite device on his title pages was a rebus of his own name. It consisted of an angel holding an R, and of a bird (nightingale), in a nest with "Jugge" "Jugge"[2]

[1] This is the motto of the Earl of Beresborough and of Lords Brougham and De Mauley.

[2] "Jug," Jug," is an imitative representation of the notes of the nightingale's song.

printed above it. This enigmatical representation occurs in the undated edition to which I must now briefly refer.

(f) *The Undated (1564?) Edition.*

There is an undated edition printed, like the 1565 edition, by R. Jugge; to this the provisional date of 1564 may perhaps be given. I have examined two copies of this edition; one of these is in the Library of the Royal College of Physicians of London, and the other belongs to Dr. J. F. Payne, the Harveian Librarian of the College. This edition has a finely executed title page, resembling in some details that of the 1565 issue. At the top is the motto of the Crown of England ("Honi soit qui mal y pense"), and at the bottom is the pelican feeding her young with her own blood. The printer's motto occurs in the ornamental border: at one side is an angel (or Cupid) holding the letter R, and at the other a bird (nightingale) in a nest with the word "Jugge" over it. The title is preceded by the paragraph sign (¶), and the setting is as follows:—

¶ The Birth
of mankynde, other-
wyse named the wo-
mans booke
Newly set foorth, corrected, and
augmented. Those con-
tentes ye may reade in
the Table of the
booke, and
most
plainely in the
prologue.
By Thomas Raynalde
Phisition.

At the end of the work is a fine colophon with the same mottoes ("Pro lege," etc.) as in the 1565 edition. Some of the capital letters are markedly ornate, such as the A in the Preface and the F in the Table of Contents. The work contains 131 leaves and 13 preliminary leaves (288 pages). The College of Physicians' copy has one folding plate giving nine figures of the organs of generation, of which the ninth is on a larger scale than it appears in the other editions I have seen; but the Birth Figures are lacking. Dr. Payne's copy, on the

Fig. XII.

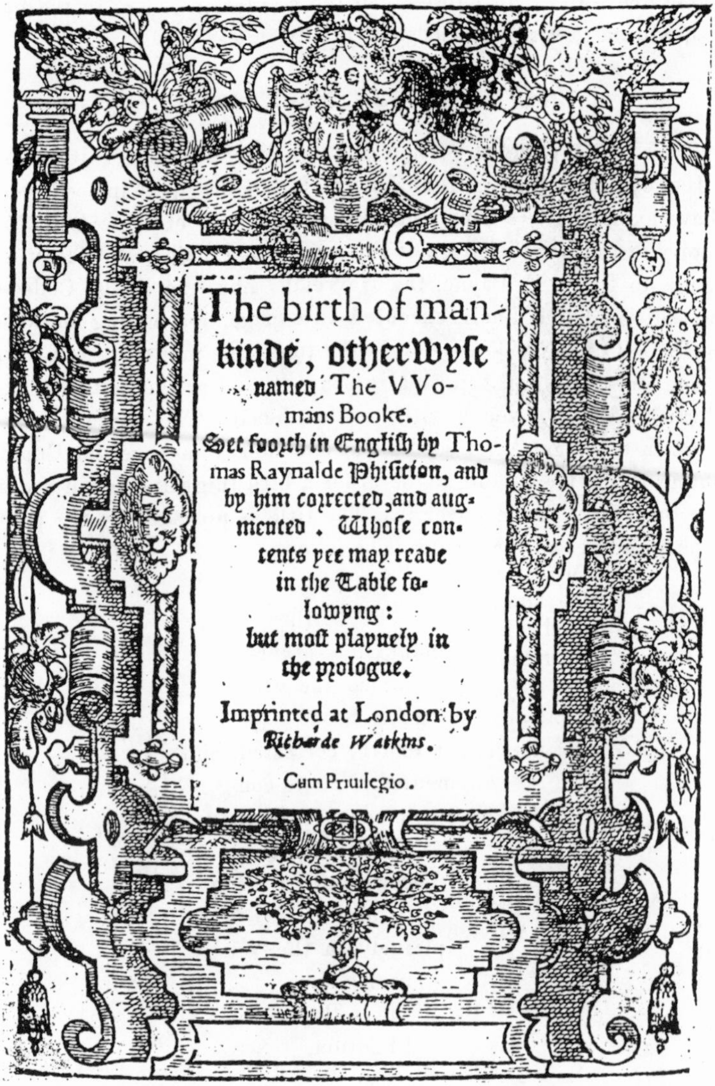

The birth of man-
kinde, otherwyse
named The VVo-
mans Booke.
Set foorth in English by Tho-
mas Raynalde Phisition, and
by him corrected, and aug-
mented. Whose con-
tents yee may reade
in the Table fo-
lowyng:
but most playnely in
the prologue.

Imprinted at London by
Richarde Watkins.

Cum Priuilegio.

Title Page of the 1598 Edition.

Fig. XIII.

L be it ſome *Ariſtarchus* may perhappes finde ſome lacke of faythfulnes and diligence in this worke:yet there is none ſo frowarde to deny, but that there is ſome fruite and profite to bee founde therein, ſeeing that it commeth now abroade much more enlarged and encreaſed, and more diligently corrected, then it was before eyther in the Latine or in the Engliſh. And where before in the other printes, there lacked matter neceſſarie to the opening and declaration of the Figures parteyning to the inner partes : it is now ſo plainely ſet foorth, that the ſimpleſt Mydwife which can read, may both vnderſtand for her better inſtruction, and alſo other women that haue neede of her helpe, the more commoditie. VVherefore my deſire is, that it may be receiued and practiſed of Midwiues, and all other Matrones, with no leſſe ſucceſſe then it is with good wyll and deſire written, to profite, and to do good to other.

A ii.

Preface to the 1598 Edition.

other hand, has the Birth Figures but lacks the folding plate of the anatomical illustrations. No doubt a perfect copy had both sets of figures, for both are described in the text.

The paragraph at the end of the second book reads as follows: "And as touching the aboue Trochiskes and emplasters described here in this place, ye shall fynde them alwayes redy made in Bucklers berie in London." It will be noted that this very closely agrees with the corresponding paragraph in the 1565 edition (q.v.).

(g) *The 1598 Edition.*

The edition of 1598 contains 204 pages and three preliminary leaves. It has an ornate title page (reproduced in fig. 12) and the title is partly in black letter and partly in ordinary type. There are five wood-cut plates containing nine figures occupying pages 83 to 88; the "Woman's Stoole" and seventeen "birth fygures" (in four plates) are found on pages 107 to 110. On the title page is the information that the work was "imprinted at London by Richard Watkins. Cum Priuilegio."; and on the last page is the statement "Imprinted at London by Richard Watkins. 1598." Further, in the scroll work near the top of the title page are the initials R.B. and T.G., in the form of monograms. The preface with a fine large initial letter is reproduced in fig. 13. In the table of contents I notice that the word "folio" is used for the odd numbers and "page" for the even numbers.

The copy of the 1598 edition which I have seen belongs to Dr. C. E. Underhill, to whom I am much indebted for permission to examine it. It is a fine and perfect copy; from the book plate, it is seen that it was once the property of F. R. MacDonald, M.D. The margins are wide, and the marginalia are not in black letter, although a large part of the work is so printed. There are practically no MS. notes in it.

It is noteworthy that in this edition the reference to "Boucklersberie" in the passage at the end of the second book has disappeared. It reads as follows: "And as touching the aboue Trochisks and emplasters described here in this place, yee shall finde them alwayes readie made in some Apotheries (*sic*) shoppes in London." This declaration concerning the Trochisks appears in practically the same form in the 1626, 1634, and 1654 editions; there was now no longer any need to specify any one shop where they were to be obtained, and doubtless "Wylliam Normeuyl" of the Three Doves had long since passed to his rest.

Richard Watkins, the printer of the 1598 edition, dwelt in St. Paul's Churchyard, and had a shop adjoining to the little Conduit, in Cheap. He had a patent, with James Roberts, for printing almanacks, and he was Warden of the Stationer's Company in 1583. He gave up his right of the sheet or broadside almanacks to the relief of the poor stationers. In 1598 he seems to have given the license for printing the "Byrth of Mankynde" by assignment to Thomas Adams, who dwelt at the White Lion, St. Paul's Churchyard. Adams appears as the printer of the 1604 edition.

It is noteworthy that comparatively few editions of Raynalde's "Byrth of Mankynde" appeared during the long reign of Queen Elizabeth. I have found no trace of any between 1565 and 1598. It is true that in the catalogue of the Radford Library in Manchester there is named an edition of Raynalde's work with the date 1593 in brackets appended; but it seems likely that this is the 1598 edition, for the date appears only in the colophon of that edition, and the 8 is rather like a 3.

(h and i) *The 1604 and 1613 Editions.*

I have not had an opportunity of examining either the 1604 or the 1613 editions. Of the former there is a copy in the British Museum and one in the Library at Washington, while of the latter the Royal College of Surgeons of London has two copies. Apparently these editions resemble the ones preceding and those immediately following them. The 1604 edition is a black letter quarto of 204 pages and was printed by T. Adams, London (*vide supra*).

(j) *The 1626 Edition.*

The edition of 1626 is a small quarto of 204 pages, including those of the Prologue but not those of the Table of Contents. The title page is very elaborately ornate, but in the copy which I have seen it is roughly printed and somewhat blurred. The arrangement and spelling of the words of the title are as follows:—

THE
BIRTH
OF
Man-kinde;
Otherwise named
The Woman's Booke,

Set forth in English by Thomas
Reynald Physitian, and by him
corrected and aug-
mented
The Contents are in the Table follow-
ing, but chiefly in the Prologue.
LONDON
Printed for A.H. and are to be sold
by James Boler.
1626.

The initial letters are poor; and at the end of the work is simply the word FINIS. This edition is only partly in black letter. The illustrations are arranged in two groups: five plates containing nine figures are placed at the end of the first book and occupy pages 83 to 87; the four plates with the seventeen "Birth Fygures" appear at the end of Chapter IV. of the second book and occupy pages 107 to 110. The first group contains as usual, the illustrations of the female generative organs, and of the fœtus and the placenta and membranes; the latter contains the picture of the "Woman's Stoole," and the figures showing the various presentations of the fœtus, including the four twin cases. The copy of the 1626 edition, which is in the Library of the University of Edinburgh, was presented to that Library in 1881 by Dr. (now Sir) Dyce Duckworth; he bought it at a sale in the Midlands. Written on the back of the title page is "1681, Mary Buxton owns (?) this Book." Mary Buxton must, I suppose, have been a Seventeenth Century midwife, but I do not find her name mentioned in Dr. Aveling's work on "English Midwives."

(k) *The 1634 Edition.*

Another edition was that of 1634. There are several copies of this edition in England; there is one in the British Museum, one in the Library of the Obstetrical Society of London, and one in the Library of the Royal College of Physicians of London. To the Harveian Librarian of the College of Physicians I am much indebted for the opportunity of examining the last named copy.

There are 204 pages and 3 preliminary leaves in this edition. The title-page has a wood-cut border of an ornate kind, well supplied with devices heraldic in appearance; in it there are figures of Fides, Humilitas, etc. The setting of the title is as follows:—

The Birth
of Man-kinde;
Otherwise Named,
The Womans Booke.
Set forth in English by Thomas
Raynald Physitian, and by him
corrected and augmented.
The contents are in the Table fol-
lowing, but chiefly in the Prologue.
London. . . Printed for A.H. and are to be sold
by John Morret, at the two Tuns
in little Britaine.
1634.

The preface is in English. The printing is partly black letter and partly ordinary type, and the marginalia are in ordinary type. In the table of contents the even numbered pages are referred to as pages and the odd numbered ones as folios; there are also directions for the placing of the plates. The interesting passage at the end of the second book is thus rendered in this edition: "And as touching the aboue Troschiskes and emplaisters described here in this place, yee shall finde them alwayes ready made in some Apothecaries shoppes in London." The whole sentence is in black letter except the word "London."

It appears to me from a comparison of this edition with the so-called 1540 edition of the Obstetrical Society of London that the latter is really the same, that it is in fact the 1634 edition. But the MS, title page has been copied, so it seems to me, from one of the editions printed by Richard Jugge, that namely which is undated but which may have been published about the year 1564. The transcribed pages at the end may have been copied from the same.

In 1634 Paré's works were translated into English, and from this time Raynalde's book declined in influence and importance.

(l) *The 1654 Edition.*

The 1654 edition is the first that is entirely printed in Roman type; the black letter which was small in quantity in the 1634 edition has disappeared altogether in this. The title-page, also, marks a new departure (Fig. 14): it is destitute of all ornamentation; the wording is different; an enumeration of the contents is given; and there is a reference to the illustrations. A third name is now given to the work, for the title reads: "The Birth of Mankind, otherwise called

Fig. XIV.

THE

Birth of Mankind,

Otherwife called,

THE WOMANS BOOK.

OR,

A Guide for VVomen,

In their { Conception. Bearing, and Suckling their Children.

CONTAINING

1. *The Anatomie of the Veſſel of Generation.*
2. *The Formation of the Child in the Womb.*
3. *What Hinders Conception; and its Remedies.*
4. *What furthers Conception.*
5. *A Guide for women in Conception.*
6. *Of Miſcarriage in women.*
7. *A guide for women in Labour.*
8. *A guide for women in their lying in.*
9. *Of Nurſing of Children.*

Illuſtrated with Figures.

Tranſlated into Engliſh by *Thomas Reynald*, Doctor of Phyſick.

The Fourth Edition Corrected and Augmented.

LONDON,
Printed for *J. L. Henry Hood*, *Abel Roper*, and *Rickard Tomlins*, and are to be ſold at their Shops in *Fleetſtreet*, and at the Sun and Bible in Pie-Corner, 1654.

Title Page of the 1654 Edition.

Fig. XV.

Directions for the nurſing of Children, and how to chooſe a good nurſe.

I Am of opinion that it is fit for every Mother to nurſe her own Child, becauſe her milk which is nothing but the blood whitened, which nouriſhed the child in the womb and of which the child was conceived, and formed, is fitteſt, and more naturall to the Child, then the milk of a Stranger, and by the womans nurſing her Child, ſhe ſhall be wholly accounted his Mother, but in caſe the Mother be ſick or weak; or hath no milk, or that her Husband will not let her nurſe her Child, then it is neceſſary to look out a nurſe, but moſt men doe know how hard it is to get a good one.

Now in chooſing a Nurſe there is foure things obſervable.

Firſt, concerning her Parentage, ſhe muſt come of good kindred, and honeſt Parentage, that is not ſtained with any vice, for often times we ſee that although the Father and Mother be healthy, and ſtrong, and ready witted, yet oftentimes the Children are weak and ſickly, or elſe fools.

2. Touching her perſon, for her Age, ſhe ſhould be at her full growth, about 22 yeare of Age, and ſhe will continue good untill ſhe be forty years, for after forty

Supplementary Matter in the 1654 Edition.

the Woman's Book, or a Guide for Women, In their Conception, Bearing, and Suckling their Children." It is also entitled "The Fourth Edition Corrected and Augmented;" but, of course, it was rather the twelfth than the fourth, unless indeed we understand the inscription to mean that it was a corrected and augmented impression of the fourth edition. The augmentation would seem to consist mainly of two and a half pages added at the end of the fourth book, numbered 191, 192, and 193; they follow the original pages 191 and 192. These pages contain "directions for the nursing of Children and how to choose a good nurse," which are additions to the statements already given in Chapter II. of the third book. One of them is reproduced in Fig. 15.

The work consists of 195 pages and 3 preliminary leaves. There is no colophon. On the title-page is the statement that it was printed in London "for J. L. Henry Hood, Abel Roper, and Richard Tomlins, and are to be sold at their shops in Fleetstreet, and at the Sun and Bible in Pie-Corner, 1654." The table of contents is numbered according to pages and not at all according to folios (cf. the 1634 edition, etc.). The anatomical plates occupy pages 81 to 86, and the Birth Fygures pages 105 to 108 (misprinted 118). The paragraph at the end of Book II. (referring to the Apothecaries of London) is practically identical with that in the 1634 edition, save that the spelling is modernized. Throughout the whole work there is a modernizing of the spelling and of the form of the words, so that this edition resembles closely other books published about the middle of the seventeenth century.

(m) *The 1676 Edition.*

A copy of the 1676 edition is in the British Museum; and it is stated in the Catalogue to be a black letter copy. So far as I have been able to discover this was the last edition to appear. Doubtless the growing popularity of Paré's works, and the appearance of "The Expert Midwife" (translation of Rueff's work), of Culpeper's "Directory for Midwives," of "The Compleat Midwife's Practice," and of P. Chamberlain's "Midwife's Practice," tended slowly to dislodge the "Byrth of Mankynde" from the position of supremacy which it had so long occupied during the sixteenth and seventeenth centuries.

The Various Editions.

I have now passed in review the various editions of the "Byrth of Mankynde," of which I have found any trace. It is possible that there may have been others. For instance, McClintock in his annotations on "Smellie's Midwifery" (vol. i., p. 53, 1876) refers to a 1606 edition which he had seen, but he adds "the 0 here has somewhat the appearance of having been surreptitiously altered with a pen and may have been printed 2." I think it is probable that it was really the 1626 edition that he had to do with. There are, therefore, thirteen known editions, or fourteen if we admit the Radford Library 1593 edition. Of these I have been able to examine nine. I have also seen four other copies, but they were duplications of the 1560, 1564 (?), and 1654 editions In all I have traced forty copies to their present resting places either in public libraries or in the private collections of medical men; and on the following page I give in tabular form the editions and where they are at present to be found.

I gladly take this opportunity of acknowledging my indebtedness to the following gentlemen for facilitating my researches: Drs. Cullingworth and J. F. Payne (London); Professor Stephenson (Aberdeen); Drs. W. L. Reid, Lawrence Oliphant, and John Ferguson (Glasgow); Dr. Underhill, Dr. Barbour Simpson, and Mr. J. R. N. Macphail (Edinburgh); and the Librarian of the Army Medical Museum and Library, Washington, U.S.A.

Summary of the Editions of the "Byrth of Mankynde."

Edition of 1540. Jonas's Translation of Rösslin. British Museum.

Edition of 1545. Raynalde's Translation. Royal College of Physicians, London; Hunterian Library, University, Glasgow.

Edition of 1552. Raynalde's Translation. Royal College of Physicians, Edinburgh; and London Obstetrical Society.

Edition of 1560. Raynalde's Translation. Royal College of Surgeons, London; Dr. J. W. Ballantyne, Edinburgh; University of Aberdeen.

Edition of 1565. Raynalde's Translation. British Museum; University of Edinburgh; R. College of Surgeons of London; Washington Library; Hunterian Library, University, Glasgow.

Edition of 1564 (?). Raynalde's Translation. British Museum; University of Glasgow; Royal College of Physicians, London; Dr. J. F. Payne, London.

Edition of 1593 (?). Raynalde's Translation. Radford Library, Manchester.

Edition of 1598. Raynalde's Translation. British Museum; Royal Medico-Chirurgical Society of London; Dr. C. E. Underhill; Washington Library; Hunterian Library, University, Glasgow.

Edition of 1604. Raynalde's Translation. British Museum; Washington Library.

Edition of 1613. Raynalde's Translation. Royal College of Surgeons of London (2 copies).

Edition of 1626. Raynalde's Translation. British Museum; University of Edinburgh; London Obstetrical Society; Faculty of Physicians and Surgeons of Glasgow.

Edition of 1634. Raynalde's Translation. British Museum; Royal College of Physicians, London; London Obstetrical Society; Washington Library.

Edition of 1654. Raynalde's Translation. British Museum; Washington Library; Dr. W. L. Reid, Glasgow; University of Aberdeen.

Edition of 1676. Raynalde's Translation. British Museum.

[The Second Part of this article, dealing with the CONTENTS of Raynalde's *Byrth of Mankynde* will appear shortly.]

The Journal of Obstetrics and Gynæcology of the British Empire

Vol. XII. SEPTEMBER, 1907. No. 3.

The "Byrth of Mankynde."

(Its Contents.)

By J. W. Ballantyne, M.D., F.R.C.P. (Edin.), F.R.S. (Edin.), *Lecturer on Midwifery and Diseases of Women, Surgeons' Hall and Medical College for Women, Edinburgh; Physician to the Royal Maternity Hospital, Edinburgh; and President of the Edinburgh Obstetrical Society.*

I have already* dealt with the author and the editions of the remarkable book known as the *Byrth of Mankynde*; and I now intend to complete the consideration of the subject by discussing the contents of the work and the light thus thrown upon sixteenth century midwifery. Before I do so, however, let me remind the reader that Raynalde's *Byrth of Mankynde* is really a composite work, for it contains an English translation (by Richard Jonas) of Rösslin's *De Partu Hominis*, along with new matter added by Thomas Raynalde. It has, therefore, three component parts: Rösslin's *De Partu Hominis* as translated by Jonas, his additions thereto, and those which came from the pen of Raynalde. In such a late edition as that of 1654, a fourth element is found in the supplementary matter added by a seventeenth century editor whose name we do not know; but it is small in amount, consisting chiefly of directions for the nursing of children, and no further reference will be made to it.

For purposes of description I might have taken any one of the first four Raynalde editions (those of 1545, 1552, 1560 and 1565), for it is probable that they all appeared during the lifetime of Raynalde;† but I have chosen that of 1560, for it represents what was practically the final recension of the work, and does not differ (save only in spelling and minor details) from so late an edition as that of 1654. The quotations which occur throughout this article

* *Journ. of Obstet. and Gynæcol. of the Brit. Empire*, vol. x., pp. 297—325, 1906.

† Raynalde is not likely to have been alive when the 1598 edition appeared.

are, therefore, to be regarded as taken from the 1560 edition except when it is otherwise stated. At the same time, I shall now and again refer to differences existing between the various editions, which may appear to have an interest other than the merely typographical and orthographical interest; and I shall draw attention to the very marked differences which are revealed by a collation of Jonas's edition of 1540 with the editions of Raynalde from 1545 onwards. Further, when I reach, in the description, the part of the book which is a translation of Rösslin's *De Partu Hominis,* I shall devote some sentences to the consideration of this German obstetrician's life and work.

The contents of Raynalde's *Byrth of Mankynde,* it need hardly be said, have a special interest for obstetricians. The book stands, so to say, between the old and the new in obstetrics: empiric midwifery was, in 1540, beginning to give way before the advance of scientific obstetrics, although many years had still to pass before the knowledge of the anatomy and physiology of labour and the principles of anæsthesia and asepsis were to sweep away the practices begotten of ignorance and superstition. The year 1550, in which Paré introduced podalic version into obstetric practice, has usually been regarded as marking the separation between the midwifery of the past and that of the present; if this be so, then Raynalde's *Byrth of Mankynde* appeared at the very time when the great transition from the empiric methods of the ancients to the scientific procedures of the moderns was being brought about. It cannot fail to be a matter of the most intense interest to know what were the practices and principles of midwifery in England in the middle of the sixteenth century, and we find them both revealed to us in Raynalde's book.

The Preliminary Matter.

The preliminary matter in a typical Raynalde edition of the *Byrth of Mankynde* (such as that of 1560) consists of the title page, the Aristarchus preface, and the table of contents. In the single Jonas edition of 1540 are the title page, the religious admonition, the dedication to Queen Catherine, the table of contents, and a table of weights and measures.

The *Title Page* differs little, so far as wording is concerned, in the various editions prior to that of 1654. That of the 1560 edition reads as follows:—"The Byrth of ‖ mankynde, otherwyse na ‖ med the womans ‖ Booke. ‖ Newly set foorth, corrected and ‖ augmented: whose contentes ye ‖ may reade in the Table of the ‖ Booke, and most playne ‖ ly in the Pro ‖ logue. ‖ By Thomas Raynalde ‖ Physition ‖ 1560." The titles of the 1545 and 1552 editions are similar, and no alterations of note appear until 1598, when the title reads: "The birth of man ‖ kinde, otherwyse ‖ named the Wo ‖ mans Booke. ‖ Set foorth in English by Tho ‖ mas Raynalde Phisition, and ‖ by him corrected, and aug ‖ mented. Whose con ‖ tents yee may reade ‖ in the Table fo ‖ lowyng: ‖ but most playnely in ‖ the prologue." One or two slight differences are to be noted in the editions of 1604, 1613, 1626, and 1634, and then there occurs the radical change of the 1654 edition (see Plate XIV. in my previous article, *loc. cit.*, p. 322).

The wording of the title page of the Jonas edition is, as might be expected, somewhat different: "The Byrth of Mankynde, newly translated out of Laten into Englysshe. In the which is entreated of all suche thynges the which chaunce to women in theyre labor, and all suche infyrmities whiche happen unto the Infantes after they be delyuered. And also at the latter ende or in the thyrde or last boke is entreated of the Conception of mankynde, and howe manye wayes it may be letted or furtheryd, with diuers other fruytefull thynges, as doth appere in the table before the booke. Cum privilegio Regali, ad imprimendum solum."

The ornamental border of the title page differs greatly in the different editions which were published before that of 1654, in which it entirely disappears; but with this matter I have already dealt fully in my first article (*loc. cit.*).

The reverse side of the title page contains (in the 1560 edition) a short *Preface*, in English, commending the work in modest terms to midwives and matrons. The same preface appears in Latin in the 1545 and 1552 editions, and in English in all the others, but it is not always printed on the reverse side of the title page. I have reproduced it in Plates II. and XIII. of my former article, both in its Latin and in its English form. Since it begins "Albeit some Aristarchus," I have named it the *Aristarchus Preface*.

The Aristarchus Preface is absent from the Jonas edition of 1540; but, occupying the corresponding position there is a *religious admonition* which reads as follows (in ordinary English): "Unto the Reader. An admonition to the reader. For so much as we have enterprised the interpretation of this present book, offering and dedicating it unto our most gracious and virtuous Queen Katherine only, by it minding and tending the utility and wealth of all women, as touching the great peril and dangers which most commonly oppresseth them in their painful labours, I require all such men in the name of God, which at any time shall chance to have this book, that they use it godly and only to the profit of their neighbours, utterly eschewing all ribald and unseemly communications of any things contained in the same, as they will answer before God, which, as witnesseth Christ, will require a count of all their words, and much more then of all ribald and uncharitable words. Everything, as saith Solomon, hath his time, and truly that is far out of time, yea and far from all good honesty, that some use at the common tables and without any difference before all companies, rudely and loudly to talk of such things, in the which they ought rather to know much and say little, but only where it may do good, magnifying the mighty God of nature in all his works, compassionating and pitying our even-Christians,* the women which sustain and endure for the time so great dolor and pain for the birth of mankind and deliverance of the same into the world. Praise God in all His works."

The rest of the preliminary matter of such an edition as that of 1560 consists of the Table of Contents, as we should now call it, or, as it it is here termed, simply the Table. I have collated the Contents of the 1545, 1552, 1560 and 1654 editions, and I find them to be almost identical, except, of course, with regard to the spelling, type, and pagination. There are, however, one or two small divergences; thus, in the editions of 1545 and 1654 there is a heading in the third book ("Remedy for the Cramp, or distention of members") which does not appear in the other two editions referred to, but the presence of the heading does not indicate the addition of new matter in the text; again, in the 1545 and 1552 editions, eleven Anatomical Figures are named and nine only in the 1560 and 1654 editions. In the 1560 edition (in my copy at least) the "byrthe fygures" (seventeen in number or eighteen, if the 'stoole' be counted) follow after the Table of Contents; but in several of the other editions they are inserted at the end of the fourth chapter of the Second Book, their proper place according to the modern rules of binding.

Before we reach the Table of Contents of the Jonas edition of 1540, we find additional preliminary matter consisting of the Dedication. It is so interesting that I reproduce it here in full. Some parts of it, it may be added, reappear in the Prologue to the Women Readers which is peculiar to the Raynalde editions. In ordinary English it reads as follows:—

"Unto the Queen. Unto the most Gracious, and in all goodness most excellent virtuous Lady Queen Katherine, wife and most dearly beloved spouse unto the most mighty sapient Christian Prince King Henry the VIII. Richard Jonas wisheth perpetual joy and felicity. Whereas of late (most excellent and virtuous Queen) many goodly and proper treatises, as well concerning holy scriptures, wherein is contained the only comfort and consolation of all godly people: as other profane arts and sciences right necessary to be known and had in use, have been by the painful diligence of such clerks which have embusied them in the same very earnestly and circumspectly set forth in this our vulgar English tongue, to the great enriching of our mother language, and also the great utility and profit of all people using the same, and among all other things, out of the noble science of Physick, have been divers proper and

* *Even-Christen* or *even-Christian* meant *fellow Christian.*

profitable matters compiled and translated from the Latin tongue into English, by the reading of the which right many have confessed themselves to have received great light and knowledge of such things, in the which they have found no small comfort and profit. And in this behalf there is in the Latin speech a book entitled *De Partu Hominis*: that is to say, of the birth of mankind, compiled by a famous doctor in Physick, called Eucharius, the which he wrote in his own mother tongue, that is, being a German, in the German speech, afterward by another honest clerk at the request and desire of his friend transposed into Latin: the which book for the singular utility and profit which ensueth unto all such as read it, and most specially unto all women (for whose only cause it was written) hath been in the Dutch and French speech set forth and imprinted in great number, so that there be few matrons and women in those parts but (if they can read) will have this book always in readiness: considering then that the same commodity and profit which they in their regions do obtain by enjoying of this little book in their maternal language, might also ensue unto all women in this noble realm of England, I have done my simple endeavour for the love of all womanhood, and chiefly for the most bound service which I owe unto your most gracious highness to translate the same into our tongue. Most humbly desiring first your grace's highness, and then consequently all noble ladies and gentlewomen with other honest matrons to accept my pains and goodwill employed in the same: the which thing as I do not doubt for the wont and incomparable benignity goodness and gentleness inset and planted in your grace's nature, so shall it be no little encouraging unto me hereafter with farther deliberation and pains to revise and oversee the same again, and with much more diligence to set it forth. For considering the manifold daily and imminent dangers and perils, the which all manner of women or what estate or degree they be in their labour do sustain and abide: yea many times with peril of their life, of the which there be so many examples needless here to be rehearsed. I thought it should be a very charitable and laudable deed, yea and thankfully to be accepted of all honourable and honest matrons if this little treatise so fruitful and profitable for the same purpose were made English, so that by that means it might be read and understanded of them all, for as touching midwives, as there be many of them right expert, diligent, wise, circumspect and tender about such business, so there be again many more full undiscrete, unreasonable and far to seek in such things the which should chiefly help and succour the good women in their most painful labour and throngs. Through whose rudeness and rashness only I doubt not, that a great number are cast away and destroyed (the more pity). For this cause and for the honour of Almighty God, and for the most bound service the which I owe unto your grace, most gracious and virtuous queen, I have judged my labour and pains in this behalf right well bestowed, requiring all other women of what estate soever they be, which shall by reading of the same find light and comfort, to yield and render thanks unto your most gracious highness, wishing greatly that it might please all honest and motherly midwives diligently to read and oversee the same, of the which although there be many which do know much more peradventure than is here expressed, yet am I sure in the reading of it their understanding shall be much cleared and have somewhat farther perseverance in the same. It is no small charge which they take upon them, for if when any strange or perilous case doth chance, the midwife be ignorant or to seek in such things which are to be had in remembrance in that case, then is the party lost and utterly perished, for lack of due knowledge requisite to be had in the midwife. Wherefore I beseech Almighty God that this my simple industry and labour may be through your grace unto the utility, wealth, and profit of all English women, according to my utter and hearty desire and intent, to whom also I daily pray long to preserve and prosper your most gracious highness, both to the continual comfort and consolation of our most redoubted and without comparison most excellent Christian prince, and also the joy and gladness of his loving subjects. Amen."

Of course the Table of Contents of the 1540 or Jonas edition differs much from those of the Raynalde editions of 1545, 1552, 1560, etc. There is no reference in it to a Prologue, for no such thing exists in the 1540 issue, unless we regard the Religious Admonition and the Dedication as jointly constituting one. Then, we note the absence of the first thirteen chapters of the first Book of all the Raynalde editions; and the sixth chapter of the fourth Book of the Raynalde editions, containing the cosmetic suggestions or "bellifying receipts," is not represented in the Table of Contents of the 1540 edition. Roughly speaking, the Jonas or 1540 edition contains the second, third and fourth Books of the Raynalde editions, plus the last chapter of the first Book and minus the last chapter of the fourth Book of these later editions. But I have already, in my former article, set forth in detail the differences between the Contents of the two issues (*loc. cit.*, p. 310).

In the 1540 edition the preliminary matter closes with two interesting paragraphs dealing with weights and measures and with drugs and apothecaries; they are peculiar to this edition. The former is introduced in these words: "For because that in this book many times be found certain measures and weights of physic, not known peradventure to all such as that chance to read it, therefore here briefly I have set them forth, showing the value and estimation of them so far as they shall be requisite to the better understanding of such things the which ye shall read in the same treatise." Then follows the table of the weights and measures, the pound being stated to contain 5,762 grains. The second paragraph refers to the obtaining of the drugs named in the text of the book, and reads thus: "Ye shall also note here that many times ye shall happen upon strange names of such things the which are occupied about infirmities spoken of in this book, for the which theyr is no English but are used in their own proper names of Greek or Latin: and they are such for the most part which are to be had only at the apothecaries, being of them right well known; wherefore when ye shall need any such thing if ye send the same names in your bill to the apothecaries they will soon speed your purpose: neither do this if ye may without the advice of some expert and learned physitian."

I have now enumerated all the parts of the preliminary matter, both as found in the Raynalde editions of the *Byrth of Mankynde* and in that first edition with which the name of Richard Jonas is associated. I now pass to the Prologue, which is to English readers perhaps the most attractive part of the work.

The Prologue.

To the English obstetrician, as well as to the student of the manners and customs of the sixteenth century, the *Prologue to the Women Readers* will be by far the most interesting part of the book. It is not a translation of anything in Rösslin;* it is not indeed a translation of anything at all. We may regard it as a piece of original writing fresh from the mind of Raynalde, giving the history of the work and throwing an important sidelight upon the way in which obstetric matters were looked upon in England in the middle of the sixteenth century. There are, it is true, a few passages in it which are reminiscent of some parts of the *Dedication to Queen Katherine* in the 1540 edition; but the greater part of it must be ascribed to Raynalde.

The intent of the author (" the entent of thauctour ") is to recite the sum and chief contents of the book, for it is " a great pricke or allurement, entising and meuinge a man, to reade any boke, when he is somwhat first admonyshed of the matters comprehended and contayned therein." Then, without giving any names, Raynalde tells how the studious and diligent clerk [Jonas, to wit] made the English translation of the Latin work [by Rösslin] entitled *De Partu Hominis*, and called it " the byrth of mankynde "; it is now to be named " the womans boke." That translation is now to be corrected and augmented, revised " from top to to," as the writer quaintly says; and there are to be " set forth and evidently declared al the inward partes of women, and that not onely in wordes, but also in ſyvely and expresse figures." Raynalde beseeches the midwives who

* There is a " Prologue " in Rösslin's *De partu hominis* (edition of 1538, Paris) but it has nothing in common with Raynalde's.

will read his book to pay special attention to these anatomical matters, for, he adds, "when a person is sycke or dyseased in any part, it is halfe a comfort, yea halfe his helth, to understand in what part the dysease is, and howe that parte lyeth in the bodye."

The second part of his book ("the seconde booke") is to concern itself with labour, "with the byrth of mankind and al the daungers, perels, and other cases happenyng to the labourynge woman at that season." This portion of the work is to be illustrated with the *Byrth Fygures* and the picture of the *Womans Stoole*; the first part had the anatomical figures from Vesalius to elucidate the text.

The third book considers the choice of a wet nurse. "Item medicines encreasyng, deminishinge, attenuatyng, engrossinge, and amendynge the mylke in the nources brestes. Also remedies for manye and sundry diseases, which oft tymes chaunce unto infantes after theyr byrth."

In the fourth book the author proposes to discuss the question of conception and the overcoming of sterility. "And farther in this last booke shall be uttered and set forth certayne embelleshinge receptes concerning onely honeste and helthsome decoration and clenlynes." The writer evidently feels that he is on dangerous ground, for he adds that he is to teach "nothinge in that place but that onely whiche may make to the honest, comely, and commendable conservinge and maintaininge of the inset and natural beautie in a woman, utterly abhorring and defying all farding, paynting, and counterfeit cast coulors, which of some dampnable and misproude people be dayly used, such as by all meanes possible, seke and search more the abhominable and divilish painting and garish setting forth of their mortal carcases (the better therby to commend it unto the eyes of foolish and fond men) than by honest, sober, debonayre and gentil maners, so to demene their life, that they may therby rather obtayne the loue, amitie, and hartie perpetual favour first of god, and then of al honest, discrete, and godly wise men."

After giving this brief summary of the contents of the book, Raynalde asks his women readers ("for whose sake and only respect it is set forth") to give it their benevolent favour and good acceptation. He is quite sure, however, that to some the work will not be acceptable. Not even an invocation of the gods and goddesses ("great Apollo, wyttye Mercury, and sweet Suada") will suffice to convince them who give so "precipitat and heady judgementes in all maner of matters," that the book is useful and good. Some will allege "that it is shame, and other somme, that it is not meete ne fyttynge such matters to be entreated of so playnly in our mother and vulgare language, to the dyshonoure (as they say) of womanhed, and the derision of theyr wonte secretes, by the detection and discoverynge whereof, men it readynge or hearing, shalbe moved thereby the more to abhorre and loothe the company of women."

But it is of no use to attempt to convince such people. Nothing is so good but it may be abused. Fire and water, meat and drink, the Bible, even the blessed Sacrament may be abused; but "to them that be good theimselfe, everye thinge turneth to good, whatever it be is to them a sufficient matter and occasion therein to seke the glory of God, and the onely profyte of their even Christen." "Wherefore," the writer concludes, "consydering that there is nothinge in this world so necessary, ne so good, holye, or virtuous, but that it maye by wyckednesse be abused, it shalbe no great wonder though this lyttle booke also, made, written, and set foorth for a good purpose, yet by lyght and leude persons be used contrary to godlynesse, honesty, or thentent of the wryter thereof."

The only possible dangers, so far as the writer can see, are that some of the medicines referred to may be employed for a criminal purpose ("some divelishe and lewde use"), and that the book falling into any "lyght marchauntes handes" may be used for the derision of women. That men by reading such a book should "conceave a certayne lothsomnes and abhorrynge towardes a woman" is answered by the fact that if this were so then "Physitians and chyrurgians wyves should greatly be abhorred and mysbeloved of theyr husbandes;" and this is not so. "And I my self likewise, which wryteth thys booke, should mervaylouslye above many other abhorre or lothe women." Such "tender reasons" are petty and trifling: "but to be short, there is no such thyng, neither any cause thereto why." In fact, it is rather to be expected that if, by any chance, a husband read the book, he may, if of a gentle and loving nature, do his wife good.

Knowing as he does what the perils of childbirth are, the writer thought "it should be a verye charitable and laudable dede, and right thankefully to be accepted of al honourable and other honest matrones, yf by my paynes this lyttle treatyse were made to speake Englyshe, as it hath been longe syth taught to speake dutch, frenche, spanyshe, and dyvers other languages." The *Byrth of Mankynde* may, if properly read and consulted, supply the "roume and place of a good mydwife;" as a matter of fact, the writer knows that it has been so used by "many honourable Ladies and other worshipfull Gentlewomen" and with much profit. But again it has to be confessed that there are some midwives who would have the book forbidden, "forsomuch as therein was descried and set foorth the secretes and privities of women, and that every boy and knave had of these bookes, reading them as openly as the tales of Robin Hood, etc." "But here nowe let not the good midwives be offended with that, that is spoken of the badde. For verely there is no science, but that it hath his Apes, Owles, Beres, and Asses." The good midwives, however, were glad to get the book. "And thus I conclude and make an ende of this rude Prologue, requyryng the gentle readers therof, that yf they shall fynde any thing therin interpretable to dyuers senses, to accepte onely that which may make to the best, accordynge to my meaning."

The First Book.

Like the Prologue, the First Book of the *Byrth of Mankynde* contains matter which is not to be found in Rhodion's *De Partu Hominis*. Its contents are mainly anatomical descriptions. The writer is very sure of the "utilitie of the first boke;" it is "as a key, openyng and clearyng the matters to be intreated of in the seconde." It deals with the "fourme, maner, and situacion of the inwarde partes of a woman," with "the campe or fielde of mankynde to be engendred therein."

An interesting paragraph deals with the relative importance of the sexes in the matter of procreation: "And although that man be as principall mouer and cause of the generation: yet (no displeasure to men) the woman doth conferre and contribute much more, what to the encreasement of the child in her womb, and what to the noryshment thereof after the byrth, then doth the man. And doubtlesse yf a man woulde demaunde to whom the chylde oweth most his generation, ye may worthily make aunswere that, to the mother: whether ye regarde the paynes in bearynge, other els the conferrence of most matter in begettyng."

The organs are then described in detail. First, "the principal coates of the body" are referred to: they consist of the superficial skin or cuticula, of the "fleshye" skin or membrana carnosa, and of the third coat or adeps, which lies between the other two. "Immediately under the fleshye skin be conteyned the Muskles." Chapter III. of the First Book is specially concerned with the "Muskles," and particularly with those of the "bellye." These are the musculi obliqui descendentes or "the Byaswyse descendyng muskles;" the musculi obliqui ascendentes or the "Byaswyse ascending muskles;" the musculi recti or "the ryght muskles;" and the musculi transversi or the "overthwart muskles." "All these foure Muskles be to the entrayles and bowelles within the belly, as foure seuerall coates: by the vertue and helpe of whom, together with the ayde of the midwiffe, all expulsion both upward and downewarde in the guttes, in the stomacke, in the matrix of the woman in the tyme of labour, and also in the bladder in tyme of makyng of water, is wrought: and yet besides this utilitie, they clothe (as I have saide) defende, fortifie, and strength the inwardes of the belly."

Chapter IV. of the First Book speaks of the "kell, called Peritoneum," a certain "thin rime;" it "yeldeth unto eche entrayle a coate and webbe of the cloth of his owne body: by the whiche his livery, they be the more arctly and straightly affixed or fastened unto hymselfe." Chapter V. gives the declaration of the names and nature of the Matrix. "The Matrix, the Mother, and the wombe, do signifie but one thing, that is to saye: The place wherein the seede of man is conceaved, fetified, conserved, nourished, and augmented, unto the tyme of deliuerance, in Latin named Uterus and Matrix." From the description which follows, it is evident that the writer intends by "Cervix Uteri" the vulva and vagina. Chapter VI. deals with "the wombe and his partes." "Nowe ye shall understande, that the founde or bottome of the matrix is not perfectly round bowlwyse, but rather lyke the forme of a mans heart, as it is paynted, sauinge that the particion or clifte in the matrix betwene both corners, the ryght and the lefte, is not so profoundlye dented inwardes as the clyfte in the hearte." From this description it would almost appear as if the uterus of one of the Mammalia were intended, or if the writer had come across a case of minor malformation of the organ (uterus septus) in the human subject. He is quite sure, however, that there is only one "holonesse" in the womb; he does not believe in the seven "selles" said to be therein. "In tymes passed, dyverse Clarkes haue written, and many other haue beleued, that there shoulde bee seuen selles, or seuen distinct places in the Matrix, in thre of the whiche, on the ryght syde shoulde onely men chyldren be conceyued, and in the other three on the lefte syde women chyldren, and yf it chaunced that the seede were conceaued in the seuenth sel, whiche was the myddelmoste, then that shoulde become a monster, halfe a man, and halfe a woman. The whiche all is but lyse, dreames, and fonde fantasyes: for the womans Matrix, as I haue saide, is euen as a stronge bladder, hauinge in it but one uniuersall holones, and the chylde when it lyeth in it, lyeth euer on the one syde more then on the other, the head beynge towardes one of the corners or angles, and not upryghte towarde the myddle brydge."

Chapter VI. (VII. correctly) speaks of the "Mother port." This is the Cervix as we nowadays call it. "It is of the forme of a haukes bell, or other lyttle mores belles" (*i.e.*, morris bells). At certain times, "the Matrix beynge apte and dysposed thereto, and other conditions requisite, thys wombe porte do naturally open it selfe, attractinge, drawing and suckinge into the wombe the sede by a vehement and naturall desyre." During pregnancy it remains closed, "untyll the tyme of delvueraunce, at what tyme agayne it delateth and openeth it self, in such amplytude and largenesse, that it is wonderfull to speake of."

Chapter VIII. contains a description of the vessels of seede, called the woman's stones, *i.e.*, the ovaries as we now know them, "wherin is engendred the seede and sparme that commeth from the woman, not so strong, ferme, and myghtie in operation as the seede of man, but rather weake, fluy, colde, and moyste, and of no great firmitie."

But, the writer tells his readers, the woman's seed is just as proper for its purpose as the man's. "These stones be nothynge so bygge as the stones of man, but lesse, flatter, much fashyoned after the shape of a great and brode almonde."

Chapter IX. has to do with the "sede bringers," not, let it be borne in mind, the Fallopian tubes, but the "two vaynes and twoo artyres which come to these two stones." Here we find a description of the blood-vessels of the pelvis along with the views then held as to the origin of what was called the "woman's seed," which are set forth at length in Chapter X. These views have now only an historical interest, for the physiological knowledge on which they rested has long since been replaced by more correct information. To the curious, however, the description given of the four "mines," shops, or workhouses existing in the body cannot but be attractive. "Of this sort of mines, there be foure principall in the bodye of man. The first is the mine of bloud, which is the lyuer, in whom the iuyce of meate, before of colour whyte, is transmuted into red, made apt and fitte to nourishe all partes of the body, attract and drawen out of the stomacke and guttes, thorow verye small and infinite lyttle vaynes into the lyuer. The seconde mine is the heart, which of the bloud attracte and drawen from the great maister vaine, proceading out of the foresayde lyuer, into his parlers, doth engendre vehement and liuely spirite, conmixed with depured and greatly elaborated bloud, within the selles of the heart, from thence sent forth throw the artires, into all partes of the bodye, being in colour yealowyshe, thinne, and hoote bloud. The thyrde mine is the brayne, of whom all the sinewes take thyr originall. In whom the wyttye spirite, the spirites of mouyng, and the spirites of al sensibilitie be engendred, and thorow the sinewes sent to all partes of the bodye. For all suche partes as moue and feele, haue that by reason of sinewes derived unto those places from the head. The fourth mine is the stones, in whom by commixtion of al the other thre foresaid metalles of the body, that is to say, vayne bloud, arteriall bloud, and liuely spirites engendred in the head, is engendred and produced sede, which bestowed in his due place becommeth like in perfection to the creature from whence it came; that is to say of mankynde, man." The writer then goes on to explain how "the seede is receaued into the stones," how the colour of the seed is transmuted, and how the seed in woman is not so firm as in man, etc.

Chapter XI. tells how the seed (of the woman) is sent from the stones to the angles or corners of the Matrix by means of a "wormye bodye," evidently the Fallopian tube. The seed in woman is supposed to be for the purpose of moistening the genital passage as "with a dewe." The writer then proceeds to moralize on what he terms the "prickes of nature." "For yf that the God of nature had not instincted, and insette in the body of man and woman, such a vehement and ardent appetite and luste, the one lawfully to company with the other: neither man ne woman woulde neuer haue ben so attentfe to the workes of generation and encreasement of posteritie, to the utter decaye in shorte tyme of all mankynde. For ye shal heare some women in tyme of theyr trauayle, meued through great payne and intollerable anguyshe, forswere and vowe them selfe, neuer to companye with a man agayne; yet after that the panges be passed, within short whyle, for entyre loue to theyr husbandes, and singular naturall delyte betwene man and woman, they forget both the sorow passed and that that is to come. Suche be the privie works of God, and suche be the prickes of nature, which neuer createth no special pleasure unaccompanyed with some sorow: neither is there for the most part any sorow, but that it hath annexed some ioy or comforte, lesse or more, to alleuiate and lyghten the burthen and weyght of displeasure."

Chapter XII. deals with the bladder in women, with stone ("but women be not so prone ne apt to engender the stone in the bladder as men be"), and with the reason why the urine when it has reached the bladder does not revert again.

Chapter XIII. is a very interesting one, for in it are considered not only the "vaynes which resort to the Matrix," but also "the termes and theyr course with the causes thereof." "Nowe to come to the declaration of the nature of termes, ye shal understand that thei be called in Latine Menstrua, for because that ons in a moneth they happen alwayes to womankynd, after XIII. or XV. yeares of age passed (beynge in theyr perfect health): In Englyshe they bee named Termes, because they retourne eftsoones at certayne seasons, tymes and termes." Having described, as best he knew, how the blood was poured into the Matrix, the writer goes on to tell the cause of the terms, that they are really intended to serve as nourishment for the fœtus ("feature"); for "prudent Lady nature" has wisely so provided; "yea, although the woman do neuer conceaue . . . . yet is there no faute in nature, who hath prepared place, and foode to be at altymes in readynes." It is noted, also, that women that have no terms cannot bear children; that the terms do not follow the waxing and waning of the moon exactly; and that the duration of the flow varies in different women.

In Chapter XIV. are considered at some length the three cauls or wrappers "wherein the infant is lapped" in the uterus. The innermost caul is named the Amnios, "in Latine Agnina, for cause it is as dilicate as lambes bee." "The mydwifes commonly call it the coyfe or byggyn of the chylde, and some call it chyldes shert, the which also many times proceadeth alone with the chylde, eyther uppon the chyldes head, or one of the armes or legges. And then the women reserve it as a thynge that shoulde betoken some great lucke to the chylde in tyme to come." "The seconde wrapper or caule in Greek is called Allantoides, in Latine Farciminosa, in Englyshe these two termes do signifie haggiswyse for because that it is fashioned much after the shape of the outwarde skynne or bagge of an haggisse puddinge." This second caul of Raynalde is our modern chorion apparently, while the third which he names Chorion or Secundina (or hoop caul) would seem to correspond to the placenta; but the description is vague and shows no evidence of close observation of nature. The description of the vessels of the umbilical cord is more exact. "Thorow these Artires, liuely spirite and freshe ayre is deriued out of the mother into the chylde, wherwith the naturall heate of the chylde is viuified and refreshed. And these two Artires with the foresayde nauyll Vayne, when the childe is borne, begin to wyther and drye, euery day more and more, and become much like a harpstryng, without any holownesse or cauitie." The urachus ("another vessell") is also described, and it is said that by it the urine passes from the bladder to the space between the first and second caul without the child's body. The placenta ("chorion") is compared to the spleen or melt "in a man or beast;" "so that to be short, Chorion is the immediat receptacle and receauer of al the vaynes and artires, to be deduced from the Matrix to the chylde, and the chylde receaueth onely at his hand the two Vaines and Artires, whiche by the way as they passe and perse thorow the other two caules, towardes the chyldes Nauyll, they sende into eche of the caules innumerable small eye vaynes and artires, whereby the caules be sustayned and encreased also."

The terms, when there is a fœtus in the uterus to be nourished, are no longer superfluous but are used in supplying nourishment to the infant *in utero*, and that part which is not needed goes to the breasts to become milk. It is not right to regard the terms as a purgation, for the blood of which they are composed is as pure and wholesome "as all the reste of the bloud in anye part of the body els." "Yet much more are to be detested and abhorred, the shameful lyes and slaunder that *Plinie, Albertus magnus de secretes Mulierum*, and diuers other mo haue written, of the venimous and daungerous infective nature of the womans Flowres or Termes: the which all be but dreames and playne dotage. To rehearse theyr fond wordes here, were but losse of inke and paper, wherfore let them passe with theyr auctours." It is not a little amusing to read such denunciations of Pliny and Albertus Magnus following so closely after some of the anatomical descriptions that have gone before; but in his views upon the functions of the placenta as set forth in the following paragraph the writer is far in advance of his time. "Forbecause that she (Nature) woulde that the pure bloud commyng from the Matrix vaynes, should be made yet purer, she suffereth not the same to entre immediately into the infante, but first useth another meane, and sendeth it into Chorion or the hoope call (as I haue sayd before), where truely it hath a certayne circulation, and another digestion, wherby it is desecate, and clensed very exquisitly, by the diligentis of nature attenuated and fined, and so at the laste sent foorth into the infant, leauyng all the grosser part in the spungye bodye of the hoope caule." (It is to be remembered that the hoop-caul is the placenta.)

The Fifteenth Chapter is concerned with some curious considerations regarding which of the three Matrix veins contain the Terms and how the milk comes to the woman's breasts. The importance of knowing which of the veins contain the menses is, the writer thinks, at once evident when we have to deal with too much or too little monthly flow; to put these anomalies right we have to apply medicines, and if the menses come only from the veins of the neck of the womb there will be no use in applying medicines to those of the fundus. The writer is of opinion (for reasons which it is unnecessary to discuss now) that the terms come from the veins at the fundus only. It is in this connexion that the writer relates the history of two cases in his practice, the one in London and the other in Paris, to which reference has already been made (*Journ. of Obstet. and Gynæcol.*, Vol. x., 1906, p. 306).

There is much else in this chapter about clots in the terms, about the "white flowers," about retention of the terms, and about the manner in which "the mylke which commeth to the brestes is engendred of the Termes (accordyng to moste mens opinions)." There is not wanting evidence, it is pointed out, of the "great familiaritie betwene the Matrix and the brestes, for so much as the ebbyng of the one is the flowinge of the other."

Such are the matters dealt with in the First Book of the *Byrth of Mankynde* in the 1560 edition; it remains for me now to examine the differences which exist between this and other editions. The 1545 and 1552 editions call for no special comment, and those that were published later than 1560 also show none other than trifling alterations; but the Jonas or 1540 edition differs widely from the rest.

The First Book of the 1540 edition is really the Second Book of the Raynalde editions (with some exceptions to which reference will be made), and the First Book of the Raynalde editions finds no counterpart, or almost none, in the 1540 edition. Almost none, for the First Chapter of the latter contains an account of "how many caules the birth is compacted and wrapped in," and the fourteenth chapter of the Reynalde editions deals with "the three caules or wrappers wherein the infant is lapped." Chapters I. to XIII., Chapter XV., and part of Chapter XIV. of all the Raynalde editions find no counterpart in the 1540 edition of Jonas; practically the whole of the First Book of the Raynalde editions, therefore, is new material added to what was in the 1540 edition. To trace the source or sources of the new material found in all the Raynalde editions would be an interesting literary investigation, but it is one which

I am not now in a position to make.* I may, however, state that no part of the added chapters is in the two editions of Rösslin's *De Partu Hominis* which I have been able to examine (those of 1538 and 1556); and in the meantime it is only reasonable to ascribe it to Raynalde, and to regard it as part of the " augmentation " mentioned on the title page of the 1545 and subsequent issues of the *Byrth of Mankynde.*

THE ANATOMICAL FIGURES.

At the end of the First Book is " The declaration by letters of the fygures folowing, wherein be set forth to the eye euery parte in woman mentioned in thys boke before: Which in the former Printinges hath ben corrupted, but nowe truely set forth." These figures are nine in number, and they have all been reproduced in Plates IV., V., VI., and VII. of my former article.

No indication is given in the letterpress as to the source of the illustrations; indeed from the reading of the accompanying descriptions it would seem that they were original and that the objects depicted in them had been seen by the writer. A little investigation, however, soon shows that this was not the case. None of them is to be found in the 1538 edition of Rösslin's *De Partu Hominis*, although four of them make their appearance in the 1556 edition; but, then, as we shall see, they had already appeared in the 1545 edition of the *Byrth of Mankynde* and had been repeated in the 1552 edition. Evidently, therefore, their original source is not Rösslin's work. As a matter of fact, they have been taken, descriptions and all, from Vesalius's book *De Humani Corporis Fabrica,* in the " first " or 1543 edition of which they are all to be found.

I have carefully compared the plates in the 1560 edition of Raynalde's *Byrth of Mankynde* with those in the 1543 edition of Vesalius's work, with the following results: The first figure in Raynalde is the twenty-fourth of the Fifth Book of Vesalius, and is found on p. 377 of that work, with the descriptive letterpress on pp. 376, 377 and 378; the second figure is the twenty-fifth of Vesalius, and is found on p. 378, with its description on pp. 379 and 380; the third figure is the twenty-sixth of Vesalius on p. 380; the fourth figure (IIII.) is also to be found in Vesalius's work at the end of his Third Book, on p. 313, with its description; Figures V., VI., VII., and VIII of Raynalde's work are the first, second, third and fourth separate figures in the thirtieth plate of Vesalius's Fifth Book, and are to be seen on p. 382 with descriptive notes on p. 383; and the ninth figure of Raynalde's *Byrth* is the twenty-seventh of the Fifth Book of Vesalius's *De Humani Corporis Fabrica,* where it is to be found with its description on p. 381.

* One would require to have access to the various editions of Rhodion's work, both in Latin and in German. Some of the new material, however, is undoubtedly derived from Vesalius's work.

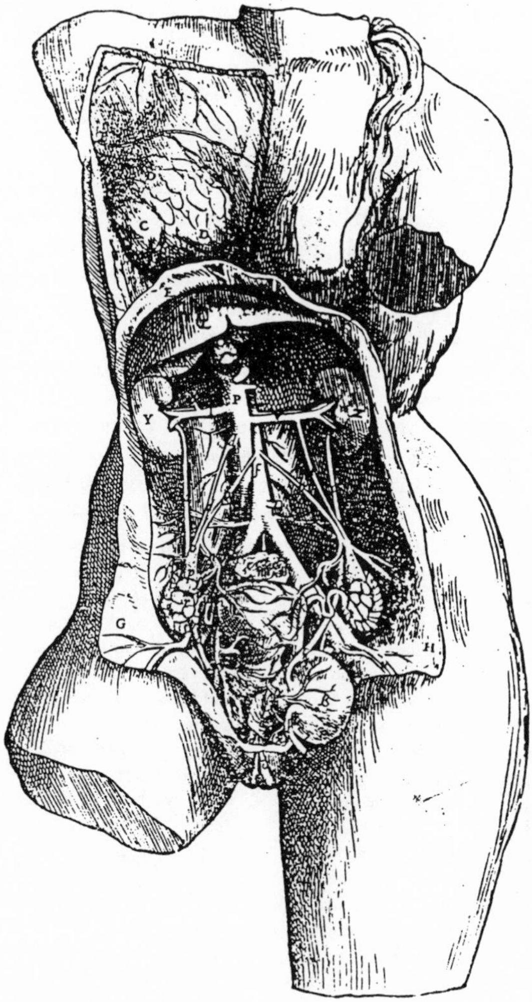

Fig. I.—Illustration taken from Vesalius's *De Humani Corporis Fabrica* (edition of 1543), where it is the twenty-fifth figure of the Fifth Book; it appears as the second figure in the *Byrth of Mankynde*.

All these figures, as they appear in the 1560 edition of Raynalde have been reproduced in my former article, where they are numbered Figs. IV., V., VI., and VII., but, in order to prevent confusion, their proper numbers (I., II., III., IIII., V., VI., VII., VIII., and IX.) are also attached to them (although some of them were reversed in the original printing, *e.g.*, IV., IIV., and IIIV. for VI., VII., and VIII.). In order that the reader may compare the plates as they appeared in Vesalius's work with their reproductions in Raynalde's 1560 edition of the *Byrth*, I give here Vesalius's twenty-fifth figure corresponding to Raynalde's Fig. II. (Fig. I.): it can be studied alongside of the Raynalde reproduction (Fig. V. of my former article). Further, in a later edition of Vesalius's work (that of 1642), all the nine figures used by Raynalde were grouped together in one plate (on p. 96), and I have thought it worth while to reproduce this also (Fig. II.). It represents in graphic form Raynalde's unacknowledged indebtedness to Vesalius.

The descriptions of the figures are literal translations of the Latin text which accompanied the illustrations in Vesalius's *De Humani Corporis Fabrica*. I give here in parallel columns the Latin description of Figure II. and the English translation of it:—

A praesentis figurae dextra mamilla cutem abstulimus, ut quàm fieri posset proximè mamillarum natura hic oculis subjiceretur. Deinde ventriculum, et cum intestinis mesenterium et lienum resecuimus, recto interim intestino non secus quàm in mox praecedente figura relicto. Ad haec, uterum suo extimo quod peritoneum ipsi porrigit involucro quodammodo spoliavimus, omnes membranas quàm licuit accuratissimè passim, ideo amputantes, ut seminis materiam testibus deferentia et rursus semen ab his utero deducentia vasa in conspectum venirent. Vesicam verò deorsum in sinistrum latus reflеximus, unà meatum à dextro rene ipsi urinam deferentem abrumpentes, ut urinam vesicae deprentium meatuum insertio appareret, ipsaque vesica uteri inspectionem non occuparet. Postremò pubis ossium portionem ab hac figura exsecuimus, quo uteri cervix ac vesicae etiam collum appositè viderentur.

We have here taken away the skyn from the ryght teate of this present figure, that the nature of the teates mighte as nygh as may be, be set before the eyes, and afterwards we have cut awaye the ventricle with the bowels, and also *Mesenterium* and the splene, leaving the strayte entrayle in thys place unmedled with, as well as we dyd in the fygure before. And moreover, we have as it were taken awaye from the uttermooste cote which *Peritonium* gave unto it, cutting away also al the pannicles, that the vessels caryinge forth the substaunce and matter of sede to the stones, and also the vessels carying away the sede from thence to the Matrix shoulde appeare and bee seene. Also we have tourned over the bladder downewarde on the lefte syde, lykewyse breaking the way or conduite which beareth foorth the urine to it from the ryght kydneye, that the insertion of the wayes of bearyng forth the urine to the bladder myght appeare, and that the bladder shuld not let the inspection or sight of the Matrix or Wombe. Last of all we have cutte away from this fygure a portion of the bones above the privie membres, thot the neckes of the matrix and of the bladder might the more commodiously be sene.

The comparison of the Latin description with the English translation leaves no doubt that Raynalde boldly appropriated both the plates and their accompanying text from the work of Vesalius; he was not even at the trouble of altering personal details which referred to Vesalius, such as the allusion to his work at the University of Padua which occurs in the explanation of the ninth figure. In this respect, however, he did not differ from the editor of some of the later editions of Rhodion's *De Partu Hominis* (*e.g.*, that of 1556), who also borrowed some of Vesalius's plates without acknowledgement.

What has ben said above refers to the 1560 edition of Raynalde's *Byrth of Mankynde*. When we turn now to the 1545 and 1552 editions we find two other anatomical figures (making eleven in all), named the fyrst and second fygures, along with several differences in the accompanying descriptive letterpress. The wording of the Declaration differs slightly, the reference to corruptions in the printing not appearing. It reads: "The declaration by letters of the fygures folowyng, wherein be set forth to the eye every part in woman mencioned in thys Boke before." The first and second figures, however, represent dissections of a *man's* body, and at the end of the descriptive letterpress the editor somewhat ingenuously says: "Here ye shal be advertysed that although these ii fyrst fygures be made principally for ye man, yet may they serve as wel to expresse the woman: for the man and woman differ in nothyng but in the pryvie partes." These two illustrations also are borrowed from Vesalius's *De Humani Corporis Fabrica*, where they appear on pp. 355 and 356 of the 1543 edition, and are named the first and second figures of the Fifth Book. They are reproduced here as Figures III. and IV. I place, again in parallel columns, the Latin and the English, and it will be noted that the latter is not so strictly a translation of the former as in the descriptions of the 1560 edition:

Praesenti figura tanta humani corporis portio delineatur, quanta ad peritonaei sedes ostendendas sufficit: exprimitur itaque hac figura anterior peritonaei sedes, sectionis serie ab octo abdominis musculis libera, nullaque ex parte dissecta.

A,A,B,C,D. His characteribus peritonaeum insignitur, quodamodoque hac figura terminatur.

E,E. Linea a mucronata pectoris ossis cartilagine ad pubis usque ossium commissuram procedens, cui oblique descendentium et ascendentium, et transversim procedentium abdominis musculorum nervosae tenuitates pertinacissime connas cuntur.

In the fyrst fygure is set forth so moch of a man's body as may be sufficient to show the forme of thee kell called Peritoneum: spoken of in the iiii Chapter, Whose compasse is here noted wyth A,B,C,D.

A,A. Noteth the grystell, nether ende or poynt of the brest plate, in the pyt or pitch of the brest agaynst the stomacke.

E.E. is a lyne descendyng upon Peritoneum from the sayde grystle, downe to the myddle joynt of the share bone.

F. is the navell.

G. sygnifieth the sede vessels of the lefte syde in men descendynge out of the amplytude of the bellye.

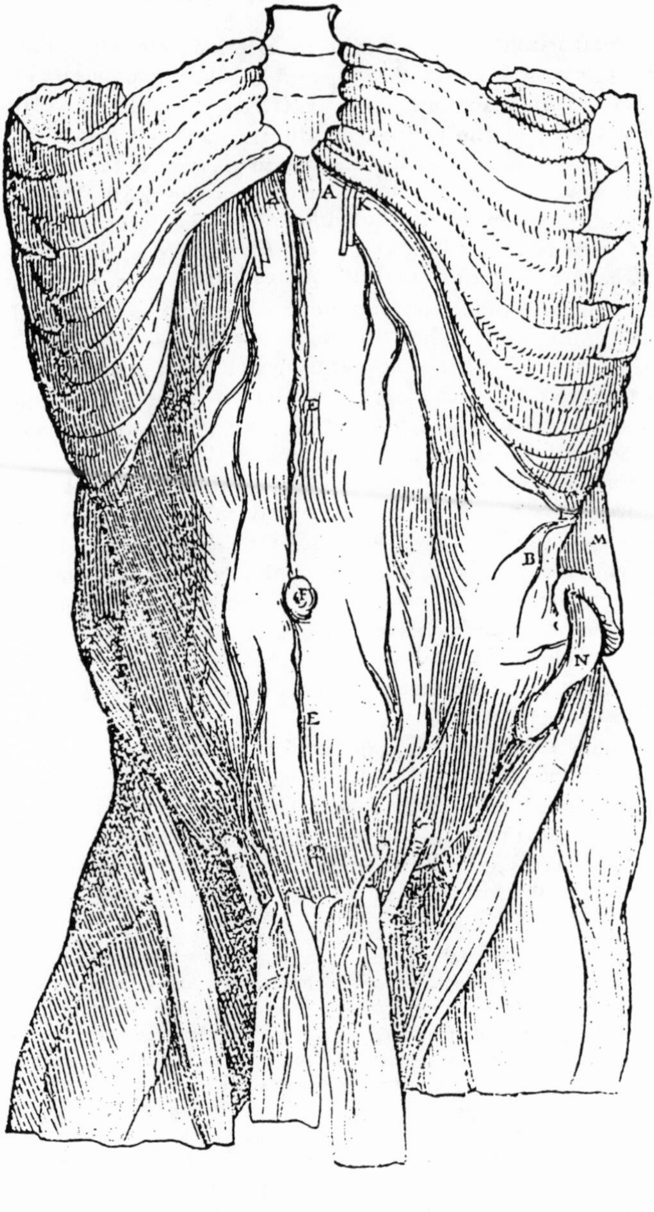

FIG. III.—Figure from Vesalius's *De Humani Corporis Fabrica* (edition of 1543) representing the dissection of a man's body. It appears as Fig 1 of the Anatomical Figures of the 1545 and 1552 editions of Raynalde's *Byrth of Mankynde*, but is omitted from that of 1560 and from all subsequent editions.

F. Umbilicus, quem inter dissecandum etiam adeptis abdominis musculis, gratia opportunae umbilici vasorum demonstrationis, reservare solemus.

G. Seminaria sinistri lateris vasa suis membranis, quas a peritonaeo mutuantur, adhuc obvoluta.

H. Seminaria dextri lateris vasa.

I. Vena ac arteria quae potissimum inferiori sedi rectorum abdominis musculorum exporriguntur, quorum et hic quoque propendet portio.

K. Vena et arteria, quae sub osse pectoris exporrectae, in anteriorem abdominis sedem prolabuntur, praecipuè rectis abdominis oblatae musculis, ac superiorè abdominis sedem universam quoque implicates: quemadmodum illae quas insignivimus, humiliorè et pubis ossibus viciniorè implicat.

L. Venarum soboles in peritonaei latera excurrentium, ac ab illis venis deductarum, quae aut à conjuge carente vena, aut geniculatim à cava pronascuntur, qua ipsius caudex lumborum vertebris colligatur, etc.

H. is the ryght seede vessell: but thys G. and H. hath no place in the women.

I. sygnyfieth the ascendinge vayne and artyre mencyoned in the last chap.

K. the descending brest vaynes and artyres spoken of in the same Chapt. as for other letters that be in this figure I wyl make no further declaration of them, for because they serve nothing to this present purpose.

Other differences between the edition of 1560 and those of 1545 and 1552 remain to be noted. These consist chiefly of verbal differences in the descriptions of the figures. Figures 3, 4, 5, 6, 7, 8, 9, 10, and 11 of the earlier editions correspond to Figures 1, 2, 9, 3, 4, 5, 6, 7, and 8 of the 1560 edition. To show the extent of the verbal changes I place here in parallel columns the description of the fifth figure of the 1552 edition and the ninth figure of the 1560 edition; they represent the same specimen, but, as will be seen, differ considerably:—

1552 Edition.

This 5 fygure is pourtrayed after ye quycke, bothe in length and bredth, according to the length and bredth of the matrix of a woman which was cut open for the same purpose by phisitions. But ye must understand that here ye founde or body of ye wombe or matrix is devyded in ye myddes: the forepart of the which, is turned up, for because that ye maye the better perceave ye cavite of the matrix signed, the uppermost with A.A.C. The nethermoste halfe wyth B.B.D. Item. C. in

1560 Edition.

And the nynth figure sheweth the Matrix cut forth of the body, being of that bygnesse as it was sene taken foorth of a woman at the laste Anothomye, which I dyd se at the universitie of Padua in Italy. And moreover we haue so devyded and cutte a sunder the bottome of the Matrix by the myddle, that the concavitie and hollowe bought within the same myght be perceaved, and the thicke substaunce also of both the coates of the Matrix in women, when they be not with chylde.

the uppermost halfe and D. in the nethermost halfe show the seame or lyne spoken of cap. vi. E.E. both in the upper and also in the nether betoken ye crassenes or thickenesse of ye inner coate, wall or skyn of the matrix in wemen not beinge with chyld, through the contraction thereof as ye shall farther rede in the sayd. vi chapt. F.F. the propendynge or heldynge parte of the seme in the matrix spoken of. cha. vi. G.G. is the porte, oryfyce, or gate of the wombe. H.H. is the second and utter coate of the matrix geven to it from Peritoneum. I.I. on both sydes of the necke of the matrix, do sygnifie, part of the kel called Peritoneum, sticking yet to the sydes of the Matrix and the necke thereof. K.K. is the place where the matrix is fastened to the upper part of the privy passage. ca. vii. L. signifieth the stub of ye bladders necke, wher it entreth into ye forepart of ye privy passage.

A.A.B.B. The concavitie and holowe bought of the bottome of the Matrix.

C.D. A line somwhat after the maner of a seame called in Latin *Scortum*, which doeth belonge to the place wherein the testycle doo lye, whiche swelleth somewhat foorth into the bought of the bottome of the Matrix.

E.E. The thickness of the inner and proper coate of the bottome of the matrix.

F.F. A portion of the innermore bottome of the Matrix, swellyng foorth downeward from the hygher seate of the Matrix, into the holownes and bought of the bottome.

G.G. The beginning of the necke or the opening place of the bottome of the Matrix.

H.H. The seconde or uttermore infolder of the bottome of the Matrix, descended from *Peritoneum*.

I.I. Here we have reserved a portion on bothe the sydes of the thinne coverynges, descended from *Peritonium*, and conteynyng the Matrix.

K. Here is also sene the substaunce of the necke of the Matrix, because the cuttyng wherwith we devyded the bottome of the Matrix, was begunne at this place.

L. A part of the necke of the bladder, implanted into the necke of the Matrix, castynge foorth into it the urine. The swellyng partes of *Abdomen* and whatsoever is els to be considered therof, they may be sufficiently knowen without derection of Karacters.

On comparing these two descriptions with the original Latin inscription found in Vesalius's work I find that the first is a free and the second a literal translation thereof. In the first (that of 1552) all mention of Padua is omitted, while in the second (that of 1560) it is referred to with the addition of the words "University of" and "in Italy" which are not in Vesalius's text. On the whole, we must accept as substantially correct the statement made in the 1560 edition, that "the declaration of the fygures . . . . in the former Printinges hath ben corrupted, but is nowe truely set forth," if by that is meant a closer adherence to the text of Vesalius's work.

The Jonas edition of 1540 is supplied with no anatomical figures.

(To be Continued.)

The Journal of Obstetrics and Gynæcology of the British Empire

VOL. XII. OCTOBER, 1907. No. 4.

The "Byrth of Mankynde."

(ITS CONTENTS.)

By J. W. BALLANTYNE, M.D., F.R.C.P. (Edin.), F.R.S. (Edin.), *Lecturer on Midwifery and Diseases of Women, Surgeons' Hall and Medical College for Women, Edinburgh; Physician to the Royal Maternity Hospital, Edinburgh; and President of the Edinburgh Obstetrical Society.*

(Continued and Concluded from Page 194.)

THE SECOND BOOK.

THE contents of the Second Book (of all editions subsequent to that of 1540) are of less interest to English obstetricians, for they are simply a translation, rather free perhaps, of Rhodion's book *De Partu Hominis.* I shall first enumerate the subjects dealt with in this Book, taking again the edition of 1560 as the standard Raynalde one, and I shall then compare the matter as it appears in the various editions, and consider the character of the translations with which Jonas and Raynalde have furnished us.

The first chapter of the Second Book of the 1560 edition (and of all the others, except that of 1540) corresponds to the second chapter of the work of Rhodion (which is not divided into "Books," but simply into twelve chapters).

Chapter I. begins with a short paragraph summarizing the contents of the First Book, and then proceeds to deal with "the tyme of byrth, and which is called naturall or unnaturall." The premonitory signs of labour are named: "first certaine dolours and paines begin to growe about the guttes, the Navyll, and in the raynes of the backe, and lykewyse about the thyghes, and the other places beynge neare to the privie partes, which lykewise then beginneth to swell and to burne, and to expell humours, so that it geveth a plaine and evident token that the labour is nere." Then comes a definition, which at least does not err by entering too much into detail: "Naturall byrth is when the chylde is borne both in due season and also in due fashion." The due season is "most commonly after the ninth moneth, or about fortie wekes after the conception;" and then follows that oft-repeated and widely-believed statement about the poor chances of survival which an eighth month child has as compared with one born at the seventh month (when "the chylde proveth very well"). The writer is somewhat in error when he describes the "due fashion": "first the heade commeth forwarde, then foloweth the necke and shoulders, the armes with the handes lying close to the body towardes the face and forepart of the chylde, beyng towardes the face and forepart of the mother, as it appeareth in the first of the byrth figures." This definition makes, therefore, a face-to-pubes case the natural one, which is, of course, an error. The author (Rhodion) is on safer ground when he states that "yf the byrth be naturall, the delyveraunce is easy without longe tarying or lokynge for it." The "byrth not natural is, when the mother is delyvered before her tyme, or out of due season, or after anye other fashion theń is here spoken of before: As when both the legges proceade fyrst, or one alone, with both the handes up, or both down, other els the one up and the other downe, and dyvers otherwyse, as shalbe hereafter more clerely declared." The "other els" as stated in the next chapter is "sidelonge (the which is most perellous) or arselonge, or backlonge, other elles (havinge two at a byrth) both proceade with their feet fyrste," etc.

Chapter II. deals with "easy and uneasy, difficult, or dolourous deliveraunce, and the causes of it: with the signes howe to knowe and foresee the same." "Verye manye," says the writer, "bee the perylles, daungers and thronges, which chaunce to women in theyr labour, which also ensue and come in dyvers wayes, and for dyvers causes, such as I shall here declare." Among the causes of delay in labour several are enumerated which are nowadays little accounted of or not considered at all, while others are omitted which are of importance. Few, for instance, will agree with the statement that the "byrth of the man is generally easier then the byrth of the female." Some of the causes are curiously set forth, such as if the mother be "too spare or leane, or that she never had chylde before, or that she be over timorious and fearefull, dyvers, waywarde, or such one that wyll not be ruled, removyng her selfe from one place to another." The old belief in the birth of the child by its own efforts appears in such statements as: "yf the childe be so faynt, weake and tender, that it cannot tourne it selfe or doth it very slowly;" "also if the child be dead in the mothers belly, it is a very perellous thing, for so much as it cannot be easely turned, neither can it welde or helpe it self to come forth, or if the chylde be sicke or weakned so that it cannot for feblenes helpe it self." Then, various signs are enumerated to help the midwife to tell in what cases the unborn infant is weak or sick, such as if the pregnant woman has been "sore lasked,"[1] if she have had "dayly and unwontly her flowres," if "straygtht after one moneth upon the

1. Lasked, purged.

conceptyon her brestes yelde any milke." Causes of delay due to teratological states were not unknown to the writer for he says that there will be difficulty if that with which the woman laboureth "be a monster, as for example, yf it hath but one bodye and two heades, as appereth in the XVII. of the birth figures such as of late was sene in the dominion of Werdenbergh."[1] Faults in the "secondine or latter birth" (membranes), such as firmness or slenderness are also named, and the delay due to loss of the humidities from early rupture is referred to. "And farther if the woman have used to eate commonly suche meate or fruytes, which do exiccate or drye, and constrayne or bynde, as Medlars, Chestenuts, and al sowre fruites, as Crabbes, Chokeperes,[2] Quinces and suche other, with over muche use of Vergeus,[3] and such lyke sowre sauces, with Rise, Mill,[4] and many other thynges, all thys shall greatly hynder the byrth." "Also," the writer continues, "the use of colde bathes after the fyrst moneth folowing the conception, or to bathe in such water where Alome is, Iron, or Salt, or anye suche thynges whiche do coarcte and constrayne, or yf he have bene often tymes heavye and mourninge, or ill at ease, or yf she have bene kepte over hungry and thursty, or have used over much watche and walkinge, eyther yf she used a lyttle before her labour things of great odour, smel, or savour, for suche thynges (in manye mens opinions) attract and drawe upward the Mother or Matrix; the which is great hinderaunce to the byrth."

The chapter closes with an enumeration of the tokens of an easy labour. "Nowe sygnes and tokens of an expedite and easie delyveraunce, be suche as be contrarye to all those that have ben rehearsed before. As for example, when the woman hath bene wonte in tymes passed, easely to be delyvered, and that in her labour she feele but little thronge or dolour, or though she have great paynes, yet they remayne not still in the upper partes, but descend alwaies downewardes to the nether partes or bottome of the belly. And to be short, in all paynefull and troublesome labours these signes betoken and signifie good spede and lucke in the labour: unquietnes, much styringe of the chylde in the Mothers bellye, all the thronges and paynes tomblynge in the forepart of the bottome of the belly, the woman stronge and mightie of nature, such as can wel and strongly helpe her self to the deliveraunce of the byrth. And agayne, evyl signes be those, when she sweteth colde sweate, and that her pulces beate and labour over sore, and that she her self in the labouring faint and sowne, these bee unluckie and mortall signes."

1. Werdenbergh, a town in Switzerland on the Rhine, in the Canton of St. Gall. The monstrosity referred to appears as fig. x. near the end of this article.

2. Choke-pear, any "rough, harsh, and unpalatable variety of the pear, used for perry," a sort of crab-pear.

3. Vergeus, or verjuice, a liquor expressed from crab-apple, sour grapes, etc.

4. Mill, millet.

The third chapter of this Book is, in some respects, the most important and the most interesting of the whole work, for it deals with "howe a woman with childe shal use her selfe, and what remedies be for them that have harde labour." If there should be any disease, swelling, or apostumation (abscess) about the uterus, vulva, or bladder (such as stone or strangury), then "in these cases it behoveth such thynges to be loked unto and cured before the time of laboure commeth, by the advise of some expert Surgion." Her diet[1] before labour will be different from that during labour. If there be constipation, she must use "suche thinges, the whiche may lenifie, mollifie, dissolve, and lose the belly: as apples fried with suger taken fastynge in the mornynge, and after that a draught of pure wyne alone, or elles tempred with the juyce of swete and very rype apples. Also to eate figges in the mornynge fasting, and at nyght, looseth well the bellye. If these profite not, *Cassia fistula* taken iii. or iv. drams one halfe hour before diner, shal loose the belly without parel." The woman also must refrain from taking constipating things ("hard egges," etc.), and it may be necessary for her to get a clyster," but it must be very gentle and easye." An easy and temperate purgation (as by mercury) may be needed, "or elles a suppositar tempered with sope, larde, or the yolke of egges." If she be faint or sickly just before her labour "then must ye comfort her with good comfortable meate, drinke, holsome and noble electuaries." Various ointments and baths are to be used before labour: "Annointmentes wherewith ye may sople the privie place, be these. Hennes grese, Duckes grese, Goose grese, also oyle Olife, Linsede oyl, or oyl of Fenegreke, or the viscosite of holyoke." She is to bathe in water in which have been seethed "Malowes, Holyoke, Camomel, Mercury, Maidenhaire, Lyneseede, Fenegreke seede, and such other thynges which have vertue to mollifie and sople." If she be not able to take such baths, she must sponge herself with the water and apply ointments locally. Sweet fumes also are useful: "it shalbe also very profitable for her, to suffume the nether places with muske, Ambre, Gallia, Muscata, which put on embres, yelde a goody savoure, by the whiche the neather places open theymselfe, and drawe downwarde." She must also "exercyse the bodye in doing some thinge, styring, moving, goynge, or standinge, more then otherwise she was wont to doe."

Hints are given as to what must be done when labour pains come on. "To withstand, defend, and to put away so neare as mai be the instant and present dolours. And as touchinge this poynt, it shalbe verye profytable for her, for the space of an houre to syt styll, then (rysynge agayne) to goe up and downe a payre of stayres, crying or reaching so loude as she can, so to styre her selfe."

Here follows the oft-quoted passage about the "womans stoole"

1. Diet here means course of living and not simply the food or drink taken.

or "obstetric chair" which is represented in the Birth Figures. "Nowe when the woman perceaveth the Matrix or Mother to ware laxe or loose, and to be dissolved, and that the humours yssue foorth in great plentie, then shall it be mete for her to sit downe leaninge backwarde in maner upright. For the which purpose in some regions (as in Fraunce and Germany) the Midwyfes have stoles for the nonce, whiche beynge but lowe, and not hye from the grounde, be made so compasse wyse and cave or holowe in the middes, that that mai be receaved from underneth which is looked for, and the backe of the stole leaning backeward, receaveth the back of the woman, the fashion of the which stole, is set in the beginning of the birth figures hereafter. And when the tyme of laboure is come, in the same stoole ought to be put many clothes or cloutes in the back of it, the which the Midwife may remove from one syde to another accordinge as necessitie shall require. The Midwyfe her selfe shall syt before the labourynge woman, and shall diligentlye observe and wayte, howe much, and after what meanes the chylde styreth it selfe, also shall with her handes fyrste annoynted with the oyle of Almondes, or the oyle of whyte Lyllies, rule and dyrecte every thynge as shall seme beste. Also the mydwife muste enstructe and comfort the partie, not only refreshing her with good meate and drink, but also with swete woordes, gevynge her good hope of a spedefull delyveraunce, encouraginge and enstomakinge her to pacyence and tolleraunce, byddynge her to holde in her breath so much as she may, also strekinge gentilly with her handes her bellye above the Navell, for that helpeth to depresse the birth downewarde."

If the patient, however, be fat, the writer recommends that she "lye grovelynge," and if necessity require it "let not the midwife bee afrayde ne ashamed to handle thĕ places, and to relax and loose the straites (for so muche as shal lye in her), for that shal helpe wel to the more expedite and quicke labour." She is warned, however, against interfering too soon "before the byrth come forwarde," and she ought not to allow the patient to expend her strength before the proper time. When the bag of membranes appears, "then maye ye knowe that the labour is at hand." If the bag do not burst of its own accord," it shalbe the Mydwyfes part and office, with the nayles easely and gentellye to breake and rent it, or yf that may not conveniently be done, then rayse up betwene your fyngers a peece of it, and cut it with a payre of shieres, or a sharpe knyfe, but so that ye hurt not the byrth with the cut." If the membranes have ruptured or been ruptured too early a dry labour results requiring the application of "oyle of whyte Lyllyes or some of the greses spoken of before" to the parts; "but chiefely in these difficulties should profite the whyte of an egge together with the yolke powred into that same place: which shoulde cause it to be most slyppery and slydynge, and supplye the roome of the naturall humidities spent before."

The birth of a child with a large head or of twins is to be assisted by the midwife, who is to "helpe all that she maye, with her hande fyrst annoynted with some oyle openying and enlargyng the waye that the issue maye be the freer."

So far the writer has been dealing with the "natural byrth when that first proceadeth the head," as is represented in the first of the Birth Figures. He now describes in turn the various ways in which the infant or infants may present and in what manner the midwife is to treat them. In the second of the Birth Figures a child is represented coming feet first (the attitude of the fœtus is wrongly represented, as it is indeed in most of the figures), and in such circumstances the midwife is apparently directed to perform cephalic version! Here, at any rate, are the directions: "Sometime it chaunced the child to come the legges and both armes and handes downwarde, close to the sydes fyrst foorth, as appeareth in the seconde of the byrth figures. In this case the Mydwyfe must do all her payne with tender handlynge and annoyntyng to receave foorth the chylde, the legges beynge styll close together and the handes lykewyse remaynynge as appeareth in the seconde figure. Howbeit, it were farre better (yf it may be done by anye possible wayes or meanes) that the Mydwyfe shouldbe tourne these legges commyng fyrst foorth, upwardes agayne by the bellywarde, so that the head myght descende downewarde by the backe part of the wombe: for then naturally agayne and without peryl might it proceade and come forth as the fyrste."

In the third of the Birth Figures the fœtus is represented as coming by the feet with the arms displaced upward alongside of the head. "This is the perylloust maner of byrth;" and the direction is that the midwife must do what she may "to turne the byrth (yf it may be possible) to the first figure," but no directions are given as to the way in which this is to be done. If she cannot do this, she is directed to convert it into the second figure by bringing down the hands; and "if this also wyll not bee, then receave the feete as they come foorth, and bynde them with some fayre linnen cloth, and so tenderly and very softly lose out the byrth tyll al be come foorth, and this is very jeoperdous labour." When one foot only presents (*Birth Figure* 4) version by the postural method ("the labouring woman's head to be the lower part of her body") is to be followed; but if it do not succeed, the midwife is to bring down the other foot. The fifth *Birth Figure* represents, rather crudely, a transverse presentation; the sole direction for its management is: "then must the Mydwyfe do so, that it may be returned to his naturall fashion, and so to come foorth." The sixth and seventh cases call for no special comment. The eighth Birth Figure shows descent of one of the arms alongside of the head; under these circumstances the midwife is told to thrust the birth in again, and if this fail she is to try postural treatment; the same procedure is to be adopted in the ninth mode of presentation when both hands come down. The directions for the management of a breech case (*Birth Figure* X.) are surprising: "Then must the Mydwife with her handes returne it agayne, untyll such tyme that the birth be turned, the legges and feete forwarde." A shoulder presentation is shown in Birth Figure XII. (described erroneously in the text as XI.), and the direction is "then must ye fayre and softly thrust it back agayne by the shulders, tyll suche tyme as the heade come forwarde." Twin cases are shown in the Birth Figures XIV., XV., and XVI.; in the first are two heads, in the second two breeches, and the third is a head and a breech presentation. In the description appended to the last-named mode of labour there is the suggestion that the possibility of head-locking was not unknown to the writer.

On the whole, it must be admitted that the management of labour as set forth in this chapter falls far behind modern practice, not to say theory. The notions regarding the attitude of the fœtus *in utero* were erroneous, the distinctions between the various presentations were incomplete (*e.g.*, face cases are not figured or named), and the management not infrequently consisted in interfering in the cases which we should now leave alone and in using ointments and posture under circumstances in which more radical methods would now be adopted. One cannot help wondering also how the midwives carried out the instructions given to them; certainly they were not burdened with details.

The fourth chapter (wrongly described as the fifth in this edition of 1560) deals with "the remedies and medicines by the which the labour may be made more tollerable, easy, and without great payne." The posture of the patient, the temperature of the lying-in room, the provocation of sneezing ("and that eyther with the powder of *Eleborus* [1] or els of pepper") and the use of ointments are all referred to. Of the oils, ointments, perfumes, washes, drinks, pills, and plasters mentioned in this chapter, we need mention two only. Here is a perfume: "Take yelowe brymstone, Myrre, Mader, Galbanum, Oppoponacum, of eche lyke much, and tempre all those together, makynge of them pylles, and with those also ye maye make fume, to be receaved underneath." The chapter closes with the prescription of "a plaster to provoke the birth." Here it is: "Take wylde Gowarde,[2] and seeth it in water, in the same water temper Myrre, the juyce of Rue, and Barlye meale so much as shalbe sufficient, stampe these thynges together, and make it plasterwyse, then laye it to the womans bellye betwene the Navyll and the nether parte. This plaster shall helpe marveylously."

Chapter V. is concerned with "howe the secondine or seconde

1. Eleborus, Hellebore, White Hellebore (*Veratrum album*) was used as a sternutatory.

2. Gowarde, Wild Gourd, Colocynth.

byrth shalbe forced to issue foorth, if it come not freely of his owne kynde." Various causes of the non-expulsion of the placenta and membranes are enumerated, such as lack of strength from prolonged labour, "entanglement" of the secondines within the uterus, and swelling of the parts. The dangers of placental retention are also named, and include "suffocation and chokinge of the Matrix" and putrefaction of the after-birth. "The seconde birth retayned and kept within will soone putrifie and rot: whereof wyll ensue yll noysom and pestiferous vapoures ascendinge to the heart, the braynes and the midriffe, through the which meanes the woman shalbe short wynded, faynte harted, often soundinge and lyinge without any maner of movynge or styrringe in the pulces: yea, and many tymes is playnely suffocated, strangled and dead of it." The remedies proposed for non-expulsion of the placenta were founded to some extent upon the causal conditions so far as these were understood. If weakness from long labour were the cause, then must the "labourer" be "recomforted and strengthed with good comfortable meates and drinkes, which maye enhart her, as broath made of the yolcke of egges, or with good olde wine, and good fat and tidie fleshe, or Byrdes, Hennes fleshe, Capons, Partrige, Pigins and such like." If the cause were contraction of the passages then oils and ointments are recommended to "make the waye slypper, sople, and easy for it to proceade." Perfumes, also, and vapours are said to be efficacious. "But if the retencion of the secondine come by reason that it is entangled or fastened in some place of the Matrix, so that it wyll not resolve ne loose; then make a fume underneath of Brimstone, Ivie leaves, and Cresses, or elles of Cresses and fygges." There are some curious prescriptions, of which the following is an instance: "Also of all odoriferous and sweete smellinge thinges, as Ambre, Muske, Frankencense, *Gallia Muscata*, and confection, neare the which savoures and perfumes put on the embers muste be so closely receaved underneth, that no part of the smell do ascende to the nose of the woman. For to the nose shoulde the savoure of nothynge come, but onely of suche thinges, the which stinke or have abhominable smell, as *Asafetida*, Castorium, mans hayre or womens hayre burnt, Pecockes fethers burnt." "Item let her be provoked to sneese with the powder of Eleborus or Pepper put in the nose, holdinge her mouth and nose so close as maye be."

The following directions for the removal of the after-birth are interesting: "And yf it bee so that any parte of the secondine do appeare, let the Mydwife receave it tenderly, losynge it out fayre and softly, least it breake, and if ye doubt that it wil breake, then let the Midwife tye that parte of the whiche she hath handfast to the womans legge or fote, not very straight, least it breake, nether very lax, least it slip in agayne, and then cause her to sneese. Nowe yf the secondine tarye or stycke, so that it come not quickly forewarde, then loose it a lyttle and a lyttle very tenderly, wrethinge it from one syde to another, tyll such time as it begotten out, but ever beware of violent and hasty movynge of it, leaste that with the seconde byrth ye remove the Matrix also." The danger of inversion of the uterus is doubtless alluded to in this last passage. The chapter closes with directions for a fumigation and a plaster, and with the following somewhat despairing instruction: "If for al this the secondine come not forewarde, then leave it, and use no more medicines ne remedyes to that purpose, but let it alone, for within fewe days it wyll putrifie and corrupt, and dissolve into a watery substaunce, thick like bryne, or other fex myxed with water, and so yssue foorth. Howbeit in the meane whyle it wyll put the woman to great paine in the head, in the heart, and in the stomacke, as we touched before."

The sixth chapter is a long one, dealing with such important matters as "howe many thinges chaunce to the women after theyr laboure, and how to avoyde, defende, or to remdye the same." Among the "many thinges" are "the fever or ague, or swelling, or inflation of the bodye, other tumblynge in the belly, or elles commotion or settelinge out of order of the Mother or Matrix," and the cause of these is sometimes "lacke of due and sufficient purgation and clensynge of the flowres after the byrth, or els contrarywyse over muche flowinge of the same, whiche sore doth weaken the woman, also the great labour and styrrynge of the Matrix in the byrth."

The "ague" we may shrewdly suspect was septic poisoning, "for that commeth of like cause by retention of the flowres." The patient is then to "drinke water in the whiche is decocte Barley beaten, or Cicer [1] and Barley together, or water in whiche be sodden *Tamarindi*, or whaye of mylke, and let her eate Cullis [2] made of a Cocke, and sweete Pomegarnates, for these thynges do provoke the flowres," etc. Various remedies, resembling those already referred to, are to be given in such conditions as swelling of the body, "frettinge and knawynge of the guttes," "paine in the privie partes," "outragious flux of flowres," "coming forth of the fundament gut," and the like. It is unnecessary to describe in detail the curious plans adopted and mixtures administered in these cases, one instance must serve for all: "To stynte and restrayne the outragious fluxe of flowres, it shalbe verye good to binde the armes strayght and strongly, and not the feete or handes, as some unwyse men doe teache, and then to set a ventose boxe, or cupping glasse with fyre (which is called borying) under the brestes without anye scarification." Here is one of the plasters: "Take of the bloud stone called *Emathites*, Bole armeniacke of eche halfe an ounce, *Sanguis draconis*, *Licium*, of eche twoo drammes, *Karabe*, otherwye named Ambre, the

1. Cicer, a chick-pea.
2. Cullis, a strong broth, a beef-tea.

cuppes of Acornes, Cipres tree nuttes, flowres of Pomegranade, of eche one dram, of the scales of Iron one dram and a halfe, Turpintine and Pitch lyke quantitie, or so much as shalbe sufficient to make a softe and somewhat liquid plaster."

The next chapter (the seventh) deals with "aborcementes or untymelye byrthes, and the causes of it, and by what remedies it may be defended, holpen and eased." Many curious causes are enumerated, among which is "a disease called *Tenasmus*, the which is when hath ever greate desyre and luste to the stool, and yet can do nothynge." Other causes are coughing, bleeding at the nose, "to be let bloud," strong purgation, hunger, cold, heat, etc. "Therefore ought women with chylde to eschewe much bathyng or going to the hotte houses in theyr teming" (teeming). "Item, the intemperancie and mutation of the ayre and weather may be cause of aborcement," and after this statement there follows an interesting paragraph on meteorology in its relation to health, as it was then understood. "Dancing and leaping" are also named as possible causes, and so are "sodayne anger, feare, dread, sorowe, or some sodaine and unloked for joy."

To the modern reader the signs of abortions enumerated by the writer of the *Byrth of Mankynde* will appear most astonishing and unconvincing; among them he will find "a great ache in the inner part of the eies toward the braynes," redness of the face, "ventositie or wynde runnynge from one syde of the bodye to the other." At the same time there is a reference, but a very brief one, to the really important sign of "greate paynes and dolours of the Matrix." The means of diagnosis given are hardly such as to justify the author's confident assertion "thus have I sufficiently declared evident sygnes, whereby may be provyded and foreseene the aborcement before it come." The methods of treatment are those which we now have come to look for from our author,—baths, fumigations, plasters, ointments, odours, and such like; but he gives the midwife one good piece of advice: "Howbeit, in all thys matter, let not to make some expert Phisition of youre counsaile, yf ye may have suche one: for because that manye such thinges come, and not all by one way or meane."

The eighth chapter (wrongly named the ninth in the edition of 1560) is concerned with "dead byrthes, and by what sygnes or tokens it maye be knowen, and by what meanes it may also be expelled." The signs are twelve in number; but they are not very convincing, as may be gathered when it is noted that the twelfth sign is if the mother's "handes put into very warme water, and then layde on the belly, and the childe steare not." There is evidence of sound knowledge, however, in the statement that "of all these sygnes nowe, the more that come togeather of theym at one tyme and in one person, the surer maye ye be that the byrth is dead."

The prognosis, grave or favourable, in cases of dead-birth labours is set forth: "Whether the Mother shalbe in parell or no, by these thinges shall ye knowe. If the woman beynge in the laboure sowne or feare as though she were in a transe: yf her remembraunce fayle her, and she were feble and scante able to moue or styre herselfe, yf she (called with a loud voyce) canne aunswere nothinge at all, or elles verye lyttle, and that verye softely, as though her voyce began to fayle her: if she be invaded or taken among in the labouring with convulsion or shrinkelynge together: if she refuse or cannot brooke meat: yf her pulces beate very faste, the which signes when ye se in the woman labouringe, it is an evident token that she shal not lyve longe after her delyveraunce, wherefore commit the cure of her to the handes of almyghtie God."

The treatment consists in getting ride of the "dead burthen" either by "medicines expulsyve "or else by certain instruments made "for the nonce." Here, again, we find described a long list of fumigations, containing such things as the hoof or dung of an ass, the skin of an adder, "hawkes' dung" or "oxe gall," of suppositories or pessaries, of drinks ("yf the woman drynke the mylke of another woman, it will styre and expell the byrthe"), and of plasters. "But yf all these medicines profyt not, then must be used more severe and harde remedyes, with instrumentes: as hokes, tounges, and suche other thinges made for the nonce." From the wording of the directions it is evidently intended that the midwife herself shall fix the hooks into the eyes, or mouth, or shoulders, or ribs of the dead fœtus and make traction, other women keeping the patient down. Arms and legs are to be cut off, if need be, and the head is to be opened with a sharp penknife if it be much swollen. Both the head and trunk may have to be broken up into pieces with "such instrumentes as the Chirurgions have readye and necessarye for suche purposes."

The last paragraph of this chapter must be quoted entire: "But contrary to all this, yf it chaunce that the woman in her labour dye, and the chylde having lyfe in it: then shal it be meete to kepe open the womans mouth, and also the nether places; so that the chylde may by that meanes both receave and also expell ayre and breath, which otherwyse myght be stopped, to the destruction of the chylde. And then to turne her on the lefte syde, and there to cut her open, and so to take out the chylde. They that be borne after this fashion are called Cesares, for because they be cutte out of theyr mothers belly: whereupon also the noble Romayne Cesar the fyrste toke his name." Assuredly the directions given here for a post mortem Cæsarean section are not too explicit!

Chapter IX. (by error called Chapter X. in the Raynalde editions) has no representative in the 1540 edition or in Rösslin's *De Partu Hominis*, and we must, therefore, ascribe it to Raynalde. It contains a list of medicines, ointments, and plasters ("suche as hath ben wel experimented and practysed"), to be used to quicken delivery and to expel the afterbirth. The reader is, by this time, able to foretell pretty accurately what kind of drugs will be in these medicines, and I need only refer to certain "trochiskes" upon which the writer evidently places much reliance. "Item, of Saffron dried by the fyre tyl it be blackyshe, of *Cassia lignea*, fine Reubarbe, Savine dryed, Myrrhe, of eche of these seven scruples, of pure muske, xvi. graynes, every of these simples exquisitely by them selves powdred, and then perfectlye myxed in one, with vi. or vii. droppes of Malvesey, temper the hole mase into lyttle roundels or trochiskes, eche waying a dram. And in tyme of neede at the womans labour, geve her hardly the wayght of vi.d. of these trochiskes beaten into fine powder, with foure sponefulles of Isope water, and other foure of good wine secke." The chapter closes with a paragraph (to which I have referred in my previous article as the "Bucklersbery paragraph") telling where the "trochiskes" are to be obtained.

Such are the contents of the Second Book as they appear in the 1560 edition. There are slight verbal differences in some of the other editions, and these specially affect the "Bucklersbery paragraph." More distinct variations separate the 1540 or Jonas edition from that of 1560 and from the rest. The whole of the ninth chapter (erroneously called the *tenth*) is absent from the 1540 edition; there is a difference in the wording of the commendation of the "plaster to provoke the birth," Jonas being less certain about its efficacy than Raynalde, and throughout the whole Book Jonas is more in the habit of introducing such phrases as "Avicenna saith" or "Hippocrates writeth" than Raynalde (*e.g.*, in Chapter VIII.). The differences are simply due to the fact that Jonas translated Rösslin's book literally, whereas Raynalde gave a more free rendering and supplemented the work here and there. For this reason, also, it comes about that the Second Book of the Raynalde editions is part of the First Book of the Jonas edition, for it really represents Chapters II. to IX. of Rösslin's *De Partu Hominis*. Jonas's First Book corresponds to Chapters I. to XI. of Rösslin.

The "Byrthe Fygures."

The "Byrthe Fygures" including the "Womans Stoole" belong to the Second Book of the 1560 and of the other Raynalde editions. They are all taken from Rösslin's *De Partu Hominis*, but they are not placed in the same order. Further, there is one in Rösslin's book which does not appear in the 1560 edition (or, so far as I know, in any of the Raynalde editions); this has been reproduced in Fig. V. It will be noted that in some respects it more nearly represents the true attitude of the fœtus *in utero* than any of the others. I have reproduced five of the Rösslin figures (including the "Stoole") which

Fig. V.

De partu ho
MINIS,& quæ circa ipſum accidunt,
Libellus D.Eucharii Rhodionis,Medici

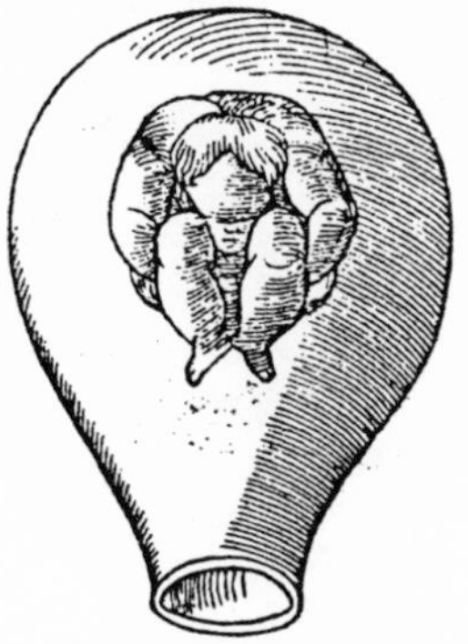

PARISIIS
Apud Ioannem Foucher in via Iacobea
1538

Fig. V.—Title page of Rhodion's *De Partu Hominis* (edition of 1538) showing supposed attitude of the fœtus in utero; this figure is not reproduced in the *Byrth of Mankynde.*

Fig. VI.

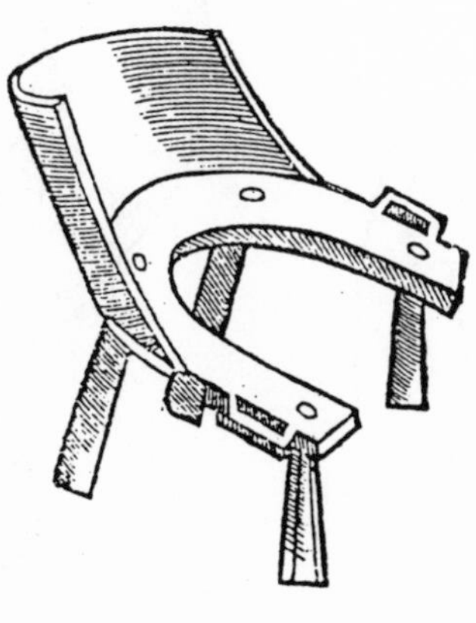

Fig. VI.—The " Woman's Stoole," taken from Rhodion's *De Partu Hominis* (edition of 1538), where it appears on folio 18.

Fig. VII.

Fig. VIII.

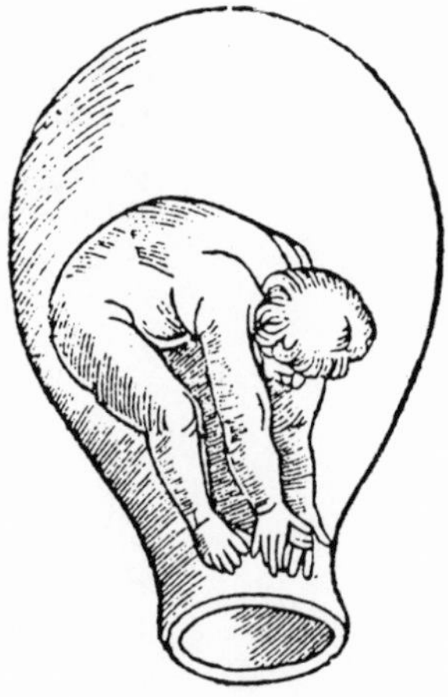

Fig. VII.—Figure showing Fœtus in Utero from Rhodion's *De Partu Hominis* (edition of 1538), where it appears on folio 20.

Fig. VIII.—Figure showing Fœtus in Utero from Rhodion's *De Partu Hominis* (edition of 1538), where it appears on folio 27.

Fig. IX.

Fig. X.

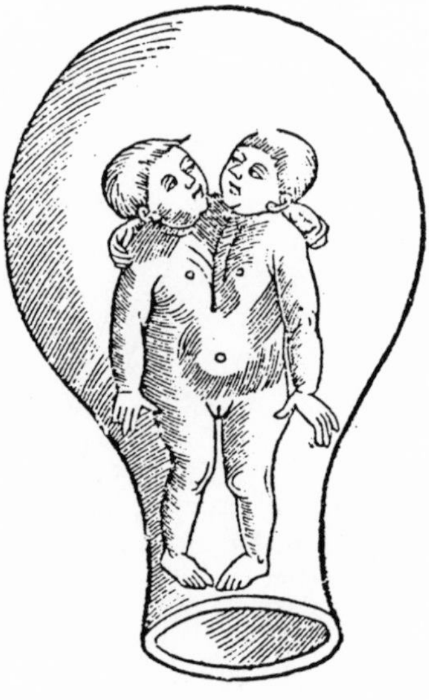

Fig. IX.—Figure of Twins in Utero from Rhodion's *De Partu Hominis* (edition of 1538), where it appears on folio 29.

Fig. X.—Figure of Double Monster in Utero from Rhodion's *De Partu Hominis* (edition of 1538), where it appears on folio 11.

have their representatives in the Raynalde editions (Figs. VI.—X.), so that the reader may compare them with the same pictures as they appear in the English translation (see my previous article, Plates VIII., IX., X., and XI.). I have already *(loc. cit.)* referred to the great interest which the "Byrth Figures" of the *Byrth of Mankynde* have excited as being the earliest, or almost the earliest, specimens of English copper plates.

The Third Book.

The Third Book of the 1560 edition of the *Byrth of Mankynde* consists of three chapters, the third being a very long one. It is devoted to the care of the new-born infant, and to its "dyverse diseases and infyrmities": the first chapter speaks of the umbilical cord and its management, the second of the nurse and her milk, and the third of the maladies of infants and the remedies required for them. "Then after that the Infant is once come to lyght, by and by the Navyll muste be cut three fyngers breadth from the belly, and so knit up, and let be strued on the head of that that remayneth, of the powder of Bole armeniacke, and *Sanguis draconis, Sarcocolla,* Myrrhe, and Cummin, of eche lyke muche beaten to pouder, then upon that bynd a peece of woll, dypped in oyle Olive, that the powder fall not of. Some use fyrst to knyt the Navyl, and after to cut it so much, as is before rehearsed." The writer mentions the belief that the length of the stump of the cord will determine the length of the "chyldes tonge," if it be a man child. He also refers to Avicenna's statement that the wrinkles on the cord betoken the number of future pregnancies the patient is to have and the intervals of time (long or short) between them; "but these sayinges be nether in the Gospell of the day, ne of the night."

The child's body is to be rubbed with oil of acorns. "After this annoyntyng, washe the Infante with warme water, and with your fynger (the nayle beyng pared) open the chyldes nosethrilles, and purge them of the fylthinesse." After the fall of the cord ("whiche commonlye chaunceth after the thyrde or fourth daye") the cicatrix is to be dusted with "ashes of a Calfes hove burnte, or of Snayle shelles, or of the powder of lead, called red lead, tempered with wyne." The proper swaddling of the child is then described, so that its limbs may grow straight ("as it is in yonge and tender impes, plantes, and twygges"); the eyes should be frequently washed, and it should sleep in its cradle in such a place that neither the beames of the Sunne by day, neither of the Moone by nyght come on the Infant." It is to be washed two or three times a day; and, after that, to put a drop or two of water into its nostrils is "very good for the eye syght."

"It shalbe beste that the mother give her chylde sucke her selfe, for the mothers mylke is more convenient and agreable to the infant, then any other womans, and more doth it nouryshe it, for because that in the mothers belly it was wont to the same, and fed with it, and therefore also it doth more desyrouslye covet the same, as that with the which it is best acquainted." Apparently the nursings are not to be frequent: "As *Avicenna* writeth it shalbe sufficient to give sucke twyse or thryse in a daye." If the mother be unable to suckle her child, then a "holsome Nourse" is to be sought out, five or six essential qualities are enumerated which she must possess, and a method of testing the milk upon the thumb nail is described. There is a long list of remedies which are said to increase the quantity of the milk. Two instances must serve: "Item, to eate shepes brestes, and the mylke of them is good;" "Item, take two drams of Crystall beaten into fyne powder, and devyde that in foure equal partes: one of these partes geve unto the Nourse, the space of foure dayes to drynke, with broth made eyther of Cicer,[1] or elles of peason."[2] The child is not to be put to the mother's breast for a day or two after birth "because that the creme (as they cal it) straight after the byrth, the first day in al women doeth thicken and congile." This, it need hardly be said, is not the rule of procedure at the present time. Weaning should take place at the end of a year, and it is not to be carried out suddenly but gradually; the infant is to be given "lyttle pylles of bread and sugar to eate" until it be able to "eate all maner of meate."

The third chapter of this Book is taken up, as has been said, with the diseases of infants and their treatment. It is of interest rather to the pediatric physician than to the obstetrician; but I may enumerate some of the subjects dealt with: "loosenesse of the bellye, cough and distillation,[3] short winde, wheales on the tounge, apostumation and runninge of the eares, bolnynge[4] of the eyes, often sneesinge, whelkes in the body, swelling of the coddles,[5] unslepinesse, yexinge or the hyckate,[6] terrible dreames, wormes in the belly, the fallynge syckenes, the palsey, and gogle eyes[7] or loking squint." It is unnecessary to quote the means recommended for the treatment of these various maladies, but the following prescription for the falling sickness (epilepsy) may be given by way of sample: "Item, to hange *Viscum quersinum*,[8] which is gathered in Marche the moone decreasynge, about the Chyldes necke, is very good."

The Third Book as it appears in the 1560 edition differs little from what it is in the other Raynalde issues, earlier or later. I have

1. Cicer, chick-pea.
2. Peason, pease.
3. Distillation, a catarrh or defluxion of rheum.
4. Bolnynge, swelling or a tumour.
5. Coddes, testicles.
6. Hyckate or yexing, the hiccup.
7. Gogle eyes, staring eyes or squint eyes.
8. *Viscum quercinum*, mistletoe of the oak.

found a few verbal differences between it and the 1552 edition (*e.g.*, in the paragraph on "Unsleppynesse"), and in the 1654 edition there is a new chapter (placed quite at the end of the work) amplifying what has been said about the nursing of children and "how to choose a good nurse." The 1540 edition differs more markedly: the chapter on "unsleepiness" is shorter, that on swelling of the coddes is not the same; there are two additional short paragraphs (the one "against the mother" and the other of short breath, hoarseness, or whistling in the throat), and there is an additional sentence on infantile constipation.

The Third Book of the Raynalde editions corresponds to Chapters X. and XI. of the First Book and to the whole of the Second Book of the 1540 or Jonas edition. It forms, also, the tenth, eleventh, and twelfth chapters of *De Partu Hominis*, from p. 61 to the end (in the edition of 1538).

The Fourth Book.

The Fourth Book of the 1560 as well as of the other Raynalde editions consists of six chapters; these are not found in Rösslin's *De Partu Hominis*, but five of them are present in the 1540 or Jonas edition of the *Byrth of Mankynde*, so that only one (the sixth) chapter is peculiar to the Raynalde editions. The Fourth Book of the 1545 and of all later editions corresponds to the Third Book of the 1540 or first impression.

An idea of the subjects dealt with in the Fourth Book can best be obtained from the short summary contained in the first chapter. I quofe (in this instance) from the 1552 edition: "Here in this fourth Boke (by ye leave of God) shal brefely be declared soch thinges which may farther or hinder the conception of man, whych as it may be by dyvers meanes letted and hyndered, so also by many other wayes it may be farthered and amended. Also to knowe by certayne sygnes and tokens whether the woman be conceaved or no, and whether the conception be male or female, and finally certayne remedies and medicines to farther and help conception: and there after we wyll (accordynge to our promyse in the prologue) set forth certayne bellyfying receptes, and so make an ende of this hole treatyse."

The second chapter gives the author's views as to the necessary conditions for the growing of corn between which and human generation he draws a parallel "Ther be in al maner of generation thre principal partes concurrent to the same: ye sower, the sede sowen, and the receptacle or place receaving and contayninge the seede." The third chapter applies this principle to the consideration of the causes of sterility, and enumerates faults in the mother receiving the seed, faults in the sower, and faults in the seed itself. The mother's womb is fancifully compared to the ground; it may be too hot, too cold, too moist or too dense. The following paragraph may be quoted to show how the author persuades himself that coldness of the matrix is a cause of sterility: "For yf corne be sowen in over cold places, soch as be in the partes of a countrey, called Sithia, and in certayne places of Almayne, or in soche places where is contynual snow or frost, or wher the sunne doth not shyne: in these places the sede or grayn sowen, wyl never come to profe, nor fructyfy, but through the vehement coldnesse of the place in the which it is conceaved, the lyfe and quickenes of the grayne is utterlye destroyed and adnihilat." The man's seed also may be defective as to heat, cold, thickness, etc. Even more fanciful is the fourth chapter which pretends to give ways of finding out whether sterility is due to defect in the woman or the man. "Let eche of theim take of wheate and barleye cornes, and of beanes of ech vii., the which they shal suffre to be steped in theyre severall uryne: the space of xxiiii. houres: then take ii. pottes, soch as they set gylyflowres in: fyl them wyth good earth: and in the one let be set the wheat, barlye, and beanes, styped (steeped) in the mans water, and in the other the wheat, barly, and beanes styped in the womans water: and everye morninge the space of viii. or x. dayes, let eche of them with theyr proper urine water the sayd sedes sowen in the foresayd pottes and mark whose pot doth prove, and the sedes therein contayned doth grow, in ye partye is not the lack of conception, and so yf ther come no other water or rayne on the pottes." This marvellous test ends with the wise remark, "but trust not moch this farfet[1] experiment." Other tokens are given, taken from the works of Hippocrates (which are often quoted in this Book), but the writer warns the reader that "these tokens, although they have a certain reason and apparence, yet be they not alwayes unfallyble, but onely lycklye." The signs of pregnancy are described (menstrual suppression, changes in the breasts, "longings," and thickness of the urine), and directions are given to enable the midwife to tell whether the unborn childe is male or female.

The fifth chapter contains various prescriptions supposed to be efficacious in curing sterility, but they call for little comment and no commendation, being founded upon the etiological theory of lack of heat or cold or moisture in the woman or in her uterus.

The sixth and last chapter of the Fourth Book may be called the "cosmetic" one, for it deals with what the writer (Raynalde) calls "dyvers bellyfying" medicines and remedies. It is proposed to show how certain blemishes ("as it were weedes of the body") are to be removed, such as "dandraffe" of the head, "hayre in places where it is unsemelye," "frekens or other spottes in the face," warts, and "pymples." There are also instructions how to keep and preserve the teeth clean, and how to prevent "stynckynge breath" and "ranke

1. Farfet, far-fetched.

savour of the armeholes." I need only quote the last paragraph: "Item, auctors do wryte the ye rootes of artichauts (ye pithe pyked oute) soden in whyte wyne and so dronke, doth clense the stenche of the arme holes and other partes of the body by the uryne: for (as Gallen also doeth testyfye) he provoketh copy and plenty of stinkynge and unsavery uryne, from all partes of the body, the whych propertye it hath by specyall gyft and not only by his hote qualyte. And thus here I make an ende of thys fourth and last boke."

The sixth chapter is wanting in the Jonas edition of 1540, but is to be found in all the Raynalde editions from that of 1545 onwards.

No part of the Fourth Book is to be found in either of the editions of Rösslin's *De Partu Hominis* (to which I have had access), and we must conclude that Chapters I. to V. were written by Jonas, Raynalde adding the sixth. Both Raynalde and Jonas seem to have gone to Hippocrates for their facts (if facts they can be called) or to some work which quotes Hippocrates. The latter is the more probable explanation of the source of the Fourth Book.

Rhodion's "De Partu Hominis."

As we have seen, Jonas translated Rösslin's or Rhodion's *De Partu Hominis* into English, adding a few chapters thereto, and Raynalde revised the translation and added new material. But who was Rösslin and what was the history of his work?

Eucharius Rösslin, Röslin, or Rhodion was a medical man practising first in Worms and then in Frankfort-on-Main. The date of his birth is unknown, and for his death-year two dates have been given—1526 and 1553. The earlier of the two is most probably the correct one, the later date being that of the death of his son. He published his work entitled *Der Swangern Frawen und Hebammen Rosegarten* (by Imperial Privilege) in 1513, and so gave to the world the first separate work on midwifery. It was dedicated to Catharine Princess of Saxony and Duchess of Brunswick and Lüneburg; and it was a compilation from the works of Hippocrates, Galen, Avicenna, Albertus Magnus, Aëtius, Gordon, and Savonarola. The earliest edition (that of 1513) had the same Birth Figures[1] as were found in Jonas and Raynalde; they were printed from woodblocks. The work was divided into twelve chapters corresponding to the first and second Books of the Jonas edition of the *Byrth of Mankynde*, and (speaking generally) to the second and third Books of the Raynalde editions of the same work. Several German editions of the *Rosegarten* appeared (in 1522, 1529, 1571); then it was translated into Latin, and, as *De Partu Hominis*, editions were brought out in 1532, 1535, 1536, 1537, 1538, 1551, 1554, 1556 and 1563; Dutch translations, under the title of *Den Rosegaert van den bevruchten Vrouwen*, came out in 1540, 1555, 1670, 1685, 1701, and 1730; and there were French

1. See Figs. vi. to x.

versions in 1536, 1540, 1563 and 1577. The English translations we have already described.

In order that the reader may judge of the accuracy of the Jonas and Raynalde rendering of Rhodion's work I place here in parallel columns the Latin and the English of two passages, one referring to the "Womans Stoole" and the other to Cæsarean section:—

From Rhodion's De Partu Hominis.

"Ad quam quidem rem, in quibusdam regionibus ut in Gallia et Germania superiori, obstetrices peculiaria sedilia habent, quae et ab humo non non alte distant, et excavata ita sunt, ut facile, quae debent, transmittant, et reclinantem tergo accipiant: quarum forma fere est talis, qualem hic adpinximus."

"At vero si diverso modo pariens emoriatur inter enitendum, id quod signis, quae morientes de se praebent, facile deprehendi potest, et partus in utero superstes, spem vitae ostendat, principio convenit morientis os, et infra genitalia cum matrice aperta et reclusa servare, quo per ea et vitalem spiritum partus capere, et anhelitum reciprocare possit, id quod mulieres mediocriter peritae satis norunt. Deinde reclinatae latus sinistrum recto vulnere novacula incidi atque aperiri (nam dexterum latus non ita liberum, propter hepar quod in eo situm habet, ingressum incidenti praebet) et inde partus inserta manu per vulnus eximi atque educi debet. Quo pacto qui nascuntur, cesares dici solent, ut etiam Romae ille a quo primo cesarum familia nomen adepta fuit, primusque caesar, eo quod caesa est matre natus, appellatus est."

From Raynalde's "Byrth of Mankynde."

"For the which purpose in some regions (as in Fraunce and Germany) the Midwyfes have stoles for the nonce, whiche beynge but lowe, and not hye from the grounde, be made so compasse wyse and cave or holowe in the middes, that that mai be receaved from underneth which is looked for, and the backe of the stole leaning backeward, receaveth the back of the woman, the fashion of the which stole, is set in the beginning of the birth figures hereafter."

"But contrary to all this, yf it chance that the woman in her labour dye, and the chylde havyng lyfe in it: then shal it be meete to kepe open the womans mouth, and also the nether places; so that the chylde may by that means both receave and also expell ayre and breath, which otherwyse myght be stopped, to the destruction of the chylde. And then to turne her on the lefte syde, and there to cut her open, and so to take out the chylde. They that be borne after this fashion are called Cesares, for because they be cutte out of theyr mothers belly: whereupon also the noble Romayne Cesar the fyrste toke his name."

It will be seen that the translation is not strictly literal, neither is it exact; for instance, the reason why the abdomen (in Cæsarean section) is to be opened on the left side rather than on the right is given in the Latin version but does not appear in the English. Possibly some divergences may be explained on the supposition that Jonas and Raynalde used editions of *De Partu Hominis* which I have not been able to see.

I have now concluded my survey of this remarkable book—*The Byrth of Mankynde*—both as regards its contents and in respect to its authors and editions. Although its precepts may bring a smile to the face of the obstetrician of the present day and merit his contempt, yet it was the most potent factor in establishing the popular customs which cluster round the practice of midwifery in these Islands, customs which can be traced and recognized even now.

Addenda et Corrigenda.

Since I wrote my article on the Author and Editions of the *Byrth of Mankynde* I have been informed of several other copies of some of the editions, and have been led to alter some of the statements made. For instance, I am doubtful of the existence of an edition of 1676. Dr. C. Nepean Longridge, to whom I am greatly indebted for a series of researches made for me in the British Museum, finds that the so-called copy of 1676 named in the catalogue of the Museum, is really that of 1626. I have, therefore, removed this edition from the list.

I have now had an opportunity of examining a copy of the 1604 edition. It belongs to Prof. H. R. Spencer, to whose kindness I am indebted for the privilege of inspecting it. It is very similar in all respects to the edition of 1598. The ornamental title page is exactly the same, with the exception of a few differences in the typography of the title. At the foot of the inscription is "Imprinted at London for Thomas Adams," instead of "Imprinted at London by Richarde Watkins." which appears in the 1598 edition. The colophon reads, "Imprinted at London for Thomas Adams, 1604." There are 204 pages in this edition, and three preliminary leaves; the plates are the same, although two plates of the "Byrthe Fygures" happen to be missing in the copy which I am describing; and the type is black letter mostly. It would seem, therefore, as if Watkins, the printer of the 1598 edition, had assigned the blocks as well as the license for printing the *Byrth of Mankynde* to Thomas Adams.

I must correct the statement made by me on p. 315 of my former article *(loc. cit.)* about the ninth figure of the Anatomical Plates. It does *not* make its first appearance in the edition of 1560; it was present (as Figure 5) in the 1545 and 1552 editions. The altering of the order of description of the figures in the 1560 issue misled me.

Here follows the revised list of the editions. I may take this opportunity of thanking Dr. R. Wilson Gibson, of Orton, Tebay, Westmorland, for kindly giving me a perfect copy of the 1654 edition.

Summary of the Editions of the "Byrth of Mankynde."

Edition of 1540. Jonas's Translation of Rösslin. British Museum.[1]

Edition of 1545. Raynalde's Translation. British Museum; Royal College of Physicians, London; Hunterian Library, University, Glasgow.

Edition of 1552. Raynalde's Translation. Royal College of Physicians, Edinburgh; London Obstetrical Society;[2] Dr. W. Blair Bell, Liverpool.

Edition of 1560. Raynalde's Translation. Royal College of Surgeons, London; Dr. J. W. Ballantyne, Edinburgh; University of Aberdeen.

Edition of 1565. Raynalde's Translation. British Museum; University of Edinburgh; R. College of Surgeons of London; Washington Library; Hunterian Library, University, Glasgow.

Edition of 1564 (?). Raynalde's Translation. British Museum. University of Glasgow; Royal College of Physicians, London; Dr. J. F. Payne, London.

Edition of 1593 (?). Raynalde's Translation. Radford Library, Manchester.

Edition of 1598. Raynalde's Translation. British Museum; Royal Society of Medicine of London; Dr. C. E. Underhill; Washington Library; Hunterian Library, University, Glasgow; Medical Institution, Liverpool.

Edition of 1604. Raynalde's Translation. British Museum; Washington Library; Dr. H. R. Spencer, London.

Edition of 1613. Raynalde's Translation. Royal College of Surgeons of London (2 copies).

Edition of 1626. Raynalde's Translation. British Museum; University of Edinburgh; London Obstetrical Society;[2] Faculty of Physicians and Surgeons of Glasgow.

Edition of 1634. Raynalde's Translation. British Museum; Royal College of Physicians, London; London Obstetrical Society;[2] Washington Library.

Edition of 1654. Raynalde's Translation. British Museum; Washington Library; Dr. W. L. Reid, Glasgow; University of Aberdeen; Dr. J. W. Ballantyne, Edinburgh.

1. In a letter which I have received from Dr. J. F. Payne, he writes "A copy of this edition, presumably perfect, was sold at Sotheby's in 1905 for £40 or £45; what became of it I do not know."

2. Now amalgamated with the library of the Royal Society of Medicine of London.

Sixteenth Century Journal
XXI, No. 2, 1990

"These griping greefes and pinching pangs": Attitudes to Childbirth in Thomas Bentley's *The Monument of Matrones* (1582)

*Colin B. Atkinson**
University of Windsor
and
William P. Stoneman
Princeton University

Thomas Bentley's *The Monument of Matrones* (1582) is one of the earliest and largest devotional books published for women in England. Its over 1,500 pages contain prayers and meditations, extracts from the Bible, and brief lives of model women. Some items by women writers have received attention recently, but this book has not been studied for what it can disclose of the lives and attitudes of sixteenth-century English women. For example, a subsection dealing with childbirth recognizes a need for prayers specifically for women in labor and for midwives. Some of these prayers, because of their intensity of emotion and female point of view, are particularly useful documents for women's history.

THE STUDY OF THE LITERATURE FOR WOMEN and by women during the English Renaissance has increased greatly in the last decade, but there is still a wealth of material which needs closer examination. In *Chaste, Silent and Obedient*, Suzanne Hull lists 163 English books for women written by both sexes published between 1475 and 1640, 85 percent after 1570.[1] Of the 18 books she classifies as devotional, the second (chronologically) is Thomas Bentley's *The Monument of Matrones* (1582),[2] one of the earliest and largest–over 1,500 quarto pages, containing prayers and meditations for a variety of circumstances, extracts from the Bible, and brief lives of biblical and other model women. Hull has aptly commented that, "In fact, *The Monument of Matrones* comes close to being an entire female library between two covers."[3]

*An earlier version of this paper was read at the annual meeting of the Canadian Society for Renaissance Studies in 1988.

[1]Suzanne Hull, *Chaste, Silent and Obedient: English Books for Women, 1475-1640* (San Marino, Ca.: Huntington Library, 1982).

[2]*THE MONUMENT OF MATRONES: conteining seuen seuerall Lamps of Virginitie, or distinct treatises whereof the first five concerne praier and meditation: the other two last, precepts and examples, as the worthie works partlie of men partlie of women; compiled for the necessarie vse of both sexes out of the sacred scriptures, and other approoued authors by Thomas Bentley of Graies Inne Student.* Printed by H. Denham. [1582]. B1. STC[2] 1892-93.

[3]Hull, *Chaste, Silent and Obedient*, 92.

Perhaps because of its size and complexity, this substantial work has been only partially studied by modern scholars. The title pages to its parts and some prayers have been analyzed for their iconographic significance, and some excerpts by women writers have been anthologized.[4] But there is an immense amount of material which has not been adequately studied, and detailed analysis of any part of the *Monument* will illuminate the lives of English women in the Renaissance. We will focus here on a subsection which concerns childbirth, but before examining it in detail, we must briefly survey the whole *Monument* to understand the context of the childbirth prayers.[5]

The *Monument* is divided into seven "lamps" and is clearly intended to cover a variety of situations. Bentley claims that he gained great spiritual help "by the reading and perusing of divers verie godlie, learned and divine treatises, of meditationes and praier, made by sundrie right famous Queenes, noble Ladies, vertuous Virgins, and godlie Gentlewomen of al ages," works which had been "dispersed into severall pamphlets, and in some part some thing obscured and worne cleane out of print, and so out of practise." Bentley has undertaken to collect these works in one volume together with works from "the holie Bible and manie other good bookes as well of praier, as of other divine matter, such as from time to time have beene penned by divers godlie men."[6] We have traced some of Bentley's sources, but other items could be original.[7] Nonetheless, whether they are reprinted from earlier works or written especially for this book, all the items are directed specifically to women and are therefore important documents for women's history.

Lamps One and Seven provide a frame for the *Monument.* Lamp One is a collection of twenty excerpts from the Bible, beginning with the prayer of Hagar in Genesis and ending with that of the five foolish virgins in

[4]The *Monument* has been excerpted recently in collections of writings by women. See Betty Travitsky, *The Paradise of Women: Writings by English Women of the Rennaissance*, Contributions in Women's Studies, No. 22 (Westport, Conn.:Greenwoood Press, 1981), and Elaine Beilin, *Redeeming Eve: Women Writers of the English Renaissance* (Princeton: Princeton University Press, 1987). Rivkah Zim in *English Metrical Psalms: Poetry as Praise and Prayer, 1535-1601* (Cambridge: Cambridge University Press, 1987) studied some of the psalms. In "The Godly Woman in Elizabethan Iconography," *Renaissance Quarterly* 38 (1985): 41-84, especially 70-79, and more recently in *Tudor Royal Iconography: Literature and Art in an Age of Religious Crisis* (Princeton: Princeton University Press, 1989), 243-56, John N. King has used the title pages of the various lamps of the *Monument* in his analysis of an iconographic tradition.

[5]An accurate bibliographic description has never been undertaken. Our forthcoming article, "'Devout and Godlie Women in Our Time': Thomas Bentley's *The Monument of Matrones* (1582) and the Tradition of Literature for Women," will survey the *Monument* as a devotional book. We are also preparing a scholarly edition.

[6]See Bentley, *Monuments,* B1, for these remarks.

[7]Nothing is known of Bentley other than that he was a student of Gray's Inn. He may have worked for Henry Denham, the printer, and selected works from Denham's stock or that of allied printers. He might have added prayers of his own composition.

Matthew 25. Lamp Seven contains descriptions or lives of various female figures found in the Bible as well as of women mentioned in the third book of Maccabees and in Josephus's *History*. The devotional material of Lamps Two through Six is thus placed in a context of women and religion going back to what was considered the beginning of Creation. In the first lamp we are given their prayers and in the seventh their lives.

Lamp Two contains prayers and meditations by such illustrious Renaissance women as Queen Margaret of Navarre, Queen Katherine Parr, and Lady Jane Dudley.[8] From devout women of his own period, Bentley moves in Lamp Three to that paragon of virtue and piety, Queen Elizabeth herself. This lamp contains psalms, prayers, and meditations pertaining to the Queen, for general use, but especially, he adds, for use on the anniversary of her accession day. There are two sets of acrostic prayers on ELIZABETH REGINA; and the King's Heast and the Queene's Vow, both compiled from the psalms.[9]

Lamps Four, Five, and Six contain prayers and meditations arranged for the various stages in a woman's life such as girlhood, motherhood, and widowhood, as well as for times of the day such as morning, evening, and before and after church. Though some (with minor changes in wording such as from "she" to "he" or "handmaid" to "servant") could as easily be used by men as well as by the women for whom they are intended, others speak with a female voice. Indeed, in some prayers, the tone is so sensitive to women's feelings that women who have read these have asked if we have discovered female authorship or at least influence. To date, we have not discovered any.

The *Monument* is a huge book containing a variety of genres, from daily household prayers to longer theological meditations. Bentley organized the book to place the daily prayers of women, to which Lamps Four and Five in particular are devoted, in a wider context of female devotion. How women actually used the book is another matter. For example, Lamp Five

[8]Included, for example, are such known works as "A Godlie Meditation of the inward loue of the soule towards Christ our Lord: composed first in French by the vertuous Ladie Margaret Queene of Navarre: aptlie, exactlie, and fruitfullie translated by our most gratious Soueraigne Ladie Queene Elizabeth, in the tender and maidenlie yeeres of hir youth and virginitie, to the great benefit of Gods Church, and comfort of the godlie" (STC 4827?) and "Godlie Praiers and Meditations, wherein the mind is stirred, patientlie to suffer all afflictions heere; to set at naught the vaine prosperitie of the world: and alwaie to long for euerlasting felicitie: collected out of holie works, by the most vertuous and gratious Princesse KATHERINE, Queene of England, France, and Ireland" (STC 4818). There are 252 pages in this lamp, and some of the prayers are attributed to unnamed "virtuous gentlewomen." We are currently tracing sources for our edition of the *Monument*.

[9]Lynn Johnson, in "Elizabeth, Bride and Queen: A Study of Spenser's April Eclogue and the Metaphors of English Protestantism," *Spenser Studies* 2 (1981): 79-81, assumes Bentley to be the author both of the epistle to the Queen in the front matter and of "The Kings Heast, or Gods familiar speech to the Queene," which Bentley claims to be drawn from the psalms as they are "learnedlie expounded by Theodore Beza."

would make a portable volume useful in many daily situations; Lamp Two, on the other hand, is an anthology of religious works by major Renaissance women more suitable for private study. It seems possible, therefore, that the *Monument* was intended as a collection of books and that individual lamps might have been bound separately.[10]

* * *

Lamp Five contains 115 prayers for virgins, women about to be married, wives, mothers, mistresses and servants, and widows. Roughly one third of the material in the lamp (38 of the 115 prayers or 62 of the 213 pages) is devoted to the subject of childbirth. This is not surprising: childbirth was a perilous event in any woman's life and may even have been more hazardous to upperclass women for whom such a major publication as the *Monument* was intended. Dorothy McLaren suggests that whereas lowerclass women might give birth only every two years (because of postpartum infertility resulting from suckling the baby), upperclass women usually utilized wet nurses and thus were more likely to have annual pregnancies and possibly more difficult deliveries.[11] Whether or not this proves to be the case, prayers centering on childbirth form a major part of Lamp Five.

The childbirth group of prayers contains only two headings, "Praiers to be said of women with child, and in childbed, and after their deliuerie," and "Thanksgiuings for women after deliuerance of childe," but headings in the *Monument* are sparse and not always helpful. A more accurate division of the group would be:

1. Ten prayers by the woman in labor (nos. 42-51; pp. 95-119)[12]
2. Four prayers of thanksgiving after childbirth (nos. 52-55; pp. 119-25)
3. Seven prayers for a woman in labor (nos. 56-62; pp.125-33)
4. Four prayers for midwives (nos. 63-66; pp. 134-41)
5. Nine prayers for difficult birth or death (nos. 67-75; pp. 141-52)
6. Two prayers for the baby (nos. 76-77; p. 151)
7. Two prayers of thanksgiving (nos. 78-80; pp. 153-56)

Many of the prayers simply ask forgiveness for sins and the grace to endure sufferings and to accept God's will. They offer praise for his mercy and love, and thanks for the safe delivery of a healthy child. Some, on the other hand, have an immediacy and a vigor that suggest a writer with a

[10]The British Library copy, however, does not indicate such use. It is bound in three volumes, but these are eighteenth-century bindings.

[11]Dorothy McLaren, "Marital fertility and lactation 1570-1720," *Women in English Society 1500-1800*, ed. Mary Prior (London: Methuen, 1985), esp. 43-46.

[12]Entries in the *Monument* are usually titled, although often the title is very brief, e.g., Another prayer. For the sake of convenience (and as part of the preparation of our edition), we have numbered the prayers in Lamp Five.

deep concern for the woman's feelings. A closer look at specific prayers in each section will illustrate these two types of prayer and demonstrate the *Monument*'s awareness of the many concerns of women during childbirth.

1. Prayers by the woman in labor

The prayers in the first group (nos. 42-51) praise God for the miracle of conception, pray for God's aid in carrying and delivering the child, and affirm a trust in God's providence. All prayers ask for less pain or for the strength to endure the pangs resulting from human sin, which women inherit from Eve.

The contrast between the two kinds of childbirth prayers can be seen by comparing nos. 50 and 51. Although both are prayers to be said during labor, no. 50 has the ring of firsthand experience. The speaker is at the end of her tether. She has kept silent until she can no longer stand the pain. She expresses the cause of her agony in specific terms:

> The babe is come vnto the place of the birth, and lo, it seemeth that thou for my sins hast shut vp the dores of my wombe, and caused the babe to stand still like to be stiffled (p. 113).

Later in the same prayer:

> How long Lord shall my bowels thus sound like an harpe, my bones and sinewes be racked asunder, and mine inward parts be thus greeuouslie tormented for my sins (p.115).

Not only is she specific about her physical complications, but the speaker also powerfully conveys her anguish. Fear of the child's death is expressed through a series of rhetorical questions:

> Shall I be the graue of my child: Shall I giue death the fruit of my bodie, for the sins of my soule; and my first (second, or third) borne for the transgressions of my youth? Alas shall that perish in the wombe, which is conceiued by thee; or shall it die in the birth, and not be borne, which thou hast so mightilie fashioned? Hast thou shut vp the doores of my wombe in thy displeasure, and couered me with the shadowe of death, that the birth may not come out (pp. 115-16).

There is an urgency and immediacy to the repetition and the suffering of the speaker not found in the second of these two prayers.

Prayer no. 51 begins with the general cry: "O God make speede to saue me, O Lord make hast to helpe me." The rhetoric throughout the prayer is rarely more explicit. The petitioner asks for relief from "these griping greefes and pinching pangs of deadlie throwes"; but the language could describe any bodily or spiritual pain. It is tempting to postulate two different writers for these prayers, the first of whom, if not a woman, is at least a man who is sensitive to female experience.

2. Prayers of Thanksgiving after childbirth

Two of these prayers (nos. 52 and 55, pp. 119-25) are for use immediately after delivery and, two are for the purifying (or churching) of the woman after childbirth. One of these prayers is noteworthy. We have characterized the voice in a number of prayers in Lamp Five as female; no. 55, on the other hand, is the only prayer in the lamp that exhibits what appears to be an anti-female voice. The speaker expresses a disgust with the "vncleannes" of women (particularly in the gratuitous image of the "fowle menstruous cloth") that is unique in the entire childbirth group. The prayer opens with thanks to God who:

> . . . didst make diuers lawes concerning vncleannes, and the keeping of women with child; and an other for the first borne, and that not onelie to put vs in mind of order in this life, and thankfulnes towards thy Maiestie: but also to teach, that this our giltie and polluted nature, like the fowle menstruous cloth of a woman, is washed by the blood of thy sonne . . . (p. 123).

The speaker later refers to herself as "thy most defiled and polluted hand maid" and bewails her "vnworthines, vilenes, and vncleannes." Through Christ she asks to be "adopted into thy children, & find fauour in thine eies."

The prayer is followed by an alternative: "Or you may giue thanks and praie as in the booke of common praier, and saie to yourselfe, or with the Minister, as followeth," and the Churching of Women section of the Book of Common Prayer is printed. Bentley follows the order and adapts the words of the 1559 version, changing the prayers to first person. For example, where the Book of Common Prayer has "ye shall therefore giue hearty thanks vnto God and pray," the *Monument* has, "I therefore heere in the Church and congregation, giue hartie thanks vnto thee, O God, and saie." (Psalm 121 follows in both cases.) The Prayer Book, after the Lord's Prayer, has a set of responses for a priest and the woman; the *Monument* combines the two into a first-person prayer. Bentley is not fundamentally altering, but merely adapting, the 1559 Book of Common Prayer for private female use.[13]

[13]Similarly Bentley adapts biblical quotations for female use, sometimes with odd results. For example, in Lamp Six, he quotes Ecclus. 9:6. The Geneva Bible has: "Cast not thy minde vpon harlots [in anie maner of thing,] lest thou destroye [both thy self and,] thine heritage." Bentley's version substitutes "ruffians" for "harlots," an interesting but simple change. Ecclus. 9.4, however, has a greater change. The Bible has "Vse not the companie of a woman that is a singer, [& a dancer, neither heare her,] lest thou be taken by her craftines." Bentley has two versions. First he quotes it: "Maidens, damsels, or women that are singers and dansers, are not to be harkened vnto, or kept companie withall"(p. 4). Then on the next page he uses it again: "Ye maids, damsels, and virgins, vse not the companie of them that are singers, dansers, and riotous men: neither herken vnto them, least you be snared, and taken by their craftie allurements." Bentley's adaptations and use of the Bible need much more study.

3. Prayers to be said for a woman in labor

Although these prayers (nos. 56-62, pp. 125-33) are provided for the use of other women attending the woman in labor, most do not particularly stress a female point of view. One (no. 56), in fact, is specifically intended for the husband. (Indeed, in the entire childbirth section, only one other prayer, no. 78, is a prayer for a husband to say.) One prayer in this group (no. 61) is, however, of interest, because it is the only one which presents the woman's point of view, this despite the fact that the child is assumed to be a man-child. The petitioner's prayer that women's agony be changed to joy expresses women's hope that labor pains will ultimately be forgotten:

> O almightie God, conuert now the sorrowes of women in trauell into gladnes, that they remember not their paines with mourning, for ioie that a man-child is borne into the world . . . (p. 131).

The prayer quickly shifts from concern for the woman in labor to the joys and concerns of motherhood. The word "mother" is repeated as the prayer asks that:

> [all infants] may come foorth into this world sound and perfect without fault or deformitie: and grant that they being borne vpon their mothers sides, may be ioifull vpon their mothers knees, and sucke the milke of comfort out of the breasts of hir loue, to the reioising of the mother. Staie the furiousnes of wicked spirits, that they shewe not their tyrannie vpon yoong infants. Keepe all women with child, that no waie being terrified, frighted, or troubled extreemelie they be vntimelie deliuered (p. 131).

The prayer concludes with the hope that women be restored to full health, and that should there be untimely births or other sorrows, the parents be strengthened to accept God's will.

4. Prayers for the Midwife

The status of the midwife in late sixteenth-century England is difficult to assess, although Thomas Rodgers Forbes asserts that it was in general a lowly one, especially in rural areas.[14] Church regulations focused on the concern that any infant in danger of death be baptized and therefore that midwives be godly. They must not practice witchcraft. Although Forbes concludes that the majority of midwives did not, he feels that the widespread concern that midwifery was allied with witchcraft suggests that many did.[15] (Sexism as the cause for the concern does not seem to have occurred to him.) Under the Tudors, medicine and surgery, including midwifery, were

[14]Thomas Rogers Forbes, *The Midwife and the Witch* (New Haven: Yale University Press, 1966), 112.

[15]Ibid., 132.

brought under government control by an Act of 1512.[16] This act was designed more to prevent sorcery and witchcraft, the substitution of children, or the destruction of fetuses than to enforce professional training. It was administered by ecclesiastical authorities, hence applicants for a license (other than university graduates) had to provide character witnesses and swear an oath before the local bishop's court. None of this, however, indicates much about the social status and skills of the midwife.

Jean Donnison states that in London (thus among the women to whom the *Monument* is addressed), "professed" midwives "enjoyed the highest status, and had undergone the longest training, in many cases as much as seven years."[17] Indeed, by the seventeenth century gentlemen's daughters attended the aristocracy and royalty during childbirth.[18] However, as the man-midwife began to appear in the late sixteenth and early seventeenth century, the ensuing struggle for power led to charges against midwives that may have been more political than accurate. How are we, then, to judge the social status of midwives in 1582? Were they ignorant superstitious women? Were they schooled in women's mysteries by midwife mothers and older wise women? If the prayers in the *Monument* reflect general opinion, we see a high degree of pride and professionalism.

The first and longest of these prayers (nos. 63-66, pp. 134-41) for midwives, no. 63, is worth examining in detail. The opening sentence sets the tone for all the prayers and affirms the importance of women and of childbirth in God's plan:

> O Heavenlie Father, which at the beginning didst appoint woman to bee the organe, instrument, and vessell to receiue, nourishe, & bring foorth man through thy woonderfull workmanship without anie labour, paine, or helpe of flesh and blood: but afterward through the sinne and disobedience of our grandmother Eue, hast made that which before was easie, now to become hard and painefull, yea dangerous and impossible (if thy helpe were not) for them to be brought to passe, and now as a meane hast ordeined man to be a god to man, and made woman through thine vnspeakable power working togither with him, to helpe & ease woman, which through their own imperfection and feeblenes are not able in this case and time of their sore labour, to helpe themselues (p. 134).

The phrase "man to be a god to man" implies that men must help each other in a postlapsarian world. By paralleling this with women helping and easing women, the dignity of the midwife's calling is asserted.

[16]"An acte concernynge the approbation of phisycyons and surgions," *Public General Statutes of the Realm, Henry VIII*, 3,. cap. 11. See Jean Donnison, *Midwives and Medical Men* (New York: Schocken Books, 1977), 6-11.

[17]Donnison, *Midwives*, 8-9.

[18]Ibid., 9-10.

The vocation of midwife is spelled out in a very proud and dignified manner. There is no apology for the role:

> For as much as it hath pleased thee among manie other more worthie, to call me thy most vnworthie handmaid to the office of a midwife among the daughters of Israel, and made me an instrument and means vnder thee to comfort them in their sorrowes, cherish them in their child-beds, ease them in their pains, and to further them in their deliuerance, with all possible paines and conuenient speede that I may, and can, in thee and for thee. I most hartilie thanke thee, and praise thy holie name and maiestie; beseeching thee of grace and vertue to walke worthie of this my calling, in all holie obedience to thee, and thy word; to thy glorie, others profit, and mine owne saluation in Christ (pp. 134-35).

Later in the prayer, the sins to be eliminated are "ignorance, negligence, slouthfulnes, slacknes, and disdaine . . . vnmercifulnes, rough handling, hardnes of hart, contempt of others, falshood, crueltie, and bloudgiltines," sins which are the human, not the female, lot. The virtues or strengths prayed for are an interesting mixture:

> make me wise-harted, skilfull, louing, gentle, tender, pitifull, cherefull, comfortable, helpfull, painefull, watchfull, strong, able, readie, willing, carefull, diligent, & faithfull, euen for thy sake onelie, without respect of filthie lucre, to pleasure all women at all times in my calling, to the full discharage of my conscience and dutie, both before thee and the world (p. 135).

This somewhat long list of virtues affirms both traditionally feminine qualities and others more usually associated with men. There are the feminine ones, such as "louing, gentle, tender, pitifull," but here they are in the service not of men but of other women. Qualities such as "strong, able, readie, willing," are virtues which, if not masculine, are at least applicable to either sex.

The midwife finishes by praying for all lawful success and for help in never hurting mother or child by negligence or ignorance. She cites Shiphra and Puah and the other midwives of the Hebrew women in Egypt as her models. All in all, this prayer projects an image of a strong capable woman, proud of her calling.

5. Prayers for difficult births or death.

These prayers (nos. 69-77, pp. 141-52) plead for forgiveness of sins, a speedy release from pain, and heavenly salvation. Very few of these prayers suggest special composition from a specifically female point of view. While Bentley has inserted proper initials in one or two prayers in other sections, it is here that they appear most frequently. M. B. and S. B.

are both near death. In one prayer (no. 74), S. B. is the woman who has died in childbirth. The prayer concludes with the belief that her soul is in "thine euerlasting glorie, where as Abraham, Isaac, and Jacob, Sarra, Rebecca, Rahel and the blessed virgin Marie are continuallie" (p. 149).

6. Prayers focusing on the child

This section (nos. 75-76, p. 151) contains only two short unremarkable prayers.

7. Thanksgiving for women after delivery

These prayers comprise nos. 78-80, pp. 153-56. Prayer no. 78 is actually for the father to say after hearing that the child is safely delivered. (As we have observed, this is one of only two for fathers in Lamp Five of the *Monument.*) He thanks God for calling him to the holy state of matrimony, thanks him for the child and the safe termination of his wife's labor, and asks for grace on himself and "this my yoke-fellowe or wife thy seruant" that they may live chastely and honestly and holily and that they may raise the child as a good Christian.

The other two prayers are for women's use and are standard thanks to God, first for bringing the woman unharmed through delivery, then for the safe arrival of the baby. Both close with pleas to deliver the mother henceforth from dangers.

* * *

The childbirth section of Lamp Five is an important document for the study of the history of women. It addresses most of the situations concerning childbirth and does so with great sensitivity to the woman's point of view. Of the thirty-nine prayers, only two are for the baby and four are prayers of thanksgiving; the rest focus on the act of childbirth itself–the agonies, the fears, and the dangers, including death, that women might expect to suffer. This contrasts strongly with the Book of Common Prayer which deals only with the churching or purification of women after childbirth. The Book of Common Prayer is of course not primarily a manual for private devotion, but nonetheless it is significant in that it offers nothing more to women at their greatest moment of need.

Although much of the *Monument* supports the belief that women were considered souls in female bodies and that therefore they should use the same prayers as men, there are some in Lamp Five that suggest an awareness that women have needs specific to their sex which require their own prayers. Women suffer pains and fears concerning childbirth seldom addressed by men. Midwives have a sense of dignity and pride in their calling and need to affirm this and ask God's help in prayer. Some of the prayers for childbirth seem only slightly adapted for female use. Others, however, in their intensity of emotion and their point of view suggest special composition either by a woman or by a man showing some sensitivity to a woman's

This examination of the childbirth section of the *Monument* (only 63 pages of a work containing over 1,500) illustrates how significant the *Monument* is to an understanding of the nature and status of women in the late sixteenth century.[19] Although much work remains to be done both on the book as a Renaissance prayerbook and on its significance for women's history, it is clear that *The Monument of Matrones* is of major importance both as a Tudor devotional book and as an source of knowledge about Renaissance English women and their concerns, and those women, biblical and contemporary, who were held up as models.

[19]The *Monument* casts light on other areas of female experience. For example, in Lamp Five there are prayers for mothers-in-law and daughters-in-law, for widows, and for old women, many of which show insight into the particular problems of these types of women.

Historians as Demonologists: The Myth of the Midwife-witch

By DAVID HARLEY*

SUMMARY. The belief that midwives were commonly persecuted as witches is widespread in the history of witchcraft and the history of medicine. Although the midwife-witch can be found in the writings of some demonologists, influenced by the *Malleus Maleficarum*, in few of the vast numbers of trials were midwives accused. The practice of midwifery required them to be respectable and trustworthy. Those who dabbled in medicine were occasionally accused but midwives were generally immune from witchcraft prosecution unless they fell foul of a zealous magistrate or there was some special local belief. Historians have been led astray by a tradition that derives from the discredited work of Margaret Murray. A few spectacular cases have been mistaken for a general pattern and midwife-witches have been seen where none exist. The history of witchcraft has been distorted but the history of midwifery has been completely unbalanced by this modern stereotype, which has served either to justify the rise of the men-midwives or to create a multitude of imaginary martyrs for the modern women's health movement. The myth of the midwife-witch is an obstacle to serious study of the history of midwives, women's health and the relationship between popular medicine and religion.

KEYWORDS: midwives, witchcraft, magical medicine, Europe, England, Scotland, New England, historiography.

Though there be numerous company of Authors that have written of Magick, Witchcraft, Sorcery, Inchantment, Spirits, and Apparitions, in Sundry ages, of divers Countrys, and in various languages: yet have they for the most but borrowed one from another, or have transcribed what others had written before them. So that thereby there hath been no right progress made truly to discover the theory or ground of these dark and abstruse matters, nor no precise care taken to instance in matters of fact, that have been warrantably and sufficiently attested.[1]

John Webster, writing towards the end of European witchcraft prosecutions, was justifiably annoyed that demonologists simply took their anecdotes and arguments from previous authorities without checking the facts. Since the existence of the phenomena was generally accepted, discussion turned mainly on the precise details of apportioning blame and identifying culprits. Unfortunately, modern historians of witchcraft tend to behave in precisely the same way.

One of the few things about late medieval and early modern midwives that almost everyone knows is that they were ignorant old crones. Moreover, it is a truth universally acknowledged, that midwives were frequently prosecuted for witchcraft. It is asserted by those who approve of the rise of the men-

[1] J. Webster, *The Displaying of Supposed Witchcraft* (London, 1677) sig. A2. This essay arises from the encouragement of Margaret Pelling and Charles Webster and has benefited from discussions with Bob Bliss, Robin Briggs, Audrey Eccles, Ann Hess, Michael MacDonald, Lyndal Roper, and Adrian Wilson.

* *17 Arlington Drive, Old Marston, Oxford OX3 0SH.*

midwives and those who deplore the decline of the midwives. It is asserted by those who believe witchcraft to have been the remnant of a pagan religion and those who believe its prosecution to have been the expression of social tension. It has become as much a part of popular historical knowledge as the hunchback of Richard III, yet it hardly ever features in the work of scholars who have engaged in detailed archival research. True or false, the belief that midwives were prosecuted as witches is clearly a powerful myth and worthy of examination in its own right, since it shapes much discussion of both midwifery and witchcraft. If it is as false as the assertion that midwives were universally poor and illiterate, one must consider why historians continue to propagate it.[2]

There can be no question that references to the existence of midwife-witches occur in the writings of demonologists. The *Formicarius* of Johannes Nider, printed in about 1473, 40 years after its composition, mentions an example and this was elaborated by Sprenger and Kramer, the authors of the notorious *Malleus Maleficarum* in 1487, into a full-blown explanatory theory. The midwives obtained the bodies of infants for magical purposes. It has been suggested that the obsession of the *Malleus* with children, impotence and infanticide arose from either deep-seated fear of the power of women or concern about the widespread medieval practice of family limitation. To those possible motives for the authors' hostility to midwives, a recent writer on the *Malleus* added the suggestion that a high level of abortion and still births, caused by social conditions, led to popular suspicion of the midwife.[3] Like many other attempts to explain witchcraft beliefs, such suggestions are helpful but rather overlook the extent to which the target of witch-hunters and the demonologists was the Devil himself and his minions on earth. Historians have a tendency to wish to explain away theological aspects of past society as though one can simply transmute them into social and political conflicts. In singling out elements of witchcraft belief that lend themselves to such explanation, the separate power of ideology is neglected. Much of the force of the *Malleus* and its successors derives from the appropriation of the alleged crimes of the Jews by authors who sought to build up the fantasy of demonic feasts. Whereas Jews supposedly had to kidnap Christian infants for their sacrifices, witches could obtain them more easily through the agency of midwives.[4]

[2] For a discussion of the literacy and affluence of English provincial midwives, see D. N. Harley, 'Ignorant Midwives: a Persistent Stereotype', *Bulletin of the Society for the Social History of Medicine* xxvii (1981) 6–9. It is not necessary in this context to consider their technical expertise, as challenged by A. Wilson, 'Ignorant Midwives: a Rejoinder', *Bull. SSHM* xxxi (1983) pp. 46–9. For a detailed study of Nuremberg midwives, see M. E. Wiesner, 'Early modern midwifery: a case study', in B. A. Hanawalt (ed.), *Woman and Work in Pre-industrial Europe* (Bloomington, 1986), pp. 94–119; id., *Working Women in Renaissance Germany* (New Brunswick, NJ, 1986), pp. 55–73.

[3] L. Dresen-Coenders, 'Witches as Devils' Concubines: on the Origin of Fear of Witches and Protection Against Witchcraft', in *Saints and She-devils: Images of women in the 15th and 16th centuries* (London, 1987), 59–63; M. Nelson, 'Why Witches Were Women', in J. Freeman (ed.), *Women: a Feminist Perspective*, 2nd edn. (Palo Alto, Calif., 1979), 463–4; E. Camerlynck, 'Feminité et sorcellerie chez les théoriciens de la démonologie à la fin du Moyen Age: Etude du *Malleus Maleficarum*', *Renaissance and Reformation*, ns 7/1 (1983), 13–25.

[4] V. J. Newall, 'The Jew as Witch Figure', in *The Witch Figure* (London, 1973), 95–124.

Such explanations also neglect to investigate the extent to which the supposed popular suspicion really existed or the clerical antagonism led to actual persecution, preferring to assume such phenomena on the evidence of a single text. The publication of the *Malleus* and other demonological works did not lead immediately to large-scale witch hunts, which mostly occurred in a later period. The influence of the *Malleus* on popular belief is very doubtful but it became a potent authority for later demonologists, especially when it began to be frequently republished in the late sixteenth century, as a response to the resurgence of prosecutions after the lull during the Reformation. Mazolini, writing in 1575 under the name 'Silvester Prierias', bases his discussions of midwives and infanticidal witches entirely on the *Malleus* and the work of Nider, generally published with it. Bodin's reference to a midwife sacrificing infants to Satan is taken from Sprenger, as is the discussion by Codronchius of midwives obtaining infants to make flying ointment, although the latter also refers to Prierias and others. Delrio hardly mentions abortion and infanticide in his compendious work, contenting himself with Nider's example and heavy borrowings from Sprenger.[5] Boguet cites Porta and Cardan, the Neoplatonist authors, on witches' ointments but, when he treats the topic of midwife-witches, he only discusses infanticidal parents, apart from quoting Bodin, Sprenger, and Nider. Although Guazzo cites Porta and Pliny on the magical use of infants' bodies, his only direct reference to midwives comes when he quotes from the *Malleus*. It is very noticeable that although seventeenth-century authors made efforts to broaden the scope of their citations, responding to the growth of classical scholarship and Neoplatonist writings, they produce no new examples of prosecuted midwives.[6]

If there was no debate about the authority of the *Malleus* on the guilt of midwives, there was some discussion on the efficacy of using children's bodies for magic. The Neoplatonist discussion of the reality of natural magic was one influence, as was the sceptical suggestion that witches were merely deluded. Pierre de Lancre argues that the fat of infants is not functional, being rather used by the Devil to dupe the witches into murder. His book contains a lurid depiction of the witches' sabbat by Ziarnko, including their cannibalistic feast:

> Voila les Conuiues de l'assemblee, ayant chacune un Demon pres d'elle: Et en ce festin, ne se sert autre viande, que charoignes, chair de pendus, coeurs d'enfants non baptisez, & autres animaux immondes, du tout hors de commerce & usage des Chrestiens, le tout incipide & sans sel.

> (There are the merry-makers of the gathering, having each a demon near her: and in this festival, no other meat is served apart from corpses, flesh of hanged men, hearts of unbaptized infants, and unclean animals, totally outside the trade and usage of Christians, the whole insipid and without salt.)[7]

[5] S. Prierias, *De Strigimagarum, Daemonumque Mirandis* (Rome, 1575), pp. 154–6, 172–4; J. Bodin, *De la Daemonomanie des Sorciers* (Paris, 1580), bk. 2, ch. 5, p. 93; G. B. Codronchius, *De Morbis Veneficis ac Veneficiis* (Venice, 1595), bk. 3, ch. 8, pp. 130, 132; M. Delrio, *Magicarum Disquisitionum*, vol. 2 (Louvain, 1600), pp. 38, 70–3.

[6] H. Boguet, *Discours des Sorciers*, 2nd edn. (Lyon, 1608), ch. 5 and 33, pp. 165, 204–8; F. M. Guazzo, *Compendium Maleficarum* (Milan, 1608), bk. 2, ch. 2 and 3, pp. 105–9.

[7] P. de Lancre, *Tableau de l'Inconstance des Mauvais Anges et Demons* (Paris, 1613), p. 107 and illustration between pp. 118 and 119.

That de Lancre and his contemporaries should describe such nocturnal meetings of witches and demons as 'sabbats' or 'synagogues' indicates the extent to which they are employing traditional myths of Jewish ritual murder. Persecutions of Jews based on the blood libel are documented from the twelfth century onwards, predating the organized witch-hunt by two centuries. When midwives were involved in such outbreaks, it was as representatives of respectable society. Thus when ritual murder was suspected in 1584, after a Worms midwife had delivered twins in the ghetto hospital, she gave expert testimony suggesting infanticide.[8] This role as expert witness, in a wide range of court proceedings, suggests an apparent paradox. As Erwin Ackerknecht remarked when discussing the forensic activities of midwives,

> in the late Middle Ages, midwives have attained an ambivalent position: on the one hand, they enjoy a number of privileges such as state salaries; on the other hand, they are favored victims of the witch-hunting craze of the 15th, 16th, and 17th centuries.[9]

Since midwives were trusted to give reliable evidence in cases of rape, bastardy, and infanticide, and even in cases of witchcraft itself, why should they be suspected of being in league with the Devil? It might be that demonologists had a logical need to see them as the source of flesh for the sabbat that was not shared by the general populace. One needs to establish how often they were prosecuted and under what circumstances.

Contemporary accounts by practising witch-hunters, apart from the *Malleus*, are rather disappointing on this point. Nicolas Rémy, the energetic Lorraine witch-hunter, discusses the vile uses made of foetal or newborn flesh by witches, citing classical and Neoplatonist authors such as Pliny and della Porta. He is indignant that directions for such preparations are to be found in the Neoplatonist works of 'Agrippa, Petrus de Abano, & Picatrix tres damnatae Magiae' (three damned witches). He provides examples from his own examination of accused witches in the period 1586–90, insisting that the practice is common in Lorraine, but none of the cases appear to involve midwives. His only reference to midwives occurs when he cites Pliny. Although Rémy abstracted the records he used, the surviving material suggests that, while the theologians and lawyers of Lorraine confused magical healers with witches, midwives did not feature prominently among those prosecuted. Since the custom of electing midwives was widespread in Lorraine, this is perhaps what one should expect.[10]

[8] R. Po-Chia Hsia, *The Myth of Ritual Murder: Jews and Magic in Reformation Germany* (New Haven, 1988), p. 206.

[9] E. H. Ackernecht, 'Midwives as Expert in Court', *Bulletin of the New York Academy of Medicine* lii (1976), 124–8; cf. Wiesner, *Working Women*, pp. 61, 70–1.

[10] N. Rémy, *Daemonolatreiae* (Leyden, 1595), bk. 2, ch. 3, pp. 209–12; E. Delcambre, *La Concept de la Sorcellerie dans le Duché de Lorraine au XVIe et au XVIIe siècle*, vol. 3 (Nancy, 1951), pp. 205–19. I am grateful to Robin Briggs for confirming my impression. He has examined hundreds of Lorraine cases without finding a single midwife prosecuted for witchcraft. For an excellent account of the social dynamics of Lorraine prosecutions, see R. Briggs, *Communities of Belief: Cultural and Social Tensions in Early Modern France* (Oxford, 1989), pp. 7–105. F. Hacquin, *Histoire de l'Art des Accouchements en Lorraine des temps anciens au XXe siècle* (St Nicolas-de-Port, 1979), pp. 36–8, 48. A good account of the selection of a Languedoc midwife is provided by the diary of a parish priest, Michel Lalande, quoted by W. Gibson, *Women in Seventeenth-Century France* (Basingstoke, 1989), pp. 118–19.

In the absence of much detailed contemporary discussion, one must turn to reputable modern historians for an assessment of the frequency of midwives being prosecuted. Erik Midelfort, in his classic study of German witch-hunting, published in 1972, identifies two vulnerable groups, the notorious and the well known. In his first group, the early victims of a panic, he places 'widows, spinsters, and midwives'. In the second group, drawn into the full-blown panic by the chains of confessions and accusations brought about by the use of torture, he places 'magistrates, teachers, innkeepers, wealthy merchants, and their wives'. Since he provides no examples of midwives being prosecuted, it is difficult to know how much weight to place on this assertion, especially in view of his failure elsewhere to differentiate between the reasons for accusing innkeepers and midwives. Herbert Pohl's recent study of Mainz trials offers some confirmation of Midelfort's identification of innkeepers and midwives as especially prone to accusation, the latter group providing some two per cent of suspects, although he suggests that midwives were accused because of their supposedly low social status, which he does not suggest in the case of other occupational groups. A more likely explanation is provided by a student of Bavarian cases who suggests that all those trades associated with food preparation or magical medicine were at risk, regardless of status.[11]

Richard Kieckhefer, in an excellent study of the early trials, suggests that 'many of the women prosecuted had curing as their occupation: they were beneficent magicians, practitioners of folk medicine, or perhaps midwives'. Unfortunately, he provides no examples of the last group. Norman Cohn, in *Europe's Inner Demons*, listed the various types of suspect: 'Finally there were the midwives and the practitioners of folk medicine. Infant mortality was very high—and who had better opportunities than midwives for killing babies? No doubt they often did kill them, through ignorance or ineptitude. But that was not the explanation that came to people's minds; and it is striking how often the village midwife figures as the accused in a witchcraft trial'. It would be even more striking had he provided a single example of a prosecuted midwife. Although there undoubtedly were cases of midwives being prosecuted in various parts of Germany, it is notable that German midwifery ordinances make no reference to magic or witchcraft. Especially after the Reformation, the main anxiety felt about German midwives focussed on their traditional practices of emergency baptism and the possible contamination of the young with superstition.[12]

[11] H. C. E. Midelfort, *Witch Hunting in Southwestern Germany, 1562–1684* (Stanford, 1972), pp. 172, 187, 195. H. Pohl, *Hexenglaube und Hexenverfolgung im Kurfürstentum Mainz*, Geschichtliche Landeskunde, vol. 32 (Stuttgart, 1988), pp. 222–3, 298; W. Behringer, *Hexenverfolgung in Bayern* (Munich, 1987), p. 201; cf. Behringer's article in *Hexenwelten. Magie und Imagination* ed. R. van Dulmen (Frankfurt, 1987). I am grateful to Lyndal Roper for advice and references.

[12] R. Kieckhefer, *European Witch Trials: Their Foundations in Popular and Learned Culture, 1300–1500* (London, 1976), p. 56; N. Cohn, *Europe's Inner Demons* (London, 1975), p. 249; Weisner, *Working Women*, pp. 64, 69; L. Roper, *The Holy Household: Women and Morals in Reformation Augsburg* (Oxford, 1989), pp. 264–5.

It is far from easy to find large numbers of accused midwives whereas there are plenty of cases where the accused was described as 'medica' or 'Ärtzin', although this may mean little more than the English term 'cunning woman'.[13] The many cases that involved witches allegedly eating the flesh of unbaptized infants at sabbats, or using it for magical purposes, do not generally include midwives among the accused.[14] Thus the factual case for the existence of midwife-witches in Europe rests largely on a handful of sensational cases. They cannot be regarded as typical and, when examined in detail, they appear to be the work of a zealous prosecutor rather than the result of popular outrage. The execution at Lindheim in 1661 of a group of women, including a midwife, for killing a child and using the body for magical purposes, led to a bitter feud between magistrate and community. The parents believed in the midwife's innocence and exhumed the body; the magistrate accused the parents; the father and some other prisoners escaped to accuse the magistrate before the imperial court at Speier. Although he had succeeded in executing a large number of those involved, including the infant's mother, Amtmann Geiss had to flee the town.[15] This case clearly shows the sharp divide between the midwife's neighbours and the magistrate, versed in demonological theory. If prosecutions were initiated by zealous authorities, then the normal process of cumulative suspicion and accusation was subverted. The good reputation of a midwife was essential in her trade since it was her best credential and her only advertisement. The evil reputation of a supposed witch was very different in nature. A midwife could only convincingly be recategorized through the use of torture and confession to persuade her neighbours that all had not been what it seemed.

Walpurga Haussmannin of Dillingen was elaborately executed in 1587, having confessed under torture to fornication with the Devil, apostasy, blaspheming the sacrament and cross, and dishonouring the font. She admitted killing forty-three unbaptized infants, often with the aid of her demonic salve, as well as cannibalism and the creation of hail. Hers is an apparently clear-cut case but her tenure of the office of licensed midwife for 19 years and her delivery of the infants of prominent officials suggest that the case is more complex than the bare story reported in the *Fugger-Zeitungen*. For such a local notable to be prosecuted, the likely explanation would be severe religious controversy within the town, coupled with political faction-fighting. A similarly political explanation probably underlies the even more famous case of the La Voisin poisoning and abortion group, arrested in Paris in 1679, having supposedly collected the remains of 2,500 infants for diabolical purposes. The normality of this case needs to be questioned not only because of the scale of

[13] P. Guerrini, *Le Cronache Bresciane Inedite dei secoli XV-XIX*, Vol. 1. (Brescia, 1922), pp. 182–3; J. Hansen, *Quellen und Untersuchungen zur Geschichte des Hexenwahns und der Hexenverfolgung im Mittelalter* (Bonn, 1901), pp. 527, 561.

[14] Hansen, *Quellen*, pp. 120, 453, 460, 462, 481, 497, 500, 542, 557, 570, 591.

[15] K. Baschwitz, *Hexen und Hexenprozesse* (Munich, 1963), pp. 302–4.

the alleged crime but also because the only substantial evidence was provided by the midwife herself and her daughter, no bodies being ever produced.[16]

The La Voisin case is also striking because the Paris midwives probably had more skill and status than their equivalents anywhere else in Europe. At *l'Hôtel-Dieu* in Paris, most of the midwives came from prosperous families and several had prominent surgeons as fathers or brothers, 'la dame Charonne' being a case in point since she was the daughter of Paré and a close friend of Jacques Guillemeau.[17] Recent attempts to assimilate individual disputes, such as that between the royal midwife, Louise Bourgeois, and the royal surgeon, Charles Guillemeau, to the received model, as 'an instance of the old enmity between doctors and midwives', tend to perpetuate misunderstandings. Whereas witch-hunters such as Peter Binsfeld regarded the term 'sage femme' as rather sinister, the Paris surgeons saw it in a more humorous light. Jacques Guillemeau joked about the term without suggesting anything detrimental to the integrity of women who since antiquity had specialized in obstetrics and gynaecology except a largely laudable desire to excel over men. Although in a medico-legal context he was concerned that judges and the common people too readily accepted the evidence of midwives on technical matters, this can hardly be taken as indicating an association with witchcraft.[18] This is not to say that Paris midwives were altogether immune from suspicion, of course. In 1660, 'la dame Constantin' was accused of causing the death of Mlle de Guerchi through a botched abortion. She defended herself vigorously but was tortured and condemned whereas a surgeon who was also implicated appears to have escaped serious punishment. Nevertheless, there was no suggestion that this was anything more than a straightforward crime and there were no generalized reflections on the moral standing of Paris midwives as a whole.[19]

The high status of Paris midwives was not perhaps reproduced elsewhere in France. Natalie Zemon Davis sees the provincial midwives as drawn from the 'menu peuple' and the artisans but their frequent Protestantism might suggest that they were literate and respectable. Even outside Paris, French medical authors generally adopted the attitude displayed by Théophile Bonet, being sympathetic to the difficulties of midwives and attentive to the utility of their remedies, despite cautioning against unduly vigorous methods of delivery. Local communities in France, as elsewhere in Europe, had to have midwives

[16] *Fugger-Zeitungen*, ed. V. Klarwill (Vienna, 1923), pp. 103–10; F. Ravaisson-Mollien, *Archives de la Bastille*, vol. 6 (Paris, 1873), pp. 38–9, 164–6, 178, 180, 194–8.

[17] H. Carrier, *Origines de la Maternité de Paris* (Paris, 1888), pp. 3–23; A. Delacoux, *Biographe des Sages-Femmes Célèbres* (Paris, 1834), pp. 56–7; W. Perkins, 'Midwives v. doctors: the Case of Louise Bourgeois', *The Seventeenth Century*, iii (1988), 135–57. For a more useful if modest study, cf. id. 'The Relationship Between Midwife and Client in the Works of Louise Bourgeois', *Seventeenth Century French Studies*, xi (1989), 28–45.

[18] P. Binsfeld, *Tractatus de Confessionibus Maleficorum et Sagarum* (Treves, 1591), p. 371; J. Guillemeau, *De l'heureux Accouchement des Femmes* (Paris, 1609), pp. 147–8 (the contemporary English translation inevitably loses the joke by translating the term as 'cunning woman', a confusion that did not arise in French usage); id., *Oeuvres de Chirurgie* (Paris, 1649), p. 480.

[19] *Lettres de Gui Patin*, ed. J.-H. Reveillé-Parise (Paris, 1846), vol. 3, pp. 225–6, 229, 232, 238–9.

whom they could trust and medical practitioners had to be able to collaborate amicably with them. An academic physician like Laurent Joubert might castigate village midwives as the main repository of popular errors but practical ones like James Primrose were markedly less hostile.[20]

The concern of Joubert in Montpellier about the ignorance of some midwives should alert the historian to the likelihood of different attitudes co-existing in the same period and even the same author. Midwives were not socially and educationally uniform and authors might well stress different aspects of midwives' behaviour depending on their audience and intentions. Thus the Roman physician, Scipion Mercurio, seems to have been relatively unconcerned about the dangers of witchcraft when he discussed the qualities needed in a midwife and the perils of abortion, in his book on midwifery first published in 1596. Only when he treats the topic of incubi and succubi does he cite the *Malleus*, 'dove è una frotta di questa sporcherie del diavolo' (where there is a collection of this filthy business of the Devil). On the other hand, in his book on popular errors published in 1603, he appears more concerned about the dangers of the midwife-witch, as befitted his changed audience and intentions in writing the later book.[21]

To some extent, the midwife-witch was a literary convention which passed from the demonologists into other kinds of writing without often influencing perceptions about the actual midwives who delivered one's own children. Thus the remark by Fernando de Rojas's famous character, La Celestina, that her friend, Parmeno's mother, had been both witch and midwife for 16 years, is unlikely to have led to any prosecutions before the Inquisition. John Bale, in a didactic comedy of 1538, depicted Sodomismus describing Idolatria's range of sorcerous services and she herself admits that she protected the children whom she delivered with Papist charms. This says more about the attitude of early Protestants towards the birthing room as a haven of popular superstition than indicating any desire to persecute midwives generally. Historians who have considered this question have been too quick to confuse concern about domestic sorcery with the hunt for maleficent witches, a confusion derived in part from demonological writers rather than actual practice. Ecclesiastical regulations concerning midwives were certainly concerned to prevent them from using charms but there is no suggestion that inhibitions on their practice of baptism in emergencies, at the 1577 Durham synod for example, were motivated by anything other than a desire to promote Protestant sacerdotalism and extirpate popular beliefs about the magical efficacy of baptism. After all, Calvin himself discussed the image of God as midwife and it is surely

[20] N. Z. Davis, *Society and Culture in Early Modern France* (Stanford, 1965), pp. 70, 81, 224, 258–64; T. Bonet, *Mercurius Compitalitius, sive Index Medico-Practicus* (Geneva, 1682), pp. 526–30; L. Joubert, *Erreurs Populaires au fait de le Medecine* (Bordeaux, 1578–9; Paris, 1580); J. Primrose, *De Vulgi in Medicina Erroribus* (London, 1638), pp. 176–9. The various excellent studies of childbirth and midwifery in France by Jacques Gélis and Mireille Laget mostly deal with a later period but see J. Gélis, *La Sage-femme ou le Médecin: une nouvelle conception de la vie* (Paris, 1988), pp. 15–64.

[21] S. Mercurio, *La Commare o Raccoglitrice* (Verona, 1642), bk. 1, ch. 18; bk. 2, ch. 17–21, pp. 71–4, 141–51, 208; id. *De gli Errori Populari d'Italia* (Venice, 1603), bk. 6, ch. 2, fos. 262–3.

anachronistic to suggest that his transition to discussion of God as providential father was influenced by 'the fact that in the early modern period male doctors were beginning to take over the work of traditionally female midwives'. Calvin is unlikely to have been swayed by developments that took place after his death.[22]

The European Inquisitions were certainly eager to monitor popular medical superstitions and secure repentance but they were not misled by the supposed authority of the *Malleus*. Occasional midwives may have been investigated, such as the impoverished cloth worker Diamente de Bisa delli Axcari della Mota who was tried at Modena in 1595, but midwifery as such did not normally feature in the depositions. Rather, they were denounced as practising *maleficium* when they unsuccessfully employed irregular healing methods. Folk-healers, frequently foreigners, were drawn into the processes of the Inquisition when they dabbled in magic, especially if it involved abuse of the sacramentals. One place where literary models could have exercised a disproportionate influence on practical policy was Spanish America, the subject of an intense theological and ethnological debate, as Anthony Pagden has shown. The colonizing missionaries, bringing their European demonology, lumped native midwives together with other indigenous healers as witches and tools of the Devil, although the mestizo aristocrat, Garcilaso de la Vega, gave a more sympathetic picture of their skills.[23] It is possible that demonology had a disproportionate influence on the Inquisition in New Spain, leading to the persecution of native midwives, but this is not easy to establish as it was rare for a native *curandera* to appear before the ordinary tribunal. Like other occupying authorities, the Inquisition in New Spain appears to have dismissed accusations of witchcraft as mere superstition. Such action against magical healing as occurred was simply part of the campaign against all forms of native religion. It is possible to find midwives punished by the Inquisition in Mexico, from the Negro *partera* incarcerated in a monastery in 1537, for incantations and illegal operations on newly delivered women, to the midwife sentenced to 200 lashes and banished from Puebla for ever in 1648. At present, however,

[22] [Fernando de Rojas], *Tragicomedia de Calisto y Melibea* (Seville, 1502), sig. d4v; J. Bale, *A Comedy concernynge thre lawes, of nature Moses & Christ, corrupted by the Sodomytes, Pharysees and Papystes* (1538), sig. B3r and v; 'Bishop Barnes's Injunctions to the Clergy of the Diocese of Durham', in *Reprints of Rare Tracts and Imprints of Antient Manuscripts*, vol. 6 (Newcastle, 1848), p. 17; J. Dempsey Douglas, 'Calvin's Use of Metaphorical Language for God: God as Enemy and God as Mother' *Archiv für Reformationsgeschichte* lxxvii (1986), pp. 131, 133.

[23] M. O'Neil, 'Magical healing, love magic and the Inquisition in late sixteenth-century Modena', in S. Haliczer (ed.), *Inquisition and Society in Early Modern Europe* (London, 1987), pp. 88–114; R. Martin, *Witchcraft and the Inquisition in Venice, 1550–1650* (Oxford, 1989), pp. 139–47, 180–9; A. Pagden, *The Fall of Natural Man* (Cambridge, 1982); M. de Murúa, *Historia del Origen y Genealogía real de los Reyes Incas del Perú*, ed. C. Bayle (Madrid, 1946), p. 320; id., *Historia General de Perú*, vol. 2 (Madrid, 1964), p. 101; G. de la Vega, *Commentarios Reales*, vol. 1 (Lisbon, 1609), p. 50. Some translations of Garcilaso have assimilated his account to the midwife-witch tradition, misleading unwary scholars. For a recent discussion of the complexity of this text, see M. Zamora, *Language, Authority, and Indigenous History in the 'Comentarios reales de los incas'*, (Cambridge, 1988).

it seems easier to find graduate physicians and theologians falling foul of the Inquisition for magical practices.[24]

If European midwives were indeed rarely prosecuted, despite the existence of a justifying theory, it is necessary to ask how often they fell foul of the law in countries where prosecution principally depended on accusation by victims of witchcraft rather than on the obsessions of powerful individuals. Those historians who assert the existence of persecution draw their examples from a wide range of times and countries. It seems desirable to single some out for closer examination. England is probably the most straightforward example available. Medieval English writers discussed midwives without feeling obliged to mention sorcery, a connection made more readily by modern historians. The later English demonologists rarely mention midwives and, when they do, the examples are rather exotic, as in the references of Alexander Roberts of King's Lynn to Constantinople and St. John Chrysostom.[25] The *Malleus* was never fully authoritative in England and Robert Plot, no sceptic, clearly regarded the midwife-witch as an altogether foreign phenomenon when he cited Bodin and Codronchius on 'the sacrifices of young Children (which are frequently offered by Midwife-Witches in some Countries, their fat being the chief ingredient wherewith they make the Oyntment indispensably necessary for their transportation to their Field-Conventicles)'.[26] The absence from England of an inquisitorial system of justice meant that witchcraft accusations remained firmly rooted in popular belief, which did not suspect midwives or concern itself with the sabbat, and the absence of torture meant that long chains of accusations were not created, dragging in prominent local people.

Before the eighteenth century, English midwives certainly were prominent within their communities, regardless of their wealth and education. Although childbed was an occasion for suspicions of witchcraft in England, as in the case of a Huntingdonshire shepherd who made an accusation over 20 years after his wife had died in labour, it was not midwives who were accused of the crime. They were more likely to be involved in checking the alleged witch for signs of the Devil's mark, as in the famous case of 1634 when a panel was chosen by the royal surgeons, Baker and Clowes, to investigate a group of Lancashire witches in a manner to be determined by William Harvey. Informally, such searches can be found operating as late as 1699, when an Essex

[24] A. Quirós Rodiles, 'Breve historia de la obstetricia en México', *Obstetricia y ginecología latino-americanas*, iii (1945) pp. 692–3; R. E. Greenleaf, *Zumárraga and the Mexican Inquisition, 1536–1543* (Washington, 1961), pp. 57–8, 114–15, 117–20; id., *The Mexican Inquisition of the Sixteenth Century* (Albuquerque, 1969), pp. 103–7. For an attempt to assess the trivial number of popular healers prosecuted before the regular tribunal, see S. Alberto, *La Actividad del Santo Oficio de la Inquisicíon en Nueva Espana, 1571–1700* (Mexico City, 1981), pp. 81–2. Angela Thompson informs me that the work of Ruth Bejar is likely to throw light on this issue.

[25] M. C. Seymour (ed.), *On the Properties of Things: John Trevisa's translation of Bartholomaeus Anglicus De Proprietatibus Rerum*, vol. 1 (Oxford, 1975), p. 305; *Ludus Conventriae; or, The Plaie Called Corpus Christi*, ed. K. S. Block (Early English Text Society, vol. 120, 1922), pp. 139–40; P. Biller, 'Childbirth in the Middle Ages', *History Today* xxxvi (Aug. 1986), pp. 42–9; A. Roberts, *A Treatise of Witchcraft* (London, 1616), p. 66.

[26] R. Plot, *The Natural History of Stafford-shire* (Oxford, 1686), p. 14. Scot and Webster do not mention the concept and John Aubrey fails to record any popular beliefs on the subject.

midwife was asked by a clergyman to examine the corpse of a deranged woman who had confessed to witchcraft:

Upon her death I requested Becke the midwife to search her body in the presence of some sober women which she did and assured me that she never saw the like in her life that her fundament was open like a mouse hole and that in it were two long bigges out of which being pressed issued blood that they were neither piles nor emrods for she knew both but excressencies like to biggs with nipples wich seemd as if they had been frequently sucked.[27]

Midwives might also be asked to confirm stories of providential monstrous births, as in the case of a Lancashire clergyman's widow who attended the children of a Catholic family. The wife had been cursed with the birth of a headless child as a judgement for saying she would rather bear a child with no head than a future Roundhead. Where midwives appear in such accounts, it is always in the role of agent of respectability.[28]

The midwife in England, as elsewhere in Europe, was firmly fixed in this position by her duty to investigate rape, bastardy, and infanticide, as well as by the highly moral behaviour expected by clients and enjoined by the midwife's oath. There were occasionally rumours in London that midwives were accomplices to infanticide, these fears giving rise to a famous ghost story in 1680 and being repeated in 1728 by Daniel Defoe. Nevertheless, such stories merely serve to emphasize the behaviour normally expected of midwives who were, for the most part, as well respected as the three Skipton midwives whose burials were specially noted in 1632, 1662, and 1694.[29] Prior to that distancing of respectable women from paid work and close contact with the poor which, together with the rise of the men-midwives, led to a decline in the status of midwives, the urban midwives of provincial England were often highly affluent and literate. An ecclesiastical lawyer proudly noted in his diary that his wife was delivered by the Mayoress of Chester. Southampton even provides an example of a man gaining his status as a freeman in 1601 by virtue of being married to a distinguished midwife. Even the poorest village midwife, if licensed, had been given testimonials by her patients, by the local clergyman and parish officers, or by medical practitioners. Regarded not as evidence of technical expertise, but as proof of acceptability among respectable neighbours, the process of ecclesiastical licensing clearly sets midwives apart from the marginal women suspected of witchcraft. Yet unlicensed midwives too required the confidence of patients and neighbours, they too had to give evidence in court cases, and there is no indication that they were fundamentally

[27] *The Witches of Huntingdon: Their Examination and Confessions* (London, 1646), p. 6; PRO: SP 16/270, fo. 137; SP 16/271, fo. 15; W. Gilbert, 'Witchcraft in Essex', *Transactions of the Essex Archaeological Society*, NS, xi (1909–10), p. 216.

[28] *Five Wonders Seene in England* (London, 1646), pp. 2–5; *The Ranters Monster* (London, 1652) reprinted in J. C. Davis, *Fear, Myth and History: The Ranters and the Historians* (Cambridge, 1986), pp. 190–4; H. Jessey, *The Lords Loud Call to England* (London, 1660), pp. 29–30.

[29] *Great News from Middle-Row in Holbourn: or true relation of a dreadful ghost* (London, 1679/80); Daniel Defoe, *Augusta Triumphans* (London, 1728), p. 9; *The Parish Register of Skipton-in-Craven*, ed. W. J. Stavert (Skipton, 1814–16), vol. 1, pp. 151, 288; vol. 2, p. 75.

different from their licensed sisters except that they might well be Quakers, Catholics, or charitable women who practised infrequently.[30]

Given this respectability, it is unsurprising that the case for the existence of English midwife-witches rests on only two famous examples. The earlier one is the case of Ursley Kempe, an Essex witch of 1582. Historians have relied on the casual assertion of Wallace Notestein in 1911 that Kempe was a midwife, which cannot be supported by reference to the original source, described at the time by Reginald Scot as 'a foolish pamphlet dedicated to the Lord Darcy'. Kempe is there depicted as confessing to a variety of improbable crimes in response to unfulfilled promises of mercy, a point scathingly noticed by Scot. Yet her accusers, who included her brother and her bastard son, at no point suggested she was a midwife, although wet nursing was involved in the tortuous tale.[31] The second case is that of Mrs. Pepper of Newcastle upon Tyne, an example of a woman who was identified by her accusers as a midwife but midwifery as such was not involved in the case. She rashly diagnosed a sick man as being either possessed or bewitched and, when he failed to improve after she had administered magical remedies, worsening instead, she fell under suspicion of having caused his bewitchment. This was the usual course of events when a would-be healer was accused. It was not unlicensed pragmatic medicine that led to prosecutions but failed magical medicine. Mrs. Pepper's description as a midwife is irrelevant to the case.[32] Clearly, some midwives did practice charms but these would be expected by the other women attending the children. Few would expose themselves to accusation by offering the kind of treatment that lay outside their established expertise. It would seem incautious to blur the distinction between midwife and wise woman, even at the village level, without stronger evidence than has yet been advanced. In any event, few cunning folk were persecuted in England for witchcraft, as opposed to sorcery.

One may safely assume that practice of midwifery would have been mentioned in witchcraft cases, because of the sensational implications. Midwives are identified as such, whether regular practitioners or not, in those legal records for which the identification is relevant, such as bastardy documents or poor law petitions, and not where it is not, as in most wills and parish registers.

[30] *The Diary of Henry Prescott, LL.B., Deputy Registrar of Chester Diocese*, ed. J. Addy, vol. 1 (Lancashire and Cheshire Record Society, vol. 127, 1987), p. 15; *The Assembly Books of Southampton*, ed. J. W. Horrocks, vol. 1 (Southampton Record Society, vol. 2, 1917), p. ix; Harley, 'Ignorant Midwives'.

[31] W. Notestein, *The History of English Witchcraft*, (Washington, American Historical Association, 1911), pp. 41, 542–3; W. W., *A True and Just Recorde of the Information, Examination and Confession of all the Witches taken at S. Oses in the countie of Essex* (London, 1584); R. Scot, *The Discovery of Witchcraft* (London, 1584), pp, 17, 49.

[32] PRO: ASSI 45/7/2/62/103. An inaccurate transcription is published in *Depositions from York Castle*, Surtees Society, lx (1861), p. 127. For a fuller discussion of this case and its magical context, see D. N. Harley, 'Mental Illness, Magical Medicine and the Devil in Northern England, 1650–1700', in A. Wear and R. K. French (eds.) *The Medical Revolution of the 17th Century* (Cambridge, 1989), p. 131.

It would be ridiculous, for example, to suppose that there was only one midwife, Mistress Blake of St. Magnus the Martyr, living within the walls of London in 1695, but midwifery was often a skill rather than a trade and sole source of income and status.[33] Since Mrs. Pepper appears to be the only known midwife accused of witchcraft in England, as far as one can tell from surviving records, it is clear that midwives were not prosecuted as witches in English courts, although the ecclesiastical authorities occasionally worried about child-bed charms. Despite less pious midwives using charms to reassure patients and protect against evil, they were not the kind of marginal magical practitioners who ended up before courts, poor and illiterate women who incompetently treated bewitchments they had diagnosed themselves. English diaries and letters indicate no association between midwives and witches. Popular ballads and chapbooks display no interest in retailing what was clearly a very foreign notion. Even slander cases brought by midwives in the English church courts do not involve suggestions of witchcraft, being rather concerned with rumours of technical incompetence or moral turpitude.

Scotland provides a better test than England for the reality of the midwife-witch for a variety of reasons, some purely fortuitous. Outside the towns, the distribution of wealth and education was very different from the English provincial structure. This should provide larger numbers of midwives whom the authorities would regard as dubious, especially in the absence of widespread licensing. Witch-hunting was more organized and less spontaneous than in England, so there are better records and the chains of structured confessions, produced by torture and interrogation, should yield more accusations of midwives. Moreover, Scottish authorities appear to have conflated black and white magic, to an extent unknown in England outside the works of demonologists, so that healers were dragged into major hunts.[34] This may, of course, simply reflect the dynamics of a major hunt anywhere, largely unknown in England.

By chance, the Scottish cases are unusually well documented because antiquarians were sufficiently interested to publish case-histories and because the records have been systematically searched by a team led by the late Christina Larner.[35] The resulting figure of twelve 'midwife/healers' is cited by Joseph Klaits, in his chapter 'Classic Witches: The Beggar and the Midwife', without mentioning that it was exceeded by such groups as the nobility, burgesses, craftsmen, and 'ministers/teachers'. Larner is cautious about the midwife-witch and describes the figures as 'extremely misleading' because status was only described where it was seen as noteworthy. Nevertheless, two of Larner's collaborators claim that the presence of six midwives among over 3,000 accused

[33] *London Inhabitants within the Walls, 1695*, ed. D. V. Glass (London Record Society, vol. 2, 1966), p. 31.

[34] C. Larner, *Enemies of God: the Witch-Hunt in Scotland* (Oxford, 1981), p. 9.

[35] C. Larner *et al.*, *A Source-Book of Scottish Witchcraft* (Glasgow, 1977).

constitutes a suggestive over-representation.[36] Since it would be dangerous to establish a stereotype on the basis of one in every 500 accused, or every 400 women accused, which seems to be rather low relative to the number of midwives in the community, one must return to the sources to examine the cases in detail, wherever possible, and search for additional examples.

Probably the most famous of the Scottish witches is Agnes Sampson, often identified as a midwife because she admitted, among a wide range of offences, administering magical medicines to take away the pains of women in childbirth. She was first named by an accused servant girl, during the political panic of 1590. James VI was personally involved in her interrogation but she did not confess under torture 'untill the Diuels marke was found upon her priuities, then she confessed whatsoeuer was demaunded of her', although the King attributed this change of heart to 'his especiall travell'. According to a contemporary report, her admissions 'were so miraculous and strange, as that his Maiestie said they were all extreame lyars'. There was no mention of midwifery in this trial of 'the wyse wyff of Keyth', who seems first to have been described as a midwife by David Calderwood, a Presbyterian author writing in the early seventeenth century. Dalyell picked up this reference from the unpublished manuscript while researching his 1834 book on *The Darker Superstitions of Scotland* and this is probably the ultimate source for most later writers.[37]

Taking away the pains of childbirth occurs in the trials of women who certainly were midwives where it can be seen to be not some innocent analgesia but a distinctly menacing Scottish magical belief. Margaret Clerk alias Bain, a midwife not listed by Larner, was accused in 1597 of transferring the pains of childbirth from patients to such effect that the husband of one woman was driven mad for several years until he died, although another man recovered. More significant to contemporary investigators, however, was the conspiracy to commit various devilish crimes, especially the destruction of one Thomas Forbes. His wife and daughter were only saved from the wave of prosecutions that followed Bain's confession by the intervention of James VI on the grounds that they had innocently consulted her in her capacity as midwife.[38] Not only men were believed to be vulnerable to this form of attack. Elie or Alison Nisbit, a Hilton midwife arrested by the Sheriff of Berwick in 1630, confessed to adultery but not that 'she tooke the paines off a woman in travell, by some charmes and horrible words; among which thir ware some, *the bones to the fire,*

[36] Larner, *Enemies*, pp. 89, 101; J. Klaits, *Servants of Satan: the Age of the Witch Hunts* (Bloomington, 1985), p. 187, nn. 23 and 30; H. V. McLachlan and J. K. Swales, 'Stereotypes and Scottish Witchcraft', *Contemporary Review* ccxxxiv (Feb. 1979), p. 89.

[37] *Newes from Scotland, Declaring the Damnable life and death of Doctor Fian* (London, 1592), sig. A2v–A3r, A4r; *Calendar of the State Papers relating to Scotland*, 10 (1936), p. 430; R. Pitcairn, *Ancient Criminal Trials in Scotland* (Edinburgh, 1833), vol. 1, pt. 3, pp. 230–41, 247–57; D. Calderwood, *The History of the Kirk of Scotland*, ed. T. Thompson (Edinburgh, Wodrow Society, 1844), vol. 5, p. 115; J. Graham Dalyell, *The Darker Superstitutions of Scotland* (Edinburgh, 1834), p. 26; C. Hole, *Witchcraft in Britain* (London, 1977), p. 13.

[38] 'Trials for witchcraft, MDXCVI–MDXCVII', in *Miscellany of the Spalding Club*, vol. 1 (Aberdeen, 1841), pp. 157–8, 163–4.

and the soull to the devill! and layed them on another woman, who straight died thereof . . ' The victim was Helen Park, another midwife.[39]

Many Scottish cases appear to have left little in the way of detailed records and even the verdict has been lost in the case of two midwives arrested in 1629, 'Janet Melros, midwife in Chattil, who has long been suspected of witchcraft' and Helen Beattie, midwife in Menner, who was one of a large group reported by the Moderator of the presbytery of Peebles. Bessie Gourdie, a midwife in Midlothian, seems to have left little trace except the fact of her execution in 1678.[40] Like other prominent people, midwives appear to have been vulnerable to accusation during major outbreaks of witch-hunting such as the Tranent hunts of 1659, when Marion Lynn was a central figure, and the Dalkeith hunt of 1661, when Beatrix Leslie was accused of the evil eye and relieving the pains of childbirth. Her guilt was ascertained by a successful *bier-richt*, despite her praying to God, conducted over the bodies of two pit-girls destroyed after killing her cat. Both midwives were executed during outbreaks when large numbers of women confessed real copulation with the devil and renunciation of their baptism.[41]

Midwives entangled in Scottish witchcraft allegations were not necessarily so unfortunate. Nothing seems to have been done to punish Margaret Reid, a Larnarkshire midwife who admitted using magical medicines after being accused by a confessed witch in 1644. It appears to have been fairly hard to convict a midwife for murdering an infant if witchcraft was mentioned, as it was in the case of a Corstorphine widow and her midwife, Margaret Wylie, tried and acquitted in 1661. By contrast, a midwife and her serving woman were condemned to be hung for the straightforward murder of a bastard infant in 1679. One of the accused in the case of two Glasgow girls afflicted with demonic obsession in 1699 was a midwife, Margaret Duncan, who is described as a merchant's widow in Larner's list. The girls recovered, the case did not proceed and the accused were acquitted.[42]

Inevitably, some accused witches died in prison, especially when judges had lost interest in the speedy prosecution of witchcraft cases. According to a correspondent of Robert Wodrow in 1727, Margaret Nin-Gilbert alias Gilbertson was midwife to a great lady in Caithness but, after confessing in 1718,

[39] C. K. Sharpe (ed.), *Memorials; or the memorable things that fell out within this island of Brittain from 1638 to 1684. By the Rev. Mr. Robert Law* (Edinburgh, 1818), pp. lvi–lvii; *The Register of the Privy Council of Scotland*, 2nd ser., vol. 3 (1901), pp. 583–4; *Selected Justiciary Cases, 1624–30*, ed. S. A. Gillon, The Stair Society xvi (1953), pp. 210–13.

[40] *Reg. P. C. Scotland*, 2nd ser., vol. 3 (1901), pp. 98, 170; Larner *et al.*, *Source-Book*, pp. 44, 86, 90, 244.

[41] Larner, *Enemies*, p. 106; *The Records of the Proceedings of the Justiciary Court, Edinburgh, 1661–1678*, ed. W. G. Scott-Moncrieff, vol. 1 (Scottish History Society, vol. 48, 1905), p. 5; J. G. Dalyell, *Darker Superstitions*, pp. 8, 36–7, 133–4; J. Nicoll, *A Diary of Public Transactions* (Edinburgh, 1836), pp. 233, 343.

[42] *Reg. P. C. Scotland*, 2nd ser., vol. 8 (Edinburgh, 1908), p. 157; *Records . . . of the Justiciary Court*, pp. 2–4; Sir John Lauder, Lord Fountainhall, *The Decisions of the Lords of Council and Session* (Edinburgh, 1749), p. 47; Larner *et al.*, *Source-Book*, p. 244; *Early Letters of Robert Wodrow, 1698–1709*, ed. L. W. Sharp (Scottish History Society, 3rd ser., vol. 24, 1937), pp. 5–6, 8, 18.

she was murdered by those whom she had implicated. She had confessed that her putrefying leg had fallen off after one William Montgomerie had attacked her while she was in the form of a cat. Another woman, who died of cold and poverty in prison at Christmas 1684, should be discounted although Larner describes her as a destitute former midwife. The original source reads 'once a milk-wife', rather a different occupation.[43]

The case of Bessie Aitken of Leith, also one of Larner's midwives, may be somewhat doubtful too, since midwifery was not attributed to her at the trial and she appears rather to have been both a consumer and purveyor of the magical healing of women's diseases among the Edinburgh poor. She did, however, advise the husband of a woman harmed by a midwife and she was accused when the cure succeeded, but she escaped with mere banishment after pleading her belly. She should probably be included in a list of accused midwife-witches, as should the two who were implicated by the confession in 1677 of Elizabeth Moodie of Haddington, East Lothian, according to Lord Fountainhall.[44]

By dint of such accretions, it has proved possible to bring the total of accused midwives to fourteen, almost as many as the members of the nobility. It may be that detailed local research, into Kirk Sessions papers for example, would link other witches to the regular practice of midwifery, such as the Orkney women in whose house, some 13 years before her trial, 'thair was ane powr woman that was trauelling of child'. However fruitful such research might prove, in supplying additional names and details, it is unlikely that it would provide a total number of midwives much exceeding one per cent. Scottish circumstances were relatively favourable to the prosecution of midwives since the initiative was frequently in the hands of the Kirk, many accusations derived from confessions under torture, and there was a local belief concerning the potentially murderous transfer of pain in childbirth. The rather different social structure of Scotland may also have led to a more lowly group of women being recruited into midwifery than was the case in England. Thus it is not surprising that there were more prosecutions of midwives in Scotland than in England yet it is notable that there were still very few.[45]

No work on the position of women in early New England is complete without a reference to the midwife-witch. Four examples are commonly used to demonstrate the persecution of colonial midwives. The most celebrated is the antinomian heretic, Anne Hutchinson. Her less educated follower, Jane

[43] *Law's Memorials*, pp. xcvi–civ; Larner *et al.*, *Source-Book*, p. 247; Larner, *Enemies*, p. 90; Sir John Lauder, Lord Fountainhall, *Historical Notices of Scottish Affairs*, vol. 2 (Edinburgh, 1848), p. 561.

[44] Pitcairn, *Ancient Criminal Trials*, vol. 2, pt. 1, pp. 25–9; Larner *et al.*, *Source-Book*, p. 242; *Law's Memorials*, p. 131 n.

[45] 'Trials for witchcraft, sorcery, and superstition, in Orkney' in *Miscellany of the Abbotsford Club*, vol. 1 (1837), pp. 152, 158–9; Larner, *Enemies*; R. A. Houston, *Scottish Literacy and the Scottish Identity, 1600–1800* (Cambridge, 1985). The history of women in early modern Scotland has been somewhat neglected although an admirable start has been made by R. Mitchison and L. Leneman, *Sexuality and Social Control; Scotland 1660–1780* (Oxford, 1989).

Hawkins, and two other women, Margaret Jones and Elizabeth Morse, are also cited. Hutchinson provides the most interesting case since her identification as a midwife rests solely on a remark by her bitter enemy, John Winthrop, that she was 'a woman very helpfull in the times of child-birth, and other occasions of bodily infirmities, and well furnished with means for those purposes', and on the fact that she was present when Jane Hawkins delivered the monstrous birth of Mary Dyer.[46] It is clear that she practised charitable domestic medicine, of the kind frequently dispensed in England by a parson's wife. She would have been present at births in the capacity of 'gossip'.

Jane Hawkins, by contrast, was clearly a midwife who dabbled in magical medicines such as oil of mandrakes. At the time, opinion about her was divided, Wheelwright describing her as 'a poore silly woman' who followed Hutchinson only to be fed and Winthrop saying she was 'notorious for her familiarity with the Devill'. Apart from her association with heresy and the monstrous birth, it was her magic that aroused suspicion because she required patients to have faith in her cures. Nevertheless, both in 1637 and 1641, the Massachusetts magistrates seem to have been unwilling to press matters to a conclusion, preferring to prevent her from practising medicine within the community by court order or banishment. She later petitioned to be allowed to rejoin her sons for comfort in her old age.[47]

Anne Hutchinson's reputation for witchcraft derives from her association with Hawkins. At the time of her trial, neither Winthrop nor anyone else suggested that Hutchinson was either a midwife or a witch although he later recorded his belief that her influence over a young man and her association with Hawkins 'gave cause of suspicion of witchcraft'.[48] The monstrous births, first from Mary Dyer and then from the banished Anne Hutchinson, did lead to some scandal, with suggestions that this was a judgement on their heresies. They became something of a model for English narratives. Ephraim Pagitt wrote that 'god punisht those monstrous wretches with a monstrous fruit, sprung from their wombs, as had before sprung from their braines'. This was not the same as calling them witches, deliberate agents of the Devil. Robert Baillie, the prominent Glaswegian Presbyterian who was highly critical of Congregational church discipline, writing in 1645, noted the monstrous births as signs from God but did not associate them with witchcraft. He did notice, in Winthrop's book, 'one abomination, which to me seems strange, That the Midwives, to their most zealous women, should not onely have familiarity

[46] [J. Winthrop] *A Short Story of the Rise, reign, and ruin of the Antinomians, Familists & Libertines* (London, 1644), p. 31.

[47] J. Wheelwright, *Mercurius Americanus, Mr. Welds his Antitype* (London, 1645), p. 7; Winthrop, *Short Story*, p. 44; J. K. Hosmer (ed.) *Winthrop's Journal: 'History of New England'* (New York, 1908), vol. 1, pp. 266–8; N. B. Shurtleff (ed.) *Records of the Governor and Company of the Massachusetts Bay*, vol. 1 (Boston, 1853), pp. 244, 329; Massachusetts State Archives, vol. 10, nos. 309, 310.

[48] Winthrop, *Short Story*, pp. 31–41, 45, 59–64; *Winthrop's Journal*, vol. 1, pp. 266–8; vol. 2, p. 8; *Collections of the Massachusetts Historical Society*, 4th ser., vol. 6 (1863), pp. 156, 227; 5th ser., vol. 1 (1871), pp. 96–7.

with the divel; but also in that very service, should commit divelish Malefices, which, so far as they tell us, were not onely past over without punishment, but never so much as inquired after'. The notes to this paragraph, if traced back to their sources, indicate that Baillie was thinking of Hutchinson as well as Hawkins, even though neither had been accused of *maleficium*. Since none of her New England contemporaries thought of Hutchinson as either a midwife or a witch, John Cotton's reply deals only with Hawkins, pointing out that as a non-member she could not be disciplined by the Church. 'But though no familiarity with the Devill could be proved against her; yet because of some other offences in dealing with young women, she was forbidden to stay in the Countrey.' Later writers in the controversy disagreed about the reputation of Hawkins and Hutchinson but it is clear that the former, although a midwife, was actually banished for magical medicine, probably because of her heresy, and the latter, although a herbalist and a heretic, was neither a midwife nor a witchcraft suspect.[49]

Margaret Jones of Charleston, executed in 1648, was accused of inflicting illness and of magical medicine but there was no reference to midwifery. Her case has been assimilated to the midwife-witch tradition through association with Jane Hawkins in secondary sources written this century. The last example, Elizabeth Morse, is often cited despite the fact that at no point during her trial and reprieve or the preceding poltergeist phenomena does anyone appear to have attributed midwifery or any other medical activities to her.[50] The misunderstanding appears to have arisen because her husband was reported as having been surprised 'that she should be both a healing and a destroying Witch' because she had been present at the successful delivery of her next-door neighbour. G. L. Burr in 1914 took this as a reference to midwifery and he has been followed by scholars unfamiliar with the role of a gossip, placing her in a group of 'midwives and magical healers' without ever documenting her 'reputation as a magical healer'.[51]

If there is so little evidence for the prosecution of midwives, how did the belief become so widely accepted? It did not feature in the old histories of midwifery which justified the rise of the men-midwives. Its main source is undoubtedly the works of Margaret Murray, an Egyptologist who was

[49] E[phraim] Pagitt, *Heresiography*, 2nd edn. (London, 1645), pp. 106–7; R. Baylie, *A Dissvasive from the Errours of the Time* (London, 1645), pp. 63–4, 73–4; J. Cotton, *The Way of the Congregational Churches Cleared* (London, 1648), p. 91; E. Johnson *A History of New-England*, (London, 1654), p. 100; S[amuel] G[roome], *A Glass for the People of New-England* (London, 1676), p. 11.

[50] *Winthrop's Journal*, vol. 2, pp. 344–5, and the references for Jones given by R. Weisman, *Witchcraft, Magic and Religion in 17th-Century Massachusetts* (Amherst, 1984), p. 197; J. Coffin, *A Sketch of the History of Newbury* (Boston, 1845), pp. 122–34; S. G. Drake, *Annals of Witchraft in New England* (Boston, 1869), pp. 141–9, 258–96; Boston Public Library, MS Am. 1502, v. 5, p. 45; Massachusetts State Archive, File 135, nos. 18–19.

[51] S. G. Drake, Annals, p. 281; G. Lincoln Burr, *Narratives of the Witchcraft Cases, 1648–1706* (New York, 1914), p. 31 n.; Weisman, *Witchcraft, Magic and Religion*, pp. 86, 88, 111. For a discussion of the gossips, see A. Wilson, 'Participant or Patient? Seventeenth Century Childbirth from the Mother's Point of View', in R. Porter (ed.) *Patients and Practitioners: lay perceptions of medicine in pre-industrial society* (Cambridge, 1985) 129–144.

determined to show that the isolated women prosecuted for witchcraft were really members of a pagan cult that had survived from pre-Christian times. Clearly both European popular beliefs and the learned writings of demonologists contain pagan elements but, to prove her case, she took literally the trial evidence and the statements of witch-hunters about the sabbat, while omitting such supernatural elements as the flying of witches to their meetings. All her quotations from obscure sources were carefully mangled to support her case. Although the midwife-witch idea, which she took from the *Malleus*, was peripheral to her thesis, her remarks were to prove highly influential. In her 1921 book, *The Witch-Cult in Western Europe*, she merely commented that 'in the sixteenth and seventheenth centuries, the better the midwife the better the witch'. This notion of effective midwifery as a reason for prosecution was novel and unsubstantiated. In her next book, *The God of the Witches*, first published in 1933, she performed a notable sleight of hand to equate the witch with the midwife:

> Throughout the country the witch or wise-woman, the sage-femme, was always called in at child-birth; many of these women were highly skilled, and it is on record that some could perform the Caesarian operation with complete success for both mother and child.

To this improbable statement, she added the assertion that they also eased birth pains, which was an impious act. This comment is derived from a deliberate misreading of her Scottish sources but it has been repeated by scores of authors ever since. Finally, she linked the *Malleus* to the men-midwives of the eighteenth century as if they were engaged in the same project:

> Religion and medical science united against the witches and when the law could no longer be enforced against them, they were vilified in every way that tongue or pen could invent.

Since medical authors in all periods since the Middle Ages have been concerned to attack quackery and popular errors, and have frequently been guilty of varying degrees of misogyny, there is just enough truth in this remark to convince the unwary that 'witch' is a synonym for 'midwife', both being attacked as dangerous illiterates.[52]

Although Murray's general thesis, based on her distortion of the Scottish confessions, continued to be a influential into the 1960s, it came under increasing attack, first from Elliot Rose in his book *Razor for a Goat*, published in 1962, and its authority dwindled to being the main prop for those who wished to believe in the existence of an 'old religion' of Satanism or witchcraft. At the same time, however, her slight comments about the midwife-witch took a side-step into medical historiography when the American medical historian, Thomas Forbes, published his article, 'Midwifery and Witchcraft', later incorporated into a widely cited book, *The Midwife and the Witch*. Forbes was able

[52] M. Murray, *The Witch-Cult in Western Europe* (Oxford, 1921), p. 170; id., *The God of the Witches*, 2nd edn. (London, 1952), p. 145.

to go far beyond the earlier apologists for the rise of man-midwifery by arguing that midwives had been not only ignorant but evil. Apart from a handful of cases, he relied for his unflattering identification of midwives on the writings of the demonologists and the work of Murray, whom he saw as 'a leading contemporary authority on the subject'.[53]

While it was Forbes who made the notion of the midwife-witch respectable in academic circles, the real impetus for the idea's dissemination came in the early 1970s. The critic of institutional psychiatry, Thomas Szasz, published *The Manufacture of Madness* in 1970, in which he was mainly concerned to show that defining madness was a social process of stigmatization, like witch-hunting, but he also suggested a role in the witch craze for the struggle between orthodox and unorthodox medicine. This was an idea whose time had come. Szasz was followed in 1973 by the trail-blazing work of Barbara Ehrenreich and Deirdre English, *Witches, Midwives and Nurses: a history of women healers.* This pamphlet, which asserts a continuity between attacks on witches and the difficulties experienced by modern women health workers, has been widely criticized even by authors sympathetic to its general polemical purpose, but it continues to be cited in the notes of academic works, despite its dependence on the discredited ideas of Margaret Murray, and it shapes much of popular perception.[54] Their position, unrepentantly summarized by Ehrenreich and English in 1978, was that millions of women were killed in the witch-hunts and they were predominantly pragmatic female healers whose main crimes were 'providing contraceptive measures, performing abortions, offering drugs to ease the pain of labour'. The *Malleus* is cited to prove the witch/wise woman/sage femme/midwife link made by Murray and the witches are portrayed as the only real medical scientists of their day:

> It was witches who developed an extensive understanding of bones and muscles, herbs and drugs, while physicians were still deriving their prognoses from astrology and alchemists were trying to turn lead into gold.[55]

The purpose of Ehrenreich and English was the entirely laudable one of campaigning for greater access for women to the management of their own health. Their work has been extensively used for this and related polemical

[53] E. Rose, *Razor for a Goat: A Discussion of Certain Problems in the History of Witchcraft and Diabolism* (Toronto, 1962); T. M. Luhrmann, *Persuasions of the Witch's Craft: Ritual Magic and Witchcraft in Present-day England* (Oxford, 1989), pp. 43–4; T. Forbes, 'Midwifery and Witchcraft', *Journal of the History of Medicine* xvii (1962), 264–83; id., *The Midwife and the Witch* (New Haven, 1966).

[54] T. Szasz, *The Manufacture of Madness* (New York, 1970); B. Ehrenreich and D. English, *Witches, Midwives and Nurses* (New York, 1973); ibid., introduction to 1976 English edition; V. V. and D. Ozonoff, 'Steps Towards a Radical Analysis of Health Care Problems and Prospects', *International Journal of Health Services*, v (1975), 299–314; M. Connor Versluysen, 'Old Wives' Tales? Women Healers in English History' in C. Davies (ed.) *Rewriting Nursing History* (London, 1980), 189–97; Larner, *Witchcraft and Religion: The Politics of Popular Belief* (Oxford, 1984), pp. 149–52.

[55] B. Ehrenreich and D. English, *For Her Own Good: 150 Years of the Experts' Advice to Women* (New York, 1978), pp. 31–3.

purposes. The sabbat has been portrayed as a physicians' conference, the witch hunt as a campaign to exclude women from science, the execution of witches as an attempt to eradicate the old wives' remedies that constituted 'a female-controlled reproductive care system', as Ann Oakley puts it.[56] Such polemicists would probably be better served by a real history of midwifery and womens' health care in early modern society. Such a history needs to be painstakingly constructed but the truth is likely to be more liberating than any romantic mythology of martyrs. Its creation is hindered rather than helped by the myth of the midwife-witch.

Distinguished historians casually identify midwives as the key targets of witchcraft prosecution. Undergraduate textbooks on witchcraft cite Forbes or Ehrenreich and English while thoughtlessly repeating outmoded prejudices about the murderous character of early modern midwifery. Even a careful author like Carolyn Merchant, who avoids reiterating the myth in her book *The Death of Nature*, published in 1980, carries it implicitly embedded in text and notes.[57] If historians can be so grievously misled, they have no grounds for complaint when polemicists are cavalier with the historical data. What is the process by which such a myth becomes so firmly entrenched despite being largely counter-factual? The American example is instructive, partly because the influence of Forbes or Ehrenreich and English has been at least as strong there as anywhere else but mainly because, as the detailed quality of much New England historiography demonstrates, more can be known about relatively obscure colonists than is likely ever to be known about the life of witchcraft suspects in most European countries. Since the facts about the American midwife-witch, as presented above, should have been readily available, the persistence of the myth is an intriguing historiographical problem.

The myth gathered momentum slowly in America. Two women who published biographies of Anne Hutchinson in 1930 took Winthrop's statement at face value and depict her as a charitable herbalist but Packard's 1931 history

[56] I. Strobl, 'Wir Hexen: das Matriarchat im Untergrund vom Mittelalter bis heute', *Neues Forum* xxiii, hft. 269/70 (May/June 1976), 59–60; J. Feldman, 'The Savant and the Midwife', *Impact of Science on Society*, 25/2 (UNESCO, 1975), p. 129; A. Oakley, 'Wisewoman and Medicine Man: Changes in the Management of Childbirth', in J. Mitchell and A. Oakley (eds.) *The Rights and Wrongs of Women* (Harmondsworth, 1976), pp. 26–30; M. Chamberlain, *Old Wives' Tales: Their History, Remedies and Spells* (London, 1981), pp. 31–66; B. G. Walker, *The Crone: Woman of Age, Wisdom and Power* (San Francisco, 1985), pp. 128–9; G. Heinsohn and O. Steiner, 'The Elimination of Medieval Birth Control and the Witch Trials of Modern Times', *International Journal of Women's Studies* V, pt. 3 (1982) 38–42.

[57] C. Hill, 'Science and Magic in Seventeenth-Century England', in R. Samuel and G. S. Jones (eds.) *Culture, Ideology and Politics* (London, 1982), 182; D. Hoak, 'The Great European Witch-Hunts: a Historical Perspective', *American Journal of Sociology* lxxxviii (1983), 1273; G. R. Quaife, *Godly Zeal and Furious Rage: the Witch in Early Modern Europe* (London, 1987), p. 93; B. P. Levack, *The Witch Hunt in Early Modern Europe* (London, 1987), pp. 127–8; C. Merchant, *The Death of Nature: Women, Ecology, and the Scientific Revolution* (London, 1982), pp. 155, 311 n. 15, 313 n. 2. It would require a separate study to trace the European influence of Ehrenreich and English, through translation into German, for example: H. Schmolzer, *Phänomen Hexe* (Vienna, 1986), pp. 84–91.

of American medicine turns her into a midwife. The success of the campaign against midwives in America had destroyed the memory of childbed as a social occasion. Although generally favourable to colonial midwives, Packard may have been influenced by Haggard's book of 1929, *Devils, Drugs, and Doctors*, which identifies early midwives with the Dickensian character, Mrs. Gamp. Packard also mentions Jane Hawkins as a midwife, physician, and reputed witch and Margaret Jones as a 'doctress'. The inclusion of Jones in the midwifery chapter may have been responsible for later confusion. Herbert Thoms, in 1933, uniquely succeeded in completely confusing Hutchinson and Hawkins. Hurd-Mead was more sympathetic than most to the difficulties experienced by early modern midwives and, despite all the examples she provided of their skill and high repute, she believed them to have been in danger of prosecution for witchraft. She followed Packard in seeing Hutchinson as a midwife and added midwifery to her laudatory portrayal of Margaret Jones, who thus joined Hawkins, mysteriously transported to Connecticut, as a midwife-witch. Hurd-Mead expressed astonishment at how few New England medical women were prosecuted. Harvey Graham's book, *Eternal Eve*, named Jones as a midwife and this was followed by Cutter and Viets in their 1964 history of midwifery. They followed Packard in seeing Hutchinson as a midwife but not as a witch. Meanwhile, a 1962 biography of Hutchinson portrayed her as a midwife only in its index, not in its text.[58]

In the first issue of *Feminist Studies* in 1972, Ben Barker-Benfield published a stimulating article on Hutchinson as a sexual threat to Puritan authority, but identified this with her role as a midwife. This is unfortunate, since Winthrop's comments clearly mean exactly what they say, that she was helpful and possessed medicines. Writing after her death, he saw danger in the confidence which families had placed in her skills, but he never said she had been a midwife. It was her pretensions as a spiritual mentor that aroused his hostility. As the first Quaker women missionaries were to discover, New England leaders regarded any active role undertaken by women as decidedly monstrous. Hutchinson's banishment and the surrounding furore can be explained without

[58] H. Augur, *An American Jezebel: the Life of Anne Hutchinson* (New York, 1930), p. 79; W. King Rugg, *Unafraid: a life of Anne Hutchinson* (Boston, 1930), p. 87; F. R. Packard, *History of Medicine in the United States*, vol. 1 (New York, 1931), pp. 44–50; H. Thoms, *Chapters in American Obstetrics* (Springfield, Ill., 1933), pp. 3–5; K. C. Hurd-Mead, *A History of Women in Medicine* (Haddam, Conn., 1938), pp. 313, 410–16, 487. Her source, equally enthusiastic about Jones as a progressive practitioner and victim of medical persecution, did not call her a midwife: J. Fiske, *New England and New France* (London, 1902), p. 145. H. Graham [Isaac Harvey Flack], *Eternal Eve* (London, 1950), p. 303; I. S. Cutter and H. R. Viets, *A Short History of Midwifery* (Philadelphia, 1964), p. 144; E. Battis, *Saints and Sectaries; Anne Hutchinson and the Antinomian Controversy in the Massachusetts Bay* (Chapel Hill, 1962), pp. 83–4, 177–9. The development of American ideas concerning the history of midwifery is best understood in the context of specifically American attitudes towards modern midwives: F. E. Kobrin, 'The American Midwife Controversy; a Crisis of Professionalization', *Bulletin of the History of Medicine*, xl (1966), 350–63; J. Barrett Litoff, *American Midwives, 1860 to the present* (Westport, Conn., 1978); id., *The American Midwife Debate: a Sourcebook on its Modern Origins* (Westport, Conn., 1986); J. Walzer Leavitt, *Brought to Bed: Childbearing in America, 1750 to 1950* (New York, 1986).

reference to midwifery and more recently religious historians have merely mentioned her supposed midwifery in passing.[59]

Although some medical historians have adhered to the tradition of seeing Hutchinson as only a midwife, while citing sources that make no such error, the new tradition requires her to be identified as a midwife-witch. Jane Donegan's valuable study of the rise of man-midwifery in America is led, by the authority of the *Malleus*, of Thomas Forbes, and Ehrenreich and English, into the association of midwifery with witchcraft. Despite providing examples of the high esteem in which the early midwives were held, by both English and Dutch settlers, she identifies Hawkins, Jones and, for what seems to be the first time, Hutchinson as midwives prosecuted for witchcraft, citing Packard, Cutter and Viets, and Haggard as her sources. Her account of these three cases is usually taken to be authoritative, by Joseph Klaits for example.[60] The power of the myth over the historical judgement of recent writers on New England history is so great that it is repeated even when it is contradicted by the evidence produced. Like Donegan, Lyke Koehler, in a book on the power of women in early colonial society, describes the influential position of midwives, examining women accused of witchcraft, infanticide, and antenuptial fornication. As in England, their evidence was crucial and they were well respected and necessarily respectable. Nevertheless, Thomas Forbes is cited so that an artificial link can be made between New Englanders and 'their European contemporaries', at which point Hawkins, Hutchinson, and Elizabeth Morse are provided as examples of midwife-witches.[61]

Almost inevitably, the weight of assertion built upon such slight foundations has begun to totter. John Demos has proved to be the most cautious of New England historians, at least on this subject. He can find only two suspects who 'can be plausibly associated with the regular practice of midwifery', Jane Hawkins of Boston and Hannah Jones of Portsmouth. It is not clear why he should identify Goodwife Jones as a midwife, in which he is unique. He may have had access to some relevant local records but the two sources that relate to the poltergeist phenomena blamed on her by a neighbour do not describe her as a midwife. In any event, her case is of minor importance since she was only bound over to keep the peace. Demos firmly dismisses the myth. Although

[59] B. Barker-Benfield, 'Ann Hutchinson and the Puritan Attitude Towards Women', *Feminist Studies*, i (1972), 65–96; C. Gardina Pestana, 'The City Upon a Hill: the Puritan Perception of the Quaker Threat to Massachusetts Bay, 1656–1661', *New England Quarterly* lvi (1983), 323–53; W. K. B. Stoever, *'A Fair and Easie Way to Heaven': Covenant Theory and Antinomianism in Early Massachusetts* (Middletown, Conn., 1978), p. 24; D. S. Lovejoy, *Religious Enthusiasm in the New World: Heresy to Revolution* (Cambridge, Mass., 1985), p. 69; C. Lloyd Cohen, *God's Caress: the Psychology of Puritan Religious Experience* (New York, 1986), p. 262.

[60] H. Brock and E. Christianson, in *Medicine in Colonial Massachusetts, 1620–1820* (Publications of the Colonial Society of Massachusetts 57, 1989), p. 130, citing S. Abbott Green, *History of Medicine in Massachusetts* (Boston, 1881), pp. 27–9, 53–4; J. Donegan, *Women and Men Midwives: Medicine, Morality, and Misogyny in Early America* (Westport, Conn., 1978), pp. 22–3, 35–6 nn., 89–91; Klaits, *Servants of Satan*, p. 98.

[61] L. Koehler, *A Search for Power: the 'Weaker Sex' in Seventeenth-Century New England* (Urbana, 1980), pp. 85, 117–18, 132 n., 200, 219, 269–70.

childbed was an anxious time, leading to occasional accusations against hostile participants, such as Elizabeth Morse, it was not midwives who were blamed for incomprehensible ailments but those who sought to cure them by highly dubious means. Yet so firmly implanted is the myth that even the rebuttal by Demos is cited by Klaits in support of the stereotype.[62]

Although the myth is perhaps beginning to fade from New England historiography, its usefulness makes it difficult to dispense with altogether. The thesis of Carol F. Karlsen, submitted in 1980, was markedly more emphatic in discussing midwife-witches than is the published version of 1987, but the argument of Demos is still not fully assimilated. Unsupported assertions are made that derive from the work of Ehrenreich and English. Karlsen continues to insist that 'women who healed people or relieved symptoms which doctors had unsuccessfully treated could come under suspicion of using magic in their medical practice. Similarly, a woman who safely delivered infants that were not expected to survive might find herself accused of witchcraft'. Karlsen retains Hutchinson and Margaret Jones as midwives, although Elizabeth Morse is classified as a paid healer. Such distinctions are not important to Karlsen's argument since her basic category is 'midwife/healer', enabling her to ignore the sinister character of some of the healing practices and blur the difference between magical healing and respectable midwifery or herbalism. An otherwise excellent study is thus marred by the persistent failure of New England historians, apart from Demos, to take note of the fact that not one New England midwife was tried for witchcraft and only one was suspected, for largely unrelated reasons.[63]

Since historians working with primary sources are misled by the myth, it is hardly surprising that textbook writers perpetrate it. Midwifery students are told that their predecessors were ignorant witches and history students are told that midwives were marginal members of society like beggars, natural witchcraft suspects.[64] 'Fortunate is the hunter of straw men in the field of witchcraft', wrote Erik Midelfort. 'Nothing would be easier than to line up a number of nimble-penned authors and knock them down with footnotes blending dust with acid.'[65] A few slashes with Ockham's razor are unlikely to demolish such a deep-rooted myth which serves the purposes of historians and

[62] J. Demos, *Entertaining Satan: Witchcraft and the Culture of Early New England* (New York, 1982), pp. 80–4, 430 n. 75; *Collections of the New-Hampshire Historical Society*, viii (1886), pp. 99–100; R[ichard] C[hamberlayn], *Lithobolia: or, the Stone-Throwing Devil* (London, 1698); Klaits, *Servants of Satan*, p. 187 n. 31.

[63] C. F. Karlsen, 'The Devil in the Shape of a Woman: the Witch in Seventeenth-Century New England' (Yale Ph.D. thesis, 1980), pp. 11, 18–20, 22–4, 176–81, 220–5, 245, 254, 310; *The Devil in the Shape of a Woman: Witchcraft in Colonial New England* (New York, 1987), pp. 9, 16, 19–21, 142–4.

[64] J. Towler and J. Bramall, *Midwives in History and Society* (London, 1986), pp. 21–42; Klaits, *Servants of Satan*. Professional beggars were actually too marginal to be commonly accused. Impoverished members of the community were more likely targets, thanks to the gradual accumulation of resentment.

[65] Midelfort, *Witch Hunting*, p. 164.

polemicists of widely varying persuasions. The existence of the myth, although only mildly disabling for the study of witchcraft, has pre-empted serious attention to the lives of midwives or the relationship between popular medical practices and religious authority. An assault on the myth may open up such projects, which might usefully start with the problem of why, despite the beliefs of demonologists, midwives were not generally persecuted as witches, participating instead in the prosecution as experts in the definition of the unnatural, and why they did sometimes become victims of the witch hunts, under certain unusual circumstances, as did other prominent people.[66] Perhaps the myth of the midwife-witch should be seen as simply one among the long list of inversions, from sodomy to salt-free food, most of which are ignored by historians. It certainly is not sufficient to simply accept the authority of the *Malleus* and assert, 'Obviously midwives had ready access to certain human remains much sought after by practitioners of witchcraft', without making clear whether one is referring to cauls or corpses.[67]

The *Malleus* is the main authority on this subject for historians yet its influence, outside the technical discussions of demonologists, is doubtful, especially in most Protestant countries and the lands under the Mediterranean Inquisitions. Even where it can be shown to have had an influence, as on the French demonologists, when men such as Rémy actually prosecuted witchcraft cases it seems mainly to have influenced their interrogation of suspects. Midwives were not accused in significant numbers because they were the wrong kind of women, respected and influential members of their local communities, more likely to be guilty of a strong will and a sharp tongue than the evil eye. Moreover, popular belief does not appear to have generalized from the handful of cases in any one jurisdiction and ascribed demonic connections even to distant midwives. The midwife-witch is a stereotype that has passed straight from the works of the demonologists into the works of historians with barely a glancing impact on the lives of real midwives. A few cases can be substantiated, in Cologne for example, but the onus of proof must lie with those asserting the reality of extensive prosecutions of midwives. No one has yet shown that there was a disproportionately high number accused. Historians treating this subject have behaved like demonologists, repeating old stories without checking their sources and making assertions without producing data to substantiate them. They have tortured a few facts to fit the Procrustean bed of an obsolete theory. That historians who casually accept the stereotype of

[66] R. A. Horsley, 'Who Were the Witches? The Social Roles of the Accused in the European Witch Trials', *Journal of Interdisciplinary History* ix (1979), 709–10; P. E. Tibbetts, 'The Weighted Coherence Theory of Rationality and Justification: a Critique and an Alternative', *Philosophy of the Social Sciences* x (1980), 271; L. L. Estes, 'The Medical Origins of the European Witch Craze: a Hypothesis', *Journal of Social History* xvi (1983), 271–84.

[67] Hansen, *Quellen*, *passim*; S. Clark, 'Inversion, Misrule and the Meaning of Witchcraft', *Past and Present* lxxxvi (1980), 98–127; V. J. Newall, *The Encyclopedia of Witchcraft and Magic* (London, 1974), p. 122.

midwives as ignorant crones should assume them to be victims of the witch-hunt is unsurprising but this myth also mars some excellent work in women's history. Campaigners on behalf of women's access to health care would be well advised to abandon this double-edged weapon since, like most myths, it ultimately produces only mystification.

Elizabeth Cellier in 1688 on Envious Doctors and Heroic Midwives Ancient and Modern

Elizabeth Cellier, a Restoration woman admired for her compassion and courage during the Exclusion Crisis (1678–81) and for a brilliant self-defense at her trial for treason, as recorded in *Malice Defeated* (1680), also deserves to be admired for her pamphlet of 1688 meant to instill pride in the London midwives. In *To Dr. —, An Answer to his Queries concerning the Colledg of Midwives*, Cellier combines satirical taunting of the doctor, her opponent, with a heroic narrative in defense of midwifery, citing ancient precedents in "Sacred and Profane Histories" for her proposed institution of learning and charity to be governed entirely by women.[1] She rallies the midwives of London, whom she considers to be oppressed by the physicians, around her plan for a college and orphanage already approved by James II four months earlier but still needing to be implemented.

Cellier's little tract seems to have been written in the course of her quarrel with an unnamed doctor, possibly Dr. Hugh Chamberlen, a Whig man-midwife who frequented the court. Although she had been sentenced in 1680 to a large fine, the pillory, and imprisonment for writing *Malice Defeated*, and although she had been incarcerated for at least three years from her first arrest in 1679, as appears from her petition to the king in 1682, Cellier had been finally vindicated by being received into the service of Mary of Modena after her release from prison.[2] She mentions, in January 1688, that she has been serving Queen Mary Beatrice for four years and encouraging her in the hope of bearing children. Her pamphlet seems to have been hastily composed, for she asserts that she promised her antagonist on 12 January to answer his "Query" about "*Whether ever there were a Colledg of Midwives in any part of the World*," and she dates her reply 16 January. She boasts she will "now prove" that such a college indeed existed "some Hundreds, if not Thousands of Years before you can prove one of Physicians" (p. 1). Like other Tories writing political works in the 1680s, Cellier asserts she is no innovator but is following good biblical and classical precedents by incorporating the midwives. She will argue that an ancient constitution of midwifery existed from time immemorial, and that divine inspiration established the first colleges of all-female midwives.

In June 1687, the previous year, Cellier presented to James II a paper entitled, "A Scheme for the Foundation of a Royal Hospital, and Raising

Eighteenth-Century Life 14 (February 1990): 24–34 © 1990 by The Johns Hopkins University Press

a Revenue of Five or Six Thousand Pounds a Year, by, and for the Maintenance of a Corporation of skilful Midwives and such Foundlings, or exposed Children, as shall be admitted therein."[3] The king approved her plan in September. Nevertheless, she probably had trouble launching her college: the London midwives may have balked at paying five pounds a year to enter such an unheard-of association. Cellier, it should be added, was regarded in high circles as trustworthy in the management of money, for in 1679 the English Catholic nobility and gentry had channeled their contributions to her for the relief and bailout of two thousand or so Catholics imprisoned in the London area because of Titus Oates' fabricated plot.[4] When she received the sneering rebuke of her antagonist on 12 January 1688, she may have decided to publicize it in this witty and imaginative defense of the London midwives. After all, she had been dubbed "*Joan of Arque*" in the heyday of the Exclusion Crisis and knew how to turn her enemies' verbal attacks back on themselves, as witness her *Matchless Rogue* (1680), a mordant mock-heroic satire against Thomas Dangerfield, a career criminal and her former accuser.

In her proposed college of midwives, Cellier planned on having female governors and instructors, the more experienced midwives imparting their skill to, as well as supervising, the others. Although the ruling midwives would willingly receive a periodic "visitation" from someone the king would appoint, as did Oxford and Cambridge, and nominate a physician or man-midwife to read a public lecture to them once a month, probably on anatomy, the college was otherwise to be free of the governance and even the collaboration of men (*Scheme*, 9:192).

Cellier's antagonist, however, has trespassed on just the province she believes should be reserved for women. He has presumed to teach midwifery while understanding "nothing" of that art. She reminds him that Hippocrates swore not to perform surgery for stones, but to leave that for physicians more skilled in the area. Her opponent has broken the Hippocratic oath by attempting to teach an "Art much more difficult" than cutting for stones, an art about which he has only book knowledge (p. 3). Since the prevailing norm of modesty in the seventeenth century kept most English women from calling in men-midwives, except when death threatened, these men often lacked sufficient experience of ordinary deliveries.[5] Hence, Cellier taunts her antagonist, saying that the women of "this Age" are too "impatient of their Pain" to "bear it, in Complement to the Doctor, *while he fetches his Book, studies the Case, and teaches the Midwife to perform her work*, which she hopes may be done before he comes" (p. 7). Far from being willing to allow males to have the experience of day-to-day deliveries, Cellier states that midwifery "*ought to be kept as a Secret amongst Women as much as is possible*" (p. 3). She evidently endorses the norm of "Modesty."

Possibly her sharply satirical tone is due to her knowledge that the College of Physicians in England, as far back as 1616, had blocked the self-governance of the women-midwives, even when James I favored it. In 1616, too, Peter Chamberlen, the great-uncle of Dr. Hugh Chamberlen, whom I suppose to be Cellier's opponent, tried to become the "Governor" of the midwives, so as to have a clear monopoly over the licensing and teaching of them. It is not known if he had the consent of the midwives or just took advantage of their petition to the king, in which they desired to be incorporated. This man's nephew, also named Peter Chamberlen, attempted to become "Governor" of the midwives in 1634, but this time it was evidently against their consent, for they protested to the king against his plan, alleging that his having the sole licensing and instructing of them would result in the "Inslaving of your petitioners."[6] This Peter Chamberlen, according to David S. Katz, brought Charles II into the world and also founded a congregation of Saturday-Sabbatarians in the 1650s, remaining a Saturday worshipper after the Restoration.[7] The third in this line of men-midwives was his son, Dr. Hugh Chamberlen, an ardent Whig who published a pamphlet in 1694 called *A Few Queries* containing a proposal for the reform of midwifery, one of a long line of reform schemes by the Chamberlens. This scheme rivalled Cellier's and was probably prepared in opposition to hers, since he states he presented it to the parliament in 1689.[8] The similarity of his title to hers seems to underline the relationship between their proposals. Later, in the 1690s, Chamberlen proposed still another scheme, this time of a financial nature, to help King William's government.

Cellier begins the defense of her proposed college of midwives by citing Exodus 1:15–21. She argues that "no Rational Person" can think that Shiprah and Puah, the Egyptian midwives who disobeyed Pharaoh's orders and let the male Hebrew children live, "could in their own Persons act as Midwives to all the Women of that *mighty People*, who about 100 Years after went up out of *Egypt* 600000 Fighting Men, besides Women and Children" (pp. 1–2). They must have been "Governesses and Teachers of other Midwives." And besides, Cellier consulted "a Learned *Rabbi*, now in Town," who told her that their very names signify they are teachers (p. 2). She adds the clincher, that the Church Father Origen, in his eleventh homily on Exodus, explains that "Shiprah and Puah were not only the *Governesses of the Midwives*, but also Women of Great Learning, and excellently skill'd in Physick, which was then practised by Women to Women." Like the "*Magi* and Priesthood" in Egypt, she concludes, these chief midwives must have had "some certain place for consulting in, from whence they might issue their Directions. Which, by your leave, Doctor, without Absurdity in the Language of these Times, might well be called *a Colledg*." Women thus had a college of midwifery in the age of patriarchs, long before male physicians did: "nor was Physick then a regular Study, nor brought under Government in that Learned Nation of *Egypt*, in *Herodotus* his time, which put together, proves the Antiquity of the Midwives Government so much antienter than that of the Doctors" (p. 2).

By claiming that the "Midwives Government" in Egypt was more ancient than that of the doctors, Cellier gives the midwives an important edge institutionally. In that era, church historians each gave their religion the same sort of advantage by claiming their faith derived from primitive or apostolic times, while insisting that the faith of others had a later origin, and was the result of encroachments on, or innovations in, ancient Christianity.

Besides giving midwives precedence in antiquity, Cellier invests their ancient Egyptian college with a sacred aura. She affirms that Shiprah and Puah had true biblical "*Piety*," for they had "*learned the knowledg of the only true God*" from the Hebrews, and so they disobeyed Pharaoh's "*bloody Orders of destroying the Males*" out of fear of offending the God of Israel. As a result, God blessed them and their descendants: "*God built them Houses, and blessed them.*" She taunts her antagonist by adding: "such a favour from God *of building Houses for them*, we do not read the Physicians ever received." By "*Houses*" Cellier means for us to understand both the physical descendants of the midwives and their professional descendants. Cellier has heard, probably from the "learned *Rabbi*" in London mentioned earlier, that the Egyptian midwives were "Incorporated into the Body of the Jews" and that their descendants "*continue in Honour*" among the Jews of Thessalonica "*to this Day*" (p. 2). Moreover, the professional descendants of those ancient midwives seem to have carried on their "*Piety*," as we shall see, even in Pagan Greece.

Moving from biblical to profane antiquity, Cellier mentions that Hippocrates used to swear by the goddesses Hygea and Panacea as much as by the gods Apollo and Aesculapius. She found this information in Jacques Guillemeau's popular work on midwifery, to which she alludes later,[9] but she draws her own conclusion: the gods of antiquity were only "Men and Women, who first found out and taught Arts and Mysteries" and were divinized because people thought a "Divine Spirit" guided them to the "knowledg of things so useful and so far above the Vulgar Capacities." Thus, women were among the first inventors of medicine since the Hippocratic Oath does not "part the *Gods* and *Goddesses*," but treats them with "equal Veneration" (p. 3).

In addition, where Guillemeau states simply that Socrates' mother was a midwife, Cellier makes a legendary heroine of Phanarota, calling her "a Woman of great *Learning* and *Skill*" and a unique monotheist in a "Learned Idolatrous City," a "*Midwife that knew and feared the true God*" (p. 5). Perhaps Cellier had read that Socrates called himself a midwife's son who practiced his mother's art (*Theatetus*, 151 C), but she goes far beyond this comparison of the arts of philosophy and midwifery. She claims that Socrates drew inspiration from his mother and was a "Martyr" only "for professing there was but *one God*," a "Wisdom" he himself admitted "he learned of his Mother" (p. 5). Cellier alludes to St. Paul's visit to Athens to hint at a parallel between the "Apostle" and Phanarota as lonely monotheists. Like Shiprah and Puah in Egypt, the Athenian Phanarota has a prophetic dimension. Cellier implies that some midwives form a lineal descent of inspired heroines in the divine plan of history: Shiprah and Puah, Phanarota and Agnodicea.

Both Guillemeau and Cellier recount the oft-told tale of Agnodicea, an Athenian woman-doctor, but Cellier alone uses biblical rhetoric to make it a romance of oppression and deliverance out of the Book of Judges. In Cellier's version, the envious physicians are just as much villains in ancient Athens as they are in London of 1688. Her story starts with some physicians among the "*subtile Athenians*" getting into the government and deciding to take advantage of some unfortunate "Miscarriages" among the "Noble Women." They use these miscarriages as an excuse to take over the art of midwifery, which had been practiced from time immemorial by women. A law is passed "that for the future no Woman should study or practise any part of Physick on pain of Death" (p. 3).

It is interesting that Cellier does not even consider the possibility that the midwives and noble women of Athens might have been inducing abortions.[10] Cellier ignores that charge and focuses instead on the envy of the physicians, because, in her mind, the same envy stands in the way of her midwives' college. However, Cellier does not have a totally idealized view of women. In her 1687 proposal, she speaks of "wicked and cruel mothers" who murder their newborn children (*Scheme*, 9:191). She explains again in 1688 that her lying-in hospitals and orphanage would help prevent these "*many Murders*" and "*the Executions which attend them*," since these mothers are the very poor who lack "fit ways to conceal their shame" and cannot "provide" for their offspring (p. 7).

Cellier shows that the physicians' oppressive law did not increase the number of their female clients in ancient Athens, because the "Modesty" of the women would not permit them "to admit of Men either to Deliver or Cure them." At this point her narrative takes on biblical characteristics. Just as God raised deliverers for the oppressed Jews in Judges, so "God stirred up the Spirit of *Agnodicea*, a Noble Maid, to pity the miserable condition of her own Sex, and hazard her Life to help them" (p. 3). Not being able to bear seeing the women of Athens die "both in Child bearing, and by private Diseases," Agnodicea, disguised as a man, studied under the famous physician Herophilus (355–280 B.C.), returned to Athens, told the women her secret, and "in a short time" became the "Successful and Beloved Physician of the whole Sex," curing them with "Skill and Industry" (pp. 3–4). Here we see Agnodicea and the women of Athens breaking the civil law in order to preserve both their lives and their modesty.

Cellier then pointedly connects her heroine's trial on a trumped-up charge of adultery to her own trial for treason in 1680. The envious physicians of Athens who "conspired" to ruin the supposed young man are parallel to the Whig plotters who conspired to ruin Cellier. Both she and Agnodicea were faced with perjured witnesses who had been bribed: "*and there being Witnesses to be found then* (as of late Years, that would swear any thing for Money) she was upon their Testimony, condemned to Death" (p. 4). Cellier hints that the envious physicians of Athens and of London, 1688, are not that far removed from Titus Oates and the Plot Witnesses of 1680.

In her account of the subsequent rescue of Agnodicea by the women of Athens, Cellier creates a veritable myth of female solidarity. At her trial for adultery, Agnodicea revealed her sex to save her life. But "the enraged Physicians accused her of transgressing the Law, which forbid [sic] Women to Study or Practise Physick" and got the death penalty. The Athenian women, however, rose up against such an injustice and entered the place of judgment, declaring to all the men that "they would no longer account them for Husbands or Friends, but for cruel Enemies" if they dared kill the person who had "restored them to their Healths." This female revolution goes much farther in Cellier than in other accounts, for her Athenian women threaten mass suicide: "they would all die with her if she were put to Death" (p. 4). This romantic gesture seems to come out of Cellier's imagination and reflect her yearning for greater unity among the London midwives against the envious male physicians waiting in the wings, she fears, to take over the governance and licensing, if not the entire practice of midwifery.

Cellier makes use of the Agnodicea story also to spell out a myth of a college of midwifery full of women of high birth. She wants to establish the prestige of her profession by insisting that it was appropriate to gentlewomen in classical Greece. When Agnodicea was acquitted in response to the women's rebellion, the magistrates revoked the hateful law hatched by the physicians and allowed the establishment of a women's college of midwifery. A new law "gave Gentlewomen leave to Study and Practise all parts of Physick *to their own Sex, giving* large Stipends *to those that did it well and carefully, and imposing severe Penalties upon the unskilful and negligent*: And there were many Noble Women who studied that Practise, and taught it publickly in their Schools as long as *Athens* flourished in Learning" (p. 4). Here Cellier hints that one of the measurements of a city's level of civilization is the prestige and care it gives to the women-midwives' mastering of their skill. If Athens in its best period allowed learned women like Agnodicea to give public lectures to other midwives, then it follows that London, which claims also to be a center of culture, should do the same.

Cellier's emphasis on women of high birth becoming midwives is interesting since midwifery was indeed a profession that English ladies who had fallen on hard times could enter. She herself claimed to come from landed gentry, from a family impoverished in the Civil Wars, and she ministered chiefly to an aristocratic and royal clientele.[11] In mentioning "*severe Penalties*" to be imposed on ignorant and negligent midwives, she shows her desire that high standards be maintained in her chosen profession.

Cellier mentions the sources she consulted on midwives in Greek antiquity, admitting that her learning does not go farther than "*Ambrose Perre*, Counsellour and Chyrurgion to the King of France, and his Ingenuous Disciple *Gulielmus*." She is referring to Ambroise Paré (1510–90) and his student Jacques Guillemeau (1550–1613), two famous French surgeons and men-midwives whose books had been published in England.[12] The details she took from their works, however, have been transformed under her hand; she has created a myth of a line of heroic women-midwives, all following divine inspiration.

Cellier then challenges her antagonist to go read "*Hebrew, Greek, Latin and Arabick* Books" and find out for himself that in the earliest times, "*Midwifery, Chirurgery*, and *the making up and administring of Medicines*, were all one." Cellier implies here that since women were the first midwives and since midwifery and the rest of medicine were originally all one, then women were the first doctors as well. She points out that the women of London are asking now for a mere portion of the healing arts they were once entitled to practice as a whole, and they are asking for it by right of ancient precedents already mentioned and others the doctor himself will uncover by research. She points out that pharmacology was once the most "servile part" of medicine, but has "now usurp'd" upon midwifery and surgery. However, the women of London are asking only to be left in charge of midwifery, "the *most Antient, Honourable and Useful Part*" of medicine (p. 5).

When she refers to books in Greek or Arabic, Cellier appears to be taunting her opponent, perhaps highlighting his ignorance of languages. Speaking to him in the name of all the midwives, as if she were already at the head of their corporation, she declares to him and other physicians aspiring to control the London midwives: "we desire you not to concern your selves, until we desire your Company, which we will certainly do as often as we have occasion for your Advice in any thing *we do not understand, or which doth not appertain to our Practice*" (p. 5). In other words, she declares that the midwives can work independently of the physicians unless there are complications which require the physicians' intervention.

Turning to British antiquity, Cellier declares that only the bards' colleges were as ancient as the midwives': "before the *Romans* came hither here were Colledges of Women practising Physick, dedicated to some of the Female Deities" (p. 5). The word "Physick" in this passage implies that women practiced all the healing arts in Druidic Britain. She adds that in ancient London itself there were "*Colledges of Women about the Temple of* Diana, *who was Goddess of Midwives here as well as at* Ephesus" (p. 6). As proof that Diana was active in midwifery, Cellier mentions that her Ephesian temple was destroyed because she was away a very long time at the labor of Queen Olympia for the birth of Alexander the Great. Again, this story is meant to raise the prestige of midwifery by showing that even a goddess performed this task.

But looking at English midwifery for the past century or so, Cellier admits that there is need for reform. She traces the problem to the bishops' licensing of midwives without an exam proving proficiency, a system which began under Bishop Edmund Bonner (1500–69) and continued until 1642. Between 1642 and 1662, the midwives were "licensed at *Chirurgions-Hall, but not till they had passed three Examinations, before six skilful Midwives, and as many Chirurgions expert in the Art of Midwifery*." But in 1662, with the Act of Uniformity, the Midwives returned to bishops' licenses, which they took in Doctors Commons, "where they pay their Money, (*take an Oath which is impossible for them to keep*) and return home as skilful as they went thither" (p. 6). Without a test of skill, the system of bishops' licenses, which continued after 1688, merely raised money without insuring high standards for the profession.

Cellier charges that episcopal licenses have brought about vast numbers of deaths of London mothers and infants. The reader senses her indignation as she mentions the "Bills of Mortality," which the "curious" may consult in "*Clerks-Hall.*"[13] She has evidently looked at them, since she affirms that between 1642 and 1662, "there did not then happen the eight part of the Casualties, either to Women or Children, as do now" (p. 6). Like Agnodicea, Cellier seems stirred up to thoughts of deliverance by the sight of all these needless deaths in childbirth, to thoughts of reforming midwifery. The alternative, she fears, is that male physicians will undertake the reform and pretend "to teach us Midwifery, especially such as confess *they never delivered Women in their Lives*, and being asked *What they would do in such a Case?* reply *they have not yet studied it, but will when occasion serves*" (p. 6). Such doctors, who rely merely on book learning, are as ridiculous as Phormio, "who having never seen a Battel in his Life, read a Military Lecture to *Hannibal* the *Great*" (p. 7). With keen wit, Cellier parallels the midwives with Hannibal, a heroic figure, because they too are the veterans of many a great "Battel."

Cellier concludes with the triumphant public announcement that the queen she serves is pregnant. She sees a connection between James II's approving of her plan to "unite the Midwives into a Corporation, by His Royal Charter, and also to found a Cradle-Hospital" in September 1687, and his queen's conceiving a child shortly afterwards. In her view, God may well have given the king a "suitable Return" for his "pious Design" to establish an orphanage for the London poor and preserve "so many *Innocents* as would otherwise be lost" (p. 7). Letting her romantic imagination soar, Cellier dreams that the males she will raise in the royal orphanage will someday make up an army led by James II's little Prince of Wales, who, "like another *Moses*," will lead these new Israelites, rescued from death, to a better world.

Cellier seems unaware of how the English are already counting on William and Mary to be the heirs of James II, since Mary of Modena has suffered a number of miscarriages. Such public gloating on her part over a future Catholic heir seems imprudent, to say the least. As a known Catholic convert and one formerly imprisoned in the Popish Plot, she may have contributed by means of this provocative announcement to the raising of fears and jealousies about Queen Mary Beatrice's pregnancy. How could Cellier be so sure, the king's enemies would ask themselves, that the queen was pregnant with a son? Perhaps it was as a result of her unwise remarks that Cellier was not among the midwives around Queen Mary Beatrice when she delivered the Prince of Wales in June 1688. A chief midwife who was Protestant and a Whig was thought a safer choice for the occasion. However, the absence of Cellier and the presence of dozens of witnesses, many of them Protestant, at the birth of James II's son, did not convince those predisposed to consider that son spurious or suppositious, a "warming pan" baby. Bishop Gilbert Burnet made much of the absence of Doctor

Chamberlen at the birth of the prince, but as the Whig doctor himself explained in a private letter to Electress Sophia of Hanover later on, he was sent for in time but happened to be visiting a patient.[14]

One reason Cellier cannot resist publicly gloating over the queen's pregnancy is that her opponents the doctors had "often Laughed" at her and accounted her "a Mad Woman these last four Years" for insisting that Queen Mary Beatrice "was full of Children, and that the *Bath* would assist her Breeding." Since it was indeed during the queen's Fall 1687 journey to Bath that she conceived, Cellier can take partial credit. Not content with her boast about foreseeing the effects of the Bath waters, Cellier declares that the only thing now standing in the way of the queen's producing "many other Royal Babes," is the presence of "over Officious" doctors, who, she hopes, "will but be pleased to let them live" (p. 7). Cellier has a point here, for the doctors nearly killed James II's newborn son in June 1688, by forbidding him the drinking of milk! Her thrust at the doctors suggests that her scheme to incorporate the midwives was opposed by the College of Physicians. According to Harold J. Cook, the monopolistic, regulatory activity of the College had reached a peak in 1687–88, after the doctors received a new charter from James II.[15]

If Cellier's opponent was the Whig Dr. Chamberlen, then her final words would be a palpable hit, for she asserts that it ought to be the "hearty Desire of all Loyal Subjects, of what Persuasion soever" to see the sons of James II "*Muster their little Soldiers*" out of the royal orphanage (p. 8). Indeed, her foundlings, according to the scheme presented in June 1687, would have been supported in part by contributions "from every church, chapel, or publick place of divine service, of any religion whatsoever" within the "limits of the weekly bills of mortality," and so hers would have been the first truly interdenominational orphanage in England. The orphans, free from any "reproach, for their innocent misfortune," would have been trained "in proper learning, arts, and sciences, according to their several capacities" and then made "free members, of every city and corporation" in England and Wales (*Scheme*, 9:192–93). Cellier was not content with saving these orphans' lives but dreamed of making them some sort of elite, like the Turkish Janizaries, that the future princes born to Queen Mary Beatrice would be proud of. Such a humane way of raising foundlings might well win the approval of persons "of what Persuasion soever."

But the probability that Cellier is also taunting a Whig doctor seems enhanced by her final quotation of Ecclesiastes 8:4, one of the most royalist verses in Scripture: "*Where the Word of a King is, there is Power: and who shall say unto him, What doest thou?*" (p. 8). This passage expresses her reliance on royal authority and her assurance that the king's commitment to her proposed women's college and orphanage will make vain whatever opposition an envious doctor can raise to her plan.

Curiously, Cellier's pamphlet, with its premature and politically unwise celebration of the queen's pregnancy, was licensed by Lord Sunderland in 1688, possibly because he or his aides did not read the work to the end.

At the close, Cellier indicates that she is writing from her "House in Arundel-street, near St. Clement's Church in the Strand," the place from which she wrote *Malice Defeated* in 1680. Her readers in 1688 might possibly recall that it was at this fine house that she had lodged the thirty Catholic boys from St. Omer's who crossed the sea to be witnesses for the five Jesuits accused by Titus Oates and tried in 1679.

Perhaps in the turmoil preceding the revolution of 1688, Cellier's plans for a midwives' college and orphanage were pushed aside, or perhaps they had already been crushed by the opposition of the physicians. Nevertheless, her scheme presented to the king in 1687 and her defense of it in 1688 survive as evidence of her compassion, courage, creative energy, and persuasive power—traits remarkable in a person of any era, but especially to be admired in a gentlewoman-midwife of the seventeenth century. In her final tract, Cellier tried to prevent the physicians' taking over midwifery in England, as was happening then in well-to-do circles in France: she wanted to keep open for English ladies one of the very few professions they could enter if they fell on hard times, and despite resistance from physicians, she also wanted to place it under the supervision of learned female midwives.

Anne Barbeau Gardiner
John Jay College of Criminal Justice, C.U.N.Y.

NOTES

1. Cellier's last tract is very rare. A copy in Burke Library, Union Theological Seminary, New York City, has no title page and consists of 8 pages. She probably had it printed for herself in 1688 and sold it from her house, as she had her previous tracts in 1680. This tract will be cited hereafter in my essay.

2. For accurate details of her life, see my intro. to *Malice Defeated* and *The Matchless Rogue*, Augustan Reprint Society, nos. 249–250 (Los Angeles: William Andrews Clark Memorial Library, 1988). See also Antonia Fraser, *The Weaker Vessel* (N.Y.: Knopf, 1984), pp. 454–63, 469–70, and J. H. Aveling, *English Midwives, Their History and Prospects* (London: Churchill, 1872), pp. 63–83.

3. Cellier's *Scheme* is printed in *The Harleian Miscellany*, 12 vols. (London, 1808–11), 9:191–98.

4. Roger Palmer, earl of Castelmaine, was Cellier's link with the nobility. She seems to have been regarded as saintly by Catholics of the era. See "A True relation of some judgments of God against those who accused the priests and other Catholics after the pretended plot in England," *Stonyhurst MSS*, printed in Henry Foley, *Records of the English Province of the Society of Jesus*, 7 vols. in 8 (London: Burns & Oates, 1877–83), 5:74–75. Those who stoned her on the pillory were said to be punished by a special providence.

5. Jean Donnison, *Midwives and Medical Men: A History of Inter-Professional Rivalries and Women's Rights* (N.Y.: Schocken, 1977), p. 20.

6. J. H. Aveling, *The Chamberlens and the Midwifery Forceps* (London: Churchill, 1882), p. 36. See also Donnison, pp. 13–14.

7. *Sabbath and Sectarianism in Seventeenth-Century England* (Leiden: Brill, 1988), pp. 48–89.

8. *The Chamberlens*, pp. 152–53. Aveling regards Hugh Chamberlen as a quack, despite his serving at court, p. 125.

9. *Child-Birth or, The Happy Deliverie of Women* (1612; rep. Amsterdam: Theatrum Orbis Terrarum; N.Y.: Da Capo, 1972). Cellier borrows many details from Guillemeau, pp. 79–82, but reshapes them by her satirical wit and heroic imagination.

10. Kate Campbell Hurd-Mead, M.D., in *A History of Women in Medicine* (Haddam, Conn.: Haddam Press, 1938), p. 45, mentions that the charge levelled against women doctors in this period was that of performing abortions. Hurd-Mead states the story of Agnodicea was told by Hyginus, Pliny, Martial, Celsus, Diodorus Siculus, Quintilian, Apollonius of Rhodes, and Apuleius, as well as by the 16th-c. lawyer Tiraquellus.

11. Her daughter Mary Cellier married Charles Dormer, whose son in 1712 became the fifth Baron Dormer. Cellier's husband is buried at Great Missenden Church with the Dormers, but Cellier disappears from view after 1688.

12. See the unpaginated preface on the origin of medicine in *The Collected Works of Ambrose Paré*, trans. out of the Latin by Thomas Johnson (1634; rep. Pound Ridge, N.Y.: Milford House, 1968). Paré mentions no women at all at the origin of medicine, but Guillemeau states that female modesty forced them into it and that their envy of men urged them on: "For women are of this disposition, that they desire to excell men, or at least to seem to go beyond them" (*Child-Birth*, p. 82).

13. Donnison, p. 12, tells of a 1662 study of the London Bills of Mortality; Cellier may have seen this work.

14. Dr. Hugh Chamberlen's letter is reprinted in T. B. Howell, *A Complete Collection of State Trials* . . . , 34 vols. (London, 1816–28), 12: cols. 155–57. Howell prints the depositions of 40 witnesses taken on 22 Oct. 1688 before the landing of William of Orange, pp. 123–44. Mrs. Judith Wilks or Wilkins, the chief midwife there, was a Protestant Whig. Sir John Dalrymple, himself a Whig, called this accusation of a spurious heir "a mere lie of party" and said it was an "affront" to the revolution of 1688 to insist on it (*Memoirs of Great Britain*, 3 vols. [1790; rep. Farnborough, England: Gregg, 1970], 2: appendix to Book V, p. 184). Sir Godfrey Kneller, at Corpus Christi College, on 28 Sept. 1697, said positively it was "a lye" that the Prince was not the true heir of James II, for Kneller recognized his every feature, James II and Queen Mary Beatrice having "sat to me about 36 times a piece; & I know every line and bit in their Faces" (Carte MSS, Bodleian Library, vol. 198, fol. 9, signed by Dr. Charlett, Dr. Gregory, and George Harbin at Dr. Wallis's).

15. *The Decline of the Old Medical Regime in Stuart London* (Ithaca: Cornell Univ., 1986), pp. 92, 203–06.

THE PERILS OF EARLY MODERN PROCREATION: CHILDBIRTH WITH OR WITHOUT FEAR?[1]

For the twentieth-century historian it is always open season on early modern attitudes. Much has been published on such topics as the magical belief systems of early modern men and women; their lack of romantic love; and their indifference to children. Such studies seem to suggest a radical difference between our own sensibilities and those of our early modern ancestors.

The attitudes of women towards childbirth have often been approached within this framework. A consensus has developed, according to which women in the seventeenth and eighteenth centuries were racked by fears of giving birth. As we shall see in a moment, the historiography is actually more complex than this. Yet I believe that one recent study reflects mainstream opinion in asserting that 'even a natural event such as childbirth, which, as the epitome of Creation itself, should have been a cause for joy, terrorized a mother's heart'.[2] This picture I shall refer to as the 'fear thesis'. In this paper I shall be assessing the validity of this thesis, concentrating on the period from about 1650 to 1750. This time-frame embraces much of the best-known evidence for maternal fears, and enables us to consider the subject both before and during the rise of 'man-midwifery' in the eighteenth century.

We can find in the literature a spectrum of views from optimism to pessimism, and these fall into three groups. The first view, which may be termed 'cautious optimism', has been enunciated by Schofield and Loudon. Schofield observes that the actual risk of dying in childbed in early modern England – around 1 per cent – resembles today's level of infant mortality, and comments:

> While most people know of cases of infant deaths, the incidence is so low that I suspect that most parents do not consider that their child is seriously at risk of dying in infancy. Similarly, in the distant past women will have known of others who died giving birth to a child; but they may also have considered it such a rare event that there was little risk that the tragedy would befall them.

Loudon, taking a slightly different tack, expresses reservations about one formulation of the 'fear thesis', and observes: 'Women in the eighteenth and nineteenth centuries had only slightly more reason to fear childbirth than our mothers and grandmothers did in the 'twenties and 'thirties of this century.'[3]

The second and clearly the majority view is the pessimism of the 'fear thesis' – advanced in different ways by some seven or eight historians. Porter and Porter have already been quoted as asserting that childbirth 'terrorized a mother's heart'; elsewhere they write that birth was 'extremely dangerous', that pregnancy 'was frequently painful, exhausting and fearful', and that 'giving birth could be a nightmare'. At one point they express puzzlement that eighteenth-century mothers do not often mention such fears in the surviving sources, but they do not doubt that the fears were there.[4] Macfarlane long ago drew attention to Jane Josselin's fears of giving birth; Beier observes that 'Jane's fears about her deliveries were understandable and, probably, very common'.[5] Crawford states that 'Many women approached childbirth with fear'. Fraser entitles a chapter on seventeenth-century childbirth 'The Pain and the Peril', and asserts therein that 'None at this time doubted that childbirth [...] was both a painful and a dangerous ordeal.' Mendelson writes that 'Women's anxieties about pregnancy and childbirth form a leitmotif which appears in various guises throughout their diaries and letters'; and Pollock (responding to Schofield) suggests that 'Notwithstanding the relatively low maternal mortality rate, it is still possible that childbirth was imbued with dread.'[6]

A further twist to the 'fear thesis' is the claim of Mendelson (supported by Crawford)[7] that:

> Not only did most women run the gauntlet themselves many times, but they were continually called upon to witness their friends' agonies in like circumstances. And so these sociable childbed gatherings – which were intended to provide maximum support for the woman in labour – became a means of equitably distributing her terror to the rest of her female acquaintance.

In other words fear was collective; the popular ritual of childbirth, though devised and maintained by women themselves, only made matters worse for them.[8] This particular interpretation renders somewhat more explicit a widespread assumption underlying the 'fear thesis': that whatever the attitudes of mothers, these were in some sense universal, generalized. When addressing this theme, historians have been looking for uniformity, not variety, amongst early modern mothers.

Thirdly, and standing alone, we have what can be called the 'radical pessimism' of Shorter, who argues not only that mothers were afraid of childbirth but also that this fear cast a pall over their whole lives, especially over their sexuality. This is linked to a wider thesis about women's history – in a nutshell, that women in the early modern period were oppressed more by bodily conditions than by social ones. The precondition of female emancipation was gynaecological progress; obstetricians (mostly male) become the unsung heroes of female history. This is of course an anti-feminist history. It should be noted that women's posited terror of childbirth lies at its heart.[9]

In considering the attitudes of mothers, we need to make two distinctions. First, we should distinguish between fear of *pain* and fear of *death*. A mother might be terrified of the pain of childbirth, while being quite confident that she would survive the experience. Second, with reference to death, we must separate analytically the *risk* of death from the *fear* of death. Let us suppose that the rate of maternal mortality was high, by comparison with some standard or other. This establishes a high risk of death associated with childbirth. But did this high risk produce a high level of fear? A modern analogy might be the risks of motoring – a comparison made by Irvine Loudon in a recent review.[10] While we all know that dozens of people are killed on the roads every week, very few of us (I suggest) set out on our daily car journeys in terror. Thus risk does not translate in any simple way into fear.

Advocates of the 'fear thesis' do not always make these distinctions. Instead they tend to combine pain, danger, and the fears of these into a single *gestalt*. As we shall see, this picture tends to break down when a more analytic approach is deployed.

We may usefully begin by considering what may be called the circumstantial evidence. This concerns, firstly, the official religious associations of childbirth, and secondly, the attendant difficulties and risks (to the extent that these can be reconstructed). After considering these lines of enquiry we may turn to more direct evidence bearing upon the 'fear thesis'.

Those mothers belonging to the Church of England were reminded officially that childbirth was dangerous. After every delivery, the 'month' of lying in having been completed, the mother would attend the ecclesiastical ceremony of *churching*. This ritual was a distinctively Anglican construction: it had been transformed at the Reformation from a rite of 'purification' into one of thanksgiving. The heart of this short service was the moment when the minister enjoined the mother to give hearty thanks to God for her 'safe deliverance [...] from the great danger of childbirth'; thus the rite of passage provided by the Church of England explicitly asserted that giving birth was dangerous. If death was a risk, pain was a near certainty. The fact that childbirth was painful was common knowledge: indeed one of the terms for giving birth was 'to cry out'.[11] Men midwives, in the eighteenth century, referred laconically to the contractions of the uterus as 'the pains'. Here too there was a religious rationale, derived from the oft-cited text of Genesis 3. 16: 'Unto the woman he [god] said, I will greatly multiply thy sorrow and thy conception; in sorrow thou shalt bring forth children; and thy desire shall be to thy husband, and he shall rule over thee.' We might imagine that these religious connotations of childbirth – the one ceremonial, the other textual – rammed home to women a message of fear and dread. Yet texts and rituals alike were open to interpretation, and it is by no means clear what effect was exerted upon mothers either by the words of the churching ceremony or by the curse laid upon Eve. As we shall see, different individuals interpreted these cultural resources according to their own specific interests.

Three examples may serve to illustrate the variety of early modern ways of dealing with these religious interpretations of childbirth. First, the two occasions for fear (pain and death) were fused together by one early modern mother, Elizabeth Viscountess Mordaunt, who in 1659 thanked God for 'my safe deliverance from *the pain and the peril* of childbirth'.[12] Second, however, one of Elizabeth's male contemporaries – the obstetric surgeon Percival Willughby of Derby – explicitly separated these. In an interesting gloss on Genesis 3. 16 he wrote: 'God was displeased with Eve, therefore he said, "In sorrow thou shalt bring forth children" – not that he would destroy her'. In other words, Providence was ultimately benign.[13] This is all the more striking coming from a man who had great experience of managing difficult births, including probably dozens of cases where the mother had died either during or after the delivery. (There was nevertheless a reason for Willughby's optimistic appraisal: it supported his wider argument for non-intervention on the part of midwives.) A third example comes from the mid-eighteenth century (1748), when we find Edward Cobden (chaplain to George II), while preaching a charity sermon to the Middlesex Hospital, explaining the pains of childbirth by Eve's sin and giving the dangers of childbirth as the reason for the churching service. So far, so traditional. Yet Cobden immediately went on to praise the recent establishment of lying-in wards at this same hospital (the Middlesex), on the grounds that this would provide opportunities for training more skilful midwives.[14] Thus the religious associations of childbirth were here being used not as fatalistic arguments, but the very reverse. (Particular interests can again be inferred: Cobden was exhorting his hearers to support the Middlesex Hospital's new initiative.)

But mention of the skill of midwives, and of moves to increase that skill, serves to remind us of another strand of the 'fear thesis' – the fact that *some* births did turn out very badly indeed. In fact, it would be easy to draw from a modern audience a gasp of horror at some of the descriptions of difficult births we find in seventeenth- and eighteenth-century sources. The technology of delivery remained, by modern standards, primitive until the introduction in the mid-eighteenth century of new techniques which could save the lives of both mother and child. Until this revolution in obstetric methods, obstructed labours usually entailed the death of the child, and posed much greater risks to the mother than did live births. Hence we can find real horror stories of births from the seventeenth and eighteenth centuries. Such accounts occur particularly in a certain genre of obstetric texts, that is, those written by male practitioners and including case-histories of difficult births. (And we would expect to find them – both because this was the nature of such men's practical experience, and because these stories supported the criticisms of midwives which these male authors were making.) But other sources such as diaries and correspondence confirm that these descriptions were not exaggerated.[15]

The typical case of this kind comprised an obstructed birth coming by the head. (The obstruction itself could be produced in various ways: for instance by mal-rotation of the foetal head in the delivery, or by disproportion between foetal head and maternal pelvis – to which maternal rickety pelvis made a minor contribution.)[16] It is a piquant twist that this form of presentation was *both* the commonest kind of normal delivery *and* the commonest source of difficulty and danger. As a result, it was by no means easy to tell when, or whether, a birth had passed over from being a normal but protracted labour into a case of genuine obstruction. Similarly it was very difficult to tell whether the child was alive or dead. Popular midwifery treatises expounded at some length the various 'signs that a child is dead in the womb'; the very diversity of these signs underscores the difficulty of diagnosis.[17] Eventually, if the birth was truly obstructed, the child would die and would be agreed to have died; and it was at this point that our horror stories would begin, for the method of choice was to extract the dead child with hooks. Such instruments (they took a variety of forms) appear cruel to us, and have been so described,[18] since if they were used on a living child they would usually kill it. But in fact their use was compassionate, for the motive was to save the life of the mother, the child's life having been given up. It was this unpleasant operation which formed the routine work of the obstetric surgeons of the seventeenth century and much of the eighteenth century, both in London and in the provinces.

I should mention that there was a formal alternative, in the operation of Caesarean section. Moreover we need to observe that that operation could be performed in either of two ways – on the living mother (in the hope of saving both lives, those of mother and child), or *post mortem* (to rescue the life of the child, and particularly to rescue its soul by means of baptism). However, while it was found in books and was occasionally practised, this was extremely rare in England before the advent of anaesthesia and antisepsis in the nineteenth century. To all intents and purposes the Caesarean operation was not a workable alternative. The reason for this seems to have been, interestingly, the resistance of women. So far as can be ascertained, early modern women – who controlled the management of childbirth – preferred the mother's right to life over that of the child; and it was probably realized that the Caesarean operation on the living mother was likely to be fatal. Thus, while formally there was a choice of actions, in practice the obstetric surgeon – and those instructing him – could only choose between delivering the dead child or not acting at all.

Hence the fact that case-histories such as those of Willughby, Cooke, Chapman, Giffard, and Smellie abound in deliveries of dead babies using the instrument known as the crotchet or sharp hook.[19] There are also tales of midwives' practices which seem to bespeak cruelty and ignorance; that was the impression the stories were meant to convey, and they succeed in that intention. By piling up such stories one can surround early modern childbirth with an atmosphere of horror. However, we need to attend to the incidence of such cases. It seems to be agreed that difficult births occurred at a rate of around 2 per cent: this is similar to today. The best estimate of maternal mortality comes from the painstaking work of Schofield, who finds a typical level of around 1 per cent of all births: this is perhaps 100 times the modern rate. I can confirm the maternal mortality from more modest researches of my own, using Dickenson's Nonconformist register of the early eighteenth century from the Halifax circuit. The method is tedious and the results are conjectural.[20] The result I have found is around 15 maternal deaths per 1000 births, which corresponds well with Schofield's national findings.

The question which then arises concerns how these rare events affected the attitudes of mothers towards the remaining 98 or 99 per cent of births. Here we have in the literature a radical disagreement, and no apparent method for its resolution. Schofield suggests that a 1-per-cent risk need not have been translated into fear, and thus argues that, on the whole, women were not afraid of giving birth. By contrast Shorter had earlier claimed that fear of difficult births radiated outwards to cast a shadow over women's whole experience of childbirth, and indeed sexuality. There seems to be no clear way of adjudicating between these contradictory interpretations; historians will proceed according to their own inclinations. Thus two advocates of the 'fear thesis' – Pollock and Crawford – have recently argued against Schofield: while not explicitly endorsing Shorter's claims, they have suggested that (in Crawford's words) 'no pregnant woman could be sure that she would be among the fortunate survivors'.[21] It would be easy to invert this reasoning: no pregnant woman could be sure that she would be amongst the small number of unfortunate non-survivors. Hence it seems clear that the risks associated with childbirth are potentially consistent with almost any assessment of women's attitudes.

In short, just as the religious connotations of childbirth give us no clear guidance on the issue, so too there is no agreed way of translating statistical risks into the domain of subjective experience. We must turn to explicit documentation in order to assess the 'fear thesis'.

If women were as terrified of childbirth as the historiography suggests, then the surviving source materials pose a puzzle. For the fact is that attitudes of fear have so far been documented for only a handful of mothers. Admittedly, testimony as to female experience is in general relatively sparse for our period; yet even taking this into account, the evidential basis for the fear thesis appears rather thin.

Some seventeenth-century mothers, on the eve of childbirth, gave explicit instructions to their husbands on how to take care of their surviving children should the birth prove fatal. Three examples have been cited in the literature: Elizabeth Josceline, whose advice was actually published in 1622 as *The Mothers Legacy to her Unborn Child*; Anne Halkett (*née* Murray), who around 1666 wrote a *Mother's Instructions*; and Anne Bradstreet, who wrote a poem 'Before the birth of one of her children'.[22] These mothers were well aware that death in childbed might be their lot, for they were making explicit provision for that event. (Elizabeth Josceline went even further, secretly buying a winding sheet for her burial as soon as she felt the foetus move within her.) It seems obvious that fear of childbirth on the part of these mothers can be inferred. Yet strictly speaking these episodes attest to *awareness of the possibility of death*, which is not the same as fear. Anne Bradstreet's verses, for instance, suggest not fear but rather resignation, combined with a pragmatic solicitude for the welfare of her children, and a wish to be remembered after death.

What then of direct testimony as to fear of giving birth? I can find in the recent literature a total of eleven mothers whose own fear of giving birth is attested: Jane Josselin; Alice Thornton; the Countess of Bridgewater (whose fear was specifically of pain); Katherine Stubbes; Mary Verney; Sarah Fountaine; Jane Hook; Lady Cornwallis; the niece of Richard Chaworth; Anne Harcourt; and Margaret Godolphin (*née* Blagge).[23] (This amounts to rather less than two mothers per historian advancing the 'fear thesis'.) We also find in the literature several instances where fear is felt by others on the mother's behalf: such as the Countess of Warwick on behalf of her niece, cited by Mendelson.[24]

Have historians been justified in generalizing from these instances? Or was Anne Meautys more typical? In 1637 she wrote to her mother-in-law: 'I grow very big and full of pains *expecting continually the hour when it shall please God to send me a safe delivery and to make me a glad mother*.' Similarly, consider the words of Frances, Lady Hatton, a few decades later (*c.*1678): 'I have not stirred out of my chamber this 3 weeks for I have never been well and am troubled in a most violent manner with a great cough *but I trust in God when I am once brought abed I shall be cured of all my illness*.'[25] Here we have not fear but calm confidence – even when 'full of pains' in one case and 'troubled in a most violent manner with a great cough' in the other. These quotes both come from a historian who elsewhere advances the 'fear thesis', but who has cited these passages in a different connection and has thus not noted the evident confidence of these particular mothers. I suspect that a search of manuscript correspondence would yield many more such instances, though I have not undertaken this particular task.

In default of such a search, what we can do relatively easily is to examine systematically the best-documented series of deliveries of particular mothers. There are two such series which command our attention: the deliveries of Alice Thornton between 1652 and 1667, recorded by herself, and those of Jane Josselin, running between 1646 and 1663, written down by her husband, the diarist Ralph Josselin. Both women are amongst those cited in support of the 'fear thesis'; both sources are available for consultation in print.[26] I shall discuss them in turn, beginning with the Josselin series of which more use has been made by historians.

Of Jane Josselin's ten deliveries the first two were described only in retrospect, since Ralph began his diary only after these births had taken place; but the remaining eight births were recorded in some detail in the diary. These births, as recorded in this document, have been quoted as evidence of maternal fears, and with good reason. Consider for example the first of the detailed descriptions, that is, the third delivery: 'I had sought to god for my wife (*that was oppressed with fears that she should not do well on this child*) that god would order all providences so as we might rejoice in his salvation'. However, as we shall shortly see, this quote is far from representative of Jane's experiences. In fact, a comprehensive examination of Ralph's accounts of Jane's pregnancies and deliveries yields several novel findings. First, while we do find fear of childbirth, we also encounter its opposite. On the one hand (this example some seven months into her fourth pregnancy, in December 1647): 'my dear wife under great fears she shall not do well of this child'; and again a week later, Jane: 'lieth under many sad fears by reason of her approaching travail'. Ralph adds, in this diary which we should remember was addressed not to historians but to God: 'Teach her to trust in thee.' On the other hand, a few weeks later, with the labour approaching, and as if in answer to Ralph's prayer: 'The lord gracious and merciful to my wife, who is more cheerful'; and another week later Jane is 'still holding up her head'.[27] These were Ralph's standard expressions to describe the pregnant Jane as being in a state the opposite to fear – *cheerful*, and *holding up her head*. Their presence in the diary shows that historians have painted a one-sided picture of Jane's experiences, focusing on fear and neglecting its opposite.

Second, in crude numerical terms instances of 'cheerfulness' and the like actually outweigh those of fear – by more than two to one. That is, in all her pregnancies combined we find only seven instances of Jane's being afraid, but some nineteen occasions on which Ralph reported her as 'cheerful' or 'cheerly' or as 'holding up her head'.[28] Of course human experiences cannot be weighed in an arithmetical balance; nor can we assume that Ralph produced a complete record of either of Jane's contrasting emotional states. Yet the numerical predominance of cheerfulness suggests a *prima facie* case against the 'fear thesis' as this has generally been enunciated. Table 1 sets out the incidence of Jane's different attitudes as recorded in Ralph's diary, pregnancy by pregnancy. This shows that cheerfulness predominated in this way within most of Jane's recorded pregnancies considered separately (six out of eight); and that in two or three of those pregnancies (her sixth, seventh, and possibly fifth) fear was not mentioned at any stage.[29]

Table 1. Jane Josselin's attitudes to childbirth, pregnancy by pregnancy.

Birth number	*Date*						
3	Dec. 1646	183	81	15	8	1	-0
4	Feb. 1647/48	-68	-61	19	12	5	
5	July 1649				76	13	
6	Dec. 1651		201	40	33	5	
7	Sept. 1654				93	2	
8	May 1658				-170	-2	-0
9	Sept. 1660						-0
10	March 1663/64				11	4	-0

Key: Numbers refer to number of days before the delivery; 'o' means in the course of the delivery. A '-' sign indicates fear; absence of a sign, 'cheerful' or 'holds her head up'.

Third, attention to context indicates that some of Jane's fearful states are intelligible as responses to specific, temporary occasions. This applies to the examples quoted above, from her fourth pregnancy: on 5 December 1647, when as we have seen Jane was 'under great fears she shall not do well of this child', Ralph's account immediately continued: 'ill in her head and back, one night ill as if she should have died'. It was doubtless the seriousness of this illness which produced her fears. Correspondingly, the lifting of those fears coincided with an improvement in Jane's health; what Ralph actually wrote (from which I quoted selectively above) was: 'The lord gracious and merciful to my wife, who is more cheerful *and free from headaches* than formerly.' Thus fear of giving birth was not a continuous, diffuse state, produced automatically by pregnancy itself, but rather was contingent upon particular circumstances.

Fourth, and related to this, Ralph's diary enables us to distinguish between two different occasions for fear: that is, fear *in advance* of the birth, and fear experienced *during the birth itself*. (In Jane's case, we find three recorded instances of fear in advance, and four of fear during labour.)[30] As might be expected, fear during the birth is generally associated with a slow, painful or difficult labour – as were all of Jane's last three deliveries. The difference between these two occasions for fear may correspond to the distinction suggested earlier, between fear of pain and fear of death. Fear during pregnancy was perhaps focused upon future pain, while fear during the birth itself may have been attached to the possibility of death. However, no such distinction, nor any explicit characterization of the nature of Jane's fears, can be found in Ralph's diary.

Finally, it does not seem that a painful and difficult labour led Jane to become afraid during the next pregnancy; despite the terrible eighth labour, in which she was 'afraid and amazed' and then experienced 'fear which made her quake and tremble', Ralph's record of the next pregnancy includes no references to fear – nor indeed to cheerfulness, as if this was not an issue at the time. The ninth labour in its turn made her 'fearful', but the tenth pregnancy was free from fear, and even during its final month we find Jane 'cheerful'.

I have been following traditional practice in taking Ralph Josselin's diary as a simple and reliable record of his wife's experiences. In fact, of course, Ralph may not have been the neutral recording instrument that we as historians might wish: his own state of mind may have influenced what he wrote down about Jane's experiences. And in this connection it is worth observing that Ralph himself on one occasion expressed fear towards an impending birth. About five months into Jane's fourth pregnancy Ralph had a startling experience in the pulpit. On 17 October 1647 he was preaching – apparently on the text of Genesis 35, concerning God's dealings with Jacob (now renamed Israel) and Jacob's journey from Bethel to Ephrath. In the course of this journey Rachel fell into 'hard labour', giving birth to a boy but herself dying shortly afterwards; as she expired she named the boy Ben-oni, which means 'son of my sorrow'.[31] Ralph wrote in his diary for this day: 'Preached, and expounding, wherein, on the consideration of Rachel's calling her child Ben-oni and on the contemplation of this fact, such a passion surprised me that I could not well speak, *the Lord in mercy preserve my dearer than Rachel.*' Perhaps it was no coincidence that in this same pregnancy, some seven weeks later, Ralph recorded that Jane was 'under great fears'. The diary tells us about *Ralph's responses to* Jane's feelings; he may well have been a more sensitive observer in the wake of his own intimation of danger. And we have other possible indications of Ralph's active contribution to the record. In the next (fifth) pregnancy Ralph never mentioned fear directly but on three occasions wrote that (for instance) Jane is 'more cheerful today than divers days'.[32] Thus in the early stages of this particular pregnancy, the issue of fear versus cheerfulness was present yet was not recorded in any easily interpretable way. Further, Ralph's references to Jane's attitudes became less frequent in the later pregnancies (Table 1); this suggests that the issue was becoming less important – perhaps to Jane, perhaps to him, perhaps to them both. The extreme instance of this was the ninth pregnancy, when the diary is devoid of any reference to the matter. Does this absence mean that Jane was 'cheerful', that Ralph's own anxieties about childbirth had diminished, that Ralph had other preoccupations, or what?

We should not therefore seek to attach any rigid meaning to the precise numerical balance between Jane's contrasting stages of fear and cheerfulness. Nevertheless it seems clear that the 'fear thesis' in its simple form is not sustained by the Josselin diary. Rather, Jane Josselin's attitude was complex, involving sometimes fear but more often a 'cheerful' confidence. Since Jane was one of the eleven mothers whose fears were directly documented, the 'fear thesis' must be severely qualified by these findings. It also appears that we cannot generalize even about the experiences of one woman, let alone the several hundred women who gave birth each day in seventeenth-and eighteenth-century England.

Let us now examine Alice Thornton's experiences in her nine pregnancies, bearing in mind the distinction between fear in advance and fear during the birth to which Jane Josselin's experiences have alerted us. Alice mentions 'fear in advance' in only two (the sixth and eighth) of her nine pregnancies. The other seven seem to have been quite free of fear. Fear during the birth itself she does not mention at all – even though some of her deliveries were both difficult and dangerous, as we shall see. Once again, then, fear of birth emerges as an occasional attitude rather than a universal one.

Why did Alice become afraid in her sixth pregnancy? And why not earlier? From her very first pregnancy onwards, she had a recurrent pattern of accidents in pregnancy and difficulties after delivery, and her first child died almost immediately after the birth. However, it would appear that initially the *birth itself* was not difficult. Table 2 sets out the incidence of her various obstetric troubles pregnancy by pregnancy, and shows that the fifth birth was the first where Alice recorded any difficulty.

Table 2. Alison Thornton's difficulties in pregnancy, childbirth, and lying-in.

Birth number	*Date*	*Troubles in pregnancy*	*Fear of birth*	*Difficult birth*	*Death of child*	*Sickness after delivery*
1	Aug. 1652	+			+	+
2	Jan. 1654					
3	Feb. 1655					+
4	June 1656					
5	Dec. 1657	+		+	+	+
6	April 1660		+	+	+	+
7	Sept. 1662	+		+		+
8	Sept. 1665		+	+		+
9	Nov. 1667	+		+	+	+

Key: A '+' sign indicates that the stated problem occurred in the given pregnancy

Notes:

6th birth: (*a*) The death of the child occurred 2 weeks after the delivery, but was linked with the birth and pregnancy by Alice. (*b*) Her sickness after delivery was very slight, i.e. 'some weakness upon my body by reason of the return of my milk' after the child's death (p.127).

9th birth: (*a*) Alice's condition in pregnancy was initially (February to May) not good, but thereafter 'very well as of any other' until within a month of delivery (i.e. June to October); then, in the final month of pregnancy, she became ill again. (*b*) The child's death occurred 19 days after delivery.

Specifically, this was a breech delivery, so occasioned by a 'great fall' Alice experienced six months into the pregnancy. The midwife did not have 'skill to turn him right, which was the cause of the loss of his life [the child, a boy, died half an hour after birth], and the hazard of my own'. The delivery was also very painful: 'I was upon the rack in bearing my child, with such exquisite torment, as if each limb were divided from [the] other, for the space of two hours; when at length, being speechless and breathless, I was, by the infinite providence of

God, in great mercy delivered.' While Alice had not feared this delivery, her mother 'feared me much' – that is, was very afraid on her behalf – and therefore composed a special prayer, a copy of which appears in the autobiography. There is no indication that Alice herself shared this fear at this time. Yet as we have seen, in her next pregnancy Alice's previous apparent confidence collapsed; perhaps the confirmation of her mother's fears in the fifth birth was partly responsible. (Notice that if difficulty in the fifth birth did indeed lead to fear in the next pregnancy, then Alice was unlike Jane Josselin.)

However, Alice herself gives a quite different reason to explain the great fears which beset her in her sixth pregnancy. During this pregnancy, she had a very disturbing dream:

> that I was laid in child-bed, had the white sheet spread, and all over it was sprinkled with small drops of pure blood, as if it had been dashed with one's hand, which so frightened me that I told my aunt of it in the morning; but she put it off as well as she could, and said dreams was not to be regarded; but I kept it in my mind till my child died.

Thus we have three possible reasons, in the autobiography, to explain the fact that Alice now became afraid of birth: her mother's earlier fears on her behalf; the trauma of the previous birth; and the morbid dream in this pregnancy.

As a direct result of her fears, Alice took special steps, in advance of this delivery, to secure her property for her surviving children in case she should die in the course of the birth. (She also seems to have changed her midwife, and perhaps for that reason lay in not at home but at St Nicholas, some distance away.)[33] As for the birth itself, though this was apparently not as severe as the previous delivery had been, Alice did describe it as 'hard labour and hazardous'.

One might have expected that Alice would be afraid of the next birth (the seventh). Yet there is no indication of this in the autobiography, and indirect evidence suggests that she was not afraid.[34] However, this seventh pregnancy was a very difficult one, in which Alice seems to have had every possible trouble except fear; and the birth was accordingly very difficult, so that Alice speaks of 'great danger and peril of my life in travail'.

Perhaps for this reason, the next (eighth) pregnancy saw a return of the fears which had racked Alice during the sixth pregnancy: she became 'terrified with my last extremity', as she later described it.[35] Accordingly she took the sacrament, and (as in the sixth pregnancy) again made preparations for the welfare of her children in the event of her death. As it turned out, she was 'one day and night in travell', had 'an exceeding sharp and perilous time', and (as in the previous delivery) 'escaped very narrowly the blow of death'. However, her recovery after the delivery was unusually swift, and she adds that 'it pleased the Almighty [...] to give me great comfort in the nursing of this sweet child, enabling me with pretty strength to go through this duty'.[36]

Alice'a autobiography gives no indication of fear during her next and final pregnancy and delivery (her ninth). Thus, even after fear had begun to cast a shadow over her childbearing, that shadow could and did lift. Moreover, even when she did experience fear this did not necessarily dominate the pregnancy as a whole. We know that in the eighth pregnancy it was not until (in Alice's words) 'about a fortnight before my delivery' that she began to be afraid; thus for probably over six months when she knew she was pregnant, Alice was apparently free of fear. Whatever way we look at her experiences, we find that Alice Thornton, like Jane Josselin, was mostly not afraid of childbirth.

In short, the direct evidence for the 'fear thesis' is unconvincing. Furthermore, against the scattered cases of documented female fear we need to set rather a lot of silences on the subject. Thus the obstetric surgeon Percival Willughby, practising in the mid-seventeenth century, recorded only a few instances of maternal fears amongst over a hundred deliveries.[37] The same emerges from those early eighteenth-century male diaries and collections of correspondence which I have managed to search.

The *Remarks and Collections* of Thomas Hearne, in the 1720s and '30s, include quite a few references to childbirth – some stories of difficult births, deaths in childbed, a normal delivery and lying-in, the names of some of the midwives of Oxford, a discussion of the notorious Mary Toft case – but give no indication that birth was associated with fear.[38] Swift's *Journal to Stella* records, in August – September 1711, several details of the lying-in of Mrs Masham, including the fact that she was 'not well' some four weeks after the birth. Though he himself was concerned in advance of the birth – saying 'God send her a good time; her death would be a terrible thing', i.e. terrible for Swift's interests at court – he does not suggest at any stage that Mrs Masham herself was afraid. A little over a year later (November 1712) Mrs Masham was delivered again, and here Swift nonchalantly chronicles the progress from late pregnancy through delivery and lying-in, never hinting at any fears.[39] Similarly a trawl through one of the later volumes of Swift's correspondence yields three mentions of impending or recent births, none of them associated with fear.[40] Again, Lord Egmont, in his diary in early 1738, mentioned quite casually the delivery of his 'daughter Perceval', that is, his daughter-in-law. (He was well aware of the dangers of childbirth, having recorded less than two years earlier a craniotomy performed during the delivery of the Princess of Orange, and still more recently a maternal death in childbed.)[41]

But perhaps the most remarkable evidence against the 'fear thesis' is the testimony of eighteenth-century women as revealed in their personal correspondence with one another. For we find that some mothers, mothers-to-be, and their female friends could treat the subject of childbirth in a spirit of jest. In 1721 Mrs Campbell – apparently pregnant, and discussing the matter in a letter to her friend Mrs Howard – wrote: 'I [...] have done no one thing relating to my progeny. I think of taking you for my nurse, and Miss Hawley [another of her friends] for midwife: if strength will be of use, she can perform that office.' Two years later, when seven months pregnant, she was again writing to Mrs Howard and made a joking implicit comparison between her own condition and that of the 'four white sows big with child' which were amongst the stock of her farm in Kent.[42]

A similarly lighthearted attitude was shown by Mrs Delany. At the end of November 1745 her sister Mrs Dewes (Ann Granville) must have been almost seven months pregnant, for she was delivered of a girl on 11 February following. Yet at this time, writing to Ann, Mrs Delany could say: 'When you are brought to bed I beg I may have an account sent me by the X post and by the other also, for fear one should miscarry.'[43] Would the word 'miscarry' have been used so lightly, when referring to a letter, between women for whom birth was an object of fear? Let it be remembered that Mrs Delany was peculiarly aware of nuances of meaning; sensitive of the feelings of others; and cautious about lying-in women, believing that they should keep calm and quiet for as long as possible after the birth.[44] A few years earlier (May 1739) she jested that her great friend the Duchess of Portland was 'in a multiplying way'. The Duchess was then seven months pregnant, and around this time herself casually observed: 'I shall be kept in town for *the old occasion* all this summer.'[45] Men, too, could treat the matter in this spirit: in 1765 we find the Earl of Buckingham joking, in a letter to Lady Suffolk, about his wife's pregnancy.[46]

To sum up so far: For the seventeenth century, the 'fear thesis' rests upon rather flimsy evidence. For the eighteenth century, we can find instances which actively contradict it.

The tenacity of the 'fear thesis' perhaps results from a certain way of proceeding on the historian's part. If we are concerned to recover early modern attitudes, it is entirely natural to be struck by stark expressions of fear such as those of Jane Josselin recorded by her husband Ralph. These expressions will then be reproduced by the historian, while less striking pregnancies will be omitted. The result is a selective record, one which suppresses instances going against the 'fear thesis' – both pregnancies where fear is simply not mentioned, and instances of actual cheerfulness on the part of the mother. Purple passages of fear are selectively preserved; once sundered from context, they appear to comprise a complete or representative record, which as we have seen is very far from being the case.

The historian then interprets these results according to two principles, or rather assumptions. The first could be called 'the assumption of homogeneity': it is assumed that there was some generalized 'popular attitude' towards childbirth amongst early modern mothers. Given this assumption, and the selectiveness already described, that attitude of course has to have been one of fear, terror, dread. The second is 'the assumption of difference': the findings are read as indicating a difference between past and present. Interestingly, this actually builds into its picture of the past a value-judgement and an implicit picture of the present. The value-judgement is that things ought not to have been as they were: that childbirth ought to have been free of fear. The picture of the present – all the more potent for being left implicit – is that by and large, things are better today. (In fact, it can scarcely be doubted that some mothers approach childbirth with great fear today.[47] Nor are the attitudes of women homogeneous in the present, any more than they were in the past.)

So far we have been narrowing the scope of maternal fears in early modern childbirth. However, I shall now suggest that these fears were actually wider than has usually been suggested. Specifically, in the particular context of difficult births, there was something else which mothers feared – something different from either pain or death, albeit connected with these; namely, *the intervention of the male practitioner*.

Earlier we saw that prolonged obstructed labour led to the death of the child, after which the mother's life could only be saved by surgical intervention – specifically, by an operation which would have killed a living child. We saw, too, that women *preferred* this operation to the formal alternative – to be found in textbooks – of Caesarean section. However, this was a 'preference' exercised only *in extremis*: what mothers really wanted was to avoid any such operation. That is, they regarded the operation performed by the male practitioner with a profound fear and loathing. Of this there is ample evidence, both direct and indirect, in the writings of these male practitioners. Willughby, for instance, wrote that 'instruments be dreadful to them'; Cooke, that 'women will seldom admit of such operations'.[48] The indirect evidence comes from the fact that male practitioners were usually only summoned to difficult births after very long delays – typically of the order of three days – and when the child was dead. (We do find cases where the child was still alive; generally, however, it was believed to have died before the male practitioner was summoned.)

Surely mothers and the women attending them – midwives, friends, relatives, and neighbours – wanted at all costs to avoid believing that the birth had come to such a pass as this. What helped them to keep up hope was that the obstructed birth usually presented just like a normal birth, that is, by the head. Some births were swift, others slow; if the delivery took a long time, this did not yet mean that all hope was lost. Thus mothers protracted their hopes, to keep their fears at bay. So long as the male practitioner was not summoned, so long as the birth was in the hands of the midwife, so long as the normal childbirth ritual was continuing – a ritual centring on the exclusion of men and on collective female companionship and conviviality – then hope could be kept alive.

On this reading, fear would have started to rise only after many hours of unsuccessful labour. There would be a tension, a struggle between hope and fear, within the mother and no doubt on the part of the other women present. Conflict amongst the women might sometimes arise (we have a few documented examples of this) over whether and when to send for the male practitioner, that is, to give up all hope for the life of the child.[49] The end result was variable. Some mothers actually died undelivered, the male practitioner having been summoned too late to save them. At the opposite extreme we find some cases where the surgeon was summoned while the child was still alive – though as I have mentioned it was usually believed to have died. In the middle were the great majority of cases seen by male surgeons – that is, cases calling for the prompt delivery of a dead child, with the aim of saving the mother's life.

Here, I suggest, is the real location of fear in early modern childbirth – in and around difficult births, rather than birth in general. If this interpretation is correct, it would go to support Schofield's argument and would again cast doubt upon the 'fear thesis'.

All this was doubtless changed by the new techniques for delivering difficult births which gradually became more widely available in the early eighteenth century – the midwifery forceps, probably also the vectis, and Deventer's method.[50] Each of these methods (of which only the forceps survives today) permitted the surgeon to deliver a *living* child in cases of obstruction. The effects must have been dramatic. Once mothers knew that a local male practitioner had this capacity, they could summon him not in fear but in hope. This led to earlier calls, with the effect that more babies survived, which in turn further promoted the process of earlier calling. Hence Giffard, practising with a forceps in London during the 1720s, was soon receiving emergency calls after hours instead of days, and to deliver a living child rather than a dead one. It was precisely this change which turned the 'surgeon' into the 'man-midwife' – the man who did what midwives did, that is, delivered living children.

Mothers' fears of the male practitioner, then, lifted dramatically between about 1700 and 1740 – at first chiefly in London, and a little later elsewhere. And it is tempting to suggest that this also produced a wider shift in women's attitudes to childbirth. We have seen that seventeenth-century mothers were by no means as terrified as historians have suggested; yet it remains possible that they were more afraid than their descendants in later generations. I have depicted the struggle between hope and fear as taking place in cases of obstruction, which comprised some 2 per cent of births – but that struggle may have arisen in other deliveries which in the end turned out as 'normal'. Jane Josselin, though seldom afraid in advance of the birth, became terrified during four of her eight deliveries: in the event, no male practitioner was summoned, and each of these babies was born alive, yet it may well have been the *possible* calling of a surgeon which she feared. By contrast, from the eighteenth century, it will be recalled that Porter and Porter were puzzled to find little evidence of maternal fears; and we have encountered instances of lightheartedness and jocularity. Was there, then, a transformation of attitudes – from fear to hope – after about 1700? This hypothesis deserves to be tested against such materials as family correspondence.[51]

Attitudes, both past and present, were and are more complex and variable than is permitted by a '*mentalité*' interpretation such as the 'fear thesis'. Once we allow for this complexity, it becomes possible to see where fear was present and where it was not – and thus to assess its concrete importance. In particular, I have suggested that mothers' fears in childbirth were especially focused on the intervention of the male obstetric surgeon. This helps to explain the specific role which such surgeons could play before they metamorphosed into the 'men-midwives' of the eighteenth century. Relatedly, it suggests that women's attitudes may have changed after about 1700, a possibility which deserves further investigation. And it indicates that even in the seventeenth century, most pregnancies and births were free of fear – and thus that early modern childbirth was often, *pace* the historians, an occasion for joy.

Adrian Wilson
University of Leeds

1. For advice and encouragement in preparing this paper, I am grateful to Timothy Ashplant, Lois Chaber, Brean Hammond, and Bill Speck. Earlier versions were given at the BSECS

conference in January 1991 and to a staff seminar in the University of Leicester History Department; I thank the participants for their comments in discussion. Special thanks go to Ludmilla Jordanova, who read the draft text in full and made very helpful suggestions which have led to significant revisions. Place of publication of all books cited is London unless otherwise stated. In quotations from early modern sources, punctuation and spelling have been modernized.

2. Roy Porter and Dorothy Porter, *In Sickness and in Health: The British Experience, 1650-1850* (1988), p.100.
3. Roger Schofield, 'Did the mothers really die? Three centuries of maternal mortality in "the world we have lost"', in *The World we have Gained*, ed. by Lloyd Bonfield, Richard Smith, and Keith Wrightson (1986), pp.230-60, at p.260; Irvine Loudon, essay review of Porter and Porter, *In Sickness and in Health, Medical History* 34 (2), (1990), pp.213-15, at p.214.
4. Porter and Porter, *In Sickness and in Health*, pp.100, 77, 85, 86.
5. Alan Macfarlane, *The Family Life of Ralph Josselin, a Seventeenth-Century Clergyman* (Cambridge, 1970), p.84; Lucinda McCray Beier, 'In sickness and in health: A seventeenth century family's experience', in *Patients and Practitioners: Lay Perceptions of Medicine in Pre-Industrial Society*, ed. by Roy Porter (Cambridge, 1985), pp.101-28, at p.104. Beier's support for the fear thesis is qualified, as she also writes that Jane Josselin 'accepted her obstetrical fate with equanimity and interest'.
6. Patricia Crawford, 'The construction and experience of maternity in seventeenth-century England', in *Women as Mothers in Pre-Industrial England: Essays in Memory of Dorothy McLaren*, ed. by Valerie Fildes (1990), pp.3-38, at p.22; Antonia Fraser, *The Weaker Vessel: Woman's Lot in Seventeenth-Century England* (London, 1984), chap. 4 and p.77; Sarah Heller Mendelson, 'Stuart women's diaries and occasional memoirs', in *Women in English Society*, ed. by Mary Prior (1985), pp.181-210, at p.196; Linda Pollock, 'Embarking on a rough passage: the experience of pregnancy in early-modern society', in *Women as Mothers*, ed. by Valerie Fildes, pp.39-67, at p.47.
7. Mendelson, 'Stuart women's diaries', p.197; Crawford, 'The construction and experience of maternity', p.22. Pollock implicitly endorses this view: 'Embarking on a rough passage', p.63, note 54 to p.48.
8. For a different reading of the childbirth ritual, see Adrian Wilson, 'The ceremony of childbirth and its interpretation', in *Women as Mothers*, ed. by Valerie Fildes, pp.68-107.
9. Edward Shorter, *A History of Women's Bodies* (New York, 1982).
10. Loudon, op. cit. (note 3 above), p.214.
11. *OED*, 'cry', verb, 21c; 'crying', vbl. sb., 2.
12. Quoted in Fraser, *The Weaker Vessel*, pp.65 and 72. Here and throughout, all emphases are my own.
13. Percival Willughby, *Observations in Midwifery*, ed. by Henry Blenkinsop (Warwick, 1863; reprinted with introduction by John L. Thornton, Wakefield, 1972), p.13. This treatise was written between about 1660 and 1672; it is cited hereafter as Willughby, *Observations*.
14. Edward Cobden, *The Parable of the Talents. A Sermon*... (1748), pp.15-16.
15. For example, *The Reverend Oliver Heywood, B.A., 1630-1702; his Autobiography, Diaries, Anecdotes and Event Books*, ed. by J. Horsfall Turner, 3 vols (Brighouse, 1881-83), II, p.167: Mistress Earnshaw of York 'was in sore labour, had her child pull'd from her by piecemeal, died at last'.
16. The next three paragraphs summarize a picture developed in more detail in my *A Safe Deliverance* (Cambridge, forthcoming). See also Audrey Eccles, *Obstetrics and Gynaecology in Tudor and Stuart England* (1981); and Jean Donnison, *Midwives and Medical Men: a History of Inter-Professional Rivalry and Women's Rights* (1977).
17. See, for example, Jane Sharp, *The Midwives Book* (1671), pp.188-89, where some twelve signs were given. The last of these was that 'Her breath stinks if the child have been dead two or three days' – one indication amongst many of the long delays in calling for male practitioners.
18. Keith Thomas, *Religion and the Decline of Magic: Studies in Popular Beliefs in Sixteenth- and Seventeenth-Century England* (first published 1971; Harmondsworth, 1978), p.15.

19. Willughby, *Observations*; James Cooke, *Melleficium Chirurgiae, or the Marrow of Chirurgery* (1685 edn); Edmund Chapman, *An Essay toward the Improvement of Midwifery* (1733); William Giffard, *Cases in Midwifery* [...] *Revis'd and Publish'd by Edward Hody* (1734); William Smellie, *A Treatise on the Theory and Practice of Midwifery* (3 vols, 1752, 1754, 1764), vols 2-3 (entitled *A Collection of Cases in Midwifery*).

20. The Reverend Thomas Dickenson's register is included in Heywood's materials, ed. by J. Horsfall Turner (cited in note 15 above), vol.I. For a short period in the early eighteenth century (1705-1709) Dickenson recorded explicitly a large number of maternal deaths 'in childbed' – large enough to suggest the possibility that he was recording *all* of these which came to his notice. (Very possibly his concern over this arose for personal reasons: the period in question began just before his marriage and continued through his own wife's first three pregnancies.) To assess the incidence of maternal mortality, it is necessary to link these deaths with the appropriate 'catchment set' of births. The usual way of doing this is to use recorded baptisms as a substitute for births, but this is not possible here. Dickenson's register includes so few baptisms (compared with either burials or with marriages) as to make it certain that most of those to whom he was ministering were not calling upon him to baptize their children. (Either they were not practising infant baptism at all, or they were using the Anglican church for the specific purpose of infant baptism.) But marriages can be used as an alternative – provided that we can make a guess as to the numbers of births per marriage. Such a guess can indeed be made using the findings of E. A. Wrigley and R. S. Schofield, *The Population History of England: A Reconstruction* (Cambridge, 1981). It is on this basis that the estimate of 15 maternal deaths per 1000 births has been reached.

21. Crawford, 'The construction and experience of maternity', p.22; Pollock, 'Embarking on a rough passage', pp.47-48.

22. For Josceline and Halkett, see Fraser, *The Weaker Vessel*, pp.76, 77; for Bradstreet, Porter and Porter, *In Sickness and in Health*, pp.99-100. Josceline is also mentioned by Crawford, 'The construction and experience of maternity', p.22, and by Pollock, 'Embarking on a rough passage', p.47.

23. Jane Josselin: Beier, 'In sickness and in health', p.104, and Pollock, 'Embarking on a rough passage', p.47. Alice Thornton and the Countess of Bridgewater: Mendelson, 'Stuart women's diaries', p.106. Katherine Stubbes: Crawford, 'The construction and experience of maternity', p.22. Mary Verney, Sarah Fountaine, Jane Hook, Lady Cornwallis, and Richard Chaworth's niece: Pollock, 'Embarking on a rough passage', p.47. Anne Harcourt and Margaret Godolphin: Fraser, *The Weaker Vessel*, pp.77-78.

24. Mendelson, 'Stuart women's diaries', pp.196-97.

25. Quoted by Pollock, 'Embarking on a rough passage', pp.44, 46.

26. *The Autobiography of Mrs Alice Thornton, of East Newton, Co. York*, ed. by Charles Jackson (Surtees Society Publications, vol. LXII, 1873) (cited hereafter as Thornton, *Autobiography*), pp.84-98, 123-27, 139-51, 164-67; *The Diary of Ralph Josselin 1616-1683*, ed. by Alan Macfarlane (British Academy Records of Social and Economic History, New Series, III, 1976) (cited hereafter as Josselin, *Diary*), pp.37-50, 101-11, 145-65, 231-57, 313-25, 399-415, 453-65, 496-502.

27. Josselin, *Diary*, pp.108, 110.

28. Excluded here are cases of 'qualms', which may appear to indicate fear but in fact must refer to minor pains: see Josselin, *Diary*, p.237 (2 March 1650/51), where Ralph writes 'my dear wife apt to qualms, but pretty cheerful'. Some ambiguity attends the fifth pregnancy, as we shall see, and I have therefore excluded from the count of instances of 'cheerfulness' the three from this pregnancy. Retaining them would have made for 22 instead of 19 such instances.

29. The ambiguity concerning the fifth pregnancy is discussed below.

30. In this respect it is difficult to interpret Ralph's account of the third birth (the first in the directly recorded series), which has been quoted above. His phrase 'I *had sought* to god for my wife (that was oppressed with fears)' may seem to indicate that the fears preceded the delivery. However, we find no mention of fear before the date of delivery, and moreover Jane 'held up her head well' the very day before the delivery, just as she had 'held up' a week earlier (Josselin, *Diary*, pp.49-50, 16 and 23 November 1645). I therefore infer that Jane's fears were experienced during the birth itself. Ralph's wording probably means that his entire account

of the birth was written down a little later. Because that account recorded the success of Ralph's prayers, it was rhetorically appropriate for him to begin it with mention of those prayers and of the occasion for them, even though the fears themselves (on this reading) took place later than his wording implies.

31. Genesis 35. 16-19.

32. 23 January 1648-49. The other two such days were 27 December and 28 January: Josselin, *Diary*, pp.150, 154f.

33. The midwife for Alice's second delivery was Jane Rimer; for the sixth birth, a Mrs Hickeringill appears to have performed this office. See Thornton, *Autobiography*, pp.91, 123.

34. As we shall see, Alice took precautions for the future of her children in both the sixth and eighth pregnancies, but not in the seventh.

35. Thornton, *Autobiography*, p.145.

36. Thornton, *Autobiography*, pp.147-48.

37. Willughby, *Observations*, pp.37, 41, 111: (1) An unnamed gentlewoman, whose first delivery had been very painful, was in her second pregnancy 'much disquieted with fears, and wept to think what she had suffered and was likely again to undergo'. (2) Lady Byron, in 1661, 'was troubled with great fears, never having any child afore'. (3) A 'good woman' of Cheshire, who had already 'had several children drawn from her body by the chirurgions at London', was again pregnant in 1650. Throughout her pregnancy 'she was heartless, and dejected, and conceited [i.e. convinced] that she should die in the child bed of this child. And for that cause fitted herself afore hand for her departure by taking the Holy Communion'. (In the event, she was delivered of a dead child in December and died herself soon afterwards.)

38. See *Remarks and Collections of Thomas Hearne*, ed. by C. E. Doble and others, 11 vols (Oxford, 1885-1921), as follows: Difficult births, VI, 338; VIII, 377. Deaths in childbed, VIII, 171; IX, 271; X, 257, 414, 446. Delivery and lying-in of Mol Clements, wife of John Keil, VI, 251, 261. The Mary Toft case, IX, 229, 234, 245-46, 259-60. Oxford midwives: Mrs Tredwell, VIII, 377; Mrs Kite, IX, 375-76; Mrs Holdship, XI, 90-91.

39. Jonathan Swift, *Journal to Stella*, ed. by Harold Williams, 2 vols (Oxford, 1948), pp.333, 337, 348, 351, 352, 357, 363, 364 (I quote from pp.363, 333); and pp.569, 573, 580, 584.

40. *The Correspondence of Jonathan Swift, Vol. 4, 1732-36*, ed. by Harold Williams (Oxford, 1965), p.252 (Lady Dysart and Lady Weymouth); pp.353 and 379 (Margaret, Duchess of Portland).

41. Historical Manuscripts Commission, *Manuscripts of the Earl of Egmont: Diary of Viscount Percival Afterwards First Earl of Egmont*, 3 vols (1920-23), II, 463 (and 467); 318-19; 437.

42. *Letters to and from Henrietta, Countess of Suffolk, and her Second Husband, George Berkely; from 1712 to 1767*, 2 vols (1824) (cited hereafter as *Suffolk Correspondence*), I, 82-83 and 104-105.

43. *The Autobiography and Correspondence of Mary Granville, Mrs Delany*, ed. Lady Llanover, 3 vols (1861) (cited hereafter as *Delany Correspondence*), II, 425. The 'X post' was the cross post – i.e. the direct post, avoiding London, recently organized by Ralph Allen of Bath. The date of delivery is inferred from the fact that 19 February was 'your ninth day' (p.425).

44. For Mrs Delany's 'caution about lying-in ladies' see *Delany Correspondence*, II, 288, 299.

45. *Delany Correspondence*, II, 53, 50.

46. *Suffolk Correspondence*, II, 298-99.

47. Certainly this seems to have been so in the relatively recent past. My title is adapted from that of Grantly Dick Read's famous book *Childbirth without fear*. This was first published (as *Revelation of Childbirth*) in 1942, and acquired the title *Childbirth without fear* in 1954. Read was of course arguing that childbirth need not involve fear. The reason he was saying this was precisely that many mothers did approach childbirth with fear at this time.

48. Willughby, *Observations*, p.164; James Cooke, *Mellificium Chirurgiae* (1648 edn, pp.251-52.

49. See Willughby, *Observations*, pp.38, 55.

50. Again I draw here upon *A Safe Deliverance* (note 16 above). See also my papers 'William Hunter and the varieties of man-midwifery', in *William Hunter and the Eighteenth-Century*

Medical World, ed. by W. F. Bynum and Roy Porter (Cambridge, 1985), pp.343-69; and 'The politics of medical improvement in early Hanoverian London', in *The Medical Enlightenment of the Eighteenth Century*, ed. by Andrew Cunningham and Roger French (Cambridge, 1990), pp.4-39, at pp.34-36.

51. The main difficulty with this hypothesis is that we do not know how attitudes in pregnancy (fear or otherwise in advance of birth) were related to attitudes during the delivery (such as fear of the male practitioner during difficult births). The experiences of Jane Josselin and Alice Thornton suggest that the two could be almost independent. In addition, there are other ways of explaining the apparent reduction in fear between the seventeenth and eighteenth centuries. It might simply reflect a more general sea-change in attitudes to death, with Puritan seriousness giving way to eighteenth-century *insouciance*. (On attitudes to death see Claire Gittings, *Death, Burial and the Individual in Early Modern England* (1984); Porter and Porter, *In Sickness and in Health*, pp.245-55.) Or it might be an illusion, arising from a change in the balance of the readily available evidence. Fear, when we find it, comes chiefly from seventeenth-century diaries – spiritual records kept by serious-minded, inward-looking individuals. A more light-hearted attitude emerges from female correspondence, perhaps written by more outward-oriented women; and it happens that the collections of such letters I have consulted come from the eighteenth century. Perhaps, then, both attitudes to childbirth coexisted throughout our period, without changing greatly over time. Further investigations should attend systematically to chronology, to social class, and to the various distinctions which have emerged here – between risk and fear; between fear in advance and fear during the birth; between difficult and easy births; and between fear of pain, fear of death, and fear of the male practitioner.

Is Childbirth Any Place for a Woman? The Decline of Midwifery in Eighteenth-Century England

BARBARA BRANDON SCHNORRENBERG

For many centuries, the normal assistant for all women in childbirth was a midwife. In the eighteenth century, however, the profession of midwifery began to decline markedly in England. By the nineteenth century midwives were mostly women of little education and generally no social status; their patients were the same. For the middle and upper classes, the stereotype of the midwife was the fat, dirty, drunken old woman whose image culminated with Dickens's Sairey Gamp in *Martin Chuzzlewit*. She is an old literary type, found in stories at least since the late Middle Ages, who still appears in popular novels about old and ignorant times. This view of the midwife has also passed from fiction into fact to encompass all midwives in all periods in many serious works. The latest evidence of this can be found in Lawrence Stone's *The Family, Sex, and Marriage*.[1] How the profession of midwifery sank to this state is the subject of this paper. There were three major factors that depressed the role and reputation of the midwife in England during the eighteenth century. The least important of these was the increase in scientific knowledge and medical skill. Far more important were the professionalization of medical practitioners and the emergence of what we usually call "Victorian" ideas of the role, abilities, and status of women.

At the beginning of the eighteenth century, the position of the midwife in Britain seemed fairly secure.[2] Any woman could set herself up in practice as a midwife. She was supposed to have a license, issued by a bishop; its qualifications were concerned with good moral character rather than knowledge or experience in the job. Since the enforcement of licensing was the responsibility of the church courts, little was done to pursue it, especially after the Restoration.[3] The midwife's training was primarily through experience. The best served an apprenticeship with an established midwife, and some even received instruction from physicians. Various handbooks and treatises on midwifery and related subjects were also available, although the value of these was sometimes questionable. Certainly one of the most valid criticisms of English midwifery was its lack of any kind of required training and regulation.[4]

In the seventeenth century there had been attempts to remedy this situation. Under the Commonwealth the licensing of midwives had been placed in the hands of the physicians, but this change, which might have meant more training, lapsed in 1660. In the first half of the seventeenth century, members of the prominent family of man-midwives, Peter Chamberlen I and Peter II, attempted to establish a corporation for midwives which would have regulated itself and enforced requirements for training. But these proposals were opposed by both midwives and physicians and so came to nothing.[5] Whether licensed or not, however, the midwife who read the available literature and who received training under a reputable midwife or physician had practical education as good as most males.

The midwife was seldom involved in any kind of prenatal care. She was called in when labor began, having perhaps been consulted shortly before about supplies or equipment she might need for the delivery. At the beginning of the eighteenth century there were essentially no provisions for delivery in hospitals; only in the second quarter of the century did lying-in hospitals and wards begin to be established in London, primarily for poor but respectable (married) women.[6] If the delivery was a normal one, all went well; the labor after all in such cases is that of the mother. If, however, things began to go wrong, the midwife was limited in what she could do. The better trained could use manual manipulation, but the midwife had no instruments. If things got really bad, the only recourse was to call in the physician or man-midwife. This provided another ground for criticism of the midwife; it was often said that she waited so late that even the skill of the physician was inadequate to save the mother or child. However, an examination of midwifery literature will show that he might well know little more than the midwife. The main difference between them was that the male, whether physician or man-midwife, used instruments.

By the early eighteenth century there were several basic texts for midwifery in circulation, and their number increased markedly throughout the century.[7] They reveal, however, the deficiencies of eighteenth-century medicine in general. What could be discovered by gross observation was clear and well understood. They could chart the development of the foetus and the various presentations of the child ready to be born. The finest of anatomical drawings, William Hunter's *The Anatomy of the Gravid Uterus*, appeared in 1774. But eighteenth-century physicians did not really understand the process of conception, the relation of the menses to reproduction, ova production, the causes of miscarriages. They were beginning to make some connections between such things as diet and exercise and a successful pregnancy and delivery, though all still believed firmly in the danger of marking the unborn child through the mother's external experiences.[8] All writers assumed that the midwife would officiate at a normal birth, though they urged that in difficult situations the physician be called in earlier rather than later.

The physician had instruments. In really abnormal births, the health of the mother was more important than that of the child. As Caesarian sections were universally fatal, they were not performed on living women. When natural childbirth failed, the physician used various instruments to crush and dismember the foetus in order to achieve its expulsion. More humane and more useful was the obstetrical forceps, which came into general use in the eighteenth century.[9] There are vague references in earlier literature to the use of forceps, but it appears that the efficient and practical prototype was developed in the seventeenth century by the Chamberlen family, man-midwives who came to England as Huguenot refugees in the later sixteenth century. They served the royal family and others; as they perfected the forceps they became widely known among their fellow professionals on the continent as well as in England. The exact design of the Chamberlen forceps was a highly guarded secret, although by the beginning of the eighteenth century various versions were being tried by other men. By about 1725 to 1730 the basic design had been developed and put to use by several practitioners, and by mid-century forceps were in general use among man-midwives.

The man-midwife was in a special category in England. Owing to the peculiar system of medical education and organization, they were a varied group. Whereas on the continent and in Scotland men practicing midwifery received standard university medical training, this was not necessarily true in England. Some of the men who practiced mainly in the area of obstetrics and gynaecology were university graduates; others were trained in the same way as their female counterparts. None of the Chamberlens, for example, held an English medical degree; those who did have university training received it on the continent. By the middle of the eighteenth century, graduates of Scottish universities dominated the man-midwife practitioners, especially in London.[10]

The whole system, if it can be dignified by such a term, of English medical education and professional organization and licensing was, by the eighteenth century, in need of major reform. Licensed medical men were divided into three separate organizations, but the actual practice of each was by no means so clearly divided or separated. The least prestigious group was the apothecaries. Although their original function had been to dispense drugs, by the eighteenth century many were serving, particularly in the provinces and among the urban poor, as general practitioners and even surgeons. They were trained by the apprenticeship system, not by formal education, and officially could charge only for drugs, not for attendance on a sick person. The regulatory body for the apothecaries was the Society of Apothecaries, which was essentially a London mercantile gild. Surgery was also controlled by what had originally been a London livery company, the Barber-Surgeons. By the eighteenth century the two functions of barber and surgeon had clearly divided; in 1745 the Surgeons Company separated from the older organization. In 1800 they were chartered by the King as the Royal College of Surgeons. The surgeons were not university-trained; a man could not be licensed as both a surgeon and a physician. The surgeons were the least numerous and probably the least important practitioners, especially outside the metropolitan area. Finally there were the physicians, organized under the Royal College of Physicians. The College, based in London like the other two companies, dominated the practice of its branch of medicine throughout the country. The College of Physicians was governed by the Fellows; admission to this body was by written and oral examination, which often seemed more concerned with the candidate's knowledge of Latin and whether he had an Oxford or Cambridge degree than with his medical knowledge. The Fellows administered examinations for the licensing of members of the College; the charter restricted licentiates to graduates of the two English universities. By the eighteenth century the Fellows had permission to grant licenses to others who might qualify, thus including the graduates of continental and Scottish universities if they chose to do so. To practice medicine in London and vicinity, a physician had to hold the College's license.[11]

The problem lay particularly in the limitation of the leadership of the profession to graduates of Oxford and Cambridge. Major changes in the study and teaching and the whole concept of medicine began in the seventeenth century and continued throughout the eighteenth. These emphasized an empirical approach to medical questions and a clinical approach to medical training. On the continent and in Scotland these ideas took hold, but England lagged far behind.[12] Some hospitals were founded, and by the mid-eighteenth century there were private medical schools in London offering lectures in anatomy and other medical subjects and training their students in the hospitals. But at Oxford and Cambridge, medical education continued in almost the same fashion as it had existed in the seventeenth century.[13] Nevertheless, the graduates of these two universities provided the political and organizational leadership of the medical profession. The often better trained and more skillful foreign or Scottish graduate was thus limited in his opportunity to achieve the fullest professional recognition.

There were also social implications in the question of any change in the organization of English physicians. Society assumed that a graduate of Oxford or Cambridge was a gentleman. Since the College of Physicians was controlled by these graduates, medicine was a gentlemanly profession, though it certainly ranked behind the church, the law, the military, or the civil service in the numbers it attracted and in prestige value.[14] But by the second half of the eighteenth century, English society itself was changing. The old order was being challenged, not just by political radicals but also by increasing industrialization. Those who were making the new society valued education and empirical results.[15] They were mostly outside the traditional establishment, and as their voice grew stronger, criticisms of the ruling class increased. There was demand for political reform and for change in various aspects of society. The gentleman physician whose main qualification was his Latin and his university degree was often a target. Obvious literary examples are Dr. Slop in *Tristram Shandy* and a number of doctors in the works of Tobias Smollett, himself a Scottish physician.[16]

Midwifery training and the relation of man-midwives to women midwives became a part of the larger issue of reform of medical education and licensing and the changing society of later-eighteenth-century England. Both men and women were practicing midwifery; no woman had a medical degree, and only some had acquired adequate training. Most of the man-midwives in London were not members of the College of Physicians; many had Scottish degrees, some were members of the Company of Surgeons. The College of Physicians had apparently recognized the problem of the man-midwife; in 1726 John Maubray and five other man-midwives were summoned before its Censorial Comitia for practicing without a license. Two years later another man-midwife was summoned; he argued that his profession did not need the College's license, and nothing more was heard on the subject for some years. In 1749 lectures on midwifery for men, sponsored by the College, were begun.[17]

By the forties a number of very distinguished man-midwives were practicing in London. The most important of these was William Smellie, a Scotsman whose teaching and writing on midwifery made him probably the most influential practitioner of the century. Older than Smellie were Frank Nicholls, the College's first lecturer on midwifery, and Sir Richard Manningham, also a College Fellow and the founder of a lying-in hospital. Students of Smellie were numerous; of them the most important was undoubtedly William Hunter, holder of a Scottish degree, anatomist as well as man-midwife.[18] Both Smellie and Hunter favored giving the same sort of lectures and clinical training to midwives as to men and had themselves begun private lectures and demonstrations for women. In 1752, with their endorsement, a proposal came before the College of Physicians to set up instruction for women in midwifery and to require this instruction before a midwife could get an episcopal license, but it was voted down.[19] Meanwhile Smellie published his *Treatise on the Theory and Practice of Midwifery* in 1751, followed by case history volumes in 1754 and 1764.[20] Smellie died in 1763; his place as the leading man-midwife was taken by Hunter. Aside from his *Anatomy of the Gravid Uterus*, Hunter made no original professional contribution. He was concerned with his own advancement, both financially and socially, so mixed in the leading political and intellectual circles of London. He became the fashionable man-midwife; he attended Queen Charlotte at the birth of her first son (later George IV) and at the births of the subsequent thirteen royal children. He died in 1783, before the birth of the Queen's fifteenth child. At the royal lyings-in, however, Hunter was not allowed in the room, where a midwife actually helped at the delivery. He also attended most of the nobility, including Lady Hester Pitt at the birth of the younger William Pitt.[21]

In 1767 the question of the relation of man-midwives to the College of Physicians again surfaced. Dr. Letch, a man-midwife, was rejected as a licentiate. He then went to the Court of King's Bench, asking the College to show cause for its actions. The Court found in favor of the College, but its judgement was tempered with a warning that the College ought to reform its practices. Meanwhile a number of licentiates of the College, including Hunter, formed the Society of Collegiate Physicians. The Society was closed to Fellows of the College of Physicians; its announced aim was the reform of the College. Twenty-three of these licentiates applied for admission as Fellows and were denied. The Society tried to debate with the Fellows, but this degenerated into an exchange of threats. The Society's members hired a gang from the neighborhood of the College's quarters and attempted to use force to break down the door to gain physical as well as paper admission. This action, known as the Battle or Siege of Warwick Lane, was clearly illegal, and the matter returned to the courts. There followed several lawsuits over the next years; in every case the College won, but always with strong advice from the judges to reform. These warnings finally took effect; in the seventies the College admitted some licentiates as Fellows and opened the possibility of more to come. The Society continued in existence, hearing papers and exchanging views until 1798, but its effectiveness as a reforming body declined rapidly after the mid-seventies.[22]

In the first reforms of the College, man-midwives were specifically exempted from the possibility of becoming Fellows.[23] Hunter, for example, who was admitted to the licentiate in 1758 on the strength of his Scottish degree, never became a Fellow. By the end of the seventies, however, the College had apparently decided that man-midwifery was there to stay. In 1783 they resolved to grant licenses to those qualified in midwifery only, for an entrance fee of £20. The number of those applying was small, but ten man-midwives were admitted to the College, including Thomas Denman, who succeeded Hunter as the fashionable man-midwife and whose training was entirely clinical.[24] Only a few man-midwives finally joined the College, although it was a far more prestigious body than the Company of Surgeons, which had been and continued to be the organization for many man-midwives. Those who attended royalty and society, such as Denman, Michael Underwood who attended the birth of Princess Charlotte, daughter of George IV, and Richard Croft, Denman's son-in-law, who let Charlotte die in childbirth, were all members of the College before it ceased to license man-midwives in 1804.[25]

As man-midwifery became socially respectable, these practitioners looked for a less awkward and contradictory name for themselves. Maubray, in his book of 1724, suggested that the man-midwife be called an "Andro-Boethogynist, or Man-Helper of Women"; it is not surprising that this name did not catch on. There was considerable use of the French term "accoucheur" in the eighteenth century, but doubtless many were not happy with the connotations a French word might carry. The term obstetrician seems to have come into use about 1828 and was apparently rapidly adopted by men to put themselves further from the midwives.[26]

In winning their own professional and social respectability, the man-midwives helped to depress the status and opportunities of their female counterparts. More and more by the end of the century, midwives attended the lower classes, who could not pay adequately; therefore, the lack of rewards meant the profession attracted fewer educated and qualified women. Of course there were exceptions, mainly London midwives who actually attended the births of royal children and who wrote books of instruction for their fellow practitioners.[27] Most of these books were positive statements of what a midwife could and ought to do. The bulk of the writing about midwifery, however, was increasingly vituperative criticism of men for taking over the profession. Although by the end of the century the best midwives were also using forceps, the most obvious ground for attack was the male's use of instruments. Actually the best men, such as Smellie and Hunter, were against their indiscriminate use. In 1751 Dr. Frank Nicholls published *A Petition of the Unborn-Babes*,[28] which accused man-midwives of killing both mothers and children by the misuse of instruments. It was said that Nicholls was paid for this support by the current royal midwife, but he was clearly sympathetic to the woman, for it was he who proposed in 1752 that the College offer lectures for midwives.[29]

The most vehement attack on man-midwives appeared in 1760, Elizabeth Nihell's *A Treatise on the Art of Midwifery, Setting forth Various Abuses therein, Especially as to the Practice with Instruments: the Whole Serving to put all Rational Inquirers in a fair Way of very safely forming their own Judgement upon the Question; Which it is best to employ, In Cases of Pregnancy and Lying-in, a Man-Midwife; or, a Midwife.* Nihell was a midwife who had studied and worked for two years at the Hôtel-Dieu in Paris, where French midwives were trained. Her husband was an apothecary, and of course there were those who said he really wrote the book. Nihell's language was intemperate and her target ill-chosen, for much of her attack was against Smellie. An example of her style and views can be seen in her summary of her three essential points.

> The *first*, is that the origin of the men, insinuating themselves into the practice of midwifery, has absolutely no foundation in the plea of superior safety, and, consequently, can have no right to exact so great a sacrifice as that of decency and modesty.
>
> The *second*, for that they were reduced first to forge the phantom of incapacity in the women, and next the necessity of murderous instruments, as some color for their mercenary intrusion. And, in truth, the faculty of using those instruments is the sole tenure of their usurped office.
>
> The *third*, their disagreement among themselves about, which are the instruments to be preferred; a doubt which, the practices tried upon the lives and limbs of so many women and children trusted to them, have not yet, it seems, resolved even to this day.[30]

Her work was reviewed at length and extremely unfavorably in the *Critical Review*, probably by its editor Tobias Smollett. The review began by proposing that she adopt as a motto *Ex nihilo nihil fit*, and went on to attack her views of "the whole body of male-practitioners, as ruffians who never let slip the smallest opportunity of tearing and massacring their patients with iron and steel instruments." Among her other points were that women were more sympathetic in their attendance on those in labor, and midwives on the whole were better trained than their male counterparts. The *Critical Review* denied these claims.[31] Nihell replied in *An Answer to the Critical Review for March, 1760, Upon the Article of Mrs. Nihell's Treatise on the Art of Midwifery*.[32] The *Critical Review* noticed this refutation in a brief but sprightly call for an end to the exchange, saying "you have delivered yourself of a monstrous birth, that fully evinces your dexterity in the obstetric art: may it, however, be the last of our begetting!"[33] It is difficult to believe that the kind of attack Nihell made did much to improve the reputation of midwives.

Besides the unwarranted use of instruments, the other main attack on man-midwives was made on the grounds of propriety. Nihell raised this issue, which, as instruments were more accepted by women as well as men, became a key question in the employment of man-midwives. Especially by the end of the century, as what we usually refer to as "Victorian" attitudes about the purity and privacy of a woman's body came to be widely held,[34] the issue was raised as to whether a woman ought to let a man see and touch her. Midwives argued that they would not compromise a woman's purity. Stories of patients seduced by their male physicians, of poor women in lying-in hospitals being subjected to the harassment of examination by countless medical students, were repeated in various versions. Man-midwifery, along with boarding schools for girls, novels, and dancing (all French in origin), were ruining English women. Once women were lost, all society would crumble.[35] This Victorian attitude toward "touching," as it was usually called, did not, however, save the mid-wife. Rather it led to nineteenth-century gynaecology and obstetrics, with male physicians diagnosing and prescribing for women whom they had not really examined and whose propriety forbade them to describe their own condition adequately.[36] While concern for female modesty did not save the midwife as the primary agent of delivery care, other aspects of the "Victorian" view of women were certainly influential in her decline. A lady does not work, and no decent woman would voluntarily involve herself in such a physical (even sexual) and messy affair as childbirth. Therefore, the kind of woman who became a midwife was for yet another reason uneducated and lower-class, and thus even less likely to be engaged by the middle and upper classes.

One of the complaints of midwives against their male counterparts was that the men were taking away women's work. All through the eighteenth century this complaint was made about many occupations; midwifery is another example of women's loss of employment in pre-industrial society. The reasons men gave for their own advance were all what would now be called male chauvinism. A woman is more delicate than a man, less able to engage in strenuous activity. Therefore, was a midwife physically fit to do her job? Although the nineteenth-century millowners argued that women's delicate and sensitive fingers made them more fitted to work in the textile mills for less money than men, the man-midwives argued that these fingers were not sensitive enough to manipulate obstetrical instruments. Since women's minds could not comprehend the mysteries of science, they could hardly be qualified to be obstetricians in the modern world. These arguments were in the end the most critical in the decline of the midwife. Women were excluded from the universities, the teaching hospitals (except as patients), from the professional organizations. Medicine, like other professions, was a male preserve, to be guarded jealously against incursion from illogical, unscientific, weak women. If medicine was to be scientific, it must be male. This is an attitude which has not died easily.

NOTES

1 *The Family, Sex and Marriage in England 1500–1800* (New York: Harper and Row, 1977), pp. 72–73, 79. Even so respected a medical historian as George Rosen makes these assumptions in "A Slaughter of Innocents: Aspects of Child Health in the Eighteenth-Century City," *Studies in Eighteenth-Century Culture*, vol. 5 (Madison: University of Wisconsin Press, 1976), pp. 293–316.

2 General accounts of midwives and midwifery in the eighteenth century vary in value. Jean Donnison, *Midwives and Medical Men* (New York: Schocken Books, 1977) is concerned primarily with the nineteenth century; however, she has an extensive introductory section on the earlier period. This is also the best-documented account of midwives. Other surveys include Kate Campbell Hurd Mead, *A History of Women in Medicine* (Haddam, Conn.: Haddam Press, 1938), pp. 460–77; Gustave J. Witkowski, *Accoucheurs et sages-femmes célèbres* (Paris: G. Sternheil, n. d.); James Hobson Aveling, *English Midwives: Their History and Prospects* (London: J. and A. Churchill, 1872; rpt., London, Hugh K. Elliott, 1967); Herbert Ritchie Spencer, *The History of British Midwifery 1650 to 1800* (London, John Bale Sons and Danielson, 1927; rpt., New York, AMS Press, 1978), really about British writing on obstetrics; Thomas Rogers Forbes, *The Midwife and the Witch* (New Haven: Yale University Press, 1966); Barbara Ehrenreich and Diedre English, *Witches, Midwives, and Nurses* (Old Westbury, N.Y.: Feminist Press, 1973), a stridently feminist account; J. Elise Gordon, "British Midwives through the Centuries 3: From the 18th Century to Today," *Midwife and Health Visitor*, 3 (1967), 275–81; M. Olive Haydon, "English Midwives in Three Centuries," *Maternity and Child Welfare*, 3 (1919), 407–9; Alice Clark, *Working Life of Women in the Seventeenth Century* (London: George Routledge, 1919), pp. 265–85; Hilda Smith, "Gynecology and Ideology in Seventeenth-Century England," in Berenice A. Carroll, ed., *Liberating Women's History* (Urbana, Ill.: University of Illinois Press, 1976), pp. 97–114.

3 See Aveling, Clark, and other sources cited above. Also see Thomas Rogers Forbes, "The Regulation of English Midwives in the Sixteenth and Seventeenth Centuries," *Medical History*, 8 (1964), 235–44, and "The Regulation of English Midwives in the Eighteenth and Nineteenth Centuries," *Medical History*, 15 (1971), 352–62; Thomas G. Benedek, "The Changing Relationship between Midwives and Physicians during the Renaissance," *Bulletin of the History of Medicine*, 51 (1977), 550–64.

4 Compare what happened to midwifery in France. See Richard A. Petrelli, "The Regulation of French Midwifery during the *Ancien Régime*," *Journal of the History of Medicine and Allied Sciences*, 26 (1971), 276–92.

5 Donnison, *Midwives and Medical Men*; Aveling, *English Midwives*; James Hobson Aveling, *The Chamberlens and the Midwifery Forceps: Memorials of the Family and an Essay on the Invention of the Instrument* (London, J. and A.

Churchill, 1882; rpt., New York, AMS Press, 1977); Smith, "Gynecology and Ideology."

6 G. C. Peachey, "Notes upon the Provision for Lying-in Women in London up to the Middle of the Eighteenth Century," *Proceedings of the Royal Society of Medicine*, 17 (1923–24): Section of Epidemiology and State Medicine, 72–76; J. E. Donnison, "Note on the Foundation of Queen Charlotte's Hospital," *Medical History*, 15 (1971), 398–400; Donnison, *Midwives and Medical Men*, pp. 25–28.

7 A useful survey of the various writers on the subject can be found in Irving S. Cutter and Henry R. Viets, *A Short History of Midwifery* (Philadelphia: W. B. Saunders, 1964), pp. 10–44. Shorter and less-exhaustive summaries are John Byers, "The Evolution of Obstetric Medicine; With Illustrations from Some Old Midwifery Books," *British Medical Journal*, 15 June 1912, pp. 1345–50; Walter Radcliffe, *Milestones in Midwifery* (Bristol: John Wright, 1967); Miles H. Phillips, "Men-Midwives of the Past," *Bristol Medico-Chirugical Journal*, 52 (Summer 1935), 83–102; Ritchie Spencer, *History of British Midwifery*.

8 This summary can be detailed from the secondary works cited above as well as contemporary works. I have read those available in the Lawrence Reynolds Collection, Lister Hill Library, University of Alabama in Birmingham: François Mauriceau, *Traité des maladies des femmes grosses . . .* , 4th ed. (Paris: Chez Laurent d'Henry, 1694. The first edition of this work was published in 1668; an English translation by Hugh Chamberlen appeared in 1672, titled *The Diseases of Women with Child and in Childbed*); John Maubray, *The Female Physician . . .* (London: James Holland, 1724); William Buchan, *Domestic Medicine*, 3rd American ed. (Norwich, Conn.: John Trumbull, 1778; first published in Edinburgh in 1769). In the Rare Book Room, Medical History Section, Medical School Library, University of Rochester: Alexander Hamilton, *Outlines of the Theory and Practice of Midwifery*, 3rd ed. (London: T. Kay, 1791); John Burton, *An Essay Towards a Complete New System of Midwifery* (London: James Hodges, 1751). On microfilm from the National Library of Medicine: Paul Portal, *Midwives, or the True Manner of Assisting a Woman in Childbearing: . . .* (London: S. Crouch and J. Taylor, 1705); John Memis, *The Midwife's Pocket Companion: or a Practical Treatise of Midwifery . . .* (London: Edward and Charles Dilly, 1765). See also Kenneth Dewhurst, "Locke's Midwifery Notes," *The Lancet*, 4 Sept. 1954, 490–91; Michael K. Eshleman, "Diet during Pregnancy in the Sixteenth and Seventeenth Centuries," *Journal of the History of Medicine and Allied Sciences*, 30 (1975), 23–39.

9 Cutter and Viets, *Short History*, pp. 44–69; Aveling, *The Chamberlens*; Kedarnath Das, *Obstetric Forceps: Its History and Evolution* (St. Louis: C. V. Mosby, 1929); Donald T. Atkinson, *Magic, Myth and Medicine* (Cleveland: World Publishing Co., 1956), pp. 163–66; John H. Peel, "Milestones in Midwifery," *Postgraduate Medical Journal*, Nov. 1947, pp. 523–29; Alban Doran, "Burton ("Dr. Slop"): His Forceps and His Foes," *Journal of Obstet-*

rics and Gynaecology of the British Empire, 23 (1913), 3–24, 65–86; and "Dusée: His Forceps and His Contemporaries," ibid., 22 (1912), 119–42, 203–7.

10 William F. Mengert, "The Origin of the Male Midwife," *Annals of Medical History*, ns, 4 (1932), 453–65; R. W. Johnstone, "Scotland's Contribution to the Progress of Midwifery in the Early Eighteenth and Nineteenth Centuries," *Journal of Obstetrics and Gynaecology of the British Empire*, 57 (1950), 583–94; Howard D. King, "The Evolution of the Male Midwife, with Some Remarks on the Obstetrical Literature of Other Ages," *American Journal of Obstetrics*, 77 (1918), 177–86; George Bancroft-Livingstone, "Louise de la Vallière and the Birth of the Man-Midwife," *Journal of Obstetrics and Gynaecology of the British Empire*, 63 (1956), 261–67; Steven A. Brody, "The Life and Times of Sir Fielding Ould: Man-Midwife and Master Physician," *Bulletin of the History of Medicine*, 52 (1978), 228–50.

11 On these three companies and medical practice, see S. W. F. Holloway, "The Apothecaries' Act: A Reinterpretation," *Medical History*, 10 (1966), 107–29, 221–36; Zachary Cope, *The Royal College of Surgeons of England: A History* (Springfield, Ill.: Charles C. Thomas, 1959); George Norman Clark, *A History of the Royal College of Physicians of London*, 2 vols. (Oxford: Clarendon Press, 1964–66); Frederick N. L. Poynter, *The Evolution of Medical Practice in Britain* (London: Pitman Medical Publishing Co., 1961); W. J. Bishop, "The Evolution of the General Practitioner in England," in E. Ashworth Underwood, ed., *Science, Medicine and History: Essays on the Evolution of Scientific Thought and Medical Practice written in honour of Charles Singer*, 2 vols. (London: Oxford University Press, 1953), II, 351–57; Joseph F. Kett, "Provincial Medical Practice in England: 1730–1815," *Journal of the History of Medicine and Allied Sciences*, 19 (1964), 17–29; Bernice Hamilton, "The Medical Professions in the Eighteenth Century," *Economic History Review*, 2nd ser., 4 (1951), 141–69.

12 Lester S. King, *The Road to Medical Enlightenment: 1650–1695* (London: Macdonald, 1970) and *The Medical World of the Eighteenth Century* (Chicago: University of Chicago Press, 1958); Albert H. Buck, *The Dawn of Modern Medicine* (New Haven: Yale University Press, 1920); Michael Kraus, "American and European Medicine in the Eighteenth Century," *Bulletin of the History of Medicine*, 8 (1940), 679–95; G. S. Rousseau, "'Sowing the Wind and Reaping the Whirlwind': Aspects of Change in Eighteenth-Century Medicine," in Paul J. Korshin, ed., *Studies in Change and Revolution: Aspects of English Intellectual History, 1640–1800* (London: Scolar Press, 1972), pp. 129–59; William R. LeFanu, "The Lost Half-Century in English Medicine, 1700–1750," *Bulletin of the History of Medicine*, 46 (1972), 319–48; Arnold Chaplin, *Medicine in England during the Reign of George III* (London, Henry Kimpton, 1919; rpt., New York, AMS Press, 1977).

13 Phyllis Allen, "Medical Education in 17th-Century England," *Journal of the History of Medicine and Allied Sciences*, 1 (1946), 115–43; James L. Axtell, "Education and Status in Stuart England: The London Physician," *History*

of Education Quarterly, 10 (1970), 141–59; A. H. T. Robb-Smith, "Medical Education at Oxford and Cambridge Prior to 1850," in Frederick N. L. Poynter, ed., *The Evolution of Medical Education in Britain* (Baltimore: Williams and Wilkins, 1966), pp. 19–52; Arnold Chaplin, "The History of Medical Education in the Universities of Oxford and Cambridge, 1500–1850," *Proceedings of the Royal Society of Medicine*, 13 (1919–20): pt. 3, Section of the History of Medicine, 83–107; Charles Singer and S. W. F. Holloway, "Early Medical Education in England in Relation to the Pre-History of London University," *Medical History*, 4 (1960), 1–17; Charles Newman, *The Evolution of Medical Education in the Nineteenth Century* (London: Oxford University Press, 1957), pp. 1–55.

14 Axtell, "Education and Status"; Edward Hughes, "The Professions in the Eighteenth Century," *Durham University Journal*, ns, 13 (1951–52), 46–55; N. D. Jewson, "Medical Knowledge and the Patronage System in 18th-Century England," *Sociology*, 8 (1974), 369–85.

15 J. H. Plumb, "Reason and Unreason in the Eighteenth Century: The English Experience," in *In the Light of History* (Boston: Houghton Mifflin, 1972), pp. 3–24.

16 William White, "A Survey of the Social Implications of the History of Medicine in Great Britain, 1742–1867," *Annals of Medical History*, ns, 10 (1938), 279–300; Rousseau, "Sowing the Wind and Reaping the Whirlwind"; Doran, "Burton ('Dr. Slop')"; W. H. Allport, "Tristram Shandy and Obstetrics," *American Journal of Obstetrics and Diseases of Women and Children*, 65 (1912), 612–17; Arthur H. Cash, "The Birth of Tristram Shandy: Sterne and Dr. Burton," in R. F. Brissenden, ed., *Studies in the Eighteenth Century* (Toronto: University of Toronto Press, 1968), pp. 133–54; G. S. Rousseau, "Pineapples, Pregnancy, Pica, and Peregrine Pickle," in G. S. Rousseau and P. G. Boucé, ed., *Tobias Smollett Bicentennial Essays Presented to Lewis M. Knapp* (New York: Oxford University Press, 1971), pp. 79–109; Cecil K. Drinker, "Doctor Smollett," *Annals of Medical History*, 7 (1925), 31–47; E. Ashworth Underwood, "Medicine and Science in the Writings of Smollett," *Proceedings of the Royal Society of Medicine*, 30 (1937), 961–74; Claude E. Jones, "Tobias Smollett (1721–1771)—The Doctor as Man of Letters," *Journal of the History of Medicine and Allied Sciences*, 12 (1957), 337–48.

17 Clark, *Royal College of Physicians*, II, 502–3.

18 Cutter and Viets, *Short History*, pp. 15–38; Johnstone, "Scotland's Contribution"; George C. Peachey, "William Hunter's Obstetrical Career," *Annals of Medical History*, ns, 2 (1930), 476–79.

19 Clark, *Royal College of Physicians*, II, 504–5; William Hunter to William Cullen, London, 22 Feb. 1752, John Thomson, *An Account of the Life, Lectures, and Writings of William Cullen M.D.*, 2 vols. (Edinburgh: William Blackwood, 1859), I, 543–44.

20 Tobias Smollett was a friend of Smellie; in the fifties and sixties, while editor of the *Critical Review*, he also edited and supervised the publication of Smellie's work. Jones, "Tobias Smollett"; Lewis M. Knapp, *Tobias Smol-*

lett, Doctor of Men and Manners (Princeton: Princeton University Press, 1949); Claude E. Jones, "Tobias Smollett on the 'Separation of the Pubic Joint in Pregnancy,'" *Medical Life*, 41 (1934), 302–5; G. S. Rousseau, "Tobias Smollett: Doctor by Design, Writer by Choice," *Journal of the American Medical Association*, 216 (1971), 85–89.

21 Johnstone, "Scotland's Contribution"; Gordon, "British Midwives"; Jane M. Oppenheimer, *New Aspects of John and William Hunter* (New York: Henry Schuman, 1946); Charles W. F. Illingworth, "William Hunter's Influence on Obstetrics," *Scottish Medical Journal*, 15 (1970), 58–60; Cutter and Viets, *Short History*, pp. 33–37; Olwen Hedley, *Queen Charlotte* (London: John Murray, 1975).

22 Iwan Waddington, "The Struggle to Reform the Royal College of Physicians, 1767–1771: A Sociological Analysis," *Medical History*, 17 (1973), 107–26; Lloyd G. Stevenson, "The Siege of Warwick Lane: Together with a Brief History of the Society of Collegiate Physicians (1767–1798)," *Journal of the History of Medicine and Allied Sciences*, 7 (1952), 105–21; R. Hingston Fox, *Dr. John Fothergill and His Friends* (London: Macmillan, 1919), pp. 143–51.

23 Dr. William Watson to Dr. John Fothergill, London, 16 Sept. 1771, Thomson, *Account of William Cullen*, I, 657–60.

24 Clark, *Royal College of Physicians*, II, 588–89; Forbes "The Regulation of English Midwives in the Eighteenth and Nineteenth Centuries"; Cutter and Viets, *Short History*, pp. 41–42, 185–86.

25 Clark, *Royal College of Physicians*, II, 636–37; Cutter and Viets, *Short History*, pp. 187–88; W. J. Maloney, "Michael Underwood: A Surgeon Practicing Midwifery from 1764 to 1784," *Journal of the History of Medicine and Allied Sciences*, 5 (1950), 289–314.

26 Forbes, "The Regulation of English Midwives in the Eighteenth and Nineteenth Centuries."

27 Mead, *Women in Medicine*, pp. 472–77; Gordon, "British Midwives"; Witkowski, *Accoucheurs et sages-femmes*; Aveling, *English Midwives*, pp. 118–29.

28 *The Petition of the Unborn-Babes to the Censors of the Royal College of Physicians of London* (London: M. Cooper, 1751).

29 Clark, *Royal College of Physicians*, II, 503–5; Donnison, *Midwives and Medical Men*, p. 32; Gordon, "British Midwives."

30 Nihell, *Treatise* . . . (London: A. Morley, 1760), pp. xii–xiii.

31 9 (1760), 187–97.

32 London: A. Morley, 1760.

33 9 (1760), 412. See also Aveling, *English Midwives*, pp. 118–26; Philip J. Kluhoff, "Smollett's Defense of Dr. Smellie in *The Critical Review*," *Medical History*, 14 (1970), 31–41. Most of the other attacks on man-midwives used exactly the same material as Nihell with perhaps different emphases. For example, Philip Thicknesse, *Man-Midwifery Analyzed: and the Tendency of That Practice Detected and Exposed* (London: R. Davis, 1764), stresses the impropriety of men treating women.

34 On this point see Muriel Jaeger, *Before Victoria* (London: Chatto and Windus, 1956); Gordon Rattray Taylor, *The Angel Makers* (New York: E. P. Dutton, 1974); Keith Thomas, "The Double Standard," *Journal of the History of Ideas*, 20 (1959), 195–216.

35 Donnison, *Midwives and Medical Men*, pp. 28–31; Mengert, "The Origins of the Male Midwife." Nihell and Thicknesse make many of these points as well.

36 Regina Morantz, "The Lady and her Physician," in Mary S. Hartman and Lois Banner, eds., *Clio's Consciousness Raised* (New York: Harper and Row, 1974), pp. 38–53; Jane B. Donegan, "Man-Midwifery and the Delicacy of the Sexes," in Carol V. R. George, ed., *"Remember the Ladies": New Perspectives on Women in American History* (Syracuse, N.Y.: Syracuse University Press, 1975), pp. 90–109; John S. and Robin M. Haller, *The Physician and Sexuality in Victorian America* (Urbana: University of Illinois Press, 1974). While these works deal with the United States, the results of these practices were the same everywhere.

American Midwifery

"Childbirth-Travells" and "Spiritual Estates": Anne Hutchinson and Colonial Boston, 1634-1638

by Deborah Kuhn McGregor

Anne Hutchinson was indeed the "breeder and nourisher" of the theological division of Puritan Massachusetts which led to the Antinomian Controversy (1636-1638).[1] However, while historians have researched, documented and interpreted the Antinomian Controversy in Colonial Boston in great detail, they apparently have failed to see the vital connection between Anne Hutchinson's role as a theological adversary and her life as a "woman very helpful in the times of child-birth. . . ."[2] Her engagement as a midwife in the ritual of birth was part of her religious activity. In fact, Puritanism provided a world-view from which the reality of her daily life in general, and the biology and common culture of childbirth in particular, derived an ethos.[3]

The Antinomian Controversy was a religious dispute which obstensibly concerned scriptural interpretation and the practice of Puritanism. The discourse as it is recorded to have occurred among clergymen and magistrates dealt with the proper weighing of emphasis in theological interpretation. A high percentage of Bostonians sided with those known now as the Hutchinsonians. They argued that belief in a covenant of works to the exclusion of the covenant of grace prevailed among their adversaries. Likewise, the Armenians—as the Hutchinsonians (or Antinomians) called them—cited what they considered to be the over-emphasis on the covenant of grace among certain ministers. Although the argument was highly specific in many ways and almost technical in its religiosity, the controversy split the town, threatening political and social cohesion. For example, Governor

Caduceus: A Museum Quarterly for the Health Sciences published by the Department of Medical Humanities, Southern Illinois University School of Medicine, Springfield, Ill. VolumeV, No. 4 (Winter 1989) pp. 1-33.

Winthrop eventually disarmed certain individuals associated with Anne Hutchinson because of his fear of violent uprising.[4]

Historians have studied the controversy extensively and continue to do so today.[5] In this essay I attempt to widen interpretation to include the midwifery of Hutchinson, including the theological disputes, as a factor as the conflict emerged. Today Western thought assumes a cosmic duality derived from the writings of Descartes in the seventeenth century. This view dichotomizes human existence through a perception of mind and body as separate entities.[6] Although the Puritans did not understand such a dichotomy, Cartesian thought influences modern historians to such a degree that scholarship often separates the history of past disease and epidemics from mainstream history. On the other hand, medical historians frequently have treated social and cultural conditions and changes as subsidiary, if not irrelevant, to their interests in the history of science and medicine. A review of the literature treating the life of Anne Hutchinson brings the limitations of the Cartesian dichotomy into relief. Mainstream historians—social, political, religious, for instance, or colonial—discuss Anne Hutchinson primarily as a theological figure, whose exile eminated from her heretical opinions. In contrast medical histories portray the Puritan woman as an early New World healer, and a well-known midwife in colonial Boston's first decade.[7]

Pertaining to the separation of mind and body, presentist blinders force an interpretation of Anne Hutchinson which distinguishes her spiritual activities from her birthing and healing activities. Puritans saw no such distinction. Her work with women in preparation for childbirth helped to initiate the first gatherings of her followers. Her ideas about religious belief and practice, in turn, originated to some degree in her experience attending births.

Cotton Mather called the joining of medical practice with spiritual guidance, "the angelic conjunction." In tune with this apt description doctors in the Colonial Massachusetts Bay were most often ministers who had acquired skills in healing to answer the needs of parishioners.[8]

Although Anne Hutchinson was not a physician, her skills did include treatment of sickness. Precise distinctions among midwifery, ministry and medical practice so clear today did not prevail in the seventeenth century. In his (for the most part retrospective) histories of the Antinomian Controversy, John Winthrop used metaphors of medicine to describe the theological activities of Anne Hutchinson among her followers.

. . . this was ever their method (i.e. the Antinomians), to drop a little at once into their followers as they were capable, and never would administer their Physicke, till they had first given good preparatives to make it worke, and then stronger & stronger potions, as they found the Patient able to beare."[9]

Winthrop perceived her spiritual activities as indistinguishably intertwined with her guidance and assistance during birth as well as in the giving of medicines.

Unfortunately there is little direct evidence of the historical experience of childbirth in the seventeenth-century Massachusetts Bay colony.[10] In fact, even the sources describing Anne Hutchinson's midwifery are scant.[11] Nonetheless, close perusal of the sources from colonial Boston in the 1630s and 1640s discloses the importance of one specific birth during the Antinomian Controversy itself. The stillbirth of Mary Dyer's deformed infant happened just weeks previous to Hutchinson's first public appearance to answer charges in Newtown before Governor John Winthrop and the General Court in November of 1637. The news of the event and Hutchinson's part in it spread only months later in March, 1638, on the last day of her second series of interrogations. Winthrop recorded his reaction in several places and the details reveal the depth of his concern and the extent to which he interpreted the ritual and history of Mary Dyer's delivery as symbolic of the theological controversy.

I intend to go beyond an intellectual consideration of Hutchinson's theological dissent or a medical discussion of monster births in this essay. Using a context provided by social history, women's history and medical history, Anne Hutchinson's midwifery becomes an integral part of her dynamic and controversial presence in the Massachusetts Bay Colony. To make the connection between Anne Hutchinson as woman midwife and Anne Hutchinson as theological adversary discloses the intersection of the day-to-day life of the original New England Puritans with the intellectual and rational structure of Puritan theology.

When Anne Hutchinson first arrived in Boston in 1634, she had already given birth to fourteen children, twelve of whom survived. She had grown up in Lincolnshire (in southern England, East Anglia, where many of the New World Puritans originated) and had also spent some years in London as a child.[12] Her father had been a nonconformist preacher before his death when Anne was still a child. She had learned to read and write, presumably under his influence, and as an adult fell under the influence of Puritan beliefs.[13] Her husband, William Hutchinson, had been a successful merchant in the village of Alford. Together they hoped to find a new home in colonial Massachusetts Bay.

In Boston, as in Lincolnshire, theology was the driving force to Anne Hutchinson's life. She cultivated friendships with ministers in England as well as in Boston; in fact, she had known John Cotton, Thomas Hooker and others before their migration to the New World. In addition to an extensive knowledge of the scriptures and Puritan theology, she had developed skills in assisting at births and in nursing sickness. During pregnancy and in childbirth, one task was the spiritual preparation of the woman. For example, John Cotton, a prominent minister in Boston, noted of Anne Hutchinson:

> . . . chiefly for that I heard shee did much good in our town, in womans meeting at Childbirth-Travells, wherein shee was not onley skilfull and helpfull, but readily fell into good discourse with the women about their spiritual estates.[14]

First appearing in 1644, John Winthrop's *A Short Story of the Rise, Reign and Ruine of the Antinomians, Familists & Libertarians* was somewhat defensive in tone, written for the English Puritans to explain the unrest in Boston during the Antinomian Controversy. Winthrop described the nexus of Hutchinson's power as residing in midwifery and her theological meetings, while at the same time asserted that she had plotted from the moment of her arrival to divide Boston.

> . . . then shee began to go to work, and being a woman very helpfull in the times of child-birth, and other occasions of bodily infirmities, and well furnished with means for those purposes, shee easily insinuated her selfe into the affections of many, and the rather because she was much inquisitive about their spiritual estates.[15]

As the Reverend Cotton's description indicates, Boston Puritans first welcomed Anne Hutchinson for both her religious devotion and for her midwifery and nursing skills. By 1637, however, the community of Boston was deeply divided along lines originating in Hutchinson's meetings in her home. Following the events of 1637 and 1638, the synod of religious leaders, the birth of Mary Dyer's "monster," the trial before the General Court and the examination before the Church of Boston, Winthrop came to characterize Anne Hutchinson's help in childbirth and sickness as the hand of the Devil and a guise for wrong opinions.

Because there is scanty documentation from Anne Hutchinson herself, the pattern of childbirth customs in the seventeenth century must provide historical clues to her practice of midwifery. It is important to consider patterns established by other instances of birth and female participation in the Old World to ascertain what Anne Hutchinson might have brought with her to the colony. When she arrived, she carried skills she had learned in Old England.[16]

Because childbirth is and has always been a female activity, historians have difficulty in determining the general patterns. Women for centuries left little record of their daily experience and, specifically, of their memories of giving birth or of their role as midwife. An examination of childbirth in the seventeenth-century Massachusetts Bay colony relies on incomplete evidence. As a portrait of historical midwifery emerges, limitations persist. The practice of midwifery varied from individual to individual. In addition, there is little detail of the experience of the birth-giving woman.

In an essay on the experience of childbirth in seventeenth-century England, Adrian Wilson offers the details of the ritual surrounding birth.[17] The ritual integrates the "natural" or biological process of birth into a cultural milieu, in our case Puritanism, where the ritual takes on even greater significance than the labor and delivery. Central to the ritual activity was the gathering of female kin and neighbors, to the exclusion of men. A special room was designated for lying in. The room was darkened, candles lit and a caudle—a special drink prepared for the duration of the labor—shared by all involved. Symbolically, the word *gossip* originated with this gathering of women who attended the birth and offered support to the woman in labor. This birthing ritual helped to create the community of women in preindustrial society.

Special qualities attended the role of midwife, including the preparation and administration of herbs.[18] In England in the seventeenth century, the publication of Culpepper's lists of plants and their medicinal uses illustrated the popularity of botanical cures. Native Americans also used herbal sources in medicine and influenced the practice among the colonists, helping the newcomers to recognize new plants as possible therapies.[19] While individual families often grew and prepared their own botanicals, undoubtedly the midwife guided them in the use of specific herbs.[20]

The midwife also stood out among the other women by virtue of her title, her authority in the lying-in chamber, her payment after delivery and her supervision of the ritual. As a rule, the physician only intervened at the midwife's request, during difficult labors, when the death of the infant and of the mother were virtually foregone conclusions. In contrast, the midwife governed normal births in which infants were born alive. While Boston in its first decade possessed fewer physicians than Old England, there were men throughout Massachusetts—two Drs. John Clark and Dr. Thomas Oliver—who had studied medicine in England. We actually know much less about the historical presence of midwives and their apprenticeship with other midwives than we know about the number of physicians in the Bay Colony.[21] Anne Hutchinson was notable for her role as a midwife in the first decade of the Massachusetts Bay Colony.

Other aspects of childbirth ritual included the eventual opening of the lying-in chamber following the birth. The mother first sat up, during a period called "upsitting," and then received visitors after a period of confinement. Essential to the Anglican ritual was the churching of the woman who had given birth. She marked her first reappearance in the public and male world by a church-governed ritual of cleansing.[22] In England many Puritan women also participated in this activity. The midwife had the serious responsibility of baptizing the infant. Sometimes this included a special urgency, if the newborn seemed vulnerable to immediate death. The midwife then was mandated to secure baptism in order to protect the soul of the infant.[23]

Childbirth in seventeenth-century America was a communal event which women governed and shared. Women supervised the biological and cultural event, translating the male-dominated social, political and religious world-view into specific customs. The past also influenced their ritual—the oral tradition of midwifery passed on from generation to generation and the wisdom of magic and superstition which explained the process of gestation and birth. Although the communal and female pattern of ritual resembled the Old World culture, the structure of the New World Puritan theology also influenced the experience of childbirth in colonial Boston of the 1630s. A woman's preparation for childbirth was a preparation of piety, a spiritual readiness for her future suffering and death. Anne Bradstreet's poem to her husband, written in the 1630s, "Upon the Birth of One of Her Children" captures the anxiety and fear of death women experienced during pregnancy. "How soon, my dear, death may my steps attend. . . ."[24]

Death more commonly accompanied childbirth in the seventeenth century than it does today. Even more to the point, however, was the lack of effective intervention technique during a prolonged or stultified labor. The demography of seventeenth-century New England suggests that there was a relatively high rate of survival among all members of the population.[25] Especially compared to the colonial Chesapeake experience, members of the early Massachusetts

John Winthrop, Governor of the Massachusetts Bay Colony during Anne Hutchinson's tenure in Boston. (Photo reproduced from Robert C. Winthrop's Life and Letters of John Winthrop *(Boston: Ticknor and Fields, 1864), vi.*

7

Bay Colony had a reasonably good chance of living a full life. There were, however, periods of epidemics during which many died. Newborn infants were often very vulnerable to sudden death.[26] Without the assistance of the possible use of forceps or caesarean section, women knew that they too, or their neighbors, risked a painful and drawn-out death in the delivery of a child.

Preoccupation with death and suffering stemmed not only from the possibility of difficult labor and ensuing death. Nearly all inhabitants of the Massachusetts Bay Colony immigrated in order to practice Puritanism. In their religious faith, they were deeply concerned with life after death and the attainment of salvation. The quest for sanctification and justification in the eyes of God were at the heart of the individual's earthly existence, as they represented knowable signs of his/her potential election to eternal grace.

Childbirth customs in early Boston did not exist in isolation from Puritanism. A woman's preparation for labor and childbirth was an intensified preparation of piety, a spiritual readiness for her future suffering and death. In this context, Anne Hutchinson's skills were tremendously valuable. Not only did she have a physiological understanding of birth (which she acquired both as a mother and as a midwife), but she also offered a powerful and comforting rendering of Puritan theology. These capacities fused in her concern for the spiritual well-being of the individuals who came to her.

Scripture was the *sine qua non* of the early Puritan colonists. While women found in the Bible that childbirth was a fearful and life-threatening travail, they also found a prescription for prolific procreation. The work of pregnancy was spiritual preparation in hope of redemption.[27] Such spiritual preparation was important to every pious Puritan colonist; through close reading of Scripture and meditation, the individual struggled to reach sanctification, justification and, upon death, glorification. Because the Puritan God was inscrutable, acting in ways irreducible to human understanding, no individual could be certain he or she would win salvation after death. Thus God might inflict pain or death upon an individual as punishment for his depravity and sinfulness, yet the same individual, if he obeyed moral law and engaged in worship, might be among the chosen saints and achieve salvation. There was no certainty until the hour of death itself. Discomfort, anxiety and oscillation between feelings of assurance of election and the desperation of denial of membership among the saints typified the Puritan experience. The theological emphasis in childbirth served to maintain the woman's agonizing insecurity in the face of death and at the same time to ordain the pregnancy and labor as a time of spiritual preparation.[28]

John Cotton wrote of Anne Hutchinson that

> ... she did much good in our Town, in womans meeting at Childbirth-Travells, wherein shee was not onely skilfull and helpfull, but readily fell into good discourse with the women about their spiritual estates. ... so as these private conferences did well tend to water the seeds publikely sowen. Whereupon all the faithfull embraced her conference, and blessed God for her fruitfull discourses.[29]

Within herself, Anne Hutchinson harbored a religious confidence and an assurance of her own salvation. She described the process of her personal enlightenment midway through her appearance before the General Court in Newtown, Massachusetts, in November 1673.[30]

> The Lord knows I could not open scripture; he must by his prophetical office open it unto me. . . . I bless the Lord, He hath let me see which was the ministry and which was wrong. Since that time I confess I have been more choice and He hath let me to distinguish between the voice of my beloved and the voice of Moses, the voice of John Baptist and the voice of antichrist, for all those voices are spoken of in scripture. . . . Ever since that time I have been confident of what he hath revealed unto me.[31]

Hutchinson had found strength in her religious beliefs and sought to share this with women gathered in preparation for childbirth. Much of what she articulated in the recorded transcripts suggested diversion from literal translation of the scriptures. She attempted to address the conflict inherent in the Puritans' accepted belief in both the covenant of works and the covenant of grace. These covenants together urged the individual's submission to God's will and, in contrary fashion, suggested the potential benefit of individual action in the pursuit of salvation.

John Cotton, an eminent minister in Boston as well as Hutchinson's teacher, supported her in her religious doctrine for the three years before the controversy threatened to divide Boston. He too was disquieted by what he perceived to be an erroneous emphasis on the covenant of works.[32] Until the General Court ruled to end Hutchinson's meetings, many members of both the religious and political hierarchy upheld her status in the community, yet church elders and governmental magistrates found her to be "... the root of all these troubles."[33]

Ostensibly, the governing hierarchy ruled against Anne Hutchinson's continued gatherings because of the underlying theological dispute and the social disorder it created. The followers of Anne Hutchinson chastised, sometimes openly taunting, ministers, like Wilson, Bulkely and Shepard, for promulgating a belief in the supervening authority of the covenant of works—suggesting, for example, that Puritans should scour their lives for external signs of sainthood while engaging in the active pursuit of salvation.[34]

Scripturally, the covenant of works originated with Adam in the Garden of Eden. In exchange for a righteous and obedient life God rewarded man with salvation. After the fall of Adam and Eve, man had only the hope offered by the covenant of grace, offered to Abraham. This latter chance of salvation eminated from God's independent decision unknowable to man. While the covenant of grace reaffirmed possible eternal life to humankind, it denied the efficacy of good works per se as the source of grace. The appellation *antinomian* went back hundreds of years and designated those who while subscribing to the covenant of grace, seemed to deny the necessity of adherence to the moral law delineated by the original covenant. They were "outside the law," or *antinomian.* Winthrop labelled Hutchinson, Wheelwright and the others as such only in the 1640s.

John Cotton, John Wheelwright and Anne Hutchinson, along with the others, felt that the Reverend Wilson too heavily emphasized the importance of good works on earth as a way of gaining entrance to heaven. They called Wilson an Armenian (after a European religious sect with a similar emphasis on works), and disputed Wilson and his followers' search for visible signs of sainthood in the form of sanctification.

Winthrop, Wilson and other ministers argued that Anne Hutchinson and John Cottons' theology offered "a faire and easie way to heaven."[35] They were suggesting, so Wilson argued, that salvation had nothing to do with doing good here on earth and that one's earthly existence was no reflection of the judgment of God. By the time of the interrogation in the Church of Boston (March 1638), John Cotton and Anne Hutchinson were no longer allied. At this time Hutchinson articulated a belief that she had experienced an indwelling of God, a tangible certainty of grace. This statement was marked clearly as heresy by Cotton and others, and resulted in the exile of herself and followers.

When Winthrop called Hutchinson "the root of all these troubles," he alluded to medicine and botany as well as to theology—this time referring to the use of plant roots as medicine. Just as in the earlier quote, he saw her opinions as "bad medicine"—"stronger and stronger potions, as they found the Patient able to bare."[36]

When Anne Hutchinson first arrived in Boston, she seemed to be a "ministering angel," a woman with abilities to help in sickness and childbirth. By 1638 she had become a seductive individual, a woman who poisoned with her medicines. Her success as a midwife was dependent upon her integration of theology with the administration of medicine. When she failed in one, she failed in the other, at least in Winthrop's mind. Her initial acceptance in Boston reflected her original strengths in both theology and medicine.

In fact, the polarity embodied in Anne Hutchinson's social status in Boston paralleled the polarities in the profession of midwife. The midwife alone possessed what knowledge and skill there was to facilitate the birth and ensure the lives of the mother and infant. However, what received either condemnation or approval was the spiritual cleanliness of the midwife, not her medical skills. In the seventeenth century and earlier the midwife could be a "ministering angel," as Anne Hutchinson appeared to be at the time of her arrival in Boston.[37] At the other end of the continuum was the midwife who relied on demonology and witchcraft in her supervision of childbirth. Obviously the latter was anathema to the Puritans. Once Anne Hutchinson fell from favor, accusations of the devil's work tainted her reputation.

Anne Hutchinson's midwifery provided a clear link between the day-to-day life of her neighbors and herself within the larger frame of Puritan theology. The experience of birth in Boston of the 1630s entailed the living-out of religious belief as well as physiological cycle. The unsettling details of the one labor and delivery in particular served as metaphor and symbol for the intense theological dispute of the Antinomian Controversy.

In childbirth we encounter the area where the anatomy of reproduction and the historical specificity of gender overlap. The ritual of childbirth integrates the biological process into a cultural milieu. In many ways, the ritual takes on even greater significance than the birthing activity itself. In the study of childbirth ritual the scholar goes beyond the outline of social and cultural patterns to glimpse women not only in their *social* roles, but in their biological and specifically *sexual* roles as well. The clash between Anne Hutchinson and the powers-that-be in early colonial Boston lays bare a conflict between female-dominated childbirth customs and male-dominated Puritan ideology.

Mary Dyer's premature delivery of a deformed and dead infant occurred just weeks preceding Anne Hutchinson's trial before the General Court on October 17th of 1637. "The Midwife Hawkins" and Anne Hutchinson supervised the event with the presence as well of several other women who were eventually ". . . forced to leave. . ." because of the adverse intensity of the labor.[38] Three women—Hawkins, Hutchinson and a third unidentified individual—were left alone with Mary Dyer.

Both Mary Dyer and Jane Hawkins (and their spouses) belonged to the core of Anne Hutchinson's following in Boston. By this time the tenuoûs nature of Hutchinson's standing *vis a vis* the ministerial and political hierarchy was evident to the three. At the time of this birth, Mary Dyer was young and bearing her first child. Her challenge to theological orthodoxy was just beginning. She met her end many years later a martyr in the name of the Quakers. Jane Hawkins, on the other hand, was an older woman and the established midwife in Boston before Anne Hutchinson's arrival. Hawkins was never granted membership in the Puritan Church, but some noted in retrospect her adherence to Hutchinson's theological perspective.[39]

Despite Jane Hawkins' exclusion from church membership, Anne Hutchinson worked with her in midwifery. Although Hutchinson's contemporaries did not call her the Midwife, they did characterize her clearly as an assistant during birth. The Midwife Hawkins then was in charge during Mary Dyer's labor. Undoubtedly, Hutchinson went to the Dyer home, several blocks away, as a routine response to the needs of a woman in labor, a woman whom she knew and counselled. Probably Dyer and Hutchinson played complementary roles as they organized and guided the community of women through the ritual and experience of childbirth.

The events of the labor and birth were recreated by Winthrop, using various sources now referred to as "hearsay evidence." Apparently the infant died two hours before its birth, although how this was known is a mystery. Either the women measured the vital signs of the unborn child, or they mistook some quality of the labor itself as a sign of its death. The bed in the lying-in chamber shook and a putrid ordor pervaded the environment. Many of the group began vomiting and children evinced fits. Thus most of the women left the room.[40] Jane Hawkins recalled that the child ". . . came hiplings till she turned it."[41] The fetus had settled into a breech position, causing a difficult labor. The Midwife Hawkins had the considerable skill to reposition the fetus during labor, to perform, in medical parlance, a podalic version. Winthrop described the infant as severely deformed, perhaps anencephalic or lacking a brain. The child was female and appeared to be covered with scales and bearing talons or claws, as Winthrop saw it, in place of nails.[42]

By the time of the birth, the Antinomian Controversy had begun in earnest. Roughly two weeks previous to the delivery, October 2, 1637, a synod had met to establish agreement and ease the rift opening among the ministers, to reestablish communication between John Wilson and John Cotton and John Wheelwright. Subsequently, in November, Wheelwright, Anne Hutchinson's brother-in-law and religious associate, was banished. Although the birth happened before Anne Hutchinson's appearance in front of the General Court in Newton, Winthrop did not know of the event for five months more. He wrote about the delivery of Mary Dyer much later, after the strenuous examinations of Hutchinson and in the context of her banishment and dismissal from the church. It was from this vantage point that Winthrop attached an aura of devilry to the event, calling it a monstrous birth and a manifestation of Special Providence.

Although the women who attended births commonly were called gossips, in this case the word of the birth did not spread. When questioned (in March of 1638 after her banishment from Massachusetts Bay) Anne Hutchinson described asking John Cotton for advice regarding the burial of the child. He counselled her, she recalled, to bury the body quietly without reporting the event to any other authorities, political or religious.[43] This was the prerogative and obligation of the midwife. Part of her power and authority was to provide a liaison between the lying-in chamber and the governing authorities. John Cotton willingly accepted the responsibility for his role, once confronted by Winthrop for his actions. Winthrop recalled his apology:

> He considered, that, if it had been his own case, he should have desired to have had it concealed. . . . He had known other monstrous births, which had been concealed, and that he thought God might intend only the instruction of the parents and such other to whom it was known . . . .[44]

Besides Cotton's reluctance to publicize the disquieting qualities of Mary Dyer's delivery, the women also kept quiet and manifested a spiritual solidarity with Anne Hutchinson who was on trial before the collective governing male authorities. They appear in retrospect to have been "a communitie of woemen [sic]." The alarming and secret experience of the delivery intensified the solidarity of the women: ". . . The bed whereon the mother lay did shake, and withal there was such a noisome savour, as most of the women were taken with extreme vomiting and purging, so they were forced to depart. . . ."[45] Even though Winthrop argues that the early departure of these women caused the concealment of the birth, it is much more plausible that, experiencing diverse individual reactions, they nonetheless stood by one another in secrecy until the conclusion of the hearing before the Church of Boston in March of 1638.

Boston in 1637 was a small community with frequent encounters among its inhabitants. The Winthrops and the Hutchinsons shared the same well, for instance, and lived across the street from one another. Networks of neighbors and shared daily activities provided context for communication and shared knowledge. While Winthrop and Hutchinson may not have spoken much, certainly the group of women who attended Mary Dyer in her labor gleaned the details of the outcome of her "travail." The well-kept secret of the birth, especially its concealment from the governing authorities, suggests that the three women attending the birth, Mary Dyer, and the other women knowledgeable as to its details, recognized the pejorative and dangerous connotation that public knowledge of the birth might carry.

Puritan cosmology included an awareness of the natural world and certain inclinations to pursue a Baconian observation and recording of the occurrences in the surrounding environment. Most did not perceive a conflict between the empiricism of Bacon and the rigors of Puritan theology. They recognized instead an opportunity to establish natural laws which would further reveal the hand of God in earthly existence.[46] In the case of childbirth, as in death, however, an implicit contradiction underlay Puritan life. On the one hand, God was the first cause of all happenings, but on the other hand, Puritans did look to the intervention of medical practitioners to treat disease, at least to a limited degree. For example, when Thomas Shephard's wife lost a baby, he recorded in his diary that the ineptitude of the midwife was the cause of the death. Puritanism urged upon him the interpretation of the infant's death as a sign of God and as a message to him and his wife concerning their failings. Still Shepard saw fit to blame the midwife, a perspective similar to our own in which medical practitioners hold a great responsibility for the outcome of a childbirth.[47]

John Cotton first interpreted the misfortunes of the Dyer birth as a providential warning primarily to the parents of the "monster." Winthrop, however, saw the birth as further evidence of the evil of Anne Hutchinson's opinions and "her monstrous errours."[48] Seeing her midwifery as predominantly spiritual, Winthrop easily interpreted the death and malformation of the child as the result of theological heresy. In fact the spiritual and physical life of the stillborn child were so intertwined that he

Quaker Mary Dyer prior to being hanged on Boston Common in 1660. Her support of Anne Hutchinson in the Antinomian Controversy had led to her expulsion from Boston in 1635. (Painting by Howard Pyle reproduced courtesy of the Friends Historical Library, Swarthmore College, Swarthmore, Pennsylvania.)

15

made no distinctions. Nor did Winthrop articulate a portrayal of the infant as a separate being. Once informed of the birth, the event became tremendously important to Winthrop's judgment of Hutchinson and to his interpretations of the Antinomian Controversy.

Winthrop recorded the story of the disclosure of the birth in his *History of New England:*

> . . . and the child buried, (being stillborn,) and viewed of none but Mrs. Hutchinson and the midwife, one Hawkin's wife . . .; and another woman had a glimpse of it, who, not being able to keep counsel, as the other two did, some rumour began to spread, that the child was a monster. One of the elders, hearing of it, asked Mrs. Hutchinson, when she was ready to depart; whereupon she told how it was. . . .[49]

The elder who confronted Anne Hutchinson with the rumored monster birth threatened her with the examination of the body of the baby if she did not disclose the details. Why he manifested curiosity to see the misshapen infant is enigmatic. Winthrop, however, followed suit. He further threatened Anne Hutchinson "that he intended to have it taken up and viewed."[50] She then offered a description of the infant, but to little avail. It seems that Winthrop was still not satisfied: "With advice of some other of the magistrates and of the elders of Boston, [Governor Winthrop] caused the said monster to be taken up."[51] Of course, the body of the stillborn infant was "much corrupted," but still they found evidence particularly of the horns and claws, which symbolized the presence of a demon.[52] Not only did Winthrop proceed, in the company of "other magistrates and elders" (both religious and political figures) to unearth the remains five months after death, but he included the astounding number of "above a hundred persons" among the spectators.[53]

The relentless investigation of John Winthrop and his fellow citizens reveals important detail in the conflict between Anne Hutchinson and the prevailing authorities. The insistence on viewing the body of the infant came in part from an interest in observing closely the material evidence. His curiosity was empirical in nature, seeking to know the detail and every aspect of such a timely and unusual occurrence. Above all, the intensity of Winthrop's concern about the birth discloses its signficance for him. The birth was closely entangled with the interrogation and examination of Anne Hutchinson. Initially, he was outraged to be excluded from the knowledge of the birth and finally convinced that the unfortunate outcome was merely a providential sign of the heresies of Anne Hutchinson and the poison she spread.

At the same time, the effort to unearth the body was similar to an autopsy. There were, in fact, autopsies done occasionally in England. Common law in the Massachusetts Bay colony held that the bodies of criminals might be used for anatomical study. This law was rare, however, and generally reserved for murderers.[54] That Winthrop would gain the support of other political and religious authorities to examine the corpse before a large public audience reveals the extraordinary passions of the Antinomian Controversy and its connections with the ritual of childbirth.

The repercussions of the concealed birth included immediate restrictions placed on Jane Hawkins, and her ultimate exile from the Massachusetts Bay Colony. When the birth was made public knowledge, Jane Hawkins accepted the responsibility for its aberrations and met the resultant restriction on her medical practice. Before her departure, she was instructed " not to meddle in surgery or physick, drink, plaisters, or oils nor to question matters of religion, except with the elders for satisfaction."[55] Hawkins' punishment discloses an important aspect of colonial childbirth. The midwife had to answer to the community, and if she failed in her responsibilities, she met with punishment.

Winthrop's condemnation of all three women reflected the prevalent seventeenth-century notion that supernatural forces determined the outcome of the birth. Although Winthrop was agitated by the news of the monstrous birth (his wife had suffered numerous failed pregnancies), he resolved his uncertainty by seeing the birth as proof of Divine sanction for his punitive measures against Anne Hutchinson and her followers. He combined his scientific or empirical interest in the physical qualities of the dead infant with an embrace of Puritan theology. The story of the Mary Dyer delivery and the aftermath displays the everyday context and acting out of the social, religious and political conflict of the Antinomian Controversy and in modern parlance, also a medical conflict.

After wintering in exile from her family and from the Boston citizens at the home of Joseph Welde in Roxbury, Anne Hutchinson appeared before the Church of Boston in March of 1638. By now she was tired and brought down by the lengthy proceedings and by her absence from her family and supporters. She had seven children at home, the youngest being Zuriel who was but one year old.[56] She appeared to be pregnant. Thomas Olliver, an elder of the church and a surgeon, spoke for her.

> . . . whereas our Sister Hutchinson was not hear at the Beginninge of this Exercise, it was not out of any Contempt or Neglect to the ordinance, but because she hath bine longe (under) Durance. She is

> so weake that she conceaves herselfe not fitt nor able to have bine hear soe longe togeather.[57]

In spite of her infirmity and exhaustion, she appeared before the group of ministers to describe and defend her behavior and her renderings of Puritan theology.

Although she had had the long winter to ponder the events of the previous fall, still she had not been able to talk to anyone about the birth she attended in October. Her concern for the fate of life after death was evident in the questions and opinions she expressed almost immediately.

> Every Man consists of Soul and Body. No *Adam dies not except his soule and Body dye.* And in Hebrews 4 the word is lively in Operation, and devides between *soule and Spirit; Soe than the Spirit that God gives man, returned to God indeed, but the Soule dyes* and That is the spirit Ecclesiastes speakes of, and not of the Soule.
>
> (Luke 19.10)[58]

She sought to make a distinction between spirit, soul and body, arguing that the spirit alone experienced resurrection. This point fell on unsympathetic ears, however. The governor, ministers and elders argued with her that there was no such distinction, particularly between spirit and soul, as it pertained to moral beings. They struggled, it seemed, to grasp her confusion and to set right her erroneous opinions:

> *Governor*: As the Body remaynes an Earthly substance after Dissolution, soe the Soule remaynes a spirituall Substance after the Curse, though we see not what substance it is turned into after Dissolution.
>
> *Mrs. Hutchinson: I thinke the soule to be nothing but Light.*
>
> *Mr. Cotton*: The Sume of her Opinion is that the soules of men by Creation is no other or better than the soules of beastes which dye and *are mortall, but are made immortall* by the Redemption of Christ Jesus. . . .[59]

A few minutes later, she asked, "How doe we *prove that both soule and body are saved?"*[60]

The Reverend John Wilson, the chief antagonist and partner of the governor, declared her opinions "to be dayngerous and damnable and to be no lesse than Sudducisme and Athiisme. . . ." [61] And still the discourse went on. At one point Anne Hutchinson declared, "I doe question whether the same Bodies that dies, shall rise agayne." To which John Davenport retorted that it was "clear that he rayseth us the same Body and not another Body for substance."[62] Not long after, Peter Bulkeley, another Puritan minister, declared that she was dancing around heresies familiar to him as "that foule, groce, filthye and abominable opinion held by Familists, *of the Communitie of Weomen*. (Job 19.25. Philippians 3)" [63]

The language Hutchinson uses leaves little doubt that she was considering on a conscious, or possibily unconscious, level, the fate of the child born in October. Would this child, if touched by the grace of God, lead an immortal existence in its misshapen and horrifying body? Certainly she must not have been inured to a birth such as that of the stillborn child of Mary Dyer. The odds of encountering this type of ill-fitted infant were small.[64] All those who examined her, except Cotton, were ignorant of the birth and could not understand the context for her queries and argument. Cotton by now was withdrawing his support and had grown much more critical of Hutchinson. If he recognized an internal agonizing on Hutchinson's part over the aberrations of the Dyer birth, he could in no way reveal them in public. Perhaps when Hutchinson stayed with him briefly between this session of interrogation and the final meeting they spoke of the monster birth, perhaps not. The other ministers, for the most part, remained angry with her opinions and further convinced of her theological error.

After the trial before the church briefly adjourned, Hutchinson returned to offer a retraction of her erroneous opinions, and to apologize, especially for her behavior towards those ministers with whom she disagreed. Still, however, she denied that she had held mistaken opinions long, only since her winter in Roxbury. She went even further to express her frustrations with language itself and her ability to articulate her beliefs. "*I confes [sic] my Expressions was that way but it was never my Judgment*." [65] Certainly this argument did not impress her "judges." They began to sense the futility of the entire discourse. John Wilson declared "this you say is most dayngerous, for if your Judgment all this while be not altered but only your Expressions, whan your Expressions are soe contrary to the Truth." [66]

The condemnation of Anne Hutchinson as a familist was the symbol of a vile heresy referring back to nonconformist sects prevalent in Europe in the sixteenth and seventeenth centuries. Although Hutchinson herself denied the association, those ministers who examined her saw the possible link between her beliefs and those of the families as damning and unconscionable. The label familist was the one used by the governing authorities, while the term "antinomian" only appeared in the next decade.[67] Practicing familists, particularly in Holland, were disproportionately women who eschewed marriage and embraced a social structure which gained the name of "the community of women," condoning a form of free love.[68]

For the hierarchy in early colonial Boston, then, "the community of women" was an abominable image. At the same time, to see the parallel between the Hutchinsonians and the familists was to bring order out of chaos for the interrogators. The phrase and the accusation involved allowed them to pinpoint and identify Hutchinson's crime, a task which was otherwise troublesome and confusing. Of course, a community of women did exist at the core of Anne Hutchinson's activities. This community, however, was not familist in origin and, therefore, Hutchinson denied the existence of the community. In this historical period women created a community, as we have seen, in the ritual of childbirth. Women shared with one another the experience of giving birth, the preparations for labor (in the specific instance of the Hutchinson gatherings) and the supervision of birth (i.e., midwifery). The community of women had a felt presence for Hutchinson during her trial and examination since their solidarity in silence protected her from prosecution. Possibly governing authorities, especially the ministers associated with Wilson and Winthrop, felt excluded and cut off from the secrets held by the silent women.

The prescribed limits of gender, most particularly women, provided a focal point throughout the period of interrogation and exile to Roxbury. For instance, one of the most obvious aspects of Anne Hutchinson's heresy was her insubordination as she characterized certain ministers as misguided and ill informed. This charge appeared early in her examination and is ready testimony to the underlying issue of the proper place of women in colonial New England. Members of the General Court admonished Anne Hutchinson for typifying the great part of the ministerial community as operating under a covenant of works, while coveting the powers and sanctification of the covenant of grace for her followers, her sympathizers and her teacher and minister John Cotton. The covenant structure of Puritan New England did not imply equality but rather subordination in a hierarchy. A covenant with God naturally implied inequality in status, and governmental covenants, family covenants and covenants among congregations were no exception. Within the family the wife was covenanted to serve her husband. As in legal contracts reciprocity was specified so that the husband was covenanted to provide love and protection to his wife and children.[69] The Reverend Hugh Peters declared toward the end of the church proceeding, speaking directly to Hutchinson, "*. . . you have rather bine a Husband than a Wife and a preacher than a Hearer; and a Magistrate than a subject.*"[70] In a similar vein, not long before, Brother Wilson admonished her for "*. . . slightinge of Gods faythfull ministers and contemninge and cryinge downe them as Nobodies.*"[71]

Consciousness of gender differences abounded during Hutchinson's interrogations. Anne Hutchinson and the ministers disagreed sharply in their use of language. Hutchinson appeared to the others to be too powerful, especially among the women. In her conceptualization and wording she seemed to assume an equality for women before God.

Transcripts of these proceedings make it clear that women were at once too strong and too weak, a duality which in many ways parallels the duality of the midwife. Women were more vulnerable to the malfeasance of the Devil because they were characteristically weak and yielding. Anne Hutchinson stood at the other pole for she had stepped out of the bounds of femininity and was too proud and forward. In early Puritan Massachusetts, then, the culture status of women was ill defined.[72] Midwifery coincided with concepts of femininity so that the midwife was a ministering angel, so pure as to be immortal, or the handmaiden of the Devil, so vilified as to be damned.

Anne Hutchinson met exile and excommunication at the end of her interrogations. She was cast out "... as a leper."[73] She and her family moved first to Rhode Island. Shortly after her arrival there, she gave birth to a hydatidiform mole, instead of the child which had been expected.[74]

Modern medicine understands this event as a gestational pathology, varying in frequency, but more common among young women in their teens or women over forty. The tumorlike growth which is conceived and born, so to speak, can lead to cancer.[75] Although there is no evidence to reveal the effect of the experience on Hutchinson, there does seem to be a plentiful assortment of comments from Winthrop, who mandated John Clarke, physician and sympathizer with Hutchinson, to describe fully the appearance of the mole. When he set about to write his narrative of the Antinomian Controversy, Winthrop lumped together Hutchinson's mole and Dyer's monster, seeing them both as evidence of God's disapproval of the women.

> Then *God* himselfe was pleased to step in with his casting voice, and brings his owne vote and suffrage from heaven, by testifying his displeasure against their opinions and practises, as clearly as if he

had pointed with his finger, in causing the two fomenting women in the height of the opinions to produce out of their wombs, as before they had out of their braines, such monstrous birth as no chronicle (I thinke) hath hardly ever recorded the like.[76]

In 1638 William Bradford responded to a letter from Winthrop, beginning with "I heard since of a monsterous, and prodigious birth . . .," while Roger Williams wrote ". . . thank you for that sad relation of the monster. . . ."[77] Winthrop saw the hand of the devil in the so-called monster births and wrote to Williams and others to tell the story. For Winthrop these unfortunate pathologies symbolized the entire theological controversy. Even before Hutchinson's accident of birth occurred, the governing powers determined that the vilifying fallacy underneath the conflict was the "abominable" community of women.

The gathering of the community of women began in the meetings in Anne Hutchinson's home, within the context of childbirth and religion. The solidarity of the women grew and the strength of her followers did also. The extension of the gathering to include men further threatened the religious and political hierarchy, yet it was the "communitie of women" which encapsulated fury.

Some have argued that Anne Hutchinson and her followers represented female agitation.[78] Lyle Koehler argues that the "communitie of woemen" dared ". . . to rebel openly against the spiritual and secular status quo."[79] In "The Case of the American Jezebels," Koehler contends that the group of women together threatened the ministry, more than Hutchinson alone. He suggests that their actions in part offered a definition for the female role in the newly forming colony.

The fact remains, however, that the defiant assertion of female rights was not the conscious goal of Anne Hutchinson and her followers. Hutchinson fell into what came to be defined as heretical way with a sincere and overpowering belief in Puritan theology. Supportive testimony to her conviction appears in the course of the trial in 1638 when she clearly could not gain vindication from the Elders. In answer to Peter Bulkeley's charge of familism she protrayed herself as Christ martyred with those around her doubting her religious purity and strength.[80] Thus her identification with theology was complete.

Nonetheless, her activity within the community of women grew out of female-dominated childbirth ritual, all the while fortified by an overwhelming involvement in Puritan theology. Anne Hutchinson took a serious interest in the spiritual welfare of these women and attempted to give them religious assurance the best way she knew; through spiritual preparation and scriptural study.

In retrospect Mary Dyer's aberrational delivery provides us with important clues to early colonial childbirth and to the social matrix of the Antinomian Controversy. The communal aspect of the ritual had widespread implications not only for the midwife, but for the women in labor and the other attendants as well. They shared an experiential knowledge of a primary rite of passage. What colonial men learned about a particular birth depended upon the communication of the midwife. At the same time, the communal birth lent a solidarity to the women involved and a sharing of tension: a considerable quality in the context of physical uncertainty in the face of possible death and spiritual uncertainty before possible damnation. The Mary Dyer delivery demonstrates as well that Anne Hutchinson was actively involved in midwifery to the extent that her closest colleagues included another midwife who was not permitted membership in the church and a pregnant young woman who later inclined toward radical religious notions. The childbirth ritual cut across theological lines and united women behind its very universality.

At the same time, the Mary Dyer delivery uncovers a tension in seventeenth-century childbirth which was symbolized by the ambiguous status of the midwife. Fear of a natural yet little understood process could turn gratitude and respect into hatred and accusation. Anne Hutchinson initially brought confidence and assurance to childbirth, which attracted a following and a community of women in need of spiritual certainty. During the trial and examination, however, the governor and the ministers, elders, and magistrates did not directly question her midwifery. Rather the terms of the charges vacillated between her challenge to the hierarchy and her erroneous opinions. There was, nonetheless, an underlying conflict between the communal childbirth ritual and Puritan theology, a conflict which found expression in such a phrase as "the communitie of woemen." Anne Hutchinson found cultural power in a ritual surrounding the biological work of women, and consequently threatened profoundly the male-dominated ideological structure of that Puritan community.

This statue of Anne Hutchinson, sculpted in 1911 by Cyrus E. Dallin, stands on the grounds of the State House in Boston. (Photo courtesy of the State Library of Massachusetts, Special Collections.)

NOTES

1. John Winthrop, *A Short Story of the Rise, Reign and Ruine of the Antinomians, amilists and Libertarians* in David D. Hall, ed., *The Antinomian Controversy, 636-1638: A Documentary History* (Middletown, Conn.: Wesleyan University Press, 968), 262.

2. *Ibid.*, 263.

3. Clifford Geertz, *Islam Observed: Religious Development in Morocco and ndonesia* (Chicago: University of Chicago Press, 1968) 97.

4. Many scholars have written about the Antinomian Controversy. David Hall dited a well-selected group of documents concerning the conflict in *The Antinomian ontroversy*. George Ellis' "Life of Anne Hutchinson, with a Sketch of the Antinomian ontroversy in Massachusetts," in Jared Sparks' *American Biography* XVI, 167-376, is seful. See also Charles Francis Adams, *Three Episodes of Massachusetts History: The ettlement of Boston Bay' The Antinomian Controversy, A Study of Church and Two overnments* I & II (Boston: Houghton, Mifflin & Co., 1892). Also John Winthrop's *istory of New England from 1630 to 1649*, James Savage, ed., (New York: Arno Press eprint, 1972). See "The Trial of Mrs. Anne Hutchinson," in *Proceedings of the assachusetts Historial Society*, 2nd Series, vol. IV (October 1888) pp. 160-191. See homas Hutchinson, *The History of the Colony and Province of Massachusetts Bay*, II, dited by Thomas Mayo (Cambridge, Massachusetts: Harvard University Press, 1936). ppendix II contains the Examination of Anne Hutchinson, November, 1637. More ecently see Francis J. Bremer, ed., *Anne Hutchinson: Troubler of the Puritan Zion* Huntington, N.Y.: Robert Krieger Publishing Co., 1981); Emery Battis, *Saints and ectaries: Anne Hutchinson and the Antinomian Controversy in the Massachusetts Bay olony* (Chapel Hill: University of North Carolina, 1962). See also William K. B. toever, *'A Faire and Easie Way to Heaven': Covenant Theology and Antinomianism in arly Massachusetts* (Middletown, Conn.: Wesleyan University Press, paperback dition, 1988, first edition 1978). Larzer Ziff's *The Career of John Cotton: Puritanism nd the American Experience* (Princeton: Princeton University Press, 1962) nvestigates the issues of the religious controversy. The literature in fact is vast and eyond the scope of a single footnote. Lyle Koehler's "The Case of the American ezebels: Anne Hutchinson and Female Agitation During the Years of Antinomian urmoil, 1636-1640," *William and Mary Quarterly 31* (1974) 55-78 is an important

article. Recent articles include James F. Cooper, Jr., "Anne Hutchinson and the 'La Rebellion' Against the Clergy," *New England Quarterly 61* (September 1988) 381-397.

5. See Bremer, *Anne Hutchinson*, and Cooper, "'Lay Rebellion.'" Be Barker-Benfield, "Anne Hutchinson and the Puritan Attitude Toward Women, *Feminist Studies 1* (Fall 1972) 65-96, and Selma Williams, *Divine Rebel: The Life o Anne Marbury Hutchinson* (New York: Holt, Rinehart and Winston, 1981). See als Ann Fairfax Withington and John Schwartz, "The Political Trial of Anne Hutchinson, *New England Quarterly 51* (June 1978) 226-240. These are a few examples of ongoin research in the area.

6. See Michel Foucault, *The Birth of the Clinic: An Archaelogy of Medica Perception* (New York: Pantheon Books, 1973); Larry Dossey, *Space, Time an Medicine* (Boulder: Shambhala Press, 1982); Owsei Temkin, *The Double Face of Janu* (Baltimore: The Johns Hopkins University Press, 1977) and Carolyn Merchant, *Th Death of Nature: Women, Ecology and the Scientific Revolution* (New York: Harper an Row, 1980).

7. See for instance Perry Miller, *The New England Mind: The Seventeent Century* (New York: McMillan Co., 1939, 1961) and Edmund Morgan, "The Cas Against Anne Hutchinson," *New England Quarterly* 10 (1937) 635-649. Anothe example is Darrett Rutman, *Winthrop's Boston: Portrait of a Puritan Town, 1630-164* (Chapel Hill: University of North Carolina, 1965) and also Koehler's "America Jezebels." Examples of the latter type, medical history include Samuel Abbott Green *History of Medicine in Massachusetts* (Boston: A. Williams and Co., 1881), John B Blake, *Public Health in the Town of Boston* (Cambridge: Harvard University Press 1959) and Harold Speert, *Obstetrics and Gynecology: A History* (Baltimore: Waverl Press, 1980). Some scholarship does attempt to bridge the schism. For instance se Nancy Cott's introduction to the controversy in *Root of Bitterness: Documents of th Social History of American Women* (New York: E. P. Dutton & Co., 1972), Be Barker-Benfield's "Anne Hutchinson" or my review of Selma Williams' biography in *Se Roles* (April 1983).

8. Cotton Mather, *Angel of Bethesda* (Barre, Mass.: American Antiquaria Society, 1972). See also his *Magnalia Christi Americana* (New York: Frederick Ungar 1970). See Richard D. Brown, "The Healing Arts in Colonial and Revolutionar Massachusetts: The Context for Scientific Medicine," in *Medicine in Colonia Massachusetts, 1620-1820* (Boston: Colonial Society of Massachusetts, 1980) vol. 57 35-47; and Eric Christianson, "Medicine in New England," in Ronald Numbers, ed. *Medicine in the New World* (Knoxville: Univ. of Tenn. Press, 1987), 101-153, and "Th Medical Practitioners of Massachusetts, 1630-1800: Patterns of Change an Continuity" in *Medicine in Colonial Massachusetts,* vol. 57.

9. Winthrop, *Short Story,* 206.

10. Catherine Scholten's *Childbearing in American Society* and Laurel Ulrich' *Good Wives: Image and Reality in the Lives of Women in Northern New England*

1650-1750 (New York: Oxford Univ. Press, 1983) contain some detail on childbirth in the early colonial period. See also Claude Edwin Heaton, M.D., "Obstetrics in Colonial America," in *American Journal of Surgery 45* (July-Sept. 1939), 606-610.

11. Although historians from George Ellis in the 1830s through Charles Francis Adams in the 1890s and the biographers of the 1930s and even further on through Battis and the present day describe Hutchinson's midwifery, Winthrop and Cotton alone provide the meager phrases which simply called her a woman helpful in the time of childbirth.

12. There are several biographies of Hutchinson. Edith Curtis, *Anne Hutchinson: A Biography* (Cambridge: Washburn & Thomas, 1930); Helen Augur, *An American Jezebel: The Life of Anne Hutchinson* (New York: Brentono's, 1930); Winifred Rugg, *Unafraid: A Life of Anne Hutchinson* (Boston: Houghton Mifflin, 1930) appeared for the tricentennial anniversary of the Antinomian Controversy. Williams' *Anne Hutchinson* is a recent biography which reflects extensive research and useful detail of her life in England and the New World. See also David Grayson Allen, *In English Ways; The Movement of Societies and the Transferal of English Local Law and Custom to Massachusetts Bay in the Seventeenth Century* (New York: Norton Paperback, 1982) for explorations of settlement patterns.

13. At the beginning of her appearance before the Church of Boston, March, 1638, Hutchinson sent a note outlining her retractions and opinions. See "Report of the Trial" in Hall, ed., *The Antinomian Controversy.*

14. John Cotton, "The Way of Congregational Churches Cleared," in David D. Hall, ed., *The Antinomian Controversy*, 412.

15. Winthrop, *A Short Story*, 263.

16. Cotton and Winthrop both testify to the skills of midwifery and medicine Hutchinson brought with her. See also Williams' biography. Ronald Numbers' introduction to *Medicine in the New World* expresses current historical interpretation which lays great store by the influence of the Old World on the ways of the New World, in medicine as well as other facets of the past.

17. Adrian Wilson, "Participant or Patient: Seventeenth-Century Childbirth from a Mother's Point of View," in Roy Porter, ed., *Patients and Practitioners: Lay Perceptions of Medicine in Pre-Industrial Society* (Cambridge: Cambridge University, 1985), 129-144. See Ulrich, *Good Wives* and Scholten, *Childbearing*. See also Peter Ruston, "Purification or Social Control? Ideologies of Reproduction and the Churching of Women after Childbirth" in Eva Gomarnikow, David Morgan, June Purvis and Daphne Taylor, eds., *The Public and the Private* (London: Heinemann, 1983) 118-131. Deborah Gorhman's "Birth and History," *Histoire Social-Social History* XVII (Nov. 1984), 383-394, provides valuable insights into the role of the midwife. See also Jane B. Donegan, *Women and Men Midwives: Medicine, Morality, and Misogyny in Early America* (Westport, Conn.: Greenwood Press, 1978) and Richard W. and Dorothy C.

Wertz, *Lying-In: A History of Childbirth in America* (New York: Schocken Books, 1975).

18. See George Gifford, Jr., "Botanic Remedies in Colonial Massachusetts, 1620-1820," in *Medicine in Colonial Massachusetts*, 263-288; George Dow, *Every Day Life in the Massachusetts Bay Colony* (Boston: Society for the Preservation of New England Antiquities, 1935), Dover Reprint, 1988; Maurice Bear Gordon, M.D., *Aesculapius Comes to the Colonies* (Ventnor, N.J.: Ventnor Publishers, 1949). All these sources offer information on the use of herbs during childbirth.

19. See Gifford, "Botanic Remedies" for reference to exchange between colonists and Native Americans. Sources for the possible herbs used include Culpepper and Leighton. The relationship between Hutchinson, her followers and the Indians is another topic which deserves study. Their group was unwilling to fight in the Pequot War, which further goaded Winthrop and also suggests an affinity for interest in the Indians.

20. In *Good Wives* Ulrich suggests possible herbs used during birth. Ann Leighton's *Early American Gardens: For Meate Or Medicine"* (Boston: Houghton Mifflin, 1970) details plants that grew in the New World. *Culpepper's Complete Herbal and English Physician* (Manchester: J. Gleave & Son, 1826) is useful for details of medical application of plants.

22. See Wilson, "Participant or Patient?" and also Rushton, "Purification or Social Control?"

23. See Thomas Rogers Forbes, *The Midwife and the Witch* (New Haven: Yale University Press, 1966). The actual role of New World Puritan midwives in baptism is unknown, although they probably had no such duty. Information on baptism is in Keith Thomas, *Religion and the Decline of Magic* (New York: Charles Scribner and Sons, 1971) and Norman Pettit, *The Heart Prepared: Grace and Conversion in Puritan Spiritual Life* (New Haven: Yale University Press, 1966).

24. See Joseph R. McElrath, Jr., and Allan P. Robb, eds., *The Complete Works of Anne Bradstreet* (Boston: Twayne Publishers, 1981), 179-180. Although the exact date of the writing of this poem is unknown, Bradstreet probably was born in 1612 and died in 1672. The poem was likely written in the 1630s or 1640s. Her father was Thomas Dudley, a governor of the Bay Colony and a contemporary of Winthrop's.

25. See James H. Cassedy, "Church Record-Keeping and Public Health in Early New England," in *Medicine in Colonial Massachusetts*, vol. 57, 249-262, and *Demography in Early America* (Cambridge, Mass.: Harvard University Press, 1969); John B. Blake, "The Early History of Vital Statistics in Massachusetts," *Bulletin of the History of Medicine 29* (1955), 46-54, and *Public Health in the Town of Boston 1630-1822* (Cambridge, Mass.: Harvard University Press, 1959); John Demos "Families in Colonial Bristol, R.I.: An Exercise in Historical Demography," *William and Mary Quarterly 25* (1968) 41-57; Maris A. Vinovskis, "Mortality Rates and Trends in Massachusetts before 1860," *Journal of Economic History 32* (1972), 184-213; and

Daniel Scott Smith, "The Estimates of Early American Historical Demographers: Two Steps Forward and One Step Back, What Steps in the Future?" *Historical Methods 12* (Winter 1979), 24-38.

26. See David Stannard, *The Puritan Way of Death: A Study in Religion, Culture, and Social Change* (New York: Oxford University Press, 1977) for one interpretation of death among Puritan children.

27. See Stoever, *"A Faire and Easie Way"* and Norman Pettit, *The Heart Prepared: Grace and Conversion in Puritan Spiritual Life* (New Haven: Yale University Press, 1966) for explorations of preparation among the Puritans.

28. See Cotton Mather "Elizabeth in Her Holy Retirement" (Boston: 1710). Also John Oliver, *A Present for Teeming American Women* (Boston, 1694). These primary sources clearly show the suffering and spiritual anxiety urged upon pregnant women in preparation for childbirth. John Demos, *Entertaining Satan: Witchcraft and the Culture of Early New England* (New York: Oxford University Press, 1982) and Lyle Koehler, *The Search for Power: The Weaker Sex in Seventeenth Century New England* (Urbana: University of Illinois Press, 1980), both explore the tensions and restrictions placed on women in colonial New England.

29. John Cotton, "The Way Cleared," in Hall, ed., *Antinomian Controversy*, 412.

30. The original transcripts appear in the *Proceedings of the Massachusetts Historical Society* IV, 160-191 and Thomas Hutchinson's *History of Colony and Province of Massachuetts Bay II.*

31. "The Trial of Mrs. Anne Hutchinson," in Hall, ed., *Antinomian Controversy*, 336-337.

32. John Cotton's ambivalence is the subject of much scholarship. See for instance, Ziff, *John Cotton* and Stoever, *"A Faire and Easie Way."*

33. "The Trial of Mrs. Anne Hutchinson", in Hall, ed., *Antinomian Controversy*, 265. See Battis, *Saints and Sectaries* and Cooper "Lay Rebellion" for interesting detail of community division and the social make-up of Hutchinson's following.

34. See Winthrop's *History*. See Stoever, *"A Faire and Easie Way"* for clarification and exploration of the theology of Hutchinson and Cotton.

35. Theodore Weld in Winthrop's *Short Story,* in Hall, ed., *Antinomian Controversy,* 212-213.

36. *Ibid*., 263. See Carol Karlsen, *The Devil in the Shape of a Woman: Witchcraft in Colonial New England* (New York: W.W. Norton, 1987) for a treatment of Hutchinson, Dyer and Hawkins as witches. There are inaccuracies in the chronology of her endering, unfortunately. Demos' *Entertaining Satan* is an admirable study of many

aspects of witchcraft, downplaying in conclusion the connections between midwifery and witchcraft. The issue of a possible link remains open, and possibly irresolvable. See also Michel Foucault, "Médècins, Juges et Sorciers au XVIIe Siècle" in *Médècine de France 200* (1969), 121-128.

37. See Elizabeth Dexter, *Colonial Women of Affairs* (Boston: Houghton Mifflin Co., 1924), 58. See William Bradford, *Of Plymouth Plantation,* edited by Harvey Wish (New York: Capricorn, 1962).

38. Winthrop, *History*, 263.

39. *Ibid*. See Cotton, "The Way Cleared."

40. *Ibid*.

41. *Ibid*., 261. See also Jane Donegan, *Women and Men Midwives*.

42. See Gordon, *Aesculapius* and Harold Speert, *Iconographia Gyniatrica* (Philadephia: F.A. Davis, 1973), 75.

43. Winthrop, *History*.

44. Winthrop, *Journal,* 267-268.

45. Winthrop, *History*, 263.

46. For readings on scientific thought and the Puritans, see Charles Webster, *The Great Instauration: Science, Medicine and Reform, 1626-1660* (New York: Holmes and Meyer, 1975); Ronald Numbers, "Science and Religion," *Osiris 1* (1985) 59-80; John B. Blake, *Public Health in the Town of Boston, 1630-1822* (Cambridge: Harvard University Press, 1959); and James Cassedy, *Demography in Early America* (Cambridge: Harvard University Press, 1969).

47. See Thomas Shepard's "Memoir of His Own Life" in Alexander Young, ed., *Chronicles of the First Planters of the Colony of Massachusetts Bay, 1623-1636* (New York: DeCapo Press, 1980 reprint of 1846 original), 555.

48. Winthrop, *A Short Story*, 281.

49. Winthrop, *History*, 261.

50. *Ibid.*

51. *Ibid*., 263.

52. *Ibid*.

53. Winthrop, *A Short Story*, 282.

54. See Frederick C. Waite, "The Development of Anatomical Laws in the States of ew England," *The New England Journal of Medicine 233* (Dec. 13, 1945) 716-726; '.R. Steiner, "Some Early Autopsies in the U.S.," *Johns Hopkins Hospital Bulletin* 149 (August 1903), 201-203; and Bear, *Aesculapius*.

55. From the Massachusetts Bay Colony *Records* I. 219, in James Savage, ed., John 'inthrop's *History*, 263n.

56. Williams, *Anne Hutchinson*.

57. "A Report of the Trial," in Hall, ed., *Antinomian Controversy*, 351. See also C. elen Brock with E. Christianson, "Appendix" in *Medicine in Colonial Massachusetts* r details of Olliver's background as a surgeon.

58. "The Trial," in Hall, ed., *Antinomian Controversy*, 354.

59. *Ibid.*, 356.

60. *Ibid.*, 357.

61. *Ibid.*, 358.

62. *Ibid.*, 361.

63. *Ibid.*, 362.

64. Ambroise Paré's sixteenth century book, *On Monsters and Marvels*, translated ' Janis L. Pallister (Chicago: University of Chicago Press, 1982) does not describe an cident of birth such as that of Mary Dyer. Nor does Percival Willoughby (1596-1685) his *Observations in Midwifery* (Wakefield, England: S.R. Publishers Limited, 1972).

65. "The Trial," 378. See Patricia Caldwell, "The Antinomian Language ›ntroversy," *Harvard Theological Review* (1976), 345-367, for a consideration of nguage and articulation of self on the part of Anne Hutchinson. See also David Hall, 'he Mental World of Samuel Sewall," in Hall et al, eds., *Saints and Revolutionaries: ›ssays on Early American History* (New York: W.W. Norton, 1984), for discussion of nguage, emotion and spiritualism among the Puritans.

66. "The Trial," 378.

67. Williams, *Anne Hutchinson*.

68. *Ibid*. See also Stoever, "*A Faire and Easie Way*" and Thoms, "Women and the ıvil War Sects," *Past and Present 13* (April 1958), 42-62, for descriptions of Familism.

69. See Lyle Koehler, "The Case of the American Jezebels: Anne Hutchinson and Female Agitation During the Years of Antinomian Turmoil, 1636-1640," *William and Mary Quarterly 31* (1974), 55-78; Mary Maples Dunn, "Saints and Sisters: Congregational and Quaker Women in the Early Colonial Period," and Gerald Moron, "'Sisters in Christ:' Women and Men in Church in Seventeenth-Century New England," both in Janet Wilson James, ed., *Women in American Religion* (Philadelphia: University of Pennsylvania, 1980); Rosemary Skinner Keller, "New England Women: Ideology and Experience in First-Generation Puritanism (1630-1650)," in Rosemary Radford Ruether and Rosemary Skinner Keller, eds., *Women and Religion in America* vol. 2 (San Franciscio: Harper & Row, 1983), 132-192.

70. "Trial of Mrs. Anne Hutchinson," in Hall, ed., *Antinomian Controversy,* 383.

71. *Ibid.*, 380.

72. Mitchell Breitwieser found that Cotton Mather's third wife held a similar position in Puritan New England. See his "Cotton Mather's Crazed Wife," *Glyph* (1979), 88-113. See also Dunn, "Saints and Sisters." John Demos, in *Entertaining Satan,* has gone a long way towards deriving the sources of definitions of womanhood in this period. He suggests a connection between accusations of witchcraft, the experience of menopause among colonial Puritan women and also that of middle age.

73. "The Trial of Mrs. Anne Hutchinson," 388.

74. Dr. Paul Younge first made this diagnosis for Battis in *Saints and Sectaries.* There was, however, earlier recognition of this condition. See Beryl Rowland, tr., *Medieval Woman's Guide to Health: The First English Gynecological Handbook* (Kent, Ohio: Kent State University Press, 1981).

75. See John Sciarra, ed., *Gynecology and Obstetrics*, vol. 4 (Philadelphia: Lippincott Co., 1988); and Martin Pernall and Ralph Benson, *Current Obstetrical and Gynecologic Diagnosis and Treatment* (Sydney, Australia: Appleton & Sage, 1987) for recent medical discussion of gestational trophoblastic neoplasms. This work suggests that the connections between hydatidiform moles and menopause are less clear and direct than the appendix to Battis' *Saints and Sectaries* suggests.

76. Winthrop, *Short Story*, 214.

77. Massachusetts Historical Society, *Winthrop Papers,* vol. IV, 1638-1644 (Boston: Merrymount Press, 1944).

78. See Koehler, "American Jezebels;" Barker-Benfield, "Anne Hutchinson;" and Kammen, *The Devil*, among others.

79. Koehler, "American Jezebels," 59.

80. "The Trial of Mrs. Anne Hutchinson," 353.

AUTHOR

Deborah Kuhn McGregor received her Ph.D. in Women's History from the State University of New York at Binghamton in 1986. Her research has centered on the history of childbirth and its ramifications for society in early America. She is interested in the general history of women and the family, and their relationships to the practice of medicine.

Professor McGregor serves as a member of the history faculty at Sangamon State University, and also teaches courses in the Department of Medical Humanities at Southern Illinois University School of Medicine in Springfield. She currently is completing work on her book, *Sexual Surgery and the Origins of Gynecology: J. Marion Sims, His Hospital, and His Patients*, which will appear in the spring of 1990.

"The Living Mother of a Living Child": Midwifery and Mortality in Post-Revolutionary New England

Laurel Thatcher Ulrich

FORTY-ONE years ago, Richard Harrison Shryock could summarize the history of early American midwifery in a few sentences. "The history of obstetrics and of pediatrics," he wrote, "affords other illustrations of the way in which inadequate medical science affected the public health. Maternity cases were left, in English-speaking lands, almost entirely to midwives. . . . And since midwives lacked any scientific training, obstetrics proceeded on the level of folk practice, and with consequences which may be easily imagined."[1] The consequences could be imagined because few persons in 1948 doubted the superiority of medical science over folk practice.

The advent of "natural" childbirth, culminating in recent years in the revival of lay midwifery, has changed historical judgments as well as obstetrics. In revisionist histories of childbirth, the pleasant story of scientific progress has been replaced by a darker tale of medical competitiveness and misplaced confidence in an imperfect science. Medical science did not on the whole increase women's chances of surviving childbirth until well into the twentieth century, the new histories argue, and may actually have increased the dangers. As Richard W. and Dorothy C. Wertz explain, puerperal fever, the dreaded infection that killed so many women in the nineteenth century, "is probably the classic example of iatrogenic disease—that is, disease caused by medical treatment itself."[2]

Ms. Ulrich is a member of the Department of History, University of New Hampshire. Versions of this article were presented at meetings of the Benjamin Waterhouse Medical Society (Boston University), the Maine Society for the History of Medicine, the American Antiquarian Society Seminar in Political and Social History, and the comparative history seminar at the University of New Hampshire. Acknowledgments: I am grateful to those groups and to Worth Estes, Judith Walzer Leavitt, Janet Polasky, and Cornelia Dayton for helpful comments. Some parts of this essay will appear in my forthcoming book on the diary of Martha Moore Ballard.

[1] Shryock, *The Development of Modern Medicine: An Interpretation of the Social and Scientific Factors Involved,* rev. ed. (London, 1948), 77-78.

[2] Judith Walzer Leavitt, *Brought to Bed: Childbearing in America, 1750 to 1950* (New York, 1986), 56-57, and "'Science' Enters the Birthing Room: Obstetrics in America since the Eighteenth Century," *Journal of American History,* LXX (1983),

Although historians trace to the eighteenth century the gradual supplanting of midwives by physicians, most detailed studies have concentrated on the nineteenth century or later. The few discussions of childbirth in early America have dealt with urban centers and with the work of prominent physicians such as William Shippen of Philadelphia.[3] Almost nothing is known about rural obstetrics or about the activities and attitudes of midwives. This essay begins to fill that gap. Its central document is the manuscript diary of a Maine midwife, Martha Moore Ballard, who lived at Augusta (then part of Hallowell) from 1778 to 1812. It also uses English obstetrical literature, scattered physicians' and midwives' records from Maine and New Hampshire, and the papers of Dr. Jeremiah Barker of Gorham, Maine.

Martha Ballard performed her first delivery in 1778, though her diary does not begin until six years later. Between 1785 and 1812 she recorded 814 deliveries. The expansiveness of her record is unusual not only among midwives (few of whom left any written evidence of their practice) but among country physicians as well. Yet the diary has received little scholarly attention. Historians who have used it have relied on an abridged version published in Charles Elventon Nash's *History of Augusta.* For most, the details of Ballard's practice have seemed less important than her symbolic image as a "traditional" midwife. One work portrays her as an untrained, intensely religious, and poorly paid practitioner, who nevertheless shared some of the attitudes of contemporary physicians. Another associates her with nineteenth-century controversies between midwives and physicians, emphasizing her helplessness when accused by a local physician of "meddling by giving her opinion of a disease."[4]

Serious study of the entire diary shatters such stereotypes. Although physicians were delivering babies in Hallowell as early as 1785, Martha Ballard was clearly the most important practitioner in her town. Because her record documents traditional midwifery at a moment of strength, it allows us to shift the focus of inquiry from the eventual triumphs of medical science to the immediate relations of doctors and midwives in an era of transition. What is most apparent on close examination is the *success* of Ballard's practice, measured on its own terms or against contemporary medical literature. Although elements of the new obstetrics had begun to

---

281-304; Wertz and Wertz, *Lying-In: A History of Childbirth in America* (New York, 1977), xi, x, 128.

[3] Catherine M. Scholten, "'On the Importance of the Obstetrick Art': Changing Customs of Childbirth in America, 1760 to 1825," *William and Mary Quarterly,* 3d Ser., XXXIV (1977), 429-431, and *Childbearing in American Society, 1650-1850* (New York, 1985), chap. 2; Wertz and Wertz, *Lying-In,* chap. 2; Leavitt, *Brought to Bed,* 36-44, 263-265.

[4] Nash, *The History of Augusta: First Settlements and Early Days as a Town, Including the Diary of Mrs. Martha Moore Ballard (1785-1812)* (Augusta, Me., 1904); Wertz and Wertz, *Lying-In,* 9-10; quotation from Scholten, *Childbearing,* 45; Leavitt, *Brought to Bed,* 37.

filter into the region, the old rituals of childbirth remained powerful. In her record, it is the physicians—particularly the young physicians— who appear insecure and uncomfortable.

The diary also extends and deepens recent discussions concerning eighteenth-century modes of delivery. A number of historians have argued that English innovations, such as William Smellie's improved forceps, encouraged an interventionist obstetrics that eventually displaced the gentler practices of midwives. Edward Shorter has countered that eighteenth-century English midwives were themselves "wildly interventionist" and that physicians, not midwives, introduced the notion of "natural" childbirth. This new obstetrics, he argues, was in part a response to general cultural trends—a medical reflection of Enlightenment respect for nature—and also a consequence of the work of pioneering physicians like Charles White, whose textbook published in London in 1773 was the first example of a fully noninterventionist obstetrics.[5]

By shifting the balance of attention from obstetrical prescriptions to obstetrical results, Ballard's diary provides a new vantage point for assessing this controversy. Although it reveals little about the particulars of Ballard's methods (we do not know, for example, whether she applied hog's grease to the perineum or manually dilated the cervix), it offers compelling evidence of her skill. Maternal and fetal mortality rates extracted from the diary compare favorably with those for physicians in both England and America, countering the horror stories of eighteenth-century literature as well as the casual assumptions of twentieth-century historians. The consequences of Ballard's practice need not be imagined.

In most respects Martha Ballard's is a typical eighteenth-century rural diary—a laconic record of weather, sermon texts, family activities, and visits to and from neighbors. Obstetrical and general medical entries are interwoven with this larger accounting of ordinary life, although she gave birth records a special significance by summarizing them in the margins, numbering each year's births from January to December. Each delivery entry gives the father's name, the child's sex, the time of birth, the condition of mother and infant, and the fee collected. Many also include the time of the midwife's arrival and departure, the names of the attendants who assisted her, and the arrival of the "afternurse," who cared for the woman during lying-in (the week or two following delivery). Succeeding entries record follow-up visits or hearsay reports about the mother and child.

The account of Tabitha Sewall's delivery on November 12-13, 1790, is typical:

[5] Wertz and Wertz, *Lying-In*, 34-43; Scholten, *Childbearing*, 34-36; Leavitt, *Brought to Bed*, 38-40; Shorter, "The Management of Normal Deliveries and the Generation of William Hunter," in W. F. Bynum and Roy Porter, eds., *William Hunter and the Eighteenth-Century Medical World* (Cambridge, 1985), 371-383.

> I was Calld by Colonel Sewall to see his Lady who was in Labour. Shee was not so ill as to Call in other assistance this day. I slept with her till about 1 hour morn when shee calld her Neighbours to her assistance. Mrs Sewal was ill till 3 hour pm when shee was thro divine asistance made the Living Mother of a Living Son her 3d Child. Mrs Brooks, Belcher, Colman, Pollard & Voce assisted us . . . Colonel Sewall gave me 6/8 as a reward. Conducted me over the river.

The only unusual thing about this account is the reference to "divine asistance," suggesting that Mrs. Sewall or her midwife encountered some difficulty along the way. Everything else about the description is routine. The father or a near neighbor summoned the midwife. The woman remained "ill" for several hours. Just before the birth she called her female neighbors. The child was delivered—safely. The father paid the midwife and escorted her home. In the eight deliveries Martha Ballard performed for Tabitha Sewall, the description of one differs very little from another. Mrs. Sewall "was safe delivered at 7 hour morn of a fine Daughter and is Cleverly," Ballard wrote, or "Mrs Suall Delivard at 1 this morn of a son & is Cleaverly."[6]

Ballard performed her first delivery at the age of forty-three shortly after moving to the District of Maine from Oxford, Massachusetts. Although she had no doubt assisted in many births in Oxford (she was herself the mother of nine children), she seems not to have practiced alone until she came to Hallowell. Demographics may explain her entry into the profession. In Oxford she had been surrounded by older women; her maternal grandmother was still alive in 1777. In Hallowell she was one of the older women in a young and rapidly growing town. The diary opens in January 1785, the year she turned fifty. It ends with her death in May 1812, just ten days after she performed her last delivery at the age of seventy-seven.

The diary tells us nothing of how she acquired her skills, though genealogical data suggest that her family had something of a medical bent. Two of her sisters married doctors; a maternal uncle was a physician. Certainly, her family demonstrated an unusual commitment to education. Her uncle Abijah Moore was Oxford's first college graduate. Her younger brother, Jonathan, was the second.[7] She probably learned midwifery in the

[6] Martha Moore Ballard Diary, 2 vols., Maine State Library, Augusta, Me., Apr. 2, 1788, Dec. 31, 1786. According to the *Oxford English Dictionary,* "cleverly" means "well" or "in health" in some dialects. This is obviously the meaning Ballard intended. Henry Sewall, Tabitha's husband, also kept a diary. He mentioned Martha Ballard's presence on only four of the eight occasions, never recorded paying a fee, and only twice mentioned the presence of other birth attendants. Henry Sewall Diary, Massachusetts Historical Society, Boston, Mass.

[7] The medical tradition continued into the 19th century. Ballard's diary was inherited and preserved by her great-granddaughter, Dr. Mary Hobart, who practiced obstetrics at New England Hospital in Boston. Clara Barton, the Civil War nurse and founder of the American Red Cross, was Ballard's grandniece.

same way her husband, Ephraim, learned milling or surveying—by practice, by observation, and by working alongside someone who knew more than she.

Ballard's assurance as a midwife is the best evidence we have of her training. In almost 1,000 births she did not lose a single mother at delivery, and only five women died in the lying-in period. Infant deaths were also rare. The diary lists fourteen stillbirths in 814 deliveries and five infant deaths within an hour or two of birth. When Mrs. Claton and her infant both died in the autumn of 1787, a week after delivery, Ballard noted the singularity of the event: "I asisted to Lay her out, her infant Laid in her arms, the first such instance I ever saw & the first woman that died in Child bed which I delivered."[8] The sight was as unusual as it was affecting. Under Martha Ballard's care, a woman could expect to become "the living mother of a living child."[9]

By twentieth-century standards, of course, both maternal and infant mortality were high. The diary records one maternal death for every 200 births. Today the rate for the United States is one per 10,000. But as Judith Walzer Leavitt has demonstrated, such dramatic gains in obstetrical safety have come in the past fifty years; as late as 1930 there was one maternal death for every 150 births in the United States. A recent study of early twentieth-century births in a Portsmouth, New Hampshire, hospital gives stillbirth rates five times as high as Ballard's. The turning point for fetal as well as maternal deaths was the 1940s.[10]

The appropriate question is how Martha Ballard's record compared with those of her contemporaries, particularly with New England physicians who began the regular practice of obstetrics in the eighteenth and early nineteenth centuries. Direct comparisons are difficult, in part because physicians' records tend to be organized much differently from hers. Most

[8] Ballard Diary, Aug. 16, 20, 1787. Since the first fatality occurred during the diary period, I have included the 177 pre-diary births in arriving at the total of 991 births.

[9] The phrase was conventional, and it persisted into the 19th century. Leavitt, for example, quotes a woman who gave thanks for having become "the living mother of a living and perfect child" (*Brought to Bed,* 34). Ballard's version of the statement was usually gender specific, as in "the living mother of a living son" or "the living mother of a fine Daughter" (Ballard Diary, Dec. 30, 1789).

[10] Leavitt, *Brought to Bed,* 23-26; Helen M. Wallace, Edwin M. Gold, and Edward F. Lis, eds., *Maternal and Child Health Practices: Problems, Resources, and Methods of Delivery* (Springfield, Ill., 1973), 185; J. Worth Estes and David M. Goodman, *The Changing Humors of Portsmouth: The Medical Biography of an American Town, 1623-1983* (Boston, 1986), 298. In 2.3% of Ballard's deliveries the child was stillborn or died in the first 24 hours of life. For Portsmouth Hospital the figures were 11.4% (1915-1917), 4.8% (1925-1941), 1.2% (1954-1957), and 0.8% (1971-1983). Because methods of compiling statistics vary markedly over time, these numbers must be considered approximations. Stillbirth ratios, for example, might be affected by abortions, spontaneous or induced. On the development of obstetrical record keeping in general see James H. Cassedy, *American Medicine and Statistical Thinking, 1800-1860* (Cambridge, Mass., 1984), 80-83.

are simply a record of fees collected. Some doctors kept notes on unusual cases; a few compiled mortality tables for their towns. Account books, obstetrical case notes, and mortality tables seldom overlap, however, so that we know the numbers of deliveries performed by one physician but not the results, the management of extraordinary cases by another but not the overall caseload, and the incidence of stillbirths for a given town but not the numbers of maternal deaths or the names of practitioners. Comparison with midwives' registers is easier, since midwives typically listed all births, live as well as stillbirths, chronologically from the beginning to the end of their careers. Few such lists survive, however, and none that I have found offers the kind of narrative detail available in the Ballard diary.

Despite the difficulties, it is nevertheless possible to construct some comparisons. Table I gives stillbirth ratios derived from Ballard's diary, two physicians' records, two midwives' registers, and several published mortality tables. At first glance it is the success of Ballard's practice that stands out. Whether her record is compared to that of Hall Jackson, a prominent eighteenth-century physician, or to Lydia Baldwin's, a contemporary Vermont midwife, it is eminent.[11] Yet none of the mortality ratios is as high as impressionistic accounts would lead us to believe, nor are there clear differences between midwives and physicians.

Most obstetrical treatises published in the first three-quarters of the eighteenth century emphasized the terrors of obstructed birth. Even authors who mistrusted "man-midwifery" and the use of forceps acknowledged the problems. Sarah Stone, an English midwife writing in 1737, described a breech delivery in which it took her an hour and a half to turn and extract the fetus. When she reached for the child, it "suck'd my fingers in the Womb, which concern'd me, fearing it impossible for the poor Infant to be born alive." Writing two decades later, Dr. Edmund Chapman, an English physician, included more gruesome tales. Among cautionary examples he cited one ignorant doctor who, not knowing "the Method of *Turning* a Child, made frequent use of the *Hook* and the *Knife,* and several other shocking and barbarous instruments, even while the Child was *Living.*" Dr. William Smellie, the London physician whose improved forceps supposedly solved such problems as these, included vivid case studies in his published works, evenhandedly distributing the blame for mismanaged deliveries among superstitious midwives and

[11] The lack of detail in the other sources makes it difficult to know whether the data are precisely comparable. Ballard's diary distinguishes between stillbirths and deaths within a few minutes or hours of birth. If other records melded those two categories, her record would look better by comparison. Still, adding the five very early deaths in her practice to stillbirths results in a ratio of only 2.3, almost identical with Jackson's and slightly lower than Baldwin's. Jennet Boardman's register includes three categories: "born dead," "died," and "died at age ——— or on ———." I list all those infants described as "dead" or "born dead" as stillborn, but exclude the "died" entries, some of which deaths may have occurred immediately after birth.

TABLE I
COMPARATIVE STILLBIRTH RATES

	Total Births	*Total Stillbirths*	*Stillbirths per 100 Live Births*
Martha Ballard Augusta, Maine 1785-1812	814	14	1.8
Hall Jackson Portsmouth, N.H. 1775-1794	511	12	2.4
Lydia Baldwin Bradford, Vt. 1768-1819	926	26	2.9
James Farrington Rochester, N.H. 1824-1859	1,233	36	3.0
Jennet Boardman Hartford, Conn. 1815-1849	1,113	36	3.3
Portsmouth, N.H. 1809-1810	541	14	2.7
Marblehead, Mass. 1808	222	7	3.3
Exeter, N.H. 1809	53	1	1.9
United States* 1942			2.0

* Fetal death ratio, defined as fetal deaths of 28 weeks or more gestation per 1,000 live births.

Sources: Martha Moore Ballard Diary, 2 vols., Maine State Library, Augusta, Maine; J. Worth Estes, *Hall Jackson and the Purple Foxglove: Medical Practice and Research In Revolutionary America, 1760-1820* (Hanover, N.H., 1979), 120; A Copy of Records from an Original Memorandum Kept by Mrs. Lydia (Peters) Baldwin, typescript, Baker Library, Dartmouth College, Hanover, N.H.; James Farrington Medical Record Books, 1824-1859, Special Collections, Dimond Library, University of New Hampshire, Durham; "Midwife Records, 1815-1849, Kept by Mrs. Jennet Boardman of Hartford," Connecticut Historical Society, *Bulletin*, XXXIII (1968), 64-69; Lyman Spalding, *Bill of Mortality for Portsmouth*, Broadside (Portsmouth, N.H., 1809, 1810); John Drury, *Bill of Mortality for Marblehead, 1808*, Broadside (Marblehead, Mass., 1809); Joseph Tilton, M.D., *Bill of Mortality for Exeter, New Hampshire*, Broadside ([Exeter, N.H., 1809]); Helen M. Wallace, "Factors Associated with Perinatal Mortality and Morbidity," in Helen M. Wallace, Edwin M. Gold, Edward F. Lis, eds., *Maternal and Child Health Practices: Problems, Resources, and Methods of Delivery* (Springfield, Ill., 1973), 507.

poorly trained physicians. In comparison, Ballard's delivery descriptions are remarkably bland: "the foet[u]s was in an unnatural posetion but I Brot it into a proper direction and shee was safe delivered." Usually she said even less: "removed obstructions" or "used means."[12]

Just as striking, given the tenor of the prescriptive literature, is her independence of Hallowell's physicians. Although the English authors agreed that midwives were capable of handling routine deliveries, authorities differed on the question of their ability to negotiate emergencies. Most publishing physicians argued that the sign of a good midwife was her willingness to call for help when needed. As Brudenell Exton put it, "the more knowledge they have, the readier they are to send for timely Assistance in Cases of Danger." Sarah Stone, the English midwife, disagreed, as did Nicholas Culpeper, a seventeenth-century herbalist and astrologer whose books were still being reprinted in New England in the early nineteenth century. Culpeper told the "Grave Matrons" who followed his advice that "the Lord will build you Houses as he did the Midwives of the *Hebrews,* when *Pharaoh* kept their Bodies in as great bondage as *Physitians* of our times do your Understandings." Both authors believed that experienced midwives were better equipped to handle difficult deliveries than officious but poorly prepared physicians.[13]

Ballard's philosophy was closer to Culpeper's than to Exton's. Although she had cordial relations with Hallowell's physicians, several of whom occasionally officiated at routine births, she seldom needed their help. A handful of her patients called *both* a doctor and a midwife at the onset of labor, but even in those cases she usually handled the delivery. Only twice in her entire career did she summon a doctor in an emergency, once in 1785 and again in 1792. She was not herself responsible for the first emergency. Arriving late, she found the patient "greatly ingered by some mishap," though the midwife or neighbor who had delivered the child did "not allow that shee was sencible of it." Calling the doctor may have been

[12] Stone, *A Complete Practice of Midwifery* . . . (London, 1737), 76-77; Chapman, *A Treatise on the Improvement of Midwifery, Chiefly with Regard to the Operation* . . . , 3d ed. (London, 1759), xiv; Smellie, *A Collection of Cases and Observations in Midwifery,* 3d ed., III (London, 1764), for example, 1-69, 416-427; Ballard Diary, Aug. 29, 1797, July 19, 1794, Feb. 18, 1799.

[13] Nich[olas] Culpeper, *A Directory for Midwives; or, A Guide for Women, in Their Conception, Bearing, and Suckling Their Children* . . . (London, 1651), "Epistle Dedicatory"; Stone, *Complete Practice,* ix; Henry Bracken, *The Midwife's Companion; or, A Treatise of Midwifery, Wherein the Whole Art Is Explained* . . . (London, 1737), 146; *Chapman, Improvement of Midwifery,* vii-xiii; Brudenell Exton, *A New and General System of Midwifery* . . . (London, 1751), 11. The library of the College of Physicians and Surgeons, Philadelphia, has an autographed and annotated copy of Exton owned by Dr. John McKechnie, who emigrated from Scotland to Maine in 1755 and apparently practiced medicine until his death in 1782. Martha Ballard may have known him; three of his married daughters were among her patients (James W. North, *The History of Augusta* [Augusta, Maine, 1870], 913-914).

Ballard's way of resolving a disagreement over the severity of the injury.[14]

In the other case she described her feelings in vivid language, though characteristically she offered little obstetrical detail:

> My patients illness Came on at 8 hour morning. Her women were Calld, her Case was Lingering till 7 p.m. I removd difuculties & waited for Natures opperations till then, when shee was more severly atackt with obstructions which alarmed me much. I desird Doct Hubard might be sent for which request was Complid with, but by Divine assistance I performed the oppration, which was blisst with the preservation of the lives off mother and infant. The life of the latter I dispard off for some time.

In the margin of the day's entry she added, "The most perelous sien I Ever past thro in the Course of my practice. Blessed be God for his goodness."[15] Whether Dr. Hubbard's emergency skills included the forceps delivery of a living child or only the dismembership of a dead one, we do not know. Fortunately, in this case as in all the others, Ballard and her patient got along without him.

In difficult deliveries, she typically gave God the credit for her success. The phrases are formulaic: "Her illness was very sever a short space but Blessed be God it terminated in Safety and the infant is numbered among the living," or "She had a Laborious illness but Blessed be God it terminated in safety. May shee and I ascribe the prais to the Great Parent of the universe."[16] One should not assume from such language, however, that Ballard lacked confidence in her own ability or that she relied on faith to the exclusion of skill. She knew that God worked through her hands.

Her confidence may actually have increased with the arrival of Dr. Benjamin Page in Hallowell in 1791. Page is remembered in local history as an extraordinarily successful physician. When he died in 1844, after more than fifty years of practice, the *Boston Medical and Surgical Journal* published an eleven-page biography proclaiming his skills as a general practitioner, surgeon, and gentleman. According to the anonymous author, Page was also "unequalled in the success of his obstetric practice. . . . [H]e attended upwards of *three thousand females in their confinement, without the loss of a single life from the first year of his practice!* This is almost miraculous, and may challenge the professional records of Europe or America for anything to compare with it."[17]

This is not the picture of Page preserved in Ballard's diary. Her first encounter with the young doctor was at the delivery of his near neighbor, Mrs. Benjamin Poor, the wife of a printer newly arrived in the town.

[14] Ballard Diary, Nov. 11, 1785.

[15] *Ibid.*, May 19, 1792.

[16] *Ibid.*, June 30, 1807; Mar. 31, 1800.

[17] "Memoir of Benjamin Page, M.D.," *Boston Medical and Surgical Journal*, XXXIII (1845), 9, 173.

Perhaps the woman intended medical delivery; perhaps she was simply worried that her midwife would not arrive in time. "I Extracted the child," Ballard wrote. "He Chose to close the Loin." The language is opaque here, suggesting either a friendly division of duties or an officious take-over by the doctor. The second encounter was more troublesome. Ballard had been sitting up all night with twenty-year-old Hannah Sewall, who had recently arrived in Hallowell from the town of York. "They were intimidated," she wrote, "& Calld Dr. Page who gave my patient 20 drops of Laudanum which put her into such a stupor her pains (which were regular & promising) in a manner stopt till near night when she pukt & they returned & shee delivered at 7 hour Evening of a son her first Born."[18] Hannah Sewall's intimidation, so called, may have had something to do with the fact that she had grown up in an elite family in a coastal town and was already familiar with medical delivery. As for Ballard, she was openly annoyed. Thereafter she was unmerciful in reporting Page's mistakes.

"Sally Cocks went to see Mrs. Kimball," she wrote. "Shee was delivered of a dead daughter on the morning of the 9th instant, the operation performed by Ben Page. The infants limbs were much dislocated as I am informed." She even questioned the doctor's judgment on nonobstetrical matters. Called to treat an infant's rupture, she recommended the application of brandy. "They inform me that Dr. Page says it must be opined [opened], which I should think improper from present appearance," she added. In June 1798, while she was engaged in another delivery, the doctor again delivered a stillborn child. Her report of the event was blunt: "Dr Page was operator. Poor unfortunate man in the practice."[19]

Page was unfortunate, but in eighteenth-century terms he was also ill prepared, as his administration of laudanum at Hannah Sewall's delivery suggests. The prescriptive literature recommended the use of opiates for *false* pains but not for genuine labor; Page was apparently having difficulty telling one from the other. Experience was the issue here as in so many other aspects of midwifery. Ballard had sat through enough lingering labors to know promising pains from false ones. Her reference to the doctor's dislocation of an infant's limbs also suggests lack of familiarity with the difficult manual operation required in breech births. The English midwife Sarah Stone had warned against doctors like him, "boyish Pretenders," who having attended a few dissections and read a few books professed to understand the manipulative arts so important to midwifery. Even Henry Bracken, an author who insisted that midwives should call in

[18] Ballard Diary, Nov. 17, 1793, Oct. 9-10, 1794. For additional detail on relations between midwives and physicians in Hallowell see Laurel Thatcher Ulrich, "Martha Moore Ballard and the Medical Challenge to Midwifery," in James Leamon and Charles Clark, eds., *From Revolution to Statehood: Maine in the Early Republic, 1783-1820* (Hanover, N.H., 1988), 165-183.

[19] Ballard Diary, July 8, Aug. 14, 1796, June 14, 1798.

a doctor in difficult births, cautioned, "I would never advise any one to employ a *young* physician."[20]

After 1800, Page's misadventures disappear from the diary. Presumably, he eventually learned the obstetrical art in the way Ballard did—by experience.

Extracting the child was only part of the problem. Toward the end of the eighteenth century, English writers began to give as much attention to the dangers of the lying-in period, particularly the problem of childbed fever, as to delivery itself. Puerperal fever may in fact have been rare in England in the early years of the eighteenth century; obstetrical treatises published before 1760 rarely comment on its treatment.[21] Thomas Denman's *Essays on the Puerperal Fever* appeared in London in 1768. Four years later, Charles White appended a detailed account of puerperal mortality in British hospitals to his *Treatise on the Management of Pregnant and Lying-In Women*.[22]

Puerperal fever is a wound infection caused by bacterial invasion of the uterine cavity. The infectiousness of the disorder was first suggested in the 1840s by Dr. Oliver Wendell Holmes in the United States and Dr. Ignaz Semmelwis in Austria, though the bacteriology of the disease was not settled until the 1880s, when Louis Pasteur demonstrated the presence of what is now known as streptococcus in patients suffering from the affliction. The symptoms—elevated temperature, headache, malaise, and pelvic pain—usually do not appear until several days after delivery. With certain strains of bacteria there is a profuse and foul-smelling discharge.[23]

At least one of Martha Ballard's patients probably died of puerperal infection. Mrs. Craig was "safe Deliverd of a very fine Daughter" on March 31, 1790, but after five days finding her "not so well as I could wish," Ballard administered a "Clister [enema] of milk, water, & salt" and applied an "ointment & a Bath of Tansy, mugwort, Cammomile & Hysop which gave Mrs. Cragg great relief." A week later the woman was still "Exceeding ill." Someone (perhaps a physician) prescribed rhubarb and Peruvian bark but without effect. A day or two later Dr. Cony "plainly told the famely Mrs. Cragg must die." She expired that night. Ballard helped

[20] Thomas Denman, *An Introduction to the Practice of Midwifery* (New York, 1802 [orig. London, 1794, 1795]), 179; Stone, *Complete Practice*, 76-77, xiv; Bracken, *Midwife's Companion*, 194.

[21] Exton, for example, gives no more attention to childbed fever than to afterpains (*System of Midwifery*, 150). In addition to the English works cited above, I have read the Worcester, 1794, edition of Alexander Hamilton, *Outlines of the Theory and Practice of Midwifery*, first published in Edinburgh in 1784. It also ignores the problem.

[22] White, *A Treatise on the Management of Pregnant and Lying-In Women* . . . (London, 1772).

[23] Erna Ziegel and Carolyn Conant Van Blarcom, *Obstetric Nursing*, 6th ed. (New York, 1972), 522-526; Wertz and Wertz, *Lying-In*, 119-128; Leavitt, *Brought to Bed*, 154-155.

dress her body for burial. "The Corps were Coffined & sett in the west room," she wrote. "Purge & smell very ofensive." Meanwhile, neighbors came by turns to "give the infant suck."[24]

Although Ballard attempted no diagnosis in this case, the symptoms fit the clinical description of puerperal fever. Perhaps one or two others among the five maternal deaths in her practice can also be attributed to infection. One woman was "safe delivered," fell ill a few days later, and died two weeks after delivery. Another died four days after giving birth at a time when scarlet fever, a form of streptococcus infection, was present in the town. In the two remaining maternal deaths, however, other symptoms were apparent. One woman was suffering from measles.[25] The other was in convulsions when delivered of a stillborn daughter and was still experiencing "fitts" four days later when she died. She was no doubt a victim of eclampsia, the most severe stage of an acute toxemia of pregnancy, a condition that is still considered one of the gravest complications of childbirth today.[26]

The Ballard diary suggests that puerperal infection was present in late eighteenth-century Maine, but the random appearance of the disease shows why it was seldom identified and discussed. In contrast, contemporary English physicians were encountering a truly alarming phenomenon. Charles White reported mortality rates for several London and Dublin hospitals that at midcentury were losing one out of every thirty or forty patients to puerperal fever. In 1770, in one London hospital, one of every four women died, most from infection. (See Table II.) White was astonished that two hospitals that had been established at the same time, were an equal distance from the center of London, were directed by eminent physicians, and treated the same number of patients should have markedly different death rates. In true Enlightenment fashion he concluded that one hospital smothered patients with an artificial regimen, while the more successful one was not only less crowded and closer to fields and fresh air but obliged patients to do more for themselves.

White believed that bad habits led to childbed fevers. "Violence used either by instruments or by the hand, in the extraction of the child or the placenta," might bring on an inflammation of the womb, a condition made worse by the custom of pampering women in childbed. A woman should not be delivered in a hot room, or have her child or placenta dragged from her, or lie in a horizontal position in a warm bed drinking warm liquids for a week after delivery. Physicians and midwives were both to blame for practices that all too frequently led to maternal death. He suspected that lower-class women, who could not afford pampering, did better in childbirth than their more affluent neighbors, and he cited christening and

24 Ballard Diary, Mar. 31, Apr. 4, 5, 10, 11, 12, 13, 15, 16, 1790.

25 *Ibid.*, Oct. 18, 21, 24-29, 1802.

26 *Ibid.*, Feb. 26, 27, Mar. 1, 2, 4, 1789; Ziegel and Van Blarcom, *Obstetric Nursing*, 208-213.

TABLE II
COMPARATIVE MATERNAL MORTALITY RATES

Place	*Total Births*	*Maternal Deaths*	*Deaths per 1,000 Births*
London A			
1767-1772	653	18	27.5
1770	63	14	222.2
London B			
1749-1770	9,108	196	21.5
1770	890	35	39.3
London C			
1747-"present"	4,758	93	19.5
1771	282	10	35.4
London D	790	6	7.5
Dublin A			
1745-1754	3,206	29	9.0
Dublin B			
1757-1775	10,726	152	14.1
1768	633	17	26.8
1770	616	5	8.1
Martha Ballard			
1777-1812	998	5	5.0
1785-1812	814	5	6.1
United States			
1930			6.7
1935			5.8
1940			3.8
1945			2.1

Sources: Charles White, *A Treatise on the Management of Pregnant and Lying-In Women* (Worcester, Mass., 1793 [orig. publ. London, 1772]); Ballard Diary; Wallace, Gold, and Lis, eds., *Maternal and Child Health Practices,* 285.

death ratios from London and Manchester parish records to prove his point.[27]

Had White known about Martha Ballard, he would have had a ready explanation for her success: she practiced among frontier women who lived close to nature. In fact, Ballard was probably guilty of one of the practices White deplored—using hot drinks laced with alcohol. Still, there

[27] White, *A Treatise on the Management of Pregnant and Lying-In Women* (Worcester, Mass., 1793 [orig. publ. London, 1772]), 17-31, 219, 236-240. White's estimates for London and Manchester work out to maternal mortality rates of 13/1000 and 6/1000 respectively. For a modern effort to compute maternal mortality ratios from parish christening and death records, see B. M. Willmott Dobbie, "An Attempt to Estimate the True Rate of Maternal Mortality, Sixteenth to Eighteenth Centuries," *Medical History,* XXVI (1982), 79-90. Dobbie believes that maternal mortality in England may have been as high as 29/1000, as compared with earlier estimates of 10-15/1000.

is plenty of evidence in the diary of the kind of vigor he admired. Ballard's patients were not all as sturdy as Mrs. Walker, who was "sprigh about house till 11 [and] was safe delivrd at 12," or as courageous (or foolhardy) as Mrs. Herriman, who "wrode in a sleigh 13 miles after her illness was on her"; but few Hallowell women could afford to lie in bed.[28] Ballard's own daughter, Dolly Lambert, was "so well as to be helpt up and sett at table for breakfast" twenty-four hours after giving birth to her fourth child. Ballard generally left her patients in the care of an afternurse a few hours after delivery, but when she stayed overnight she helped to get the woman out of bed in the morning. "Got my patient up, Changd her Lining and came home," she wrote (in this case, twelve hours after delivery), and "help[ed] Mrs Williams up & maid her Bed and returned home" (twenty-four hours after birth).[29]

Modern epidemiology confirms Charles White's belief that environment affected mortality, though, of course, the theoretical explanations differ from his.[30] Because Ballard was a part-time practitioner who delivered women at home and shared their postpartum care with nurses and family members, she had little opportunity to spread puerperal infection from one patient to another. The opposite conditions existed in the London hospitals, where as White himself suspected, the use of instruments in delivery probably increased the lacerations and tears that encouraged septicemia. Higher incidence of venereal disease in London may also have been a factor.[31]

That childbearing was safer in rural Maine than in London hospitals hardly seems surprising. The more interesting question for our purposes is how the literature emanating from those hospitals affected obstetrical practice in country places. Here the writings of Dr. Jeremiah Barker of Gorham, Maine, are particularly revealing. In February 1785 Barker initiated a discussion in the *Falmouth [Maine] Gazette* over the causes of an unusual "mortality among child-bed women, which has prevailed of late." Dr. Nathaniel Coffin, whose practice was in Falmouth (now Portland), submitted an angry response that was published in the next issue of the paper. Yes, several women had died in childbed in and about the town, but since the cause was unknown there was nothing that could have been done to save them. He denied that there was an epidemic, and he accused Barker of awakening "all those fears and apprehensions, which are but too

28 Ballard Diary, Mar. 11, 1790, Jan. 19, 1800.

29 *Ibid.*, Apr. 17, 1801, May 31, 1799, Nov. 28, 1787; see also *ibid.*, June 30, 1794, June 3, 1795, Aug. 10-11, 1799.

30 Some 19th-century Americans debating the causes of childbed fever used the same environmental argument, anticipating the conclusions but not the logic of 20th-century historians (Charles E. Rosenberg, *The Care of Strangers: The Rise of America's Hospital System* [New York, 1987], 124-126, 376, n. 10, n. 11).

31 Dorothy I. Lansing, W. Robert Penman, and Dorland J. Davis, "Puerperal Fever and the Group B Beta Hemolytic Streptococcus," *Bulletin of the History of Medicine*, LVII (1983), 70-80. On the complexities of the puerperal fever debate in the 19th and early 20th centuries see Leavitt, *Brought to Bed*, chap. 6.

often cherished by the sex." The debate continued through four issues of the newspaper, Barker insisting that an excess of bile characterized all the cases of puerperal fever he had studied, Coffin retorting that Barker had misread the symptoms.[32]

Barker included additional detail on the puerperal fever controversy in "History of Diseases in the District of Maine," a manuscript that he wrote after his retirement from active practice in 1818. Taken together, the newspaper stories and the "History" tell us a great deal about how medical reforms, initiated in London, were received in America. In his letters to the *Falmouth Gazette* Barker appears as a bold empiric asserting the power of direct experimentation against the dated theories of academic physicians. In his manuscript he reveals that the source of his ideas was a work by Thomas Denman, presumably his 1768 essays on puerperal infection.[33]

According to Barker's history, the puerperal fever outbreak began at the same time as an equally troubling rash of wound infections. In the spring of 1784, he recalled, "some unusual appearances took place in wounds & bruises, even trivial ones, which baffled the skill of the Surgeon, and issued in the death of the patient. . . . Local inflammations chiefly from injuries were more frequent and untractable during the year than I ever knew them to be before or since. The subjects of these complaints were chiefly males and apparently of good constitutions."[34] At the same time, several women in Gorham, Falmouth, and adjoining towns contracted puerperal fever. Although Barker made no connection between the two phenomena, it is difficult for a twentieth-century reader to avoid doing so.

Since Barker gave no statistics on the number of men who died from infected wounds or of women who suffered from childbed fever, and since birth and death records for the region are incomplete, it is impossible to know how serious the problem really was. Barker simply tells us that few women who suffered childbed fever survived, and that he attended autopsies in three different towns. Yet his description confirms the rarity of the disorder in the region. "The ill success which attended my practice," he wrote, "induced me to write to several aged & experienced physicians in different portions of Massachusetts, for advice, as puerperal fever had never appeared among us excepting in a few sporadic cases, which yielded to common means." His correspondents had never seen such an epidemic themselves, but they referred him to the works of Denman and other unnamed British authors. It was from Denman's book, apparently, that Barker got his notions about bile and the use of "the bark" (quinine) as a remedy. He also wrote to Dr. Ammi Ruhamah Cutter of Portsmouth, New Hampshire, who had reportedly experienced high mortality from

[32] *Falmouth Gazette and Weekly Advertiser*, Feb. 12, 26, Mar. 5, 12, 1785.
[33] Jeremiah Barker, "History of Diseases in the History of Maine," chap. 3, Barker Papers, Maine Historical Society, Portland.
[34] *Ibid.*

childbed fever. Cutter suggested applying "fermenting cataplasms to the abdomen composed of flower & yeast."[35]

Barker credited none of these sources in his 1785 newspaper letters, however, nor did he elaborate on the problems in his own practice. Alluding to an unusual childbed mortality, "especially in the town of Falmouth" (where Nathaniel Coffin practiced), he offered his remedies as a disinterested effort to "secure the happiness of mankind." Although he claimed to have "taken the opinion of the Massachusetts Medical Society," he gave no names.[36] Whom was he addressing? Surely not his fellow doctors. If that had been his intent, he would have limited himself to the private correspondence he had already begun. Instead he reached beyond the medical fraternity to the literate public of his region. The very form of his argument suggests that some part of his intended audience was female.

When Barker asserted that his patients could testify to the effectiveness of his methods, Coffin countered, "I am sorry the Dr. is obliged to have recourse to the female sex for a vindication of them." He suggested that the young doctor read "Astruc, Brooks, and others" to correct his faulty diagnosis. Barker retorted that the proposed authors were not only "Obsolete" but "esteemed of less consequence, in many respects, than the opinion of some of the female sex, founded on experience, in this more enlightened age."[37] The reliance on experience was, of course, a staple of Enlightenment medicine. Whereas earlier physicians had relied on theoretical learning, English reformers like William Smellie had emphasized the necessity for practical training in the manual arts of midwifery. Ironically, the obstetrical Enlightenment encouraged physicians to assume women's work in the very act of celebrating its importance. As Thomas Denman expressed it, "A natural labour was the last thing well understood in the practice of midwifery, because scientific men, not being formerly employed in the management of common labours, had no opportunity of making observations upon them."[38]

Barker's regard for female experience was conditional. He praised enlightened women who sought his care but mistrusted traditional midwives and nurses. His case notes from 1774 describe his efforts to deliver a woman with an imperforate vagina after the ministrations of her "friends" had failed. "I found that nothing could be done but to dilate the Perineum for the egress of the Child," he wrote, "and 'tho the operation is simple, yet fearing the sensure of the Vulgar (if any misfortune should befall the patient, afterwards) advised to send for Dr. Savage as an

[35] *Ibid.*

[36] *Falmouth Gaz.*, Feb. 12, 1785.

[37] *Ibid.*, Feb. 26, Mar. 5, 12, 1785. Coffin was perhaps referring to Richard Brookes, *The General Dispensatory* . . . (London, 1753), or *The General Practice of Physic* . . . (London, 1754) and to Jean Astruc, *A Treatise of the Diseases of Women* . . . , 2 vols. (London, 1762), or *Elements of Midwifery, containing the Most Modern and Successful Method of Practice in Every Kind of Labor* . . . (London, 1766). Astruc's works were translated from the French.

[38] Smellie, *Collection*, III, 533-543; Denman, *Introduction*, I, 171.

assistant." As it happened, the dead fetus was delivered before the second physician arrived.[39] Barker's concern about the censure of "the Vulgar" suggests the difficulties many physicians had in establishing credibility in the region, not only in obstetrical but in general medical cases. One young man entering practice in Waterville, Maine, in the 1790s even signed contracts with prospective patients, promising not to charge them if his remedies failed.[40]

Like Benjamin Page of Hallowell, Barker had begun his medical career after a brief apprenticeship with a Massachusetts physician. In 1774 he was twenty-two and in his second year of practice. The newspaper debate suggests that, ten years later, he had grown tired of his practice in Gorham and adjoining towns, and perhaps hoped to attract the attention of prosperous families in the port of Falmouth.[41] Jeremiah Barker knew that women, whether vulgar or enlightened, were guardians of a doctor's reputation.

For his part, Coffin was furious at Barker for questioning the skills of other physicians. He was also dismayed that the younger doctor should invoke the authority of the Massachusetts Medical Society, even though he was not a member. When Coffin wrote to the society in 1803 recommending a number of new members from Maine, he explicitly excluded Barker, partly on the basis of the 1785 affair, which still rankled. (The society ignored his advice and elected Barker anyway.)[42] Thus a young physician moving into obstetrical practice in the 1780s and 1790s had two obstacles to overcome—folk reliance on traditional midwifery and the mistrust of older, more conservative physicians.

For our purposes, however, the more important issue is the way in

[39] Jeremiah Barker, Medical Cases, 1771-1796, Barker Papers.

[40] Loose paper dated Apr. 29, 1802, Moses Appleton Papers, Waterville Historical Society, Waterville, Me. On the larger question of lay resistance see William G. Rothstein, *American Physicians in the Nineteenth Century: From Sects to Science* (Baltimore, 1972), 128-138, and Joseph F. Kett, *The Formation of the American Medical Profession: The Role of Institutions, 1780-1860* (New Haven, Conn., 1968), 101-107.

[41] Barker was born in Scituate, Mass., began his practice in Gorham, Me., in 1772, removed to Barnstable on Cape Cod after a year, returned to Gorham in 1779, and finally went to the Stroudwater section of Falmouth in 1796 (James Alfred Spalding, *Jeremiah Barker, M.D., Gorham and Falmouth, Maine, 1752-1835*, reprinted from *Bulletin of the American Academy of Medicine*, X ([1909], 1-2). Barker had an indirect link to British medicine. In mid-career his mentor, Dr. Bela Lincoln of Hingham, Mass., had spent a year studying in London hospitals and acquiring an M.D. from King's College, Aberdeen. Spalding, *Barker*, 1-2; Clifford K. Shipton, *Sibley's Harvard Graduates: Biographical Sketches of Those Who Attended Harvard College*, XIII (Boston, 1965), 456.

[42] Nathaniel Coffin to Massachusetts Medical Society, May 8, 1803, and Jeremiah Barker to Joseph Whipple, July 12, 1803, Countway Medical Library, Boston. In the long run, Barker may have been more forgiving than Coffin. His manuscript history describes Coffin as a physician "who commanded an extensive practice in physic, surgery and obstetrics, with good success" ("History," chap. 2).

which the puerperal fever incident of 1784-1785 began to shape Barker's practice. All of the cases of childbed fever described in his history came from that outbreak, yet he used them to support a long, detailed discussion of the cure and prevention of the disorder. Even by his own account, puerperal fever cannot have been a serious problem in the region. Most of the physicians to whom he wrote had seen only scattered cases; all of them referred him to British authors for an understanding of the subject. Coffin even doubted that the deaths could be attributed to a single disease, and he questioned whether he or any other physician could have done anything to prevent them. In Barker's own practice the trouble also faded away. There were additional cases during the winter of 1784-1785, he wrote, yet the disease showed "decreasing malignancy and mortality. Since which it has not appeared among us, excepting in a few sporadic cases, which seldom proved fatal."[43] Yet by 1818 his interpretation of the 1784-1785 cases had expanded to encompass citations from medical literature published as late as 1817.[44] Barker measured his entire career against that single early disaster. Since it was never repeated, he assumed that his preventative practices were successful.

Barker combined the noninterventionist prescriptions of the late eighteenth century—better ventilation, lighter food, avoidance of alcohol—with more heroic measures. "The means of prevention may be reduced to two," he argued. First, the physician should treat the patient during labor as though she were already a victim of the disease, drawing blood, administering emetics and cathartics, debarring her "entirely from spirits," and keeping her "on a low diet, without any animal food, in a well ventilated apartment without any curtains, on a mattress or straw bed." The second method involved "facilitating or rather hastening, by artificial means, the termination of labour."[45] Presumably, this meant using forceps and possibly ergot, a powerful and dangerous drug that, when given orally, stimulates uterine contractions.[46] In this, Barker departed from the advice of his 1784 mentor, Thomas Denman, who, like Charles White, believed that forceps should be used rarely and that hastening labor led to postpartum complications.[47]

What we have, then, is a clear example of the way in which medical literature in combination with local experience came to define a practice.

43 Barker, "History," chap. 2.

44 *Ibid.* The citations on puerperal fever are, as he gave them, "Dr. Terriere, 1789; Dr. Biskell, *Medical Papers,* v. 2, 1798; *London Medical Repository,* May 1815; *New England Journal,* v. 4, 5; Dr. Channing, *New England Journal,* vol. 6, 1817; *Medical Repository,* vol. II."

45 Barker, "History," chap. 2.

46 Leavitt, *Brought to Bed,* 144-145.

47 On some things Denman had changed his own mind by 1794. Although he continued to oppose intervention in labor, he did accept bloodletting as a cure for puerperal fever, something he had dismissed in his earlier treatise, as had Barker in his *Falmouth Gaz.* letters. Denman, *Introduction,* I, 184-190, II, 253-254; *Falmouth Gaz.,* Feb. 26, 1785.

Barker's need to differentiate himself from other practitioners, as well as his desire to apply the latest in scientific knowledge to the management of his practice, made it impossible for him to see the 1784-1785 outbreak as an anomaly. Thereafter, he was convinced that it was his own intervention that had prevented a similar disaster from occurring. In contrast, Martha Ballard's nonscientific, even providential interpretation of events enabled her to treat each case on its own terms. For every patient, she did what she knew how to do and let God determine the outcome. This is not to say that she was incapable of experimentation or that she never wondered why one infant died and another lived. It is simply to argue that her craft was oriented toward practical results rather than theoretical explanation. The death of Mrs. Claton or Mrs. Craig did not destroy her confidence in the soundness of her methods. Hers was not an approach that encouraged innovation, but neither did it promote ill-considered intervention.

Adrian Wilson has estimated that, in nature, 96 percent of births occur spontaneously. Approximately 4 percent involve serious obstruction of some kind and cannot be delivered without intervention. An additional 1 percent, though spontaneous, result in complications—minor ones such as fainting, vomiting, and tearing of the perineum, or major events like hemorrhaging or convulsions.[48] Martha Ballard's records fit Wilson's typology well. Approximately 95 percent of her entries simply say "delivered" or "safe delivered." In the remaining 5 percent, some sort of complication is indicated, by explicit reference to obstructions, an oblique comment on the severity of the labor, or simply an acknowledgment that the delivery was accomplished through the mercy of God. Her records thus attest to the relative safety of childbearing as well as to her skill in managing difficult labors. Her ministrations no doubt improved the conditions of birth, but, perhaps even more important, she did little to augment the dangers.

In this regard it is interesting to compare her records with those of James Farrington of Rochester, New Hampshire, a nineteenth-century physician whose caseload was similar to hers and whose records, unlike those of his eighteenth-century predecessors, are extraordinarily complete. Dr. Farrington began the study of medicine in 1814, two years after Ballard's death. His manuscript records include a systematic register of 1,233 deliveries performed between 1824 and 1859. At first glance, his stillbirth and mortality ratios confirm the conclusions of revisionist histories—that childbirth became more dangerous in the nineteenth century. Farrington's stillbirth ratios are higher than any of the eighteenth-century practitioners and closer to those of the nineteenth-century midwife Jennet Boardman. (See Table I.) Even more striking is the number of maternal deaths at delivery. That he was occasionally called to complete someone else's mismanaged delivery is certain, though those few cases that include extended descriptions suggest that, regardless of

[48] Adrian Wilson, "William Hunter and the Varieties of Man-midwifery," in Bynum and Porter, eds., *William Hunter*, 344-345.

practitioner, nineteenth-century obstetrical practice added new dangers to the old problems of obstructed birth. Curiously, there is no indication of puerperal fever in Farrington's records. One might have expected at least a few cases of infection over such a long career. Since his tightly organized accounts, with one exception, list deliveries *only*, it is possible that such cases, usually arising a week or so after delivery, appeared in another set of more general medical records.[49]

Farrington recorded five maternal deaths. One woman, he wrote, was "enfeebled by intemperance." Another had a severe cold and "spoke but few words after delivery, but sunk away without a groan." The most dramatic case had been abandoned by another physician. Farrington described it as "preternatural labor requiring in the end the dissection of the infant," adding details that might have come from English obstetrical literature a hundred years before: "the external parts of generation much lacerated and mangled by *hooks, pincers, and knives*." The woman survived Farrington's extraction of the dismembered fetus but died five days later. A fourth woman died of bleeding after an unidentified attendant failed to extract the placenta. The fifth woman suffered a ruptured uterus: "in a few minutes the whole child could be felt expelled from the Uterus within the abdominal cavity." The woman lived about an hour.[50]

The numbers are small, however, and, without more detail on postpartum infection, inconclusive. The most striking contrast between Farrington's and Ballard's records is not in mortality rates themselves but in their characterizations of delivery. The process of labor was biologically the same, yet their descriptions differ markedly. Whereas Ballard thought in terms of the general outcome ("left mother and child cleverly"), Farrington focused on theoretical categories. Labors were "natural," "tedious," "premature," "preternatural," "complicated," or, after 1838, "instrumental," regardless of whether the mother and child survived.[51] Twenty percent of the deliveries in his records are listed as something other than "natural."[52]

49 Franklin McDuffie, *History of the Town of Rochester, New Hampshire, from 1722 to 1890*, ed. Silvanus Hayward (Manchester, N.H., 1892), I, 345-346; James Farrington Medical Record Books, 1824-1859, Special Collections, Dimond Library, University of New Hampshire, Durham. Farrington added an entry about the woman dying five days after delivery in different colored ink at the end of his delivery record. On the general pattern of listing childbed deaths under other causes see Wertz and Wertz, *Lying-In*, 125-126.

50 Farrington, Medical Record, Case #451, Sept. 9, 1835, #118, Feb. 24, 1825, #442, May 28, 1835, #292, Jan. 30, 1831.

51 Farrington used forceps before 1838; he just did not have a separate category to cover instrumental labors.

52 Joan M. Jensen's analysis of 109 deliveries by an early 19th-century Chester, Pa., physician shows no maternal deaths at delivery, 7% stillbirths, and 30% difficult labors (*Loosening the Bonds: Mid-Atlantic Farm Women, 1750-1850* [New Haven, Conn., 1986], 30-33). The low caseload of this physician, roughly 14

Here the telltale category may be his 102 cases of "tedious" labor, defined in the medical literature as lasting longer than twenty-four hours. In one case, which terminated safely at twenty-six hours, Farrington reported taking blood from the woman's arm, then giving an opiate. Four hours before the birth he gave her "Ergot in Infusion" and was pleased when he was able to deliver the child "without Instruments though for several hours no alteration was made by the force of the Pains."[53] Reading such an account, one finds it difficult not to think of Ben Page's administration of laudanum at the delivery of Hannah Sewall. Ironically, the remedy that so dismayed Martha Ballard was by now a standard part of the physician's arsenal. The three remedies—laudanum, ergot, and forceps—went together, accomplishing, as the physicians and perhaps many of their patients thought, an artificial hastening of labor.[54]

Judith Walzer Leavitt has argued that women chose medical intervention. Sally Drinker Downing, for example, sought out the services of the Philadelphia physician William Shippen, who administered opium during her 1795, 1797, and 1799 deliveries. Leavitt concludes that "the prospect of a difficult birth, which all women fearfully anticipated, and the knowledge that physicians' remedies could provide relief and successful outcomes led women to seek out practitioners whose obstetric armamentarium included drugs and instruments."[55] Leavitt may be right about Downing, yet Martha Ballard's diary adds a new dimension to the question of choice. At ten o'clock on the evening of October 21, 1794, she was summoned to the house of Chandler Robbins, a Harvard graduate and new resident of Hallowell. "Doctor Parker was calld," she wrote, "but shee did not wish to see him when he Came & he returnd home. Shee was safe delivered of a son her first Born at 10 hour 30 minutes Evening"—that is, twenty-four and one-half hours after summoning the midwife. Ballard's reward for officiating at this "tedious labor" was eighteen shillings and the satisfaction of knowing that God and the parents were pleased.

This brief survey of Martha Ballard's diary and related documents supports the reformist point that birth is a natural process rather than a life-threatening event. It suggests that rural midwives were capable of managing difficult as well as routine births, that the need for medical intervention was by no means obvious, and that puerperal infection, though present, was still only a random problem in the last years of the eighteenth century. For midwives like Martha Ballard or Lydia Baldwin, experience defined competence, yet in the years following the Revolution a number of brash young men with more confidence than experience took

---

deliveries a year, suggests the presence of other practitioners, probably including midwives.

[53] Denman, *Introduction,* 171; Farrington, Medical Record, Case #539, Aug. 8, 1839.

[54] Leavitt, *Brought to Bed,* 43-44.

[55] *Ibid.,* 40.

up the practice of delivering babies. Not content with the more restrained role of older doctors, they consulted British literature and sought advice from other physicians to solve their problems and validate their skills. That they gravitated toward works that emphasized the necessity of intervention is hardly surprising. In a competitive environment no bright young physician could embrace Charles White's advice that the less done in childbirth the better. Employing forceps, letting blood, administering opiates and ergot, they set themselves apart from the manual skills and the providential faith of the midwives.

During the earlier years of Martha Ballard's midwifery in Hallowell, however, the success of such physicians was by no means assured. In 1800, when age, ill health, and a move to a more distant part of the town forced her to cut back her practice, she was the single most important practitioner in her town, and she knew it.

"Science" Enters the Birthing Room: Obstetrics in America since the Eighteenth Century

Judith Walzer Leavitt

At the end of the eighteenth century, Dr. William Shippen, Jr., of Philadelphia attended all the childbearing women in the well-established Drinker family. The family members chose Shippen instead of a woman midwife, in spite of their ambivalence about having a man in the traditionally all-female birthing room, because they believed the physician offered the best hope for a successful outcome. The Drinkers, and many others like them, considered themselves fortunate to be living at a time when male physicians began replacing female midwives in the birthing rooms of the American urban elite. The families expected—and believed that they received—better care at the hands of physicians than they thought possible with traditional female attendants. This article examines the promises of the new physician-directed obstetrics—beginning in the 1760s, when physicians entered obstetrics, and ending in the 1940s, after they dominated the field—and evaluates the extent to which those promises were kept.[1]

Before 1760 birth was a women's affair in the British colonies of North America. When a woman went into labor, she "called her women together" and left her husband and other male family members outside. "I went to bed about 10 o'clock," wrote William Byrd of Virginia, "and left the women full of expectation with my wife." Only in cases where women were not available did men participate in labor and delivery, and only in cases where labor did not

Judith Walzer Leavitt is associate professor of the history of medicine and women's studies at the University of Wisconsin, Madison. This paper was written for and first delivered at the Seventh International Symposium on the Comparative History of Medicine—East and West, Susono-Shi, Japan, Sept. 19-26, 1982. This symposium is held by the Division of Medical History, the Taniguchi Foundation.

[1] Cecil K. Drinker, *Not So Long Ago: A Chronicle of Medicine and Doctors in Colonial Philadelphia* (New York, 1937), 50-62; Jane B. Donegan, *Women and Men Midwives: Medicine, Morality, and Misogyny in Early America* (Westport, Conn., 1978); and Catherine M. Scholten, "'On the Importance of the Obstetrick Art': Changing Customs of Childbirth in America, 1760 to 1825," *William and Mary Quarterly*, 34 (July 1977), 426-45. The scope of the subject of changing obstetric practices necessitates a synthetic analysis and does not permit full documentation here of all the points raised.

The Journal of American History Vol. 70 No. 2 September 1983

progress normally did physicians intervene and perhaps extricate a dead fetus. The midwife orchestrated the events of labor and delivery, and the women neighbors and relatives comforted and shared advice with the parturient.[2] Women suffered through the agonies and dangers of birth together, sought each other's support, and shared the relief of successful deliveries and the grief of unsuccessful ones. This "social childbirth" experience united women and provided, as Carroll Smith-Rosenberg has argued, one of the functional bonds that formed the basis of women's domestic culture.[3]

Within their own homes, birthing women controlled much of the experience of childbirth. They determined the physical setting for their births, the people to attend them during labor and delivery, and the aids or comforts employed. Midwives traditionally played a noninterventionist, supportive role in the home birthing rooms. As much as possible they let nature take its course: they examined the cervix or encouraged women to walk around; they caught the child, tied the umbilical cord, and if necessary fetched the placenta. In complicated cases, midwives might have turned the fetus—podalic version—or fortified women with hard liquor or mulled wine. They may have manually stretched the cervix or, rarely, administered ergot. Midwives spent most of their time, as the written record reveals, comforting the parturient and waiting. The other women attendants supported the midwife and the birthing woman, and the atmosphere in the birthing rooms—if everything proceeded normally—was congenial and cooperative. Parturient women, who felt vulnerable at the time of their confinements, armed themselves with the strength of other women who had passed through the event successfully.[4]

Despite the very positive aspects of social childbirth, a romantic image of childbirth in this period would be misleading. Women garnered support from their networks of companions, but they continued to fear their births because of the possibilities of death or debility that could and frequently did result from childbirth. Although statistics do not exist to measure these dangers precisely, women's fears were not unfounded. By all accounts maternal and infant mortality rates were high. Furthermore, postpartum gynecological problems

[2] Louis B. Wright and Marion Tinling, eds., *The Secret Diary of William Byrd of Westover, 1709-1792* (Richmond, 1941), 79. The phrase "called her women together" is common in early childbirth accounts. See, for example, the diary of midwife Martha Moore Ballard of Augusta, Maine, November 21, 1785—"Mrs. Cowen called her women together this evening; was safely delivered of a daughter about the middle of the night and is comfortable"—and February 18, 1786—"Mrs. Fletcher called her women again this morn, and was safely delivered of a daughter." Charles Elventon Nash, *The History of Augusta: First Settlements and Early Days as a Town, including the Diary of Mrs. Martha Moore Ballard (1785 to 1812)* (Augusta, Maine, 1904), 243, 246.

[3] Carroll Smith-Rosenberg, "The Female World of Love and Ritual: Relations between Women in Nineteenth-Century America," *Signs*, 1 (Autumn 1975), 1-29. For a discussion of social childbirth, see Richard W. Wertz and Dorothy C. Wertz, *Lying-In: A History of Childbirth in America* (New York, 1977), 1-26.

[4] On midwives' practices, see Donegan, *Women and Men Midwives*, 9-11, 25, 127; Scholten, "'On the Importance of the Obstetrick Art,'" 429-31, 432-34; Claire E. Fox, "Pregnancy, Childbirth and Early Infancy in Anglo-American Culture: 1675-1830" (Ph.D. diss., University of Pennsylvania, 1966), 122-27; and Janet Bogdan, "Care or Cure? Childbirth Practices in Nineteenth Century America," *Feminist Studies*, 4 (June 1978), 94.

resulting from unsutured perineal tears, prolapsed uteri, or vesico-vaginal and recto-vaginal fistulas caused some women extreme discomfort and disability throughout the rest of their lives. Women's fears of death and physical debility led them away from traditional birthing patterns to a long search for safer and less painful childbirths. Thus some women, especially those who were economically advantaged, tried to modify traditional births by incorporating new possibilities as they became available. Like the Drinkers, these eighteenth-century women readily invited physicians to attend them, despite their worries about the propriety of having men participate in intimate female events, in hopes that the "man-midwives" could provide easier and safer births.[5]

The entrance of physician-accoucheurs into the practice of obstetrics in America during the second half of the eighteenth century marked the first significant break with tradition. The story of this new midwifery, familiar to historians of American medicine, centers on Shippen, who in 1762 returned from his studies in London and Edinburgh and established the first systematic series of lectures on midwifery in America. Shippen initially trained both female midwives and male physicians in anatomy (including the gravid uterus), but he soon limited his lectures to male students. He established a private practice of midwifery and became a favorite of Philadelphia's established families. The Drinkers found him "very kind and attentive" during labor and delivery and noted that he remained with his patients even during protracted labors and "sleep't very little."[6]

Shippen was the most famous of late-eighteenth-century physicians who practiced midwifery, but he was not alone. Numerous doctors expanded their practice of medicine and began to attend laboring women. The transition to male attendants occurred so easily among advantaged urban women that it can only be explained by understanding the women's impression that physicians knew more than midwives about the birth process and about what to do if things went wrong. Women overturned millennia of all-female tradition and invited men into their birthing rooms because they believed that men offered additional security against the potential dangers of childbirth. By their acknowledgement of physician superiority, women changed the fashions of childbearing and made it desirable to be attended by physicians.[7]

Women had good reason to believe that physicians could provide services that midwives could not. Many of the physicians who practiced obstetrics at

[5] For more on women's perceptions of childbirth as a time of danger, see Judith Walzer Leavitt and Whitney Walton, "Down to Death's Door: Women's Perceptions of Childbirth in America," in *Proceedings of the Second Motherhood Symposium: Childbirth: The Beginning of Motherhood*, ed. Sophie Colleau (Madison, 1982), 113-36. On efforts to compute maternal mortality, see B. M. Willmott Dobbie, "An Attempt to Estimate the True Rate of Maternal Mortality, Sixteenth to Eighteenth Centuries," *Medical History*, 26 (Jan. 1982), 79-90; and Louis I. Dublin, "Mortality among Women from Causes Incidental to Childbearing," *American Journal of Obstetrics and Diseases of Women and Children*, 78 (July 1918), 20-37.

[6] Drinker, *Not So Long Ago*, 60-61; Lewis C. Scheffey, "The Earlier History and the Transition Period of Obstetrics and Gynecology in Philadelphia," *Annals of Medical History*, 2 (May 1940), 215-16.

[7] Scholten, "'On the Importance of the Obstetrick Art,'" 434-36.

the turn of the nineteenth century had trained in Great Britain, where a tradition of male accoucheurs had already developed. The men had access to education then denied to women, which provided theoretical understanding of female anatomy and the process of parturition. Whereas women midwives relied on practical experience and an appeal to female traditions—attributes to some extent taken for granted and unappreciated—men physicians had the extra advantage and prestige associated with formal learning. Even though most American practitioners had not attended medical school and were themselves apprentice trained, physicians carried with them the status advantages of their gender and of the popular image of superior education. Furthermore, birthing women perceived that the male presence had already contributed opium and forceps to obstetrics and promised even greater benefits in the future. The appeal to overturn tradition was strong.[8]

When Shippen attended Sally Drinker Downing's 1795 birth, which was complicated by a footling presentation, he administered opium to relieve her suffering. Two years later, faced with another of Downing's difficult labors, Shippen "was oblig'd to force her mouth open to give some thing with a view of reviving her." In 1799 Downing again suffered a protracted labor, and Shippen took fourteen ounces of blood and then gave her eighty to ninety drops of liquid laudanum. According to Downing's mother, when labor still did not progress, Shippen gave "an Opium pill three grains he said, in order to ease her pain, or to bring it on more violently." When this still did not produce the desired result, Shippen threatened to use his instruments, but finally Downing delivered without them.[9]

Shippen's practice of allaying painful and lengthy labors by bleeding, giving opium, and occasionally using forceps illustrates why women wanted physicians to attend them. The prospect of a difficult birth, which all women fearfully anticipated, and the knowledge that physicians' remedies could provide relief and successful outcomes led women to seek out practitioners whose obstetric armamentarium included drugs and instruments.[10]

The promise of the new obstetrics developed in part through formal physician education in midwifery. Shippen's first lectures covered pelvic anatomy, the gravid uterus, the placenta, fetal circulation and nutrition, natural and unnatural labor, and the use of obstetrical instruments. He demonstrated on manikins and on patients—poor women for whom he provided accommodation—and used drawings and textbooks. Other physicians followed his example, and courses in midwifery became available in Boston and New York as well as in Philadelphia.[11]

During the first half of the nineteenth century medical education in obstet-

[8] Bogdan, "Care or Cure?" 96–97; Donegan, *Women and Men Midwives*, 141–57.

[9] Drinker, *Not So Long Ago*, 51–53, 58–61.

[10] Jane B. Donegan argues that William Shippen practiced conservative obstetrics when compared to the "meddlesome midwifery" of others. Donegan, *Women and Men Midwives*, 118–19. For more on Shippen, see Betsy Copping Corner, *William Shippen, Jr.: Pioneer in American Medical Education: A Biographical Essay* (Philadelphia, 1951).

[11] Lawrence D. Longo, "Obstetrics and Gynecology," in *The Education of American Physicians: Historical Essays*, ed. Ronald L. Numbers (Berkeley, 1980), 205–08.

rics expanded. To many professors and students, the subject was embarrassing, and students generally did not observe women in labor but received only a theoretical education. Samuel D. Gross, a student who observed Thomas Chalkley James teaching obstetrics at the University of Pennsylvania early in the nineteenth century, noticed that "it was seldom that he . . . looked squarely at his audience. His cheeks would be mantled with blushes while engaging in demonstrating some pelvic viscus, or discussing topics not mentionable in ordinary conversation. It was often painful to witness his embarrassment."[12]

Although the embarrassment of physicians about their role in obstetrics soon disappeared, the question of a man's proper behavior in a woman's birthing room continued to influence the teaching and the practice of obstetrics. Of William Potts Dewees, who succeeded James at the University of Pennsylvania, Gross observed, "he did not hesitate to call things by their proper names. No blush suffused his cheek in the lecture-room." But Dewees, whose *A Compendious System of Midwifery* went through an influential twelve editions, remained as horrified by the idea of ocular inspection as his most modest patients. Using manikins, he taught his students how to perform unsighted digital explorations of parturient women. Even when applying forceps, Dewees taught, "every attention should be paid to delicacy . . . the patient should not be exposed . . . even for the drawing off of the urine. . . . The operator must become familiar with the introduction of the instruments without the aid of sight."[13]

Students graduating from such didactic obstetrics courses were forced to enter the birthing rooms of their first patients in relative ignorance. Never having witnessed actual births, and armed with only theoretical knowledge, they must have been somewhat apprehensive. Yet doctors knew that they had to be confident to gain confidence, and they forged ahead and delivered babies. One medical graduate wrote that when he delivered his first baby he examined the laboring woman, "but whether it was head or breech, hand or foot, man or monkey, that was defended from my uninstructed finger by the distended membranes, I was as uncomfortably ignorant, with all my learning, as the foetus itself that was making all this fuss."[14] Rejecting the practical experience of the midwives' training, those physicians who had formal medical education tried to raise the practice of obstetrics to a higher level by emphasizing anatomy and physiology.

Through their theoretical training, this physician elite had the potential to expand the practice of obstetrics beyond individual experience. Apprentice-trained midwives and physicians could develop expertise only insofar as their

[12] *Autobiography of Samuel D. Gross, M.D., with Sketches of His Contemporaries*, ed. [Samuel Weissel Gross and Albert Haller Gross] (2 vols., Philadelphia, 1887), II, 240.

[13] *Ibid.*, 247; William P. Dewees, *A Compendious System of Midwifery, Chiefly Designed to Facilitate the Inquiries of Those Who May Be Pursuing This Branch of Study* (Philadelphia, 1843), 289. See also Irving S. Cutter and Henry R. Viets, *A Short History of Midwifery* (Philadelphia, 1964), 156–57.

[14] Wertz and Wertz, *Lying-In*, 50.

mentors' or their own experiences took them, and wider perspectives remained elusive. But when some doctors took it upon themselves to study anatomy of the female pelvis and activity of the gravid uterus, they removed knowledge from its anchor in individual human experience and brought it to a more abstract level. Learning what was possible and probable in labor and delivery, what was normal and abnormal, provided these birth attendants with knowledge to make judgments on the individual cases they examined.

While the potential for this increased enlightenment may have existed, the question remains whether or not individual medical graduates related their theoretical knowledge to the actual cases they faced in their practices. Did male physicians enhance or improve childbirth, as women believed they would? Did "science," the symbol of the promises of physician-directed obstetrics, come to the aid of birthing women?

Either because of their training or because of families' expectations, male physicians, the apprentice trained and medical school graduates alike, intervened in the birth process more than midwives. Walter Channing, professor of obstetrics at Harvard, advised that a doctor, when called to attend laboring women, "must do something. He cannot remain a spectator merely, where there are many witnesses, and where interest in what is going on is too deep to allow of his inaction. Let him be collected and calm, and he will probably do little he will afterwards look upon with regret."[15]

Physicians' favorite interventions during the first half of the nineteenth century were bloodletting, drugs, and forceps, frequently all used together. Dewees advocated substantial bloodletting. In one of his difficult cases, for example, with the woman standing on her feet, he took "upwards of two quarts" until she fainted. Dewees remarked that "every thing appeared better . . . I introduced the forceps, and delivered a living and healthy child." Physicians believed that venesection could relieve pain, accelerate labor, soften a rigid cervix, ease podalic version, and reduce inflammation. They even bled patients who were hemorrhaging, using the logic that further reducing circulation would produce blood clotting and stop the hemorrhage. They also relieved puerperal convulsions by bloodletting. In fact, as one historian has concluded, "bloodletting . . . was uncritically accepted as the fashion in early American obstetric practice."[16]

The use of opium or laudanum (tincture of opium) seems to have been equally popular among physician-accoucheurs in the nineteenth century. In cases of protracted labor, as in the Shippen-Downing cases cited above, physicians tried opium to accelerate cervical dilation and ease suffering. They also employed cathartics to open the bowels, ergot to stimulate contractions, tobacco infusions to encourage the cervix to dilate, and manually breaking the waters to accelerate labor. In cases of extreme need physicians could surgically separate the pubic bones to facilitate passage of the fetus's head or they could

[15] Walter Channing, *A Treatise on Etherization in Childbirth. Illustrated by Five Hundred and Eighty-One Cases* (Boston, 1848), 229.

[16] Dewees, *Compendious System*, 354, 356–57; A. Clair Siddall, "Bloodletting in American Obstetric Practice, 1800–1945," *Bulletin of the History of Medicine*, 54 (Spring 1980), 101–10.

introduce the crochet, the instrument used for fetal dismemberment and extraction.[17]

The forceps constituted the favorite instrument of physician intervention, and women both feared and respected the "hands of iron." When Shippen referred obliquely to forceps in front of Sally Downing's mother, she "was afraid to ask him, least he should answer in the affermative" that he needed to use it. Her confidence in its benefits, however, remained intact. Women were grateful for the tool that could extricate a fetus in compacted labors.[18]

In 1812 a medical graduate of the University of Pennsylvania entered practice in a town where two aging physicians already practiced. These physicians "had never used forceps, and were in the habit of resorting to Smellie's scissors and the crotchet in all cases where the foetal head became obstructed in the pelvis." The young doctor soon established himself among the women in the community by successfully using forceps to deliver healthy babies who might have otherwise been destroyed. This sort of success story achieved a legendary quality in the nineteenth century as physicians expanded the use of forceps at the same time as they developed their practices.[19]

When well used, forceps could save lives; when misused, they could increase women's perineal lacerations and cause head injuries to the fetus. Dewees called attention to the "mischief" that forceps could cause and believed that their dangers were enhanced because physicians used them too often and unnecessarily. He cautioned: "The greatest care must be taken, before we begin our traction, that no portion of the mother is included in the locking of the blades—this must be done by passing a finger entirely round the place of union. . . . I was once called to a poor woman who had had a considerable portion of the internal face of the right labium removed, by having been included in the joint of the short forceps."[20] Physicians who had been trained to use forceps only on manikins and who were required by custom to perform the forceps operation without the benefit of sight ran considerable risk of creating new problems for the women whose obstructed labors they tried to ameliorate.

Countless stories testify to the severe lacerations nineteenth-century women suffered in childbirth, and they suggest an increase in the problem over what women had previously experienced. The accusation of "meddlesome midwifery" followed physician-accoucheurs, and textbooks cautioned against forceps misuse often enough to suggest that a significant problem existed. The midwifery professors repeatedly warned against unnecessary use of forceps. But their students or apprentice-trained physicians in practice in America's communities found forceps a very valuable tool, and because of their eagerness or their limited practical experience, they may have overused and misused

[17] Bogdan, "Care or Cure?" 96–98; Scholten, "'On the Importance of the Obstetrick Art,'" 437–39. The crochet was the classic instrument to destory the fetus in utero and permit extraction.

[18] Drinker, *Not So Long Ago*, 60–61.

[19] Robert P. Harris, "History of a Pair of Obstetrical Forceps Sixty Years Old," *American Journal of Obstetrics and Diseases of Women and Children*, 4 (May 1871), 55–59.

[20] Dewees, *Compendious System*, 290.

them. Dewees observed so many cases of misused forceps that he concluded, "The forceps . . . are nearly as fatal as the crochet itself."[21]

If forceps were at fault in the "meddlesome midwifery" of the early nineteenth century, they were not the only problem. Almost any intervention by the physician created a potential for harm. If a birth could not proceed without help, the physician provided a service not available elsewhere and necessary to saving lives. If, however, as was statistically more probable, labor was proceeding normally and physicians intervened anyway, their actions introduced dangers not otherwise present. Dewees's caution about careful use of forceps inadvertently informs us about the concurrent dangers of infection. He taught his students to check that no part of the mother was caught in the blades of the forceps by "passing a finger entirely round the place of union." An unwashed and ungloved finger could have carried a higher risk to women's lives than a perineal laceration.

H. B. Willard, a medical graduate who practiced obstetrics in Wisconsin beginning in 1849, rarely used forceps because "the idea of instruments is *horrible* to *friends* & *patient* beside there is much liability to injure the parts." But Willard had no reservations about using opium and ergot, internally manipulating the fetus to change its presenting part, and routinely rupturing the waters with his fingernails. Willard chose his therapeutic activities by weighing the effects they would have on his reputation in the community. He tried not to use emetics, for example, because he knew "the idea of being vomited at such a time is exceedingly repugnant to the patient & to friends." By the same reasoning, Willard, when called to attend a laboring woman, followed Channing's dictum to do something. He knew his patients expected action. The only cases in which Willard did not interfere were those in which the baby had been born before his arrival. In those cases he might still have had opportunity to extract the placenta manually.[22]

Physicians' obstetrics courses taught them the theoretical basis of their craft, their apprenticeships gave them tools with which to effect a successful birth, but nowhere did they receive clear guidelines for the practical application of their knowledge. Doctors had numerous techniques at their command and complete leeway in their use. If physicians used forceps too often, or if they intervened in the birth process too eagerly, it was because they were more persuaded by the faces of women in agony than by the cautions of their elders. Their decisions about intervention were made on the spot and in relative isolation. Even the professors at the medical schools and the leading textbooks taught by anecdotal example, making generalizations hard if not impossible to construct. Physicians could convince themselves easily and in good conscience that their judgment to intervene in labor was in the best interests of the patient. The majority of successful outcomes in each

[21] *Ibid.*, 286.

[22] H. B. Willard Obstetrical Journal, June 29, 1854, p. 23, H. B. Willard Papers (State Historical Society of Wisconsin, Madison); *ibid.*, Feb. 4, 1856, p. 29. For a woman's perceptions of physicians' interventions, see Harriet Connor Brown, *Grandmother Brown's Hundred Years, 1827–1927* (Boston, 1929), 93.

individual's practice underlined the truthfulness of these conclusions. Early obstetric "science" provided knowledge and wherewithal, but the principles of the practical application remained at the bedside in the hands of individual, isolated doctors such as Willard.[23]

The decision of whether or not to employ the skills of the physician remained with women, where it had traditionally been. The parturient, her midwife, and her assistants frequently decided to call a doctor after labor had begun and then gave or withheld permission for each procedure suggested. Dewees, for example, after one successful forceps operation, was called to other cases because "the influence of this case upon many of the midwives of this city, procured me many opportunities of applying the forceps." Similarly the young Pennsylvania doctor's success in delivering babies using forceps led women to refuse to let other doctors use the crochet. Although physicians had broken the gender barrier and birth was no longer exclusively a woman's event, women continued to hold the power to shape events in the birthing room.[24]

Despite the strength of tradition, however, birth had changed for the women who invited physicians to attend them in their homes. These women formed a minority of all birthing women, and they probably were limited geographically to the major cities and economically to the advantaged classes. Most Americans still could not afford doctors and employed midwives and delivered their babies in the same ways as their mothers and grandmothers had. But for those women who chose physicians instead of or in addition to midwives, birth became less a natural process and more an event that could be altered and influenced by a wide selection of interventions. Women and physicians realized that fate no longer held women in such a tight grip and that decisions could be made and actions could be taken that would determine what kind of a birth a woman would have and perhaps whether she and her baby lived or died. This mental perception of the ability to shape the birth experience became even more important in the second half of the nineteenth century when anesthesia emerged as the paramount birthing panacea.

Fanny Longfellow's exuberant description of her childbirth under ether in 1847 (the first in the United States) indicates just how ready women were to change their childbirth experiences. "I never was better or got through a confinement so comfortably," she wrote her sister-in-law. "I feel proud to be the pioneer to less suffering for poor, weak womankind. This is certainly the greatest blessing of this age, and I am glad to have lived at the time of its coming and in the country which gives it to the world."[25] By the middle of the century middle-class women had become accustomed to male birth attendants, although a large segment of the population—probably growing as immigration soared in the latter part of the century—remained faithful to female

[23] See, for example, Willard Obstetrical Journal, Oct. 27, 1849, p. 3; *ibid.*, Dec. 25, 1849, p. 2; *ibid.*, May 7, 1850, p. 4.

[24] Dewees, *Compendious System*, 281; Harris, "History of a Pair of Obstetrical Forceps," 57.

[25] Edward Wagenknecht, ed., *Mrs. Longfellow: Selected Letters and Journals of Fanny Appleton Longfellow (1817–1861)* (New York, 1956), 129–30.

midwives and traditional birth procedures. Advantaged urban women sought every obstetric improvement that male physicians could offer because they continued to fear childbirth and its attendant discomforts. As one woman wrote about her fourth birth in 1885, "Between the oceans of pain, there stretched continents of fear; fear of death and dread of suffering beyond bearing."[26]

Women who experienced their childbirths as the hour in which they "touched the hand of death" eagerly sought relief from the frightful event. Ether and chloroform promised such relief, and women like Longfellow embraced anesthesia enthusiastically. One of Channing's patients told him after her etherized birth "how wonderful it was that she should have got through without the least suffering, and how grateful she was."[27]

Women, in fact, were initially more eager than physicians to use anesthesia. Among physicians, there was some uncertainty about the safety of the new drugs for midwifery practice. Charles D. Meigs of Philadelphia rejected chloroform and ether and carried on a well-publicized campaign against their use, claiming that he had "not yielded to several solicitations as to its exhibition addressed to me by my patients in labour." Meigs relied upon women's painful contractions to help him determine labor's progress and believed that their inhibition would make him a less effective birth attendant.[28]

Walter Channing, one of ether's most ardent supporters, in 1848 surveyed forty-six Boston-area physicians' use of anesthesia in labor and found that they held back on using the drugs except in cases when "patients have demanded it with an emphasis which could not be resisted . . . in many cases, and in the practice of some physicians, it has only been used when such demand has been made."[29] A. R. Thompson of Charlestown, Massachusetts, wrote Channing the following account, which was not unusual. "This lady had informed herself fully as to the use of ether, and had made up her mind to take it. . . . When called to her, I frankly told her that I had never used the ether, nor had I ever seen it used in any case, but that I had no prejudice against it, and would consent to its administration in her case. . . . she was resolute, and demanded the ether. . . . I poured an ounce of ether into the sponge, and the lady held it to her mouth and nose." A successful outcome and a happy patient convinced this doctor to use the drug in subsequent cases. After his second trial of ether Thompson concluded confidently, "My conviction at the time was strong, that the ether had greatly diminished the sufferings of the mother, and shortened the term of her travail for many hours."[30]

[26] Elizabeth H. Emerson, *Glimpses of a Life* (Burlington, N.C., 1960), 4–5.

[27] Mrs. Hal Russell, "Memoirs of Marian Russell," *Colorado Magazine*, 21 (Jan. 1944), 35–36; Channing, *Treatise on Etherization*, 165. For more on women's fears of death and debility, see Leavitt and Walton, "Down to Death's Door."

[28] "Dr. Meigs' Reply to Professor Simpson's Letter," *Medical Examiner and Record of Medical Science*, 11 (March 1848), 148–51. See also John Duffy, "Anglo-American Reaction to Obstetrical Anesthesia," *Bulletin of the History of Medicine*, 38 (Jan.-Feb. 1964), 32–44.

[29] Channing, *Treatise on Etherization*, 300. Walter Channing did not report what proportion of these physician-attended births used anesthesia.

[30] *Ibid.*, 334–35. The woman who insisted on using ether was forty-three years old and had had five children.

Perceptions of safety and patterns of anesthesia use varied enormously among physicians in the second half of the nineteenth century. The medical journals reported both safe and hazardous results of ether and chloroform in midwifery. A Massachusetts physician, for example, wrote in the *Boston Medical and Surgical Journal* that chloroform was "a dangerous and often deadly agent," and a Detroit physician concluded in the *Journal of the American Medical Association*, "I have yet to see my first case of the least evil result to mother or child." With such variation in the literature, individual physicians relied heavily on their own experiences and on the desires of their patients. A rural Iowa doctor wrote about his five hundred obstetric cases: "I do not use an anaesthetic of any kind, especially in forceps cases. I want the patient to know what is going on." Another physician came to the opposite conclusion and recommended anesthesia "not only in troublesome instrumental labor, but in all cases where the pains of travail fall upon women."[31]

Many physicians felt pressed, if their time was limited and other patients were waiting, to use forceps, anesthesia, or both to direct labor into patterns under their control. This inclination was reinforced by women eagerly seeking relief from their suffering. A Colorado physician related that after several hours of hard labor one of his patients "implored me to do something, and, with many misgivings, at last I decided to use instruments [and] chloroform."[32]

Physicians came to adopt anesthesia so generally that one physician in 1895 observed that "the profession has come to regard the use of chloroform in parturition as almost utterly devoid of danger," so that "chloroform is given in labor often recklessly, carelessly, and copiously." As physicians delivered the babies, other attendants—"ignorant nurses, husbands, bystanders, and even . . . the patients themselves"—dripped chloroform or ether onto a sponge or cloth and held it to the woman's nose. No standard procedures guarded physicians' drug dosages.[33]

Physicians' reports of their obstetric practices in the second half of the nineteenth century indicate that use of anesthesia did not necessarily lead to increased forceps use. While doctors did come to use anesthesia routinely in forceps deliveries, the reverse was not true. Physicians did not use forceps routinely on patients to whom they gave ether and chloroform. Physicians employed forceps in approximately 8 percent of their births, whereas some used anesthesia at almost every birth they attended, and on the average about 50 percent of all physician-attended births utilized chloroform or ether.[34]

[31] B. E. Cotting, "Anesthetics in Midwifery," *Boston Medical and Surgical Journal*, 59 (Dec. 9, 1858), 369; Bedford Brown, "The Therapeutic Action of Chloroform in Parturition," *Journal of the American Medical Association*, 25 (Aug. 31, 1895), 355; A. D. Bundy, "Obstetrics in the Country," *Medical and Surgical Reporter*, 56 (Feb. 12, 1887), 201; D. M. Barr, "Anaesthesia in Labor," *ibid.*, 42 (March 13, 1880), 221.

[32] Charles Fox Gardiner, *Doctor at Timberline* (Caldwell, Idaho, 1938), 211. See also Barr, "Anaesthesia in Labor," 227.

[33] Brown, "Therapeutic Action," 354; Duffy, "Anglo-American Reaction," 38. See also J. F. Ford, "Use of Drugs in Labor," *Wisconsin Medical Journal*, 3 (Oct. 1904), 257-65.

[34] Ford, "Use of Drugs," 257-58; Bundy, "Obstetrics in the Country," 200-01. In 91,000 births in Iowa physicians used forceps in 7 percent of cases. E. D. Plass and H. J. Alvis, "A Statistical

Thus, despite the haphazard use of open-drop ether or chloroform and the consequent potential for overdosing, actual anesthesia use must have remained light. Profound anesthesia would have caused labor to decelerate and would have made forceps necessary to lift the baby that the woman was unable to push out. But with light analgesia physicians did not require forceps, and in the nineteenth century forceps abuse did not increase as a result of increased use of anesthetics. Physicians aimed at pain relief, not total unconsciousness. If a woman administered her own drugs on a handkerchief she held up to her nose, her arm would periodically drop away, decreasing the drug dose. One physician advised maintaining what he called a "dreamy sleep, in which the patient follows in her imagination the direction of the physician." While labor progressed, he conversed with his patients about scenes from their childhood or the Sunday school picnic. One of his patients "almost immediately after the first inhalation burst out in a beautiful song, and continued singing one after another until her babe, a large boy, first child, was born."[35]

Most physicians continued to intervene in the birth process in the second half of the nineteenth century as they had earlier. Their safety record when measured by mortality matched the record of midwives, who continued to follow a basically noninterventionist birth policy. Dorothy Reed Mendenhall studied births in Wisconsin at the beginning of the twentieth century and concluded that for both physicians and midwives maternal mortality was higher than it needed to be but that "we must admit that the midwife is, on the whole, probably less culpable in regard to the deaths of parturient women than physicians in the state."[36] Similar maternal mortality rates for midwife- and physician-attended births indicate that physicians, with all their expertise and intervention techniques, did not, as they had promised, enhance the safety of the birth experience for women. Medicine may have improved comfort levels and may have rescued some women from complicated labors, but it did not, on the whole, increase women's chances of survival.

There are other indications that physicians' techniques created new problems for birthing women and actually increased the dangers of childbirth. Inappropriate forceps use and the careless administration of ether and chloroform introduced serious lacerations and breathing disorders that otherwise might not have developed. Most significant, physicians carried puerperal fever, which was potentially disastrous, to birthing women. Because their medical practices included attending patients with communicable diseases, doctors were more likely than midwives to bring with them on their hands and on their clothing the agents of infection. Epidemics of puerperal fever developed

Study of 129,539 Births in Iowa, with Special Reference to the Method of Delivery and the Stillbirth Rate," *American Journal of Obstetrics and Gynecology*, 28 (Aug. 1934), 297.

[35] Barr, "Anaesthesia in Labor," 226–27. In the twentieth century, obstetricians appear to have used anesthesia with the intent to produce unconsciousness, and forceps use consequently increased.

[36] Dorothy Reed Mendenhall, "Prenatal and Natal Conditions in Wisconsin," *Wisconsin Medical Journal*, 15 (March 1917), 353. See also the comments of Dorothy Reed Mendenhall and Florence Sherbon, in American Association for Study and Prevention of Infant Mortality, *Transactions of the Seventh Annual Meeting, Milwaukee, October 19–21, 1916* (Baltimore, 1917), 67–68, 63–64.

even in the practices of physicians like H. H. Whitcomb of Norristown, Pennsylvania, who did "as little meddling as possible" in normal labor and delivery. In 1886 one of Whitcomb's parturient patients developed fever because Whitcomb brought the infection from a previous patient who had scarlet fever. That winter and spring Whitcomb transmitted the fever to thirty-two more birthing women.[37]

Physicians offered some women relief from suffering and aided others through compacted labors, but because of the variations in ability and practices of their birth attendants, most women continued to experience discomfort and to fear childbirth. One woman described her physician-attended delivery as "hell. . . . It bursts your brain, and tears out your heart, and crushes your nerves to bits. It's just hell."[38] Despite their continued suffering, those middle-class women who had chosen physicians to attend them did not want to return to midwives. They believed that if birth were to be eased, improvements would come from progress in medicine. Instead of turning to tradition, these women demanded more of their physicians and continued to hope that safe and comfortable deliveries would come to them from the medical world.

As the symbol of what science had to offer, anesthesia enhanced the place and role of physicians in birthing rooms across America. Women who could afford physicians and their new panacea demanded the advantages of painlessness. Despite the presence of men and modern technology in the birthing rooms, however, the birth experience of nineteenth-century middle- and upper-class women retained many important elements from traditional births. Anita McCormick Blaine's first childbirth in 1890 illustrates this. During her pregnancy Blaine corresponded with her mother, who was in Europe seeking a cure for a hearing loss, and sought the traditional female support network to help her through the impending crisis. "If you could but be with me now, what wouldn't I give," wrote Blaine to her mother, Nettie Fowler McCormick. Despite all of Blaine's preparations and the doctor and nurse who would attend her, McCormick yearned to be with her daughter. She wrote detailed instructions about her care, advising, for example, rubbing olive oil over the abdomen and perineal area to ease delivery. She worried that only a mother could rightfully do such an intimate job. "I don't know if you have a person you could let do it," she wrote, "but I wish I were there to do it." Blaine sequestered herself in her childhood Chicago home, where "every chair and table speaks to me of dear familiar times . . . nothing is so sweet as to feel the presence of all it reminds me of." She surrounded herself with helpers, including her old friend Harriot ("Missy") Hammond, who traveled from Virginia to be with her. Strengthened by these traditional comforts, Anita delivered her baby while under the influence of chloroform, supported by her nurse and by Hammond

[37] H. H. Whitcomb, "A Report of 616 Cases of Labor in Private Practice," *Medical and Surgical Reporter*, 56 (Feb. 12, 1887), 201. Puerperal sepsis caused the largest proportion of maternal deaths at the beginning of the twentieth century. See American Association for Study and Prevention of Infant Mortality, *Transactions of the Seventh Annual Meeting*, 67.

[38] Russell Kelso Carter, *The Sleeping Car "Twilight"; or, Motherhood without Pain* (Boston, 1915), 10–11.

and attended by a male physician. Her husband waited outside. She united the old and the new in obstetric experiences and planned the whole event to meet her expectations. The doctor managed the birth in the context that the woman created.[39]

The decision-making process in the turn-of-the-century birthing rooms reveals how the old and the new intermingled. Doctors, who may have had only minimal practical experience, were invited to attend women in their own homes in the presence of other women, many of whom had had considerable birth experience and had developed strong opinions about birth procedures. Within the birthing rooms, these attendants negotiated. There were desires and expectations on both sides. Women retained a lot of power in their own homes, and physicians bowed to it or risked damaging their reputations among a whole community of women and losing patients. As the new obstetrical techniques—from forceps to anesthetics—became available, they worked to the advantage of physicians, who held the monopoly over their use, while centuries of female traditions and the domestic environment in which they operated worked to the advantage of women.

Inexperienced physicians sometimes welcomed the knowledge of other attendants. Morris Fishbein admitted that he learned more from a laboring woman whose eighth birth he attended than from all his classroom experience. Other physicians found attending a birth in the presence of well-informed women intimidating. One physician recalled, "A young doctor, fresh from medical college, can pass many embarrassing moments in the presence of the neighborhood midwife." Another recalled that a mother's instructions to him "rattled me so that I hardly knew what I was doing."[40]

Women invited physicians to attend them in order to benefit from their expertise and technology, yet doctors could not obtain patients' compliance with all procedures. An Oklahoma physician explained why he could never try to shave a patient's pubic hair even though he thought it would help prevent infection. "In about three seconds after the doctor has made the first rake with his safety [razor], he will find himself on his back out in the yard with the imprint of a woman's bare foot emblazoned on his manly chest, the window sash round his neck and a revolving vision of all the stars in the firmament presented to him. Tell him not to try to shave 'em."[41] Another practitioner realized that a doctor "has his living to make and cannot be too insistent with his patient over whom, usually, he has no control." Yet physicians could not

[39] Anita McCormick Blaine to Nettie Fowler McCormick, Aug. 24, 1890, folder 2B, box 46, Nancy Fowler McCormick Papers (State Historical Society of Wisconsin); McCormick to Blaine, Aug. 1890, folder 1E, box 459, *ibid*.

[40] Morris Fishbein, *An Autobiography* (Garden City, N.Y., 1969), 25; Helen MacKnight Doyle, *A Child Went Forth: The Autobiography of Dr. Helen MacKnight Doyle* (New York, 1934), 322; Gardiner, *Doctor at Timberline*, 211-12. See also Leon Herman, *A Surgeon Thinks It Over* (Philadelphia, 1962), 60.

[41] S. H. Landrum to editor, *Journal of the American Medical Association*, 58 (Feb. 24, 1912), 576. See also George S. King, *Doctor on a Bicycle* (New York, 1958), 61-63; Marcus Bossard, *Eighty-One Years of Living* (Minneapolis, 1946), 39-40; and Mary Bennett Ritter, *More than Gold in California, 1849-1933* (Berkeley, 1933), 219.

come into the birthing room and do nothing. One Kansas doctor admitted, "Perhaps the best way to manage normal labor is to let it alone, but you cannot hold down a job and do that." Women and physicians negotiated the procedures that would be used, and the result represented a compromise between the two worlds of women's tradition and men's medicine. The compromises held until the twentieth century, when physician-directed obstetrics finally became master of the birthing room.[42]

At the turn of the twentieth century physicians delivered approximately half of the nation's babies. Most of these births took place in the woman's home and combined aspects of traditional and modern techniques. Although most of the medical profession accepted the need for aseptic conditions—sterile gloves, shaved perineum, antiseptic infusions, sterile instruments—they could not always achieve them. Whether because of their own abilities or their patients' demands, physicians intermingled modern surgical techniques with wide individual variations and traditional practices. One physician writing in 1912 about the management of normal labor observed that only when the fetal head was resting on the perineum was it necessary "to expose the perineum so that its condition may be constantly watched." The inference, harking back to William Dewees, was that before this stage ocular examination should remain limited. Another doctor refused to shave his patients because the attempt "would cause such a violent attack of hysteria that the physician would cease to be of any further use to that family"[43] Physician practice—susceptible to the physical environment, to the opinions of other attendants, and to the physician's own training and inclination—varied from individual to individual. Physicians acted when they were called to women in labor, but the nature of their interventions followed particular circumstances more than any predetermined science of obstetrics.

The lack of a systematic approach to the practice of obstetrics in the early twentieth century can be directly traced to the quality and emphasis of obstetric education, which had not changed substantially during the nineteenth century. After the middle of the nineteenth century, "demonstrative midwifery," or teaching students by having them observe actual laboring women, became available, but many medical schools ignored this innovation and

[42] J. H. Mackay to editor, *Journal of the American Medical Association*, 58 (March 9, 1912), 720, J. H. Guinn to editor, *ibid.* (March 23, 1912), 880. Women physicians, although small in number, provided a bridge between the two worlds of domesticity and medicine, and many women who wanted the benefits of medicine within the comforting female sphere chose women doctors as birth attendants. More research is needed in this area, beginning with women's hospitals such as the New England Hospital for Women and Children, which, one could argue, tried to provide a milieu in which traditional women's supports could be maintained. See, for example, Virginia G. Drachman, "Female Solidarity and Professional Success: The Dilemma of Women Doctors in Late Nineteenth-Century America," *Journal of Social History*, 15 (Summer 1982), 607-19; and Regina Markell Morantz and Sue Zschoche, "Professionalism, Feminism, and Gender Roles: A Comparative Study of Nineteenth-Century Medical Therapeutics," *Journal of American History*, 67 (Dec. 1980), 568-88.

[43] "Management of Normal Labor," *Journal of the American Medical Association*, 58 (Jan. 27, 1912), 274; William M. Gregory to editor, *ibid.* (Feb. 24, 1912), 577. See also F. W. MacManus to editor, *ibid.* (March 9, 1912), 720; and Landrum to editor, ibid. (Feb. 24, 1912), 576.

continued their didactic teaching from textbooks and manikins. In 1910, although most practitioners now received a formal medical education, many still graduated having witnessed few or no live deliveries. Such training at best prepared students to understand birth pathology and perhaps how to intervene to rescue a woman in trouble, but it did not prepare physicians to attend women in normal labor.[44]

J. Whitridge Williams, professor of obstetrics at the Johns Hopkins Medical School, surveyed obstetrics education in 1912 and concluded that the "average practitioner, through his lack of preparation for the practice of obstetrics, may do his patients as much harm as the much-maligned midwife." Williams found that most medical students had the opportunity during their training to watch only one woman deliver, and one-quarter of the medical schools admitted that their graduates were not competent to practice obstetrics. Yet most physicians did deliver babies as part of their medical practices, spending on the average approximately 30 percent of their time attending childbirth.[45] In the era when scientific investigations began to permeate the medical world, when medical schools increasingly built laboratories and trained their students in research methods and findings, and when specialization and hospital expansion transformed medical practice, physicians' obstetrical practices were characterized by individual variation and unsystematic application of general principles.

Williams believed that only the developing specialty of obstetrics practiced in the hospital setting could improve the marginal situation. Even as a significant portion of American women continued to be attended by midwives in their own homes, and most others continued to invite general practitioners to attend them at home, increasing numbers of women from the middle and upper classes became convinced that childbirth in the hospital offered them the safest experience. Many factors converged to encourage women to move to the hospital in the early years of the twentieth century, including the new anesthetic agent scopolamine (best monitored in the hospital), the preference of the specialists, the growth of the hospitals themselves, and the increasing technology and ease of management in the well-equipped and well-staffed hospital setting.[46] Still seeking more comfortable births and safer deliveries

[44] Longo, "Obstetrics and Gynecology," 211–14; Virginia G. Drachman, "The Loomis Trial: Social Mores and Obstetrics in the Mid-Nineteenth Century," in *Health Care in America: Essays in Social History*, ed. Susan Reverby and David Rosner (Philadelphia, 1979), 67–83; American Association for Study and Prevention of Infant Mortality, *Transactions of the Seventh Annual Meeting*, 63–64; Francis A. Long, *A Prairie Doctor of the Eighties: Some Personal Recollections and Some Early Medical and Social History of a Prairie State* (Norfolk, Nebr., 1937), 29.

[45] J. Whitridge Williams, "Medical Education and the Midwife Problem in the United States," *Journal of the American Medical Association*, 58 (Jan. 6, 1912), 1, 2, 5; R. H. Riley, "The Public Health Aspect of the Teaching of Obstetrics in Undergraduate Medical Schools," *ibid.*, 106 (April 25, 1936), 1438. See also Lawrence D. Longo, "John Whitridge Williams and Academic Obstetrics in America," *Transactions and Studies of the College of Physicians of Philadelphia*, 3 (Dec. 1981), 221–54.

[46] I have explored some of these points elsewhere; see Judith Walzer Leavitt, "Birthing and Anesthesia: The Debate over Twilight Sleep," *Signs*, 6 (Autumn 1980), 147–64. See also Morris J. Vogel, *The Invention of the Modern Hospital: Boston, 1870–1930* (Chicago, 1980); and Judy

with minimal negative aftereffects, many women decided to try the newest that obstetrics had to offer and go to the hospital to be attended by specialists. "I have placed myself in the hands of. . . . a specialist in obstetrics," wrote Lella Secor to her mother in 1918. "I have every confidence in him, and it is a great relief."[47]

In hospital deliveries women left their family and friends at the door of the labor room and faced their birthings alone, as one account of birth in the 1930s indicated:

> Arriving [at the hospital], she is immediately given the benefit of one of the modern analgesics or pain-killers. Soon she is in a dreamy, half-conscious state. . . .
>
> She knows nothing about being taken to a spotlessly clean delivery room, placed on a sterile table, draped with sterile sheets; neither does she see her attendants, the doctor and nurses, garbed for her protection in sterile white gowns and gloves; nor the shiny boiled instruments and antiseptic solutions. She does not hear the cry of her baby when first he feels the chill of this cold world, or see the care with which the doctor repairs such lacerations as may have occurred. . . . Finally she awakes in smiles, a mother with no recollection of having become one.[48]

This woman was separated from the people she loved, she was in an unfamiliar environment controlled by others, and she was unconscious during her childbirth. Women did not view the stay in the hospital as a time when they lost important parts of the traditional birth experiences, but rather as a time when they gained protection for life and health, aspects of birth that had been elusive and uncertain in the past. Women gave up some kinds of control for others because on balance the new benefits seemed more important. In seeking life and health, women relinquished consciousness and self-determination.

The hospital appealed to women because it was modern, well equipped, and staffed by experts: it represented the newest medical advance. Also, it physically separated women from their domestic chores and allowed them to relax. One wrote, "My stay in that hospital was like a lovely vacation." "I can't tell you the relief I feel as I walk out my door headed for the hospital to have a baby," wrote another. "I have nothing to worry about . . . and have only to concentrate on giving birth. All this peace of mind, plus expert medical attention, makes me wonder why anybody would consider it a 'privilege' to

Barrett Litoff, *American Midwives: 1860 to the Present* (Westport, Conn., 1978). The declining number of midwives in this period contributed to the trend toward hospital deliveries and was affected by many of the same factors. See Frances E. Kobrin, "The American Midwife Controversy: A Crisis of Professionalization," *Bulletin of the History of Medicine*, 40 (July-Aug. 1966), 350-63; Neal Devitt, "The Statistical Case for Elimination of the Midwife: Fact versus Prejudice, 1890-1935," *Women and Health*, 4 (Spring 1979), 81-96; *ibid.* (Summer 1979), 169-86. The midwife was the major victim of advancing obstetrics. Physicians took almost two centuries to replace midwives in America's birthing rooms, but by the 1940s medicine successfully superseded female midwifery. Since few women were obstetricians, when midwives "disappeared," women lost the occupation for their sex; they also lost with midwifery centuries of female tradition, ritual, and unity, which today they are seeking to reclaim.

[47] Barbara Moench Florence, ed., *Lella Secor: A Diary in Letters, 1915-1922* (New York, 1978), 170.

[48] Roy P. Finney, *The Story of Motherhood* (New York, 1937), 6-7. See also Alan Frank Guttmacher, *Into This Universe: The Story of Human Birth* (New York, 1937), 192-267.

have her baby at home." Home did not hold for mid-twentieth-century women the same comforts and sustenance that Blaine had found so reassuring. The "hospital is equipped with every modern device for the safe delivery of babies," wrote one mother. "Nursing and medical attention is available at any hour of the day or night. How much simpler—and more restful—to be in a hospital where babies are an accepted business."[49]

The move to the hospital was hastened by specialists' attempts to wrest birth away from general practitioners and to systematize birth procedures within the hospital setting. One very influential move in this direction was made in 1920 by Joseph B. DeLee of Chicago in the first volume of the new journal for specialists, the *American Journal of Obstetrics and Gynecology*. DeLee argued that birth was a "pathologic process" and that "only a small minority of women escape damage during labor." He continued, "So frequent are these bad effects, that I have often wondered whether Nature did not deliberately intend women should be used up in the process of reproduction, in a manner analogous to that of salmon, which dies after spawning?" Because of its dangers, DeLee believed birth needed careful monitoring in skilled hands. He recommended reducing birth to predictable patterns by using outlet forceps and episiotomy routinely and prophylactically in normal deliveries. DeLee sedated the parturient with scopolamine, allowed the cervix to dilate, gave ether during the second stage, performed an episiotomy, and lifted the fetus with forceps. He then extracted the placenta, gave ergot to help the uterus contract, and stitched the perineal cut.[50]

DeLee's operation achieved wide acceptance. It represented the new move in the 1920s and 1930s to make obstetrics scientific, systematic, and predictable by putting it under the control of the specialist. Not all specialists wanted to use prophylactic forceps, but hospital-based obstetricians did develop routines for managing childbirth that incorporated systematic use of pain-relieving drugs, labor inducers, and technological intervention. Mothers-to-be learned they could plan when and how they would have a baby, and doctors could predict the course of labor because they controlled it. "The old way [of having a baby] was no fun," wrote one father about the 1916 birth of his first daughter. In 1938 his second daughter "was born the new way—the easy, painless, streamlined way." His wife, in consultation with her doctor, decided which day to deliver, and after a matinee and dinner went into the hospital. Pituitrin induced her labor, Nembutal and scopolamine deadened her perceptions of it, and the doctor delivered her baby. "Why, I wouldn't mind

[49] Gladys Denny Shultz, "Journal Mothers Report on Cruelty in Maternity Wards," *Ladies' Home Journal*, 75 (May 1958), 44; M. F. Ashley Montagu, "Babies Should Be Born at Home!" *ibid.*, 72 (Aug. 1955), 52. See also Betty MacDonald, *The Egg and I* (Philadelphia, 1945), 108, 163. Betty MacDonald found the hospital a place of luxury, wonderful food, and good care—almost, she said, enough to keep her pregnant the rest of her life. Hospital stays may have looked attractive because women's own networks of home helpers were diminishing in this period. This hypothesis needs further research.

[50] Joseph B. DeLee, "The Prophylactic Forceps Operation," *American Journal of Obstetrics and Gynecology*, 1 (Oct. 1920), 40–41, 34–35.

having another baby next week," she said, "if that's all there is to it."[51] Increasing numbers of middle-class women turned to the hospital and the specialist for their childbirths, believing, as one of them put it, "the vexation of hospital routine shrinks to infinitesimal importance beside the safety of the delivery room."[52]

But was streamlined childbirth safe? Had medicine at last provided safer and more comfortable childbirth for women, as it had promised since the 1760s? Observers in the 1930s realized that while the practice of obstetrics now could relieve women of much of their discomfort, childbirth remained unsafe. Maternal mortality in the 1930s remained "unnecessarily high," according to a group of concerned physicians in New York. The physicians claimed hospital deliveries contributed to the high mortality with a "high incidence of operative interference during labor . . . undertaken when there was no indication or a plain contra-indication." Birth in the hospital encouraged interference because the equipment and staff were readily accessible, and this interference increased the number of maternal deaths. Rather than making childbirth safer, physicians in the 1920s and 1930s, according to their own evaluation, were responsible for maintaining high rates of maternal mortality.[53]

In home deliveries of the nineteenth century, increased anesthetic use had not caused an increase in forceps deliveries, but in hospital deliveries of the early twentieth century, a direct relationship existed between anesthesia and forceps. The New York doctors noted, "The frequent use of instrumentation is based upon the easy accessibility of anaesthesia. . . . the increase in the use of anaesthesia is a factor in keeping the maternal mortality rate stationary." According to the New York study, use of anesthetic agents led directly to increased instrumentation because drugged women were less effective in pushing the baby out. Approximately 25 percent of hospital deliveries were operative. "The increase in the use of instrumentation brings with it an increased hazard," the doctors found. "Clearly a reduction of the mortality rate can be achieved through a reduction in operative interference."[54]

During the early 1930s the medical profession publicly confessed the short-comings of its obstetric practices. Rudolph Wieser Holmes declared at a crowded American Medical Association meeting, "I have seen hundreds of

[51] J. P. McEvoy, "Our Streamlined Baby," *Reader's Digest*, 32 (May 1938), 15-16. For reaction to prophylactic forceps, see J. Whitridge Williams, "A Criticism of Certain Tendencies in American Obstetrics," *New York State Journal of Medicine*, 22 (Nov. 1922), 493-99.

[52] "I Had a Baby, Too: A Symposium," *Atlantic Monthly*, 163 (June 1939), 768. See also Lenore Pelham Friedrich, "I Had a Baby," *ibid.* (April 1939), 461-65. Clara Rust delivered five babies at home before going to the hospital to have her sixth in 1926. She was apprehensive about leaving home but went to the hospital on the doctor's advice. Jo Anne Wold, *This Old House: The Story of Clara Rust* (Anchorage, 1976), 207-08.

[53] New York Academy of Medicine Committee on Public Health Relations, *Maternal Mortality in New York City: A Study of All Puerperal Deaths, 1930-1932* (New York, 1933), 213-15, 125-27.

[54] *Ibid.*, 113-17, 126-27. An Iowa study found that 11.8 percent of all births and 23 percent of hospital births were operative statewide. Plass and Alvis, "Statistical Study," 293-305.

FIGURE 1
Maternal Mortality Rates.
Birth Registration States or United States, 1915-1964.

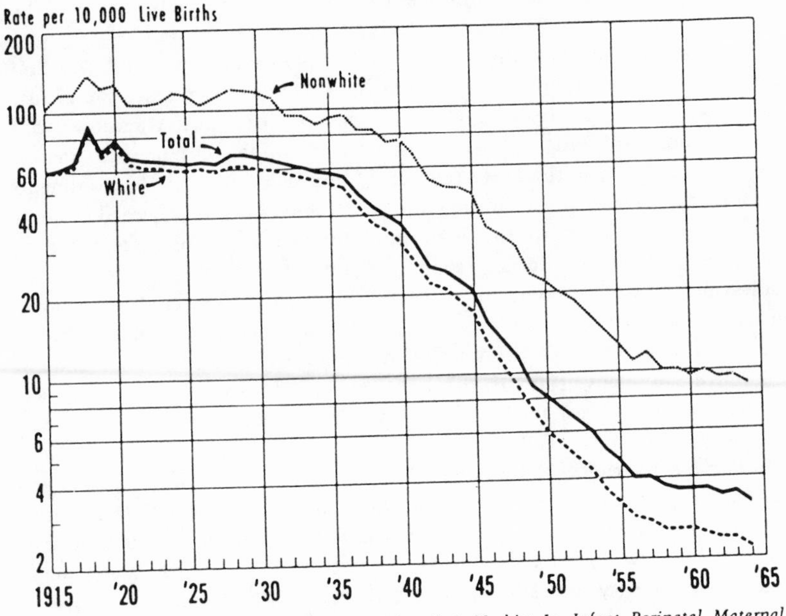

Source: Sam Shapiro, Edward R. Schlesinger, Robert E. L. Nesbitt, Jr., *Infant, Perinatal, Maternal, and Childhood Mortality in the United States* (Cambridge, Mass.: Harvard University Press, 1968), 145. Reprinted by permission.

women die on the delivery table because of the wrongful use of drugs." Holmes blamed his own role in accelerating scopolamine use for many of these deaths. The public health commissioner for the state of New York calculated that maternal deaths were actually increasing in the twentieth century and concluded that "they represent the results of meddlesome and unskillful obstetrical practice."[55] Philadelphia doctors agreed that 56 percent of preventable maternal deaths could be blamed on physician errors of judgment and technique.[56] A national study concluded similarly that "artificial delivery is becoming increasingly frequent, especially in hospital practice," and "that interference with normal labor is accompanied by some added risk to both mother and child." Although the indictment was not universally accepted by

[55] "Childbirth: Nature v. Drugs," *Time*, May 25, 1936, p. 36; Matthias Nicoll, Jr., "Maternity as a Public Health Problem," *American Journal of Public Health*, 19 (Sept. 1929), 967. See also R. W. Holmes, R. D. Mussey, and F. L. Adair, "Factors and Causes of Maternal Mortality," *Journal of the American Medical Association*, 93 (Nov. 9, 1929), 1440-47.

[56] Philadelphia County Medical Society, Committee on Maternal Welfare, *Maternal Mortality in Philadelphia, 1931-1933* (Philadelphia, 1934), 130. See also New York Academy of Medicine, *Maternal Mortality*, 32-38.

physicians in the 1930s, the statistics indicated that going to a hospital to have a baby managed by a physician and under the influence of twentieth-century drugs and instruments did not necessarily give a mother an advantage and may have put her in greater jeopardy than her neighbor who stayed home.[57]

Obstetrics by the 1930s had not achieved a record of safety commensurate with its abilities. Aided by the New York Academy of Medicine's 1933 revelations about preventable maternal and infant deaths, however, physicians moved toward regulation and standardization of hospital obstetrics practice in their attempts to decrease the overuse and misuse of drugs and operative procedures. Their success in establishing minimum standards for delivery procedures is evident in the fact that hospital births increased from 35 percent of all births in 1933 to 72 percent ten years later and "for the first time was not associated with higher rates of puerperal death." Other factors converged with hospital regulations to cause the decrease in maternal mortality after 1935. The prenatal-care movement spearheaded by middle-class women succeeded in gaining adherents and in educating the public about the importance of care during pregnancy. Probably most important in the decelerating death rates was the wide adoption of sulfonamides, blood transfusions, and, after World War II, antibiotics, which were successful in combating infection and hemorrhaging, two of the major causes of maternal deaths. After 1935 maternal mortality fell dramatically.[58] (See Figure 1.)

Even though birth became safer in the 1940s and 1950s, women who went to the hospital to deliver their babies—95 percent of American women by 1955—missed some of the features of traditional social childbirth and found hospital births sterile both in the antiseptic sense and in the human dimension. Continuing to appreciate the safety factors they acknowledged to exist for hospital births, women as early as the 1950s attacked the "cruelty in maternity wards." One woman from Elkhart, Indiana, wrote to the *Ladies' Home Journal*, which exposed these conditions: "So many women, especially first mothers, who are frightened to start out with, receive such brutal inconsiderate treatment that the whole thing is a horrible nightmare. They

[57] White House Conference on Child Health and Protection, *Fetal, Newborn, and Maternal Morbidity and Mortality: Report of the Subcommittee on Factors and Causes of Fetal, Newborn, and Maternal Morbidity and Mortality* (New York, 1933), 14, 220. For information about the effects of instrumental deliveries on the children, see Charles Edwin Galloway, "Prevention of Birth Injury and Its Resulting Mortality from the Standpoint of the Obstetrician," *Journal of the American Medical Association*, 106 (Feb. 15, 1936), 505-07.

[58] Joyce Antler and Daniel M. Fox, "The Movement toward a Safe Maternity: Physician Accountability in New York City, 1915-1940," in *Sickness and Health in America: Readings in the History of Medicine and Public Health*, ed. Judith Walzer Leavitt and Ronald L. Numbers (Madison, 1978), 386; Lawrence D. Longo and Christina M. Thomsen, "The Evolution of Prenatal Care in America," in *Proceedings of the Second Motherhood Symposium*, ed. Colleau, 29-70. During the 1920s and 1930s, when increasing numbers of women moved to the hospital to have their babies, maternal mortality rates remained as high as they had been earlier in the century, approximately 60 deaths per 10,000 live births (over 100 per 10,000 for nonwhite women). See Sam Shapiro, Edward R. Schlesinger, and Robert E. L. Nesbitt, Jr., *Infant, Perinatal, Maternal, and Childhood Mortality in the United States* (Cambridge, Mass., 1968), 144-45. When maternal mortality rates started falling after 1935, the decrease was not due to the hospital per se but to measures such as antibiotics that could have been used at home.

give you drugs, whether you want them or not, strap you down like an animal." A woman from Columbus, Ohio, concurred that a new mother was "foiled in every attempt to follow her own wishes."[59] These women noticed that the physical removal of childbirth from the woman's home to the physician's institution shifted the balance of power. Birth was no longer part of the woman's domain, as it had been during all the years it remained in the home. It had become instead a medical affair run by medical professionals. Women were no longer the main actors, but instead physicians acted upon women's bodies. Women who were attended by physicians in their own homes could still determine what would happen—could refuse to be shaved or could demand ether. They still controlled much of their own births. But women who entered the birthing rooms of medicine were captured by the routine and by the expertise surrounding them. They could not decide what kind of births they would have.

Some hospitals and some doctors did not depersonalize birth with institutional routine and were able to provide satisfying birth experiences. But many women agreed with the Bozeman, Montana, mother who wrote, "The cruelest part of [hospital] childbirth is being alone among strangers." She found nothing familiar to comfort her through the difficult hours, only the routine of hospital life that she and others described as an assembly line.[60] Physician-directed hospital obstetrics had not met the expectations of American women by protecting a woman's dignity and integrity at the same time as it insured safety. Women wanted the psychologically comforting practices of their traditional birthing rooms to be incorporated into the modern practices of the birthing rooms of science.

The two images of birthing women in America—one "brought to bed" in her own home by the women she had called together and the other drugged and "alone among strangers" in an impersonal hospital—frame the American obstetric experience. The eighteenth-century woman felt vulnerable to death and debility despite the strength she derived from her friends; the twentieth-century woman felt vulnerable to the institutional routine despite the strength she derived from encouraging maternal-mortality statistics. Neither woman had what she knew she needed at the time of her birth: the certainty of a healthy outcome with the freedom to make choices about how to conduct the important event.

For most of American history women controlled the events in their own birthing rooms. They negotiated with their birth attendants and decided, as much as their bodies and available technology allowed, how the birth would proceed. But the physical forces of birth did not always cooperate, and women felt at the mercy of fate. When physicians entered the women's birthing rooms with the promises of increased safety, women shared their traditional jurisdiction. Forceps, therapeutic aids, and especially anesthesia seemed to offer some insurance against the uncontrollable terrors of birth. Women found that

[59] Shultz, "Journal Mothers Report on Cruelty," 44–45.
[60] *Ibid.*, 44.

they could, if they wanted to, give their bodies up to labor and to the physician and return to consciousness after the event. Birthing women and their physicians, in the company of other women or family members, negotiated the particulars of birth procedures behind the closed doors of the birthing room and achieved a compromise that recognized the needs, wishes, and abilities of all participants.

But physicians could not achieve the safety that they had promised. While some women were relieved or saved by medical intervention, on balance women's experiences of birth were not significantly enhanced by the entrance of medicine. Physicians remained unable to control their own use of obstetric technology for the first 170 years they were engaged in delivering America's babies, and they frequently added to women's problems by increasing the danger of infection, overdosing with anesthesia, and mutilating with forceps.

In the twentieth century the hospital offered an alternative that appealed to women and to doctors because it promised at last the safety that both had sought. Removing birth from women's homes seemed just another step in the progression of changing practices. But women soon discovered that the birthing rooms of medicine differed fundamentally from their own birthing rooms. In hospitals women lost the ability to make choices for themselves and surrendered to the medical community the power to determine what would be done to their bodies. The early years of hospital-based obstetrics did not even provide additional safety for women. As the doctors themselves eventually realized, the hospital made interventions so easy that doctors resorted to operations and drugs in excess of what was indicated and by their own action kept the rates of maternal mortality high. Only after 1940 did medicine begin to achieve a record of safety commensurate with the promises it had held out to women centuries earlier.

Each major transition in American childbirth history—from midwives to physicians, from home to hospital, from "hands of flesh" intervention to "hands of iron" interference—represents in a different form the lure of science and its promise to improve the childbirth experience.[61] As women and their attendants came to perceive birth as an event that could be shaped and manipulated, their *mentalité* of pathology and impending emergencies led each generation to seek its own kind of ideal childbirth. Ultimately it was the promise of continual advances that pushed birth into new patterns throughout American history. Women and their birth attendants believed that medicine could solve their birthing problems and make procreation safe, and they put their faith in scientific advances years before science had significant impact on obstetric practices. Systematic application of tested principles did not characterize obstetrics until the twentieth century, yet medical activities in the birth room were valued much earlier because physicians carried an image of expertise and knowledge along with the instruments and drugs in their medical bags.

[61] The dichotomy between hands of flesh and hands of iron is used particularly effectively by Adrienne Rich, *Of Woman Born: Motherhood as Experience and Institution* (New York, 1976), 128-55.

Despite the lack of a systematic practice of obstetrics, women did not want to turn back to traditional practices. Only in very recent years has any sizable group of American women tried to change direction. Today some women are trying to alter the patterns of childbirth by seeking practices that were common in the past. These women, still a minority and predominantly middle-class, search out midwives or demand that doctors turn back the clock to a time when decision making involved all the birth participants. Women in America today who are seeking to control their own births are fighting the legacy of childbirth as structured by the medical profession and are demanding that choice be reintroduced to the childbirth agenda. They are not trying to overturn science but rather to control it within the traditional female context.[62]

The two themes of social childbirth and scientific childbirth have persisted throughout American history. In the seventeenth and eighteenth centuries, social childbirth practices prevailed. The two vied for supremacy during the nineteenth century, and not until the middle of the twentieth century did the medical model succeed in dominating American childbirth. Yet physician-directed childbirth has not met all the needs of birthing women. The pendulum swing between the two traditions has not yet found its resting place.

[62] As an example of contemporary interests, see Barbara Katz Rothman, *In Labor: Women and Power in the Birthplace* (New York, 1982).

"Called to Her at Three O'Clock AM": Obstetrical Practice in Physician Case Notes

AMALIE M. KASS

OBSTETRICAL case notes of physicians engaged in daily practice add an important perspective, hitherto neglected, to the history of nineteenth-century obstetrics. They give detailed descriptions of the physician's activities, reveal his perceptions of his role, and disclose his attitude toward his patients. Although this perspective is not always consistent with earlier interpretations, it provides a more complete picture of events in the birthing room and a more balanced view of patient-physician relationships than has heretofore been accepted. As Erwin Ackerknecht suggested more than twenty-five years ago, case histories are an important source for "gauging the actual practice of medicine as well as the social nature of healing relationships."[1]

For many years historians relied on textbooks and lectures of prominent obstetricians and on the unusual cases in medical journals to describe nineteenth-century obstetrical practice.[2] The result was an account of

1. Erwin H. Ackerknecht, "A plea for a 'behaviorist' approach in writing the history of medicine," *J. Hist. Med. Allied Sci.*, 1967, *22*, 211–14, quoted in Guenter B. Risse and John Harley Warner, "Reconstructing clinical activities: Patient records in medical history," *Soc. Hist. Med.*, 1992, 5, 183–205.

2. See, e.g., Theodore Cianfrani, *A Short History of Obstetrics and Gynecology* (Springfield, Ill.: Charles C Thomas, 1960); Irving S. Cutter and Henry R. Viets, *A Short History of Midwifery* (Philadelphia: W. B. Saunders, 1964); Palmer Findley, *Priests of Lucina: The Story of Obstetrics* (Boston: Little, Brown & Co., 1939); Harvey Graham, *Eternal Eve, the Mysteries of Birth and Customs that Surround It*, 2nd ed. (London: Hutchinson, 1960); Frederick C. Irving, *Safe Deliverance* (Boston: Houghton, Mifflin, 1942); William F. Mengert, "The origin of the male midwife," *Ann. Med. Hist*, 1932, n.s., *4*, 453–65; Harold

Research for this paper was funded in part by an ACOG-Ortho Fellowship in the History of American Obstetrics and Gynecology. An abbreviated version was presented at the annual meeting of the American College of Obstetrics and Gynecology in May 1993.

I am grateful to Allan Brandt, Anne Hecht, Nancy Kass, John Harley Warner, and Ellin Wolfe for helpful comments; to Susan Rishworth and Margaret N. Burri for their assistance; to Mrs. Ferdinand E. Chatard for information about her distinguished medical family; and to Richard Wolfe and the staff of the Countway Library of Medicine for their interest and encouragement. I also appreciate the criticisms and suggestions made by the editors and reviewers of the *Journal of the History of Medicine and Allied Sciences*.

ISSN 0022–5045 VOLUME 50 PAGES 194 TO 229

improved care by physicians who had replaced "uneducated and unqualified persons, mostly women," and had raised midwifery from "the low condition into which it had fallen, . . . placing it where it rightfully belongs, and where we hope it will long remain, in the hands of well-educated men."[3] The assumption that men trained in anatomy and physiology could assure better outcomes was implicit in these accounts and led to descriptions of "the gradual march of progress in midwifery" and to praise for "the rational practice of midwifery" in America.[4]

Beginning in the 1960s, social historians, especially feminist scholars, produced a vigorous revision of the hitherto accepted version of nineteenth-century obstetrics.[5] By examining women's journals, diaries, and letters, feminist historians gave voice to the emotions and thoughts of the women who experienced pregnancy and parturition. They criticized male physicians for replacing midwives and for destroying what had always been a female-controlled experience. Many feminists questioned the beneficial effect of the "medicalization of childbirth," asserting that delivery was neither safer nor easier than before.[6]

Speert, *Obstetrics and Gynecology in America, A History* (Chicago: American College of Obstetricians and Gynecologists, 1980); Herbert Thoms, *Chapters in American Obstetrics*, 2nd ed. (Springfield, Ill: Charles C Thomas, 1961); idem, *Our Obstetric Heritage, the Story of Safe Childbirth* (Hamden, Connecticut: Shoe String Press, 1960); J. Whitridge Williams, "A sketch of the history of obstetrics in the United States up to 1860," *Amer. Gynecol.*, 1903, *3*, 266–94.

3. William W. Wellington, "Biographical sketches of deceased members of the Obstetrical Society of Boston, with an outline of the earlier obstetrical history of Boston and vicinity," *Boston Medical and Surgical Jrl.*, 1881, *105*, 494–97.

4. Cutter and Viets, (n. 2) *Short History*, pp. 155, 172. The terms *midwifery* and *obstetrics* were used interchangeably in this period and are also used interchangeably in this paper.

5. See, e.g., Janet Bogdan, "Care or cure? Childbirth practices in nineteenth century America," *Feminist Studies*, 1978, *4*, 92–98; Charlotte Gregg Borst, "Midwives in Early New England 1620–1820: From Healer and Community Authority to Quack and Outsider" (M.A. thesis, Tufts University, 1977); Jane B. Donegan, *Women & Men Midwives: Medicine, Morality and Misogyny in Early America* (Westport, Conn.: Greenwood Press, 1978); Nancy Schrom Dye, "History of childbirth in America," *Signs*, 1980–81, *6*, 97–108; Claire E. Fox, "Pregnancy, Childbirth and Early Infancy in Anglo-American Culture 1675–1830" (Ph.D. diss., University of Pennsylvania, 1966); Pamela S. Eakins, ed., *The American Way of Birth* (Philadelphia: Temple University Press, 1986); Sylvia D. Hoffert, *Private Matters, American Attitudes toward Childbearing and Infant Nurture in the Urban North, 1800–1860* (Urbana: University of Illinois Press, 1989); Judith Walzer Leavitt, *Brought to Bed: Childbearing in America 1750 to 1950* (New York: Oxford University Press, 1986); Judy Barrett Litoff, *American Midwives: 1860 to the Present* (Westport, Conn.: Greenwood Press, 1978), chap. 1; Sally G. McMillen, *Motherhood in the Old South: Pregnancy, Childbirth, and Infant Rearing* (Baton Rouge: Louisiana State University, 1990); Catherine M. Scholten, "'On the Importance of the Obstetrick Art': Changing customs of childbirth in America 1760 to 1825," *William Mary Quart.*, 3rd ser., 1977, *34*, 426–45; Richard W. Wertz and Dorothy C. Wertz, *Lying-In: A History of Childbirth in America* (New York: Free Press, 1977). Laurel Thatcher Ulrich, *A Midwife's Tale* (New York: Knopf, 1990) is a study of obstetrics as practiced by a female midwife in Maine from 1785 to 1812.

6. In *Brought to Bed: Child-bearing in America 1750 to 1950*, (n. 5), Judith Walzer Leavitt reconciled a distinctively feminist approach with the gains in safety that accompanied medicalized childbirth. Her book has become a standard text in women's studies courses as well as in medical history courses.

Neither traditional nor feminist historians gave much attention to the physician in his day-to-day practice of obstetrics.[7] Case notes give voice to *his* emotions and thoughts as he attended delivery after delivery. They make clear that physicians were neither as confident nor as successful as had been implied by the earlier historians. Nor were they uncaring or as interventionist as the revisionist critics suggest. Case notes show that parturient women continued to control childbirth and that their female friends and relatives continued to assist them in the birthing room throughout the nineteenth century. They also give evidence that physicians were delivering the babies of poor urban women and rural women earlier than has been generally assumed.

PHYSICIAN CASE NOTES

The analysis that follows is based on the handwritten case notes of eighteen physicians, plus a manuscript summary and a published summary of two additional obstetrical practices (See Appendix for brief biographical identification). All were recorded by male physicians who practiced allopathic medicine.[8] They lived in cities, small towns, and rural areas of Massachusetts, Maryland, Pennsylvania, New Jersey, and New York State. Their records cover ninety years, from 1785 to 1875.

The sole criterion for inclusion in the study was the information they contained; that is, that they add in some way to our knowledge of "the actual practice of medicine as well as the social nature of healing relationships."[9] They differ significantly from obstetrical cases found in medical journals or in the proceedings of medical societies. Whereas the latter accounts focus on the unusual situation, a difficult delivery, unexpected complication, or monstrous birth, these case notes detail continuous day-to-day practice, most of which was uneventful (except, of

7. For studies of obstetrical practice that are based upon physician case notes, see Steven M. Stowe, "Obstetrics and the work of doctoring in the mid-nineteenth century American South," *Bull. Hist. Med.*, 1990, *64*, 540–66; Paul Berman, "Obstetrical Practice in South Central Massachusetts from 1834 to 1845," in *Medicine and Healing*, ed. Peter Benes, Dublin Seminar for New England and Folk Life, Annual Proceedings 1990, pp. 185–90; Amalie M. Kass, "The obstetrical casebook of Walter Channing, 1811–1822," *Bull. Hist. Med.*, 1993, *67*, 494–523.

8. Several physicians made disparaging remarks about homeopathic physicians and "quacks" in their notes. See, e.g., John Barnard Swett Jackson, "Medical Cases and Autopsies, 1830–69," 21 vols., MSS, Countway Library of Medicine, Boston, Massachusetts. 2nd ser., II, 29 January 1837; William Thornton Parker, "Cases," 3 vols., 1841–1851, MSS, Countway Library, Boston, Massachusetts, III, 3 September 1850.

9. Ackerknecht, (n. 1).

course, for the parents of the newborn child). Day books, account books, and prescription books were not used.[10]

Nineteenth-century American physicians were not taught to keep case notes during their medical training.[11] Once in practice, they did not necessarily keep patient records, and, when they did, there was no standardized form as might be found in institutional records. Thus, there is great variety in the form and the content, reflective of the interests and concerns of each physician.

Some kept obstetrical notes separately from the rest of their practice; others described all their cases seriatim, childbirth mixed in with broken limbs, tumors, fevers, venereal disease, vaccinations, and swallowed buttons. Some recorded only their more interesting cases. Some gave very brief descriptions, identifying the patient, date of delivery, sex of the newborn, perhaps adding a brief note if there were complications. Many counted their patients as they went along.[12] Some noted the weight of the baby; one recorded the distance from cord to head and cord to toes.

10. For studies based upon account books, prescriptions books, etc., see, e.g., Irvine Loudon, *Medical Care and the General Practitioner 1750–1850* (Oxford: Clarendon Press, 1986); J. Worth Estes, "Therapeutic Practice in Colonial New England," in *Medicine in Colonial Massachusetts, 1620–1820*, ed. Philip Cash, Eric C. Christianson, and J. Worth Estes (Boston: University Press of Virginia for the Colonial Society of Massachusetts, 1980), pp. 289–383.

11. Medical education in the United States prior to the latter part of the nineteenth century consisted of apprenticeship with an established physician and attendance at two sixteen-week series of lectures. Neither the preceptor nor the formal lectures provided an opportunity for the student to observe patients and take case notes. Students who paid an extra fee to accompany their professors to the almshouse, dispensary, or hospital might observe a few patients, but they were not required to take notes. Midwifery was also taught didactically, with a mannikin to illustrate the lectures. The lessons of practical obstetrics followed attainment of the medical degree. Kenneth M. Ludmerer, *Learning to Heal, The Development of American Medical Education* (New York: Basic Books, 1985); Ronald L. Numbers, *The Education of American Physicians* (Berkeley: University of California Press, 1980); Speert, (n. 2) *Obstetrics and Gynecology*, pp. 72–84. Five physicians in this study received additional education in Europe, where they were exposed to clinical training which included observation of patients and notetaking based on the professor's comments during ward rounds. Their case notes are more detailed than some of the other physicians, suggesting that their European training made a difference. For medical education in France and Britain, see Erwin H. Ackerknecht, *Medicine at the Paris Hospital 1794–1848* (Baltimore: Johns Hopkins Press, 1967); Charles Newman, *The Evolution of Medical Education in the Nineteenth Century* (London: Oxford University Press, 1957); Russell C. Maulitz, *Morbid Appearances, the Anatomy of Pathology in the Early Nineteenth Century* (Cambridge: Cambridge University Press, 1987), pp. 1–82; M. Jeanne Peterson, *The Medical Profession in Mid-Victorian London* (Berkeley: University of California Press, 1978), pp. 40–89; R.G.W. Anderson and A.D.C. Simpson, eds., *The Early Years of the Edinburgh Medical School* (Edinburgh: Royal Scottish Museum, 1976).

12. Numerical analysis of medical practice became popular among American physicians during the nineteenth century. John George Metcalf was among the first to publish an analysis of obstetrical practice. "Statistics in midwifery," *American Journal of the Medical Sciences*, 1843, n.s. *6*, 327–30, 334; ibid., 1847, n.s. *14*:295–329; W. Chew Van Bibber, a son-in-law of Pierre Chatard, published an analysis of Chatard's obstetrical cases. *Transactions, Medical and Chirurgical Faculty of Maryland*, 1855: 33–64. See also, James H. Cassedy, *American Medicine and Statistical Thinking 1800–1860* (Cambridge: Harvard University Press, 1984).

Another paid particular attention to the relationship of maternal pulse to cord pulse. A few specified the position with which the fetus presented.[13] Many of the physicians kept obstetrical case notes only for a few years in a long practice. Occasionally, the notes resemble short stories, with the outcome of a particularly complicated case unknown until the end of five or six pages and a compassionate reader hoping against hope that all will turn out well.

There are also differences in style, reflecting the education and personality of the writer. Pierre Chatard, educated in Montpellier and Paris, wrote in French with a tendency toward florid descriptions. Henry Hartshorne was well educated in Philadelphia, but the notes that remain from the early years of his practice are scribbled among the pages of small account books. The initial cases of John Metcalf were interspersed with his medical school lecture notes (probably to use up the paper) but were later transcribed and included in a seven-year account solely of obstetrical cases. William Parker kept meticulously neat notes of all his medical cases, including obstetrics, and added a daily meteorological record. Tyler Briggs, on the other hand, wrote his obstetrical notes in an intermittent diary that also recorded an occasional non-obstetrical case, the purchase and sale of horses, social events, and financial matters. John Barnard Swett Jackson included his obstetrical cases among many others, including autopsies, in twenty-one volumes of his complete medical practice.[14]

That these physicians kept notes suggests they believed their work was worth recording, either for their own education or for that of other physicians. Wrote William Darrach, "I make this report to present concisely what I have done, wherein I have been instrumental in curing, & releiving [sic], wherein I have been useless, and wherein I have done harm.

13. For length of cord, see George Dock, "Records of events from the commencement of my medical studies in Philadelphia, Oct. 10th, 1840, with a record of the principal Surgical Cases & Operations that I have had during my professional life, up to the date of July 1856," MSS, College of Physicians of Philadelphia, Pennsylvania. For pulse, see Jackson, (n. 8). Many American physicians categorized fetal presentation according to the system of the French obstetrician Jean-Louis Baudelocque (1746–1810). William Potts Dewees reiterated Baudelocque's system in his *Compendious System of Midwifery*, first published in 1824, which was the most widely used textbook in the United States. Pierre Chatard commented that the textbooks described too many positions to make sense of them.

14. Pierre Chatard, "Obstetrical Case Records," 2 vols., 1828–46, MSS, Medical and Chirurgical Faculty of Maryland, Baltimore, Maryland; Ferdinand Edmé Chatard, "Obstetrical Case Records," 3 vols., 1829–84, MSS, Medical and Chirurgical Faculty of Maryland, Baltimore, Maryland; Henry Hartshorne, "Account Books," 3 vols., 1848–51, MSS, College of Physicians of Philadelphia, Pennsylvania; John Geo. Metcalf, Notebook, 1826–32, MSS, Countway Library, Boston, Massachusetts; Tyler Briggs, Diary, 1817–39, MSS, Countway Library, Boston, Massachusetts; Jackson (n. 8). At the end of his career, Jackson reviewed all his case notes, added comments to some of them, and compiled a summary index. The original twenty-one volumes were later bound into six volumes.

These facts well considered will enable me to correct my ignorance . . ." Samuel Butler meant to record whatever "I could look back profitably [upon] in days to come."[15]

Pierre Chatard wrote for the instruction of his son whom he expected to succeed him in practice.[16] Walter Channing and John Barnard Swett Jackson, both professors at Harvard Medical School, probably kept their records as sources for lectures and journal articles.[17] Case notes also served as reminders of medications prescribed and the condition of patients over time. Several case books doubled as account books.

NINETEENTH-CENTURY OBSTETRICS

The obstetrical practice of these physicians coincided with the years when physicians were increasingly accepted as providers of obstetrical care. Medically trained men were assumed to have the education and experience that would make childbirth safer and less difficult, thus assuaging the fear of prolonged labor, severe pain, debilitating injury, and death that many women associated with parturition.[18] Upper class and middle class women especially were attracted to professionally trained "accoucheurs," although some continued to prefer the ministrations of a female midwife and, as will be seen below, lower class women did call for the assistance of physicians. Except for very poor women whose lack of financial resources and social supports forced them to give birth in almshouses and public hospitals, childbirth in the United States occurred at home.

15. William Darrach, "Report of my medical practice of 1830," Darrach Papers, 7 vols., MS B 324, History of Medicine Division, National Library of Medicine, Bethesda, Maryland, vol. 1; Samuel Butler, "Journal of Cases Treated, beginning June 29, 1849," 1849–55, MSS, College of Physicians of Philadelphia, Pennsylvania.

16. Many of the comments were directly addressed "mon cher fils" to whom Chatard explained that authorities could be wrong and that a father's experience was more instructive and more reliable than the textbooks. The notebooks contain precepts and essays on obstetrical topics in addition to the case notes. Pierre Chatard, (n. 14) I, "Sur les accouchements."

17. For journal articles based upon case notes see, e.g., Walter Channing, "List of Midwifery Cases," Walter Channing III Papers, Misc. Box, Massachusetts Historical Society, Boston, Massachusetts, 24 March 1813 and Walter Channing, "Observations on puerperal fever with diarrhea," *New England J. Med. Surg.*, 1813, *2*, 232–41; Walter Channing, "List of Midwifery Cases," 13 September 1821 and Walter Channing, "A case of arm presentation, with remarks," *New England J. Med. Surg.*, 1822:*11*, 30–37. The medical school at Harvard had several names during its early years: Medical Institution of Harvard College, Massachusetts Medical School, Medical School of Harvard University, etc. Henry K. Beecher and Mark D. Altschule, *Medicine at Harvard: The First Three Hundred Years* (Hanover, N.H.: University Press of New England, 1977), p. 30. I have used the name by which it has been consistently known since the middle of the nineteenth century.

18. See, e.g., Leavitt, (n. 5) *Brought to Bed*, pp. 13–35; Hoffert, (n. 5) *Private Matters*, pp. 38–43.

Delivering babies was part of a physician's general practice. Contemporary medical theories taught that disease was caused by internal imbalances of the body and that cures depended on restoring its proper equilibrium.[19] Thus, obstetrical complications were treated by commonly accepted forms of depletion, that is, bleeding, purging, and puking. Ergot might be administered to hasten labor though many physicians found its effects too unpredictable to warrant its use. Instruments were used to remove an impacted fetus, but great skill was required to do so. Cesarean deliveries were extremely rare, primarily because all surgical procedures ran the risk of infection, in addition to the horror of an unanesthetized patient.

The physician's greatest anxiety came from uncontrollable hemorrhage, convulsions, and puerperal fever, for which the etiology was unknown. Much of the physiology of labor and delivery remained a mystery. Bacteriology and endocrinology had not yet begun to affect obstetrical knowledge or practice. Anesthesia was not discovered until 1846 and the principles of asepsis were not introduced until the latter part of the century.

PHYSICIAN EXPERIENCES

On Saturday evening at 8 of the clock I was called to Mrs. Chloe K— living about 3 miles from my residence. . . . the pains were slight and seldom. As she was young and a stranger I was told I might wait in the adjoining room till called for. Having sat up an hour or two and judging I should not be wanted at present, I wrapped my cloak about me and in a few minutes dropped asleep. I was awakened about day light and introduced to my patient. John Geo. Metcalf, May 20, 1827.[20]

Here, as in most obstetrical cases, unless the physician had attended the patient in a previous labor, he first met her at the time of delivery. Occasionally, prearrangements were made by a husband who engaged the physician for his wife, by a woman who felt the need for reassurance

19. For an overview of nineteenth-century medical practice, see William G. Rothstein, *American Physicians in the Nineteenth Century, from Sects to Science* (Baltimore: Johns Hopkins University Press, 1972); Gert H. Brieger, ed., *Medical America in the Nineteenth Century: Readings from the Literature* (Baltimore: Johns Hopkins Press, 1972); John S. Haller, Jr., *American Medicine in Transition* (Urbana: University of Illinois Press, 1981); Richard Harrison Shryock, *Medicine and Society in America 1660–1860* (New York: New York University Press, 1960); Charles E. Rosenberg, "The Therapeutic Revolution," in Morris J. Vogel and Charles E. Rosenberg, eds., *The Therapeutic Revolution, Essays in the Social History of Medicine* (University of Pennsylvania Press, 1979), pp. 3–25.

20. Metcalf, (n. 14).

in advance of her labor, or because there had been medical problems during the pregnancy. There was no such thing as prenatal care per se.[21]

On the other hand, post-partum visits occurred once or twice a day for a week or more following a normal delivery and continued for many weeks if there were serious complications. Walter Channing delivered Mrs. C. of twins on 29 April 1829 and visited twice daily 30 April and 1 May. On 2 May he was called between 6 and 7AM because the patient was showing signs of puerperal fever. He was back that day at noon, 7PM, and 11PM; on 3 May at 1AM, 7AM, 11AM, 3PM, 7PM and 10PM. The following day he visited three times, continued to see her once or twice each day for more than a week, and declared her "perfectly recovered" on 1 June.[22]

Whereas Metcalf and other physicians whose patients were widely dispersed in rural areas rode long distances and stayed many hours for each case, those, like Channing, whose patients were in closer proximity might come and go. "Stayed the night and went home for breakfast . . . called again at 10AM," wrote George Snow.[23] Some doctors reveal a fair amount of impatience. "Case of confinement which detained me til afternoon" or "this case delayed me nearly 2 hours . . . causing me to miss the Germantown car . . ."[24] But, for the majority, waiting was a tolerable part of practice. In Metcalf's case referred to above, he stayed more than 21 hours, including the time he was asleep. Occasionally, he remained even longer at the scene. In one home, he claimed to have slept in the only available bed, a three-foot-long cradle.[25]

The physician might be called at the onset of labor pains or later when the second stage of labor approached. Some women seem to have been reluctant to summon the physician until absolutely necessary, a decision that could be hazardous in complicated cases.[26] Considering the time required for someone to fetch the physician, for him to leave what he was

21. For prenatal care, see Speert, (n. 2) *Obstetrics and Gynecology*, p. 142; Hoffert, (n. 5) *Private Matters*, pp. 20–22, 52 n.; McMillen, (n. 5) *Motherhood in the Old South*, pp. 25–56; Ann Oakley, *The Captured Womb, a History of the Medical Care of Pregnant Women* (New York: Basil Blackwell Inc., 1984), p. 11.

22. Walter Channing, Obstetrical Cases, 3 March 1829 to 29 July 1829, Walter Channing II Papers, Massachusetts Historical Society, Boston, Massachusetts, Temp. Box 1, 29 April 1829.

23. George Snow, "Case Book," 1865–75, MSS, Countway Library, Boston, Massachusetts.

24. Samuel Butler, "Journal of Cases Treated, beginning June 29, 1849," 1849–55, MSS, College of Physicians of Philadelphia, Pennsylvania, 27 November 1853; Hartshorne, (n. 14) II, 29 July 1850.

25. Metcalf, (n. 14), 28 May 1828.

26. For example, Mrs. Green delayed sending for Dr. Parker who arrived to find that the baby had been born in breech position and was barely alive. The infant died the following day. It is possible that Parker could have turned the fetus or hastened the delivery to avoid the stress that may have caused the fatality. Parker, (n. 8), II, 6 September 1848.

doing, at night to dress, saddle his horse or outfit the carriage (unless he returned with the summoner), and then to ride or walk to the home of the patient, it is not surprising that in many instances the physician arrived after the baby had been born. He might still be needed to remove the placenta or to treat post-partum pain, hemorrhage, or convulsions.

A few physicians developed reputations as obstetricians although specialization was uncommon. Both Pierre and Ferdinand Chatard, father and son, had large obstetrical practices. Many years they attended at more than 150 homes spread throughout Baltimore, occasionally with 2 or 3 cases in a single day, often remaining many hours, reaching their patients on foot or by carriage. There could not have been much time for a large general practice. Neither Chatard gained his obstetrical prominence as a professor, yet frequently they were called upon by other physicians to consult in difficult cases.[27] On the other hand, despite his position at Harvard Medical School and his reputation as Boston's foremost obstetrician, Walter Channing had a much smaller obstetrical practice and counted on his general practice for additional income.[28]

For most physicians, obstetrics was a small part of their practice. William Parker averaged 22 cases of childbirth in a yearly practice of about 650 cases; John Metcalf averaged 18 obstetrical cases per year.[29] Avery Skilton did not report his non-obstetrical cases but the list of obstetrical cases averages 60 per year with a minimum of 15 and a maximum of 118. It also demonstrates how his practice increased as he became more established in Troy, New York.[30] Other physicians followed the same pattern of a small beginning practice, gradual acceptance and larger numbers of cases, followed by a decline in later years.

Although Metcalf's frequent naps suggest that the physician was not busy much of the time, physical and mental strength were necessary for

27. It is possible that the European training each Chatard had experienced set him apart from other Baltimore physicians and thus made him more desirable as a consultant.

28. Kass, (n. 7).

29. An anonymous physician in Plymouth, Massachusetts, noted 436 cases in 22 years of practice, an average of 20 per year. George Snow reported 313 cases in 10 years, an average of 30 per year. Robert Thaxter was reported to have delivered 2600 babies in 52 years, an average of 50 per year. William Ingalls reported that he had delivered 2000 babies in 43 years, an average of 46 per year. N.N., "Journal of Practice," 1826–47, MSS, Countway Library, Boston, Massachusetts; Snow, (n. 23); Benjamin Cushing, "Record of the Obstetric Practice of Dr. R. Thaxter," 1801–52, MSS, Countway Library, Boston, Massachusetts; William Ingalls, "A synopsis of private obstetrical practice for forty-two years previous to January 1, 1876," *Boston Medical and Surgical Journal*, 1877, *96*, 485–98.

30. Avery Skilton, "Obstetric Memoranda," 1827–37, Ives Papers, MS fB 12, History of Medicine Division, National Library of Medicine, Bethesda, Maryland. Skilton did not give many case details, but he did include date of delivery, followed by notations indicating the nature of the delivery, i.e., abortion, premature, natural, stillborn, breech, difficult, instrumental, etc.

complicated cases, and the physician often departed much exhausted by his efforts (though surely not as exhausted as the woman whom he had attended). "Had a hard, hungry and tedious time . . . the most severe snow storm of the season on that day," commented George Snow regarding a labor that lasted many hours. Ferdinand Chatard referred to the "great exertion" needed to remove an adherent placenta and the tension he experienced in cases when danger threatened and he felt constrained not to frighten the mother or "her friends." Deliveries that required turning the fetus or instruments to remove an impacted fetus were extremely demanding. Pierre Chatard described his utter fatigue in a craniotomy that required unusual force.[31]

PHYSICIAN UNCERTAINTIES

Judith Walzer Leavitt has indicated that obstetrics was an imprecise science at best, and the case notes, written for their personal use, demonstrate that physicians were far less certain of themselves than their professional status might have predicted.[32] Physicians frequently were unable to comprehend outcomes—why a seemingly healthy newborn died shortly after birth or why a woman whose delivery had been uneventful seemed to waste away. Complicated deliveries produced even more questions. Why had convulsions occurred, why had the uterus ruptured, why should a labor continue for five or even six days? Twin births seldom were expected and the surprise, in addition to the difficulty of delivery, led to obvious consternation.[33]

Uncomplicated cases could be equally puzzling. Ferdinand Chatard pondered "the uncertainty of prognosis as to the duration of labour" when, within a brief time, he examined two patients who were simultaneously in labor. "The os tincae was but slightly dilated & pains very moderate in the case of Mrs. Stuart; in Mrs. Wilby's case on the contrary, the dilation was complete and the pains appeared sufficient to effect a speedy termination of the case & yet Mrs. Stuart's child was born about 11 o'clock PM & Mrs. Wilby's case was prolonged to about 2 o'clock AM."[34]

31. Snow, (n. 23) 1 March 1822; Ferdinand Chatard, (n. 14) III, 14 September 1859; Pierre Chatard, (n. 14) I, 1833.

32. Leavitt, (n. 5) *Brought to Bed*, pp. 62–63.

33. Twin births were often premature or one baby was significantly smaller than the other. Physicians took much interest in the placenta and any other unusual features of twin births.

34. Ferdinand Chatard, (n. 14) III, 10 December 1860.

In the privacy of their case notes, physicians also confessed their mistakes. "An oversight to be avoided in the future" is a representative comment.[35] One young physician faced with severe post-partum hemorrhage admitted his ignorance. "Forgot to try the effect of cold water . . . went home and *studied hard* [sic] on Uterine Hemorrhage." From another, "The hard labor and my hard handling have excited and may even cause inflammation."[36] Both these remarks were made by novice physicians, but experienced men also candidly described serious errors. At the end of his career, Pierre Chatard mishandled a case in which the use of ergot and forceps resulted in a torn perineum. "For a primipara for whom the progress of labor is always too long," he warned, "labor should not be accelerated."[37] Ferdinand Chatard described many mistakes, including the following.

May 18 [1860] Mrs. A. H. in labour of 6th child called to her about 9½ PM on 17th—had some slight pains—os tincae not all dilated & cervix soft & elongated. I left her with the certainty that many days would elapse before she was confined—The Pains however became decided about five AM & the child was born about 6AM but it having come feet present & nobody to assist her the child died in the birth. I did not see her for some time after.

Unsuspected complications also led to admissions of error. "I made a great mistake, from want of sufficient care in the examination," wrote a physician when an expected vertex presentation turned out to be breech, followed by the foot of a second child. "It was the feel of the ends of the toes that deceived me in the first examination of the case."[38] Some physicians revealed how frightened they were, especially those whose obstetrical experiences were limited by a small practice. "There was a spasmodic contraction of the uterus . . . first case . . . I was some frightened and told them to send for Dr. Wilder."[39]

The frequency of consultations attests to physicians' awareness of the imperfection of their knowledge. "Becoming somewhat perplexed . . . I rather hastily requested the advice of Dr. H.," is a typical comment.[40]

35. N.N., (n. 29) 30 May 1842.

36. Butler, (n. 15) 10 November 1849; Hartshorne, (n. 14) I, 23 October 1848.

37. Pierre Chatard, (n. 14) II, 7 July 1844. Chatard also told his son of a case early in his career when he had not acted soon enough in a case of hemorrhage. "I was young then and influenced by the opinions of others." I, "Accouchements avant 1829."

38. N.N., (n. 29) 16 December 1846.

39. Briggs, (n. 14) 27 December 1827. Briggs had been in practice more than ten years when he experienced this first case of spasmodic contractions.

40. N.N., (n. 29) May 20, 1839.

The small number of cases in some practices made it difficult to accumulate sufficient experience to meet every emergency. More than ten years after completing his medical training, William Darrach confronted a thirty-hour labor and the stillbirth of a large child. "Ought forceps to have been used?" he wondered. In a subsequent case, he did employ forceps, but he also called for the assistance of a more experienced physician.[41] A seventh-month miscarriage accompanied by severe convulsions and the destruction of the dead fetus by instruments led another physician to comment, "This was the first case of the kind to which I had been called in my own practice. I witnessed one of less severity in a patient of Dr. H. many years ago."[42] Textbooks often were no help. "[A]ll directions in the books . . . are so vague and indefinite," complained John Metcalf.[43]

Faced with the realities of practice and limited by the still unsolved riddles of conception, parturition, and diseases associated with childbirth, physicians reflected upon these mysteries. William Darrach wrote three pages of "Queries relative to the human placenta," in which he pondered whether there is more than one kind of blood in the placenta and lamented the difficulty of finding answers. Puerperal fever led to questions and conjectures, but no certain answers, from several physicians.[44] Pierre Chatard wrote seven pages for the edification of his son about the use of ergot, pointing out that, although initially he had opposed the drug, he had changed his mind and now recommended prudent use in appropriate situations.[45] John Barnard Swett Jackson tried to

41. Darrach, (n. 15) V, 11 January 1836, 4 October 1836, 25 December 1836.

42. N.N., (n. 29) 31 July 1847.

43. Metcalf, (n. 14) 30 July 1829. Although physicians did not hesitate to call upon their colleagues when confronted with an unfamiliar or difficult situation, those who practiced in less populous places were forced to rely on their own abilities. Samuel Butler, in practice in rural New Jersey, did not call for assistance in "one of the hardest cases ever" or in another "very severe case requiring use of catheter and forceps." It is possible that he did not want to yield anything to other physicians but also possible that there was no one to call upon. Butler, (n. 15) 1 October 1853, 23 October 1853.

44. Darrach, (n. 15) I, "Queries relative to the human placenta." For puerperal fever, ibid., 30 November 1829; Pierre Chatard, (n. 14) I, "Precepts"; Henry Hartshorne, "Is Erysipelas Contagious,—and Are Erysipelas & Puerperal Peritonitis identical? 5 mo 28th 1847" and "Case of Puerperal Fever—Notes begun 6/6/47," *Case Book at Pennsylvania Hospital*, 1846–47, MSS, College of Physicians of Philadelphia, Pennsylvania.

45. Pierre Chatard, (n. 14) I, "Sur l'ergot." Ergot or spurred rye was introduced in the United States in 1807 by Dr. John Stearns. Its powerful effect accelerated and strengthened uterine contractions but, if improperly used, ergot could cause perineal tearing and fatal stress to the fetus. Speert, (n. 2) *Obstetrics and Gynecology*, pp. 181–82. Chatard published two journal articles about ergot. In the first he recounted twelve births in which he administered ergot and his conclusion that it could be dangerous to the child. Later in his career, more specific criteria for the use of ergot enabled him to recommend it to his son.

use his skills as a pathologist to explain disease or death in childbirth. Walter Channing, too, turned to autopsies for answers to riddles he encountered. One autopsy helped to explain a maternal death by revealing a severely diseased uterus that had ruptured during labor. In another case, however, the family refused to permit a post-mortem examination, and Channing's questions regarding "a great quantity of green fluid . . . [that] poured from the vagina" remained unanswered.[46]

PHYSICIAN-PATIENT RELATIONSHIPS: CONCERN FOR THE PATIENT

Contrary to the assertion that physician-assisted childbirth meant a loss of empathy and emotional support for the mother, physicians were genuinely concerned for their patients. "She was so sick with her first child that I felt exceedingly anxious about the result of this labor," wrote George Snow.[47] "I was extremely anxious that no unusual or unnecessary interference should be made" and "she was soon removed, but with great care to her bed, everything being made dry and comfortable about her" are remarks from Channing's case notes. He recorded his anxiety when "a great and alarming hemorrhage" occurred, and his hopes "for immediate resolution of disease" in a case of puerperal fever.[48] Other physicians expressed similar reactions, especially following a difficult delivery or complicated recovery. Comments such as "to my joy" and "got safely to bed" frequently occur in Metcalf's case notes. "Happily no laceration took place," he wrote when a potential complication was avoided.[49]

Physicians made frequent references to women's fears, pain, and exhaustion. "I assured my patient she would soon get through with her troubles," wrote John Metcalf. Another time he "assured my patient that all was right, and patience and good courage was all that was wanting to ensure a speedy and safe delivery." A physician in Plymouth, Massachusetts, was acutely aware of pain, making frequent references to "uncommon suffering," and "(suffering) more than I have ever known at this period of labor." William Parker noted the suffering of his pa-

---

Pierre Chatard, "Experimental observations on the medicinal properties of the ergot or spurred rye," *Medical Repository*, 1820, n.s. 5, 11–22; "Observations on Ergot," ibid., 1821, n.s. 6: 139–65.

46. Walter Channing, (n. 22) 3 April 1829; idem, (n. 17), 23 July 1821.

47. Snow, (n. 23) 15 June 1873.

48. "no interference," Channing, (n. 17) 22 October 1816; "soon removed," Channing, (n. 22), 2 April 1829; "alarming hemorrhage," (n. 17) 6 April 1814; puerperal fever case, (n. 22) 29 April 1829.

49. Metcalf, (n. 14) 5 September 1827, 11 May 1828, 14 February 1829.

tients, his efforts to encourage them, and their fatigue after enduring long labors.[50]

Scholars critical of the intrusion of men into the birthing room have accused them of using obstetrics to ingratiate themselves with their patients, thereby to acquire future business as the family's doctor.[51] It is possible to interpret these comments as self-serving, since no physician would want the patient or her family to think he was not concerned for her well-being. Reputations were always at risk, especially in childbirth cases where the presence of a man still caused consternation to many women and where an unexpected crisis always could occur. However, there is no reason to doubt the sincerity of concern for the welfare of the woman and her child. "All the good symptoms continue" or "I gently supported the perineum" were not written to impress future generations but to record the physician's personal view of the case.[52]

Physicians could be critical of husbands who seemed to be insufficiently attentive to their wives. "Her husband being a miserable fellow, and not helping much towards getting a living, the poor woman had worked hard to get a scanty support for herself and her little ones," commented John Metcalf in the case of a woman in her sixth labor. "Her beast of a husband has been drunk since Monday," complained George Porter when he made a post-partum visit.[53] In this case, as in others, the newborn was named for the physician who had delivered

50. Metcalf, (n. 14) 25 March 1828, 11 February 1828; N.N., (n. 29) 9 November 1846, 2 December 1841; Parker, (n. 8) III, 17 January 1850, 23 February 1850, 24 October 1850, 23 February 1851. Not all physicians were sympathetic with their patients. " . . . there seems to be so strong a disposition in pregnant women, or in those who have recently passed through the ordeal of childbirth, to exaggerate in the relation of what they have experienced, and we are told tales of great hardship and many hair-breadth escapes." Ingalls, (n. 29), p. 492.

51. See, e.g., Scholten, (n. 5), p. 56. Walter Channing is accused of expressing this notion in a pamphlet published anonymously but usually attributed to him. "Women seldom forget a practitioner who has conducted them tenderly and safely through parturition—they feel a familiarity with him, a confidence and reliance upon him, which are of the most essential mutual advantage in all their subsequent intercourse as physician and patient . . . It is principally on this account that the practice of midwifery becomes desirable to physicians. It is this which ensures to them the permanency and security of all their other business." Critics do not include the subsequent remarks. ". . . the physician takes a deeper interest and feels a more intimate and personal connexion with those, whom he has attended in this scene of suffering and danger, than with patients of any other description. . . . We should serve our patients with all our heart and soul; and they should know that we do it not merely because it is our business, or because we expect to be supported, or to grow rich by the occupation, but because we feel for their welfare as friends, and as friends will strive for their advantage." *Remarks on the Employment of Females as Practitioners in Midwifery. By a Physician* (Boston: Cummings and Hilliard, 1820), p. 19.

52. Parker, (n. 8) III, 13 May 1850; Metcalf, (n. 14) 14 February 1829.

53. Metcalf, (n. 14) 7 September 1826; George Porter, "Commonplace Book," [n.d.], MS Film 11, History of Medicine Division, National Library of Medicine, Bethesda, Maryland, Case #1.

him. Perhaps this was one way a mother could express appreciation for the physician's care and support.

Not every physician was uniformly attentive. Ferdinand Chatard frequently arrived after a baby had been born or, having gone to see another patient, would return to find he had missed the birth. Some case notes are too terse to indicate the feelings of the physician. But none suggests physician disregard for the well-being of the patient.

PHYSICIAN-PATIENT RELATIONSHIPS: DEFERENCE TO THE PATIENT

Contrary to the assertion that the physician's presence in the birthing room meant the loss of control by childbearing women, physicians generally performed according to the wishes of their patients.[54]

First of all, the physician was present at the invitation of the woman. He was "called to her" when she wanted his assistance, and he was admitted to the birthing room when she was ready.[55] "I was called 7AM but was requested merely to be in the readiness when labour should be more advanced. Called again about 11AM—as before. At 7PM I saw her for the first time. She was sitting up, complaining of slight pain and would scarcely submit to an examination."[56] From another case book,

> I received a call from Mr. S. to visit his wife, she being in labour with her third child. It was about 8 of the clock when I arrived. From the attendants I learnt that her pains had gone off in some measure, and I being a stranger to Mrs. S. she concluded not to see me for the present . . . At 3AM I was awaked by one of the women and desired to come immediately as Mrs. S. was now quite willing to see me.[57]

Furthermore, decorum forced physicians to defer to their patients. One of the ironies of nineteenth-century obstetrics is the apparent conflict between men attending women in childbirth and the increased importance placed on female modesty and delicate feelings.[58] Tact and sen-

54. Judith Walzer Leavitt corrected many misunderstandings by demonstrating that parturient women and their friends continued to control childbirth throughout the nineteenth century, despite the presence of the physician, (n. 5) *Brought to Bed*, pp. 98–108.

55. Several physicians used this or similar terminology to introduce each case. See, e.g., Butler, anonymous physician, Metcalf and Parker. John Metcalf frequently began his cases with the phrase, "I was called to visit——."

56. Channing, (n. 22) 17 April 1829.

57. Metcalf, (n. 14) 1 November 1826.

58. Jane B. Donegan has written especially perceptively on this point, (n. 5) *Women and Men Midwives*, pp. 141–236.

sitivity were required of the physician. As for the woman, she had to be willing to allow him to touch, though not to see, the most private parts of her body. Case notes do not specifically describe this aspect of examination or delivery, but it seems safe to assume that the prescriptions against visual examination were observed.[59]

Several physicians sought permission to conduct an examination by asking the women in attendance to transmit the request. The presence of the male physician may well have added tension to an already stressful situation. Metcalf described several cases in which labor pains subsided upon his entrance into the birthing room. He tactfully withdrew until recalled by the attendants.[60]

Parturient women assumed a leading role throughout the delivery and post-partum care. Limited in his opportunity for physical examination, lacking instruments to measure maternal or fetal changes, to a great extent the physician depended on superficial observations and, more importantly, on what the woman wanted to tell him about the progress of her labor and any unusual sensations she might be experiencing.[61] One physician described the progress of a labor thus: "she began to have those kind of pains which are called by the women 'striving.'" Another, "found from her own account that pains had come on about 3 hours, and that the membranes were ruptured." Because the patient was not having active labor pains he did not examine her. Still another, "found the patient . . . by her own account, attended with considerable hemorrhage and in great pain." In William Parker's notes, his post-partum patients "report themselves" as "better," "worse," or "nicely," and he does or does not prescribe medication according to those self-descriptions.[62]

59. I have not been able to find many references in the case notes to "touch" or to visual examination. It would be interesting to know if the rules of decorum applied to all social classes. Snow referred to the unusually large breasts of an unmarried patient, seventeen years old, which may or may not suggest visual examination, Snow, (n. 23) 8 August 1869. Jackson described a perineal examination by candlelight, which also suggests that sight was significant, Jackson, (n. 8) n.s., I, 22 August 1837. It is difficult to imagine how physicians performed difficult fetal extractions or treated perineal tears by touch alone.

60. Metcalf, (n. 14) 27 February 1827.

61. The stethoscope, first used in France by René Théophile Hyacinthe Laennec in 1816, had been introduced shortly thereafter to the United States. It was occasionally employed to hear the fetal heartbeat. Channing mentioned it once, while Jackson and Snow made more frequent references. This does not necessarily mean that Channing did not use the stethoscope more frequently. For development of the stethoscope and its impact on clinical medicine, see Stanley J. Reiser, *Medicine and the Reign of Technology* (Cambridge: Cambridge University Press, 1978), pp. 23–44, 61.

62. Striving pains, Metcalf, (n. 14) 20 May 1827; membranes ruptured, N.N., (n. 29) 3 September 1837; hemorrhage, Metcalf, (n. 14) 26 October 1827; "nicely," Parker, (n. 8) passim.

Women decided when they were ready to go to bed.[63] Some chose the position in which they delivered, despite obstetrical textbooks which prescribed lying on the left side.[64] Some refused enemas or opium or ergot; others demanded ergot or forceps, forcing the physician to use therapies he might not think warranted. Still others simply ignored their doctor's advice. "Child had been given catnip tea—contrary to my advice," observed George Porter when he visited a patient several days after delivery.[65] Physicians listened attentively to their patients. "Patient expressing a great desire to sleep, I allowed to remain quiet," wrote Ferdinand Chatard in a situation in which the placenta had not yet been completely removed.[66]

Decorum also limited examination post-partum and forced the physician to prescribe on the basis of the patient's description. "Mrs. Cashman has had, ever since her confinement, a severe pain in pudendum with swelling . . . supposed ruptured perineum, advised poultice."[67] In another case, the nurse and not the doctor, examined for a suspected prolapse. Patients described breast abscesses, and their nurses or friends carried out the physician's instructions.

63. See, e.g., " . . . sitting up & so continued till 3 AM 5'—when a severe pn (pain) drove her to bed & in 5' (3.10') a fine boy was born." Parker, (n. 8) III, 29 December 1850. Case notes make frequent reference to placing the woman in a clean, dry bed and to a change of clothing after delivery. "The bed usually consisted of two sacks; one of feathers, the other, underlying, of straw. The upper was folded half up on itself, and the patient was placed across the exposed half of straw. After delivery, she was made to stand up, and, while the soiled clothing dropped, clean and fresh garments were slipped on her from above. She was then lifted up (by the doctor if he was strong enough), and placed in the bed, which in the meanwhile had been readjusted by the nurses. Thus 'in the straw' and 'putting a woman to bed' had a greater significance formerly than nowadays . . . The patient now [1888] on a comfortable mattress, from which she is not to be removed, taking the position she pleases in an easily-detached dress, supplied with scores of napkins readily managed and quickly removed, passes neatly and with comparative facility through the ordeal." Benjamin Eddy Cotting, "Professional reminiscences," *Boston Medical and Surgical Journal*, 1888, *118*, 1–7, 32–35, 61–63, 85–87. See also, Irvine Loudon, *Death in Childbirth, an International Study of Maternal Care and Maternal Mortality 1800–1950* (Oxford: Clarendon Press, 1992), p. 287.

64. Obstetrical texts and lectures generally advised delivering with the woman lying on her left side at the edge of the bed, which satisfied decorum because patient and physician would not see each other. See William Potts Dewees, *A Compendious System of Midwifery* (Philadelphia: H. C. Carey and I. Lea, 1824), p. 193; John Geo. Metcalf, "Lectures on Midwifery, at the Medical School of Harvard University," Boston, 1825 and 1826, MSS, Countway Library, Boston, Massachusetts, p. 71. Nonetheless, Metcalf noted "our patient placed upon her back which is a very favourite position among the women in this vicinity" and he also " . . . found my patient upon her hands and knees on the floor . . . this is her second delivery in the same position," Metcalf, (n. 14) 13 January 1827; 7 July 1829; Snow described one patient as "very uneasy changing position from bed to the floor," and in another case "delivered her in the 'Irish position' on her hands and knees," Snow, (n. 23) 19 June 1870; 9 April 1870.

65. Porter, (n. 53) 4 May 1861.

66. Ferdinand Chatard, (n. 14) I, Case #49.

67. Parker, (n. 8) III, 16 June 1850.

Even the more assertive physicians had to pay attention to the wishes of their patients. Thus Pierre Chatard, who prided himself on "boldness and savoir faire," deferred to "that sensibility which takes the part of the suffering woman as well as the family and friends."[68]

The language in some of the case books suggests a dominant role for the physician and the importance of the husband rather than the woman. "I put —'s wife to bed with a daughter/son," is the formulaic inscription in Benjamin Osgood's list of obstetrical cases. Other physicians' notes give the same picture. "Delivered —'s wife of a son," wrote Tyler Briggs. Austin Flint used an abbreviated form: "—'s W." However, since the husband paid the physician's bill, nameless women do not necessarily signify their unimportance. Midwives, too, customarily referred to their patients by the husband's name.[69]

Although physicians appear to have been gentle and supportive of their patients, unpleasant relations did occur. Pierre Chatard was replaced by another doctor and thereafter lost the confidence of the family. One of William Parker's patients was continually dissatisfied with her care and finally sent "an impudent note" of dismissal.[70] In both examples, it was the patient's wishes that led to the dismissal.[71]

CHILDBIRTH AS A SOCIAL EVENT

Despite the presence of a male physician, childbirth remained a social event with other women in the birthing room to help the mother and to assist with the delivery.[72] "The pains began to grow sharp and the women and myself were sent for at the same time," wrote Metcalf.[73] In most instances, the women were present before the physician was summoned. In addition to the female friends and relatives, there might be a nurse or a midwife. Their presence not only was expected, but essential.

68. Van Bibber, (n. 12) Case #26, p. 59; Pierre Chatard, (n. 14) I, Observation #45. Chatard regretted that he had acted so kindly toward the patient because he was replaced by another physician called in to complete the delivery.

69. See, e.g., Robert R. McCausland and Cynthia MacAlman McCausland, eds., *The Diary of Martha Ballard 1785–1812* (Camden, Maine: Picton Press, 1992).

70. Chatard, see n. 68 above; Parker, (n. 8) III, 20 March 1851.

71. Other patients or families may well have been displeased or disappointed by the outcome of a case, but since the physicians are telling the stories they do not cite many examples.

72. This has been pointed out by several historians, notably by Leavitt, (n. 5) *Brought to Bed*; Hoffert, (n. 5) *Private Matters*; McMillen, (n. 5) *Motherhood in the Old South*. Their findings contrast with some earlier studies. See, e.g., "The crowd of supportive friends and family disappeared with the arrival of the doctor." Catherine Scholten, *Childbearing in American Society: 1650–1850* (New York: New York University Press, 1985), p. 47.

73. Metcalf, (n. 14) 22 March 1827.

They played an active role from beginning to end, encouraging the woman, judging her progress, administering enemas and/or emetics, providing nourishment or warm drinks, adding needed hands for the delivery, receiving and caring for the newborn, or applying an abdominal bandage. In cases of maternal death, "the friends" as well as the family requested, agreed to, or refused permission for a post-mortem examination.

Physicians noted the rare events when the women were absent. Samuel Butler was called to a patient at 1AM and found only her husband present. "Could get no female attendant . . . the scene was rather ludicrous." Wrote George Snow, ". . . before the bed could be arranged and the women called, the child was born." Tyler Briggs was astonished to find "no person present but the patient."[74] Though some scholars have claimed that female friends were unwelcome in the birthing room because they posed a threat to medical authority, these physicians did not disparage the friends and relatives who assisted at childbirth.[75] John Metcalf frequently used the first person plural when describing procedures: "we found the child had bled," "we removed our patient to a dry and comfortable bed," "I now left our patient to rest." Neither did physicians take unnecessary actions or procedures to justify their presence. On the contrary, Walter Channing sought a second opinion to "relieve the anxiety of the friends." Even the novice physicians did not try to impress the friends or patient by taking unusual actions.[76]

PHYSICIANS AND MIDWIVES

Much controversy has centered on the rivalry between physicians and midwives, with physicians accused of active dislike for midwives who, after all, were competing with them for business.[77] Feminist scholars also have suggested that midwives, familiar with childbirth from their own experiences and those of friends and relatives whom they assisted in labor,

74. Butler, (n. 15) 19 June 1852; Snow, (n. 23) 26 November 1869; Briggs, (n. 14) 15 April 1833.

75. See, e.g., Hoffert, (n. 5) *Private Matters*, pp. 74–77; Leavitt, (n. 5) *Brought to Bed*, p. 103.

76. Metcalf, (n. 14) 27 February 1827, 20 May 1827, 22 February 1829; Channing, (n. 17) 22 October 1816. It is always possible that the physicians did not record everything they did, especially if they considered it to be routine practice. However, there is no indication that they used drugs or forceps to demonstrate their ability.

77. Walter Channing in particular has been singled out because he wrote a pamphlet, published in 1820, that denounced female accoucheurs as inferior in all respects to medically trained men, (n. 51) *Remarks on the Employment of Females*. For references to physician-midwife antipathy, see, e.g., Ulrich, (n. 5) *A Midwife's Tale*, pp. 254–56; Scholten, (n. 72) *Childbearing in American Society*, pp. 42–45.

were better able to comfort parturient women, particularly in stressful situations that produced anxiety for medically trained men.[78]

Physicians assumed that midwives could and would handle normal cases without difficulty. Samuel Butler did not understand why he had been called for a delivery that "could have been attended by the woman present." Another "was at a loss to imagine why the midwife could not have finished the labour."[79] Physicians who were called by the midwife in cases of severe hemorrhage or impaction did not normally condemn her efforts. ". . . an intelligent woman who has been in the habit for several years of conducting such cases in her neighborhood when the labors are natural" is the reference to a midwife who needed help with delivery of an encephaloid child presenting knee first.[80] Though he had written a pamphlet warning women and their husbands against the employment of female accoucheurs, Walter Channing did not criticize a midwife who called for assistance in a case of retained placenta and hemorrhage and had kind words for "an intelligent nurse" who assisted him in another case.[81] On the other hand, several physicians did refer to the midwife as "the old woman present," which suggests less than universal admiration.

Most of these physicians did not see themselves competing with midwives for business. The exception is J.B.S. Jackson who indicated very clearly that fear of being shown up by a midwife caused him to resort to extreme and unnecessary measures. It was a case of arm presentation in which he and another physician were called in consultation. When neither was able to turn the baby, Jackson extracted it with the blunt hook. "I should perhaps have left the patient to natural efforts, but an old woman undertook to put me aside & take the case into her hands. The child would at any rate have soon been born. She would have had credit of defeating the Drs."[82]

On the other hand, midwives were criticized for mistakes, or what the physician considered to be mistakes. For example, one midwife gave gin

78. Bogdan, (n. 5); Laurel Thatcher Ulrich, "'The living mother of a living child': Midwifery and mortality in post-revolutionary New England," *William Mary Quart.*, 1989, 3rd ser., *46*, 27–48; Barbara Ehrenreich and Deirdre English, *Witches, Midwives, and Nurses, a History of Women Healers* (Old Westbury, N.Y.: Feminist Press, 1973); Eakins, (n. 5) *The American Way of Birth*, pp. 17–20.

79. Butler, (n. 15) 7 July 1852; N.N., (n. 29) 26 February 1843.

80. Snow, (n. 23) 14 November 1866.

81. See n. 51 above. "Intelligent nurse," Channing (n. 22) 17 April 1829, 26 June 1829; "midwife," ibid., 9 May 1829. It is not certain whether nurse and midwife were separate categories or whether the words were used interchangeably during these years. For example, *The Boston Directory* (Boston: Charles Stimson Jr., 1829) listed one midwife and 37 nurses, but it is possible that some of the nurses were midwives.

82. Jackson, (n. 8) n.s., III, 23 April 1842.

to a woman with puerperal fever; a physician was called in and prescribed cathartics. Another midwife so mismanaged an arm presentation that the physician was unable to turn the baby.[83] Pierre Chatard was disgusted with a midwife who deserted a "colored woman" while the placenta still adhered and with another midwife and physician who allowed their patient to hemorrhage more than twenty-four hours before calling for his help.[84]

Other birth attendants also were criticized. John Metcalf described a situation in which the "friends" had erred. He had arrived after the baby was born, and, although he was told that the woman was doing well, he

> found the patient . . . attended with considerable haemorrhage, and in great pain . . . The attendants were dosing down warm drinks to ease the pain. Supposing all was not right, I observed that perhaps the placenta did not all come away well, and desired it might be shown to me. I was informed that no one had seen it, but supposing it to have come away in the sheet that was about the woman, they had folded it up . . . and laid it away with the rest of the dirty clothes. On search, however, no placenta could be found, and the cause of the flooding and pain was now revealed.[85]

Nurses, too, were cited for negligence or incorrect treatment. Parker claimed that one nurse had almost killed a newborn by giving white poppy syrup to quiet it and another did not administer "the oil I prescribed," thereby causing headache and discomfort to the patient.[86]

Physicians also criticized their male medical colleagues for mistakes or lack of attention to their patients. "Dr. in a hurry broke funis & went off leaving placenta undelivered & no swathe round abdomen!" noted William Parker who was called in when the woman began to hemorrhage severely. He arrived too late to save her. "She suffered for want of proper professional attendance," he commented about another patient who had dismissed her first physician.[87] J.B.S. Jackson wrote of a physician who "eats, drank and kept himself pretty well tipsy" while waiting for a labor to proceed. Jackson also noted that Walter Channing had insisted on use of instruments when an additional dose of ergot might have terminated the labor. "All would soon have been well," he confided to his case book.[88] John Metcalf had the temerity to criticize "Dr. W——d of U——e" for not admitting that his use of forceps had severely dam-

83. Parker, (n. 8) I, Case #6; Ferdinand Chatard, (n. 14) II, Case #447, 1840.
84. Pierre Chatard, (n. 14) I, Observation #5, Observation #37; II, 18 February 1844.
85. Metcalf, (n. 14) 26 October 1827.
86. Parker, (n. 8) I, Case 15; II, 23 July 1848.
87. Parker, (n. 8) III, 25 February 1851, 3 June 1851.
88. Jackson, (n. 8) XV, 3 October 1835, 14 January 1836.

aged the patient. "On the whole the Faculty had better be disigenious [sic] and tell the truth," he added.[89]

OBSTETRICS AMONG RURAL AND POOR WOMEN

Contrary to the thesis that middle class and upper class women, primarily urban women, were the basis for physicians' obstetrical practice, physicians also delivered many babies of poor urban women and women in rural areas.[90]

In small towns of central Massachusetts, Austin Flint was delivering babies in normal labors as early as 1785. Benjamin Osgood practiced in rural areas northwest of Boston during the second decade of the nineteenth century and John Metcalf worked in another semi-rural part of Massachusetts during the next two decades. Ferdinand Chatard delivered free black women and Native American women in Baltimore in the 1830s and 1840s. He also was called by slaveholders to deliver black women, often, but not exclusively, for difficult cases. For several years of his long practice, Chatard delivered women at St. Vincent's Asylum. Since Chatard was not a medical teacher, he could not have done so for the instruction of students. More likely, he did so as a Catholic physician.[91] Tyler Briggs noted his first "black case," and several other physicians also indicated race.

Samuel Butler's practice depended largely on poor women in small New Jersey towns, and he complained bitterly about his inability to collect his fees. Tyler Briggs was continually in debt, but we do not know if that was due to ineptitude as a horse trader or inability to collect the money owed to him. George Snow was called to many squalid homes. Channing and Jackson also had many poor patients. Some, but not all, were clients of the Boston Dispensary.[92] William Parker frequently men-

89. Metcalf, (n. 14) 24 February 1830.

90. For social class of obstetrical patients and location of obstetrical practice, see Scholten, (n. 72) *Childbearing in American Society*, p. 49; Bogdan, (n. 5), p. 92; Donegan, (n. 5) *Women & Men Midwives*, p. 4; Leavitt, (n. 5) *Brought to Bed*, p. 8. Joan M. Jensen has shown that rural women in the Brandywine Valley, Pennsylvania, were using physicians to deliver their babies at the beginning of the nineteenth century. *Loosening the Bonds: Mid-Atlantic Farm Women, 1750–1850* (New Haven: Yale University Press, 1986), pp. 30–33.

91. St. Vincent's Infant Asylum was established in 1856 under the direction of the Sisters of Charity to care for unmarried mothers and young children. The Chatard family were longstanding supporters of Catholic charities in Baltimore. Saint Elizabeth Ann Seton, who founded the Sisters of Charity, was a patient of Pierre Chatard when she lived in Baltimore. Personal communication from the Daughters of Charity, Emmitsburg, Maryland.

92. The Boston Dispensary, founded in 1796, provided free medical care to the poor who were treated in their homes or at the offices of physicians appointed by a board of managers. For young physi-

tioned the occupation of his patients or their husbands, which included servants, carpenters, factory workers, and dressmakers, as well as unemployed Irish immigrants.

The lives of the poor women are poignantly revealed. Cold damp rooms, scanty clothing, and poor nourishment were common features for many of Jackson's Boston cases. Tuberculosis, dysentery, and other underlying diseases killed women and their babies. In Newburyport, Massachusetts, George Snow described a patient who had "done most of her weeks washing after pains commenced" and another "back working at the mill the following day." More amusing, in retrospect, was the case of a poor woman whose bed collapsed during the delivery. Snow was "obliged to finish the case on the floor."[93] Several physicians commented on the fate of poor women who suffered permanent damage from abortions.

Some physicians let their personal prejudices intrude, indicating in their notes when a patient was a prostitute or foreign born, especially if she were Irish. Boston physicians held many of the anti-Catholic and anti-immigrant prejudices that characterized social relations during this period. "A miserable American married to an Irishman" is Parker's description of one patient. Jackson noted "Irish" women as opposed to women of "respectable families."[94]

Physicians were aware of the marital status of their patients, noting if the woman were unmarried or not married long enough for legitimate birth. This attention to bastardy may have been a reflection of physicians' potential roles as witnesses in legal cases regarding paternity and inheritance. The record would refresh their memories if necessary.

---

cians, an appointment as dispensary physician was an opportunity to treat many more cases than would come their way in private practice and to establish a reputation with the public as a skilled practitioner. Jackson was a district physician from 1832 to 1835 but the poor women whom he delivered were not limited to that period. Robert W. Greenleaf, *An Historical Report of the Boston Dispensary for One Hundred and One Years, 1796–1897* (Brookline, Mass.: Riverdale Press, 1898). Channing was not a dispensary physician but he did consult in difficult cases. He was one of the founders of and physician to the Boston Lying-In Hospital, intended to serve "poor but respectable women" during their final days of pregnancy, delivery, and recovery. Leonard K. Eaton, *New England Hospitals 1790–1833* (Ann Arbor: University of Michigan Press, 1957), pp. 221–26; "Boston Lying-In Hospital," *Med. Mag.*, 1833, *1*, 19–20; Frederick C. Irving, "Highlights in the History of the Boston Lying-In Hospital," typescript, Countway Library, Boston, Massachusetts. For medical care of the poor in the antebellum period, see Charles E. Rosenberg, *The Care of Strangers: The Rise of America's Hospital System* (New York: Basic Books, 1987), part 1.

93. Snow, (n. 23) 8 October 1866, 30 January 1867, 1 March 1868.

94. Parker, (n. 8) II, 25 September 1849; Jackson, (n. 8) XII, 21 May 1834; 2nd ser. II, 6 January 1837; "good family," ibid., 19 December 1837; "decent circumstances," ibid., 13 December 1837. For Boston and immigrants see, Peter R. Knights, *Yankee Destinies, the Lives of Ordinary Nineteenth-Century Bostonians* (Chapel Hill: University of North Carolina Press, 1991), pp. 106–24; Oscar Handlin, *Boston's Immigrants* (Cambridge: Belknap Press of Harvard University Press, 1959), pp. 178–229.

However, it might equally well have been a reflection of their moral standards. George Snow referred to one of his patients as a "strumpet."[95]

THERAPEUTICS

Obstetrical therapeutics varied greatly from physician to physician. Though textbook prescriptions and medical lectures suggest that there would have been uniformity of practice, individual personalities, preferences, and experience influenced what physicians actually did. Those whose case loads included consultations and complicated deliveries would be more likely to use ergot or instruments.[96] Physicians who had confidence in their skill with forceps would have been more willing to use them; physicians whose obstetrical practices were relatively small would have hesitated to do so. Predisposition for or against particular therapeutic practices also may have resulted from the teachings of a particular preceptor or professor.

Ergot, or *secale cornutum*, long had been known to increase and strengthen uterine contractions and was introduced to American medical practice at the beginning of the nineteenth century as an aid to prolonged labor, convulsions, and abortion. It also was used to expel a retained placenta and to curtail post-partum hemorrhage. However, the unpredictability of its effect and the potential for harm to mother or fetus made its use questionable and was an additional reason for the differences in usage.[97]

Although critics have asserted that the use of ergot reached "dangerous proportions" in the first half of the nineteenth century, with physicians "reluctant to sit passively by and await natural delivery," the physicians in this study do not support this claim.[98] William Parker administered ergot in fewer than 3 percent of his cases, with no fatalities. Most of the others used it even less frequently.

Walter Channing gave ergot four times in 195 cases between 1811 and

95. Snow, (n. 23) 21 September 1869. The association of obstetrical practice and jurisprudence can be traced to European and colonial American midwives who were expected to testify in legal cases regarding paternity and legitimacy. Thomas Rogers Forbes, *The Midwife and the Witch* (New Haven: Yale University Press, 1966); Borst, (n. 5); Wertz and Wertz, (n. 5) *Lying-In*, pp. 6–9. Walter Channing was Professor of Midwifery and Medical Jurisprudence at Harvard Medical School.

96. Paul Berman has written about obstetrical therapeutics in the practices of physicians in rural New England from 1800 to 1860, demonstrating that ergot and forceps were not overused. See Berman, (n. 7), and "The Practice of Obstetrics in Rural New England 1800–1860," paper presented at 66th Annual Meeting, American Association for the History of Medicine, Louisville, Kentucky, 15 May 1993.

97. See n. 45 above. Also Wertz and Wertz, (n. 5) *Lying-In*, pp. 65–66; Leavitt, (n. 5) *Brought to Bed*, pp. 144–47. As with every aspect of case notes, we only know what the physician included or what can be inferred from his words. It is possible that he was not totally accurate or that he did not consider today's important questions to be interesting or worth noting.

98. Haller, (n. 19) *American Medicine*, pp. 161–62.

1822. In each case the baby was stillborn, but we do not know whether the fetal death occurred prior to labor or during the birth process. Channing indicated that one mother thought her baby had been dead for some days. In a case from his 1829 notes, Channing described how, after four days of ineffectual labor pains, he gave his patient ergot

> in a cup of souchong tea with cream and sugar. Pains very soon came on of a forcing character, I prepared another cup of the tea as before, but before the half hour was over at which I meant to repeat it the labour was over. *This patient did not know she had taken ergot.* [italics mine] The child was large, and breathed with some difficulty for a few minutes after birth. It was in no degree asphyxiated, nor was its colour livid.

Channing gave no explanation for the rare subterfuge. It is apparent from his remarks that he was aware of the potential danger and relieved that both mother and child were well.[99]

Pierre Chatard initially opposed ergot. In 1820 and 1821, he published two articles that detailed its capricious effects and cautioned against its use. However, by the time he began to record his obstetrical cases, he had changed his mind and, as indicated above, he instructed his son that proper dosages in specific circumstances would expedite labor without harming the fetus.[100] Nonetheless, he used ergot in fewer than 2 percent of his cases from 1829 to 1846.

Among the physicians in this study, his son used it most frequently, some years in more than 10 percent of his cases. He recognized that ergot worked in some labors and was ineffective in others, but since the reasons remained obscure he continued to use it. In just about every such case, Ferdinand Chatard explained why he had done so: to expedite a tedious labor, to strengthen ineffective contractions, to hasten a miscarriage and halt bleeding, or to accede to an earnest request from the mother. In some cases in which ergot in retrospect might seem to have been appropriate, Chatard did not prescribe it. In the final years of his long career, he employed ergot less frequently. Perhaps the "therapeutic nihilism" that replaced "heroic medicine" in the second half of the nineteenth century was responsible for the change.[101]

99. Channing, (n. 22) Obstetrical Cases, 1 July 1829.

100. See n. 45 above. Chatard's experiments with ergot have been called "some of the earliest clinical trials of ergot in America." John E. Savage, "A sketch of the history of obstetrics and gynecology in Maryland, 1770–1976," *Md. State Med. Jrl.*, 1979, *28*, 53–58.

101. It should be pointed out that, in one case, Ferdinand Chatard gave ergot knowing the fetus was either breech or footling presentation, a practice that ran counter to every dictum, including that of his father. For the shift from "heroic medicine," see John Harley Warner, *The Therapeutic Perspective: Medical*

Physicians also were cautious in their use of forceps and other obstetrical instruments.[102] Some never used instruments; others had to send for their instruments when needed because they did not ordinarily bring them to a delivery. In most instances, they also sent for a colleague who gave a second opinion and assistance during a strenuous and tiring operation. Forceps frequently were employed if ergot had not been effective and the woman was losing her strength. Cases in which a deformed pelvis, unusually large child, hydrocephalic child, or abnormal presentation made delivery impossible might require use of instruments that destroyed the child, who might or might not already be dead.

As would be expected, the three physicians who gained reputations as obstetricians and were called by midwives and other physicians to assist with difficult cases recount larger numbers of instrumented deliveries. Channing used instruments nine times in 195 cases between 1811 and 1822, three times in 18 cases in 1829. In each instance, the fetus was so severely impacted within the pelvis that the mother was unable to force its expulsion.

Nonetheless, he was generally loathe to resort to instruments. As he described a delivery in which the woman was in severe pain and the fetal head was protruding, "the strength was good, pains powerful, pulse & countenance good, no headache, thirst or febrile symptoms, I recommended further delay." An hour and a half later, when no progress had occurred, he successfully used the lever to deliver a very large child. In another case, Channing arrived to find the mother already "much exhausted, the tongue dry & rough, the centre of it *brown*, the *sides white*, & thickly crusted. The body was hot & dry in some parts, moist in others, great despondence was present." He decided that "the case had been trusted long enough to the natural efforts, & that the safety of the patient was to be looked for in delivery." Delivery took another three hours and required use of the hook to bring down a dead fetus. Other cases were

Practice, Knowledge, and Identity in America 1820–1885 (Cambridge: Harvard University Press, 1986); John Duffy, *The Healers, A History of American Medicine* (Urbana, Ill.: University of Illinois Press, 1979), pp. 98–108; Rosenberg, (n. 19) "Therapeutic Revolution"; Haller, (n. 19) *American Medicine*, pp. 36–99; Rothstein, (n. 19) *American Physicians*, pp. 177–85. In the twentieth century, endocrinology and pharmacology have produced drugs that have replaced ergot.

102. Many writers have ascribed the replacement of midwives by physicians to the ability of the latter to use instruments, especially forceps. Midwives, who lacked the same educational opportunities, did not use forceps. For women who dreaded a long and painful labor, the availability of a forceps delivery was a significant reason to rely on medically trained men. See, e.g. Haller, (n. 19) *American Medicine*, pp. 152–53; Thoms, (n. 2) *Chapters*, pp. 7–14; Wertz and Wertz, (n. 5) *Lying-In*, pp. 34–44; Leavitt, (n. 5) *Brought to Bed*, p. 39; Nancy Schrom Dye, "The Medicalization of Birth," in Eakins, (n. 5) *American Way of Birth*, pp. 27–28.

even more horrific, requiring the deliberate destruction of the fetus in an effort to save the mother's life. These cases usually necessitated repeated examinations and prolonged interference and could result in puerperal fever, thus producing still another crisis. However, delay in resorting to instruments also could have dire results.[103]

Pierre Chatard admitted that he used forceps more than was necessary but added that women had begged for them, and he wanted to relieve their suffering. He believed that selective use of forceps could save lives, whereas prolonged labor could cause urinary fistulas, fatal inflammation, or puerperal fever in feeble or excessively nervous women. However, lest his candor lead to the assumption that he used forceps indiscriminately, he did so only in 141 or 3 percent of 4309 cases.[104]

His son made more frequent use of forceps, 6 percent of all cases between 1851 and 1860. However, there is little evidence to support the notion that he did so for his own convenience. As with ergot, in each instance of forceps or other instruments, Ferdinand Chatard gave an explanation: the urgency to halt severe hemorrhage, convulsions or the threat of convulsions, exhausting labor lasting many hours or days, excessive pain, an impacted child, deformed pelvis, and "at her urgent request."

The case books also give vivid examples of the injuries that occurred in some instrumented deliveries. Limited by poor illumination, often uncertain about the cause of obstruction or the condition of the fetus, there were many chances for physicians to make dreadful mistakes. Injured babies and maternal lacerations, ruptures, and infection were potential consequences of frequent examination and instrumented deliveries but often the alternative was death to mother and child. However, there is no indication that forceps were used to hasten delivery and enable the physician to be on his way, as has been charged by some critics.[105]

103. Channing, (n. 17) 22 June 1822, 15 September 1821, 12 November 1820; idem, (n. 22) 3 April 1829.

104. Van Bibber, (n. 12) gives 144 as number of forceps deliveries; I counted 141. The difference does not significantly affect the percentage of forceps cases. However, Chatard did use forceps more frequently in the final seventeen years of his practice than in previous years. For Chatard's views of the proper use of forceps, see Vol. 1, "Sur les accouchements avant 1829." Since he did not understand the etiology of puerperal fever, the notion that it might be caused by the continuous pressure and irritation of prolonged labor probably accorded with Chatard's experience. Puerperal fever also occurred in obstetrical cases in which there was no physician. William Parker described a case of puerperal fever in which the woman had been delivered by a midwife, Parker, (n. 8) II, 15 July 1849.

105. ". . . with little or great hesitation, [physicians] were using forceps to accomplish or even to speed deliveries," Bogdan, (n. 5), p. 96.

MEDDLESOME MIDWIFERY

Nineteenth-century obstetrical text books and medical journals contained frequent warnings against "meddlesome midwifery," a term that suggested unnecessary interference in childbirth.[106] Revisionist historians, basing their claims on evidence from these sources, assert that "meddlesome midwifery" was inherent in physician-delivered obstetrics, with detrimental results for women.

However, the case notes show that physicians engaged in daily practice were loathe to interfere with nature. Robert Thaxter avoided instruments as long as possible. He even opposed supporting the perineum, preferring to leave as much as possible to nature. George Snow ruptured the membranes early in labor but concluded that there was little benefit, and wrote that "meddlesome midwifery is bad." John Metcalf assured his patients that nature should never be interfered with when she has "power to accomplish her purposes alone."[107]

This is not to suggest that physicians did nothing but "catch the baby." At the minimum, they urged evacuation of the bladder and rectum before the second stage of labor and prescribed laxatives afterward. If labor were slow or the cervix were rigid, they might bleed the patient or rupture the membranes. They might administer opiates to relax the patient or to relieve pain. When convulsions threatened or occurred, they bled the patient, applied leeches and cold compresses, and did anything else they could think of to avert disaster. Hemorrhage was the greatest fear and was treated with ergot, ice, tampons, and drugs, such as copper sulfate, aluminum sulfate, and iron compounds. In cases of puerperal fever, the full array of nineteenth-century "heroic medicine" came into play, venesection, cathartics, emetics, cold applications, frictions, etc. Although these therapies were useless at best and could be harmful, they are further demonstration of the extent to which nineteenth-century medicine was limited by lack of knowledge.

On the other hand, physicians aligned the cervix when the uterus was poorly positioned and turned the fetus if it presented in an abnormal po-

106. William P. Dewees, whose obstetrical text was widely used, was one of many who warned against too much interference with natural processes, Dewees, (n. 64) *Compendious System*, p. 145. On the other hand, he also advocated bleeding to syncope to strengthen uterine contractions.

107. Cushing, (n. 29); Snow, (n. 23) 5 October 1870; Metcalf, (n. 14) 21 February 1828. Snow ruptured the membranes in nine of the remaining 176 cases recounted in his casebook. One was a case of convulsions which required turning the baby to expedite delivery. Another was a twin birth that also required version. In the seven remaining cases, which were vertex, the head immediately engaged and the child was soon born.

sition. They also resuscitated stillborn babies and delivered many babies who presented with a foot or breech and had not been turned. They prevented miscarriages, saved women whose abortions had nearly killed them, and, in one case, treated a husband for "exhaustion & excitement at his new duties."[108]

The vast majority of cases were completely uneventful, and, contrary to charges of "meddlesome midwifery," the physician "sat down to await the delivery."[109] A typical entry by Ferdinand Chatard reads "#2855, January 25 [1857], Mrs. T. in labour of 5th child called to her about 11 o'clock AM—pains slight—os tincae dilated about the size of half a dollar—vertex presenting in the 1st position—labour slow but natural—the child was born at twenty minutes of 5PM Sex a boy. The placenta was removed without difficulty." From George Snow, "#263 April 1 [1874] Delivered Mrs. A. E. Goodwin, Rings Island, of female child. Primipara. Labor set in about one o'clock A.M. and child was born at 12 o'clock. Vertex presentation. Labor natural. Had a good convalescence and good flow of milk." From another physician, "Case #243 Aug. 16 [1843] Wife of Samuel Cole Jr. in her 5th labour, which continued about 2 hours. The waters being low, I ruptured them, and the child, presenting naturally, was born by the next pain. placenta soon followed. Child, male."[110]

ANESTHESIA

Analysis of the use of anesthesia shows physicians hesitant to try new technologies. Acceptance of anesthesia was hampered by uncertainty about its effect on the mother and on the child, in addition to the belief of many people, including physicians and theologians, that pain was an essential part of childbirth.[111] Only seven of the case books cover the years after 1846 when ether was first demonstrated, and only five of those physicians recorded its use.[112]

108. Parker, (n. 8) III, 3 April 1850.

109. The term is from Metcalf, (n. 14) 30 May 1829. Several other physicians referred to "sitting" and to "yielding the seat" when another physician assisted.

110. Chatard, (n. 14); Snow, (n. 23); N.N., (n. 29).

111. The controversy surrounding the use of ether has been well described in Martin S. Pernick, *A Calculus of Suffering: Pain, Professionalism, and Anesthesia in Nineteenth-Century America* (New York: Columbia University Press, 1985); Leavitt, (n. 5) *Brought to Bed*, pp. 116–27; Wertz and Wertz, (n. 5) *Lying-In*, pp. 109–18; Speert, (n. 2) *Obstetrics and Gynecology*, pp. 136–41. See also John Duffy, "Anglo-American reaction to obstetrical anesthesia," *Bull. Hist. Med.*, 1964, *38*, 32–44.

112. George Dock did not use chloroform for his obstetrical cases, but he did use it for removal of the nail on a big toe. Dock, (n. 13) 15 April 1848. The Channing case notes do not cover the years af-

William Parker first gave chloroform in March 1848.

> . . . in great apprehension procured some chloroform [underlined by Parker] . . . the parturient effort seemed hardly to have any cessation & the relief from pn [pain] was delightful . . . This makes a memorable case . . . she has hitherto suffered much more than women in general . . . in one hour the labor was complete & that with remarkable comfort & subsequent success to the patient.

He gave chloroform again in July, the patient's pains "being very tormenting & she being willing." The third time, the patient lost her sight, seemed faint and delirious. There were no permanent effects, and a healthy baby boy soon was born. But Parker must have been badly frightened because he never gave chloroform again.[113]

Ferdinand Chatard administered "chloroform in moderate quantities" at the request of a patient in September 1849 and noted approvingly that pain was much reduced, and the patient remained conscious. He continued to use it, though infrequently, sometimes at the woman's request to relieve pain, more often to control convulsions or to produce unconsciousness during an instrumented delivery.[114] George Snow also used anesthesia sparingly, sometimes ether and sometimes chloroform, more frequently to quiet convulsions than to allay pain. He noted that the attendants objected on one case for fear "it would affect the child." The Plymouth physician employed ether in one of the final cases of his notebook. "It had some effect in lessening the sensibility, the pains being borne with less expression of suffering than before, but the trial was not as satisfactory as I could have wished."[115]

By the end of the period under study, changes in therapy can be discerned. Medicine gradually was becoming more "scientific" and new obstetrical knowledge also was developing. Ferdinand Chatard and George Snow analyzed urine, noted the presence of excessive albumen, and referred to "insipid diabetes."[116] J.B.S. Jackson examined ova and tissue with a microscope. The demise of "heroic" therapeutics in gen-

---

ter 1846. However, Channing was an early advocate of the use of anesthesia in childbirth. Walter Channing, *A Treatise on Etherization in Childbirth* (Boston: William D. Ticknor, 1848).

113. Parker, (n. 8) II, 1 March 1848, 23 July 1848, 7 August 1848.

114. Chatard referred to the use of chloroform 21 times in 1619 cases from 1850 through 1860 and 6 times in 507 cases from 1870 through 1875.

115. Snow, (n. 23) 6 Sept 1867; N.N., (n. 29) 10 November 1847. The fifth physician was Samuel Butler, whose wife "inhaled chloroform which was a great comfort to her" in a severe and lingering labor that produced a stillborn son. Butler did not speculate on the possible relationship between anesthesia and the stillbirth. Butler, (n. 15) 4 April 1855.

116. Probably a reference to "diabetes insipidus," defined as "the habitual excretion of large amounts of pale urine of low specific gravity," *Stedman's Medical Dictionary*, 15th ed.

eral medicine was reflected in the restriction of bleeding to cases of convulsions. Purging and emesis also disappeared. Technology yielded improved forceps and "Barnes dilators" to expedite labor. And patients who wanted to deliver their babies in their mothers' homes could get there more rapidly by train.[117]

CONCLUSION

Until the latter part of the eighteenth century, midwives held almost exclusive domain as obstetrical caregivers in America. Only in the most extreme cases were physicians called upon for assistance. Most midwives left no written records, although Martha Ballard is a notable exception. Ballard's diary, admirably explicated by Laurel Thatcher Ulrich, provides a rare glimpse of a midwife's daily practice.[118]

During the nineteenth century, physicians were increasingly employed in obstetrical cases. Fortunately, some physicians kept case notes and thus have provided an important source of information about the daily practice of obstetrics. In these private documents physicians described their activities, their mistakes, and their questions. They make clear that their primary concern was the safety and well-being of the patient. They consciously tried to avoid unnecessary interference with natural processes but used the drugs and instruments that could make the difference between a successful delivery and tragedy. They were acutely aware of the imperfections in their knowledge, although they also assumed that the medical knowledge they did possess was superior to that of midwives and a more certain guarantee of safe childbirth.

In the twentieth century childbirth shifted from women's homes to hospitals, where record keeping is an essential part of practice. Moreover, twentieth-century women are willing to discuss their childbirth experiences publicly, a habit their earlier sisters would never have condoned. Thus the history of twentieth-century obstetrics will be easily documented. But the full history of nineteenth-century obstetrics requires close attention to the case notes of the physicians who brought its babies into the world.

117. Urine, Ferdinand Chatard, (n. 14) III, 16 January 1871, 29 June 1873, 2 November 1874, 16 October 1875; Snow, (n. 23) 15 June 1873; microscope, J.B.S. Jackson, (n. 8) 2nd ser., IV, 17 July 1849; Barnes Dilators, Snow, (n. 23) 10 December 1872; train, ibid., 3 October 1874.

118. Ulrich, (n. 5) *Midwife's Tale.*

APPENDIX

Following are the physicians whose case notes form the basis for this study.

TYLER BRIGGS (1817–51) studied with a local physician in Hopkinton, Massachusetts, before attending medical lectures in Boston and receiving a degree in medicine from Brown University. He spent the rest of his life in Northbridge, Massachusetts, and practiced in small rural towns of the Blackstone Valley. His diary, 1817–39, contains 117 obstetrical cases along with a few other medical cases.

SAMUEL WORCESTER BUTLER (1823–74) was the son of a physician who served as Presbyterian missionary to the Cherokee Indians, first in Georgia, and then in the Indian Territory. Butler graduated from the University of Pennsylvania Medical School in 1850, began practice with another physician already established in Burlington, N.J., and started his own private practice there in 1852. The difficulties of earning a living in the small towns led him to pursue an additional career as author and editor of several medical journals and books. Butler was a religious man who sought spiritual consolation from many tribulations described in his journal, 1849–55, which includes thirty-four obstetrical cases among a general practice.

WALTER CHANNING (1786–1876) studied medicine at the University of Pennsylvania and spent a year in London and Edinburgh before returning to Boston where he spent the remainder of his life. Channing was appointed Lecturer in Midwifery at Harvard in 1815 and Professor in Midwifery and Medical Jurisprudence in 1818. He was one of the founding editors of the *New England Journal of Medicine and Surgery*, held office in the Massachusetts Medical Society, and was assistant physician at the Massachusetts General Hospital from 1821–39. Channing's "List of Midwifery Cases" includes 195 cases, many of which are fully described. His 1829 notes on midwifery refer to eighteen cases which are given in detail.

FERDINAND EDMÉ CHATARD (1805–88) was the son of Pierre Chatard. Born in Baltimore, he received a medical degree from the University of Maryland and continued his studies in Montpellier, Paris, and Edinburgh. Upon his return to Baltimore, he became associated in practice with his father. After his father's death, he too became one of Baltimore's leading obstetricians. His three-volume case books, 1829–85, contain detailed descriptions of 5265 obstetrical cases.

PIERRE CHATARD (1767–1848) was born in the French colony of St. Domingo (now Haiti) and studied medicine in Montpellier and Paris before coming to Baltimore, where there was a substantial group of French colonial émigrés, in

1800. He practiced general medicine but his many obstetrical cases gained him a reputation as "the most distinguished early obstetrician in Maryland" (J. W. Williams). Pierre Chatard was Consulting Physician at Baltimore Hospital from 1812 and a member of the faculty of Washington University, Baltimore, from 1838. He published in American and French medical journals on a variety of subjects. He was the progenitor of four generations of Chatards who practiced medicine in Baltimore. His two-volume obstetrical case books contain summary comments on 2503 cases before 1829, plus detailed accounts of fifty-one of those cases, detailed notes on 1787 cases from 1829–46, and essays and "precepts" regarding the practice of obstetrics.

WILLIAM DARRACH (1796–1865) was the son of a physician and father of three physicians. Born in Philadelphia, he received the A.B. and A.M. degrees from Princeton, studied medicine with Dr. Philip Syng Physick, became resident physician at the Philadelphia Almshouse, received the M.D. degree from the University of Pennsylvania, and spent three years continuing his medical studies in Europe. Darrach practiced general medicine in Philadelphia and was consulting physician to the Philadelphia Medical Dispensary. He was Professor of Theory and Practice of Medicine and also served as the President at the Pennsylvania Medical College. His obstetrical notes, 1835–43, describe 137 cases.

GEORGE DOCK (dates unknown) began his medical studies at the University of Pennsylvania in 1840. After graduation, he was resident physician at the Philadelphia Almshouse (Blockley Hospital). Two years later he moved to Harrisburg, Pennsylvania, where he began a private practice that seems, from his case book, to have included much eye surgery. The case book, 1840–56, describes thirteen obstetrical cases among those of a general practice.

AUSTIN FLINT (1760–1850) practiced medicine in Leicester, Massachusetts, and nearby towns. He studied with a preceptor and was admitted to the Massachusetts Medical Society in 1804. Harvard Medical School awarded him an honorary M.D. in 1825. Three generations of his descendants were also physicians: his son, Joseph Henshaw Flint (1786–1846) was a surgeon; his grandson, also Austin Flint (1812–86), was an authority on medical practice, auscultation, and medical ethics; his great-grandson, another Austin Flint (1836–1915), was an eminent physiologist. His "Memorandum of Births" (1785–1834) lists 1750 obstetrical cases.

HENRY HARTSHORNE (1823–97) was the son and brother of Philadelphia physicians. He received the M.D. degree from the University of Pennsylvania and served as Resident Physician at the Pennsylvania Hospital for two years before starting his own practice. His varied career included teaching positions in several medical schools, physicianships at numerous hospitals, Quaker missionary work in Japan, and publication of fiction and poetry. Hartshorne was a strong advocate of medical education for women and was also a founder of the American Public Health Association. His account books, 1848–51, describe eleven obstetrical cases among a general practice.

JOHN BARNARD SWETT JACKSON (1806–79) was the nephew of James Jackson, one of Boston's most prominent physicians during the first half of the nineteenth century. J.B.S. Jackson was educated at Harvard College (B.A. 1825) and Harvard Medical School (M.D. 1829) and studied in Paris, London, and Edinburgh. In addition to a small private practice, he was associated with the Massachusetts General Hospital for forty-four years and was appointed Professor of Pathological Anatomy at Harvard Medical School in 1847. Jackson developed the pathological collections for the Warren Museum at Harvard and the museum of the Boston Society for Medical Improvement. Many specimens from the autopsies recorded in his notebooks were contributed to those collections. His twenty-one volume case books contain descriptions of deliveries he performed, obstetrical cases in which he consulted, and autopsies of fetal and maternal deaths among many other medical cases.

JOHN GEORGE METCALF (1801–92) was the son and grandson of physicians. He graduated from Brown University in 1822, studied medicine with local physicians, and received a medical degree from Harvard in 1826. Walter Channing was his Professor of Midwifery. He practiced in Mendon, a town in south central Massachusetts, where he was also active in local affairs, social reform, and medical associations. Metcalf was one of the earliest American physicians to accumulate obstetrical statistics which formed the basis for publications on the practice of obstetrics. His note book, 1826–32, contains 107 obstetrical cases, of which seventy-two are fully described.

BENJAMIN OSGOOD (1781–1863) practiced medicine in Littleton, Massachusetts, and surrounding towns during the years covered by his case book. His father was also a physician and it is possible that he initially studied medicine with him. Osgood attended medical lectures at Harvard in 1806 and 1807, but did not receive a degree. He became a member of the Massachusetts Medical Society in 1822. Osgood moved to Westford, a town next to Littleton, shortly after the last case recorded in the obstetrical casebook, 1810–17, which lists eighty-seven cases.

N.N.–This physician practiced in Plymouth, Massachusetts, and surrounding towns. His anonymous case book, 1826–47, contains detailed descriptions of 436 obstetrical cases.

WILLIAM THORNTON PARKER (1818–55), the son of a physician, was born in Andover, Massachusetts, studied medicine with Dr. A. L. Peirson of Salem, Massachusetts, and received degrees from Dartmouth and Harvard Medical

School. He had an extensive practice in South Boston where he worked diligently for modest remuneration. He was a member of the Massachusetts Medical Society and the Boston Society for Medical Improvement. Parker had to give up his practice when tuberculosis began to seriously affect his health. He died at age thirty-six. Parker's three-volume notebooks, 1841–51, describe 101 obstetrical cases among many other cases of a general practice.

GEORGE LORING PORTER (1838–1919) graduated from Brown University in 1859 and received the M.D. degree from Jefferson Medical College in 1862. It is likely that he also studied obstetrics with Dr. Ellwood Wilson at the Obstetric Institute in Philadelphia. He had an illustrious career in the Medical Service of the United States Army, including service at Gettysburg and medical charge of the Lincoln assassination conspirators. Upon discharge, he practiced in Bridgeport, Connecticut, where he specialized in surgery. Porter was President of the Connecticut Medical Society in 1889 and held many other offices in medical and local organizations. His "Commonplace Book" contains notes on Wilson's lectures and three obstetrical cases, 1861.

AVERY SKILTON (— d.1857?) studied medicine with local physicians in upper New York State and was licensed by the Connecticut State Medical Society in 1827. Thereafter he lived in Troy, New York, where he had a varied and extensive medical practice. His "Obstetric Memoranda," 1827–37, lists 670 cases.

GEORGE SNOW (1835?–93). Little is known about Snow. He became a member of the Massachusetts Medical Society in 1866, the same year that he began to practice medicine in Newburyport, Massachusetts. His case book, 1865–75, fully describes 313 obstetrical cases.

EBENEZER STONE (1797–1869), A.B. Brown 1820, A.M. Brown 1822, received a medical degree from Harvard in 1824 where he studied obstetrics with Walter Channing. Stone practiced medicine for more than forty years in Walpole, Massachusetts, and surrounding towns. He was a popular family physician as well as an advocate for educational reform and temperance. His case books, 1824–36, contain nine obstetrical cases along with those of general practice.

In addition, published accounts of two other physicians' obstetrical practice were used. Both included personal recollections in addition to statistical summaries.

WILLIAM INGALLS (1813?–1903) graduated from Harvard College, B.A. 1835, and Harvard Medical School, M.D. 1836. He practiced general medicine in Boston for more than forty years. Ingalls also taught medicine in one of Boston's many private medical schools. Tyler Briggs (above) was one of his students. His synopsis of 2000 obstetrical cases was presented to the Obstetrical Society of Boston and published in the *Boston Medical and Surgical Journal*, 1877, *96*, 485–98.

ROBERT THAXTER (1777?–1852) was a "country physician" who practiced medicine in towns south of Boston for more than fifty years. He became a member of the Massachusetts Medical Society in 1803. His obstetrical practice, 2600 cases from 1801 to 1852, was summarized and analyzed by his nephew, Dr. Benjamin Cushing. The original records were lost and only the manuscript of Cushing's notes remains.

MIDWIFERY.

LABOR, OR CHILD-BIRTH.

At the end of nine months from the time of conception, the natural period of pregnancy is accomplished. Some women are a day or two earlier, others a day or two later; but they usually have their child in forty weeks from the time of conception. When a woman is about to be confined, or approaching this period, it is not unusual for her to be troubled with irregular pains for several days previous to her labor; these are called false pains, and are usually most troublesome during the night. They do not bear down like real labor pains, but are sharp, last for a few moments, and then pass off. In real labor, there is a discharge from the vagina or privates. When the pains are suspected to be false, it will be necessary to relieve them by warm drinks, such as hop tea, sage tea, balm tea, poppy tea, or a little camphor and hot water mixed. If these do not succeed, give two teaspoonsful of paregoric, or twenty or thirty drops of laudanum; and this may be repeated, if necessary, every three or four hours. When the pains come on at regular intervals, or gradually increase in severity, and are attended with a bearing down sensation, a looseness of the bowels, frequent inclination to make water, and a slight discharge, or moisture of the organs of generation, it will be presumable that true labor has begun, however slow or feeble it may be at the commencement; the discharge, which is called a "show," is often colored, and sometimes there is an appearance of fresh blood. Should the case prove tedious, the pains not strong and irregular, then bathe the feet in hot water, and give some warm tea; this will often be followed by a good effect, and the pains will become more regular. If the bowels are costive, a dose of castor oil, or an injection of salt and water, will produce an operation, relieve the lower bowels, and advance the process of the labor; this ought never to be neglected when the bowels are not open, or costive. Drinking freely of cold water often increases the pains, and may be used with perfect safety.

As soon as labor commences, the belly sinks, the dress becomes loose, and there seems to be a descent of the womb into the bottom of the belly; the motions and weight of the child are felt to be lower than formerly, and the head of the fœtus, or child, falls down to the orifice of the womb, and presses upon it. The physical causes that determine the exit of the fœtus, or child, are the contraction of the uterus and that of the abdominal muscles. By their force the mucous secretion takes place; this lubricates the parts, and the mouth of the womb gradually undergoes dilatation, or opening, increasing at each bearing down. As the process of labor advances, the finger directed into the vagina, will find a considerable tumor; this contains the uterus; the pains now gradually increase, the pulse becomes stronger and more powerful, the face is flushed, perspiration flows in abundance, and the whole body is in extreme agitation. The process still goes on; the pains are now more frequent, stronger, and lasting; presently a strong pain comes on, and suddenly the water, medically called *liquor amnii*, gushes out, wetting the bed. There is now a longer interval between the pains, and on the midwife introducing her finger, she will discover a great change: a large round hard substance is felt, which she at once knows to be the head of the child. The pain returns again, the woman becomes more distressed, the face looks intensely red, accompanied by a trembling of the lower limbs; the pains are much longer; she lays hold of a towel which is commonly fastened to the bedpost for the purpose, and she bears down with all her might, exclaiming, "Oh! when will all this be over?" It is impossible to express the impatience, to depict the suffering of this awful and critical moment. The head of the child descends lower and lower into the *vagina*, or water-passage, till it firmly presses upon the perineum, and shows itself at the outward or external orifice. In some instances, if it be the first child, it is apt to remain some time before it is expelled; if she has had the second or third child, or more, a few pains will generally complete the work.

But in every instance, let me impress on your mind *patience;* and let nature alone, and she will accomplish the labor.

The parts of generation during labor should always be well oiled or greased with lard, as it greatly assists and mitigates the suffering, and lubricates the parts or passage. When the infant gradually advances, enlarging the passage so that the crown of the head may be felt, the birth is then advanced one-third. When advanced forward as far as the ears, it is then in the vagina, or passage; if the membranes have not already burst, they may now be opened, and the waters by their effusion will render the vagina slippery, and promote the birth of the child. After the child is born, there is a freedom from pain, and the mother feels rejoiced, and "thanks God" for the sudden transition from pain and severe suffering to comparative ease.

The navel-string must be divided as soon as the child is born, or a few minutes afterward, giving time to establish fairly by breathing or crying the new mode of life. I generally let it remain five or ten minutes, until the cord ceases to beat, before its connection with the mother is severed, or, in plain language, the navel-string is cut. If separated immediately after birth, it may do an injury, and therefore it is advisable to let it remain on the bed for a few minutes. As soon as you see the signs of life fully established in the child, the navel-string must be tied with a small cord made of a few threads well twisted together, or waxed so as to make it strong. First, tie the navel string two inches from the belly of the child, and then again two inches from the first, so as to leave a space between, and with a sharp pair of scissors cut the cord or navel-string so as to separate the child from the mother; and remember to tie the cord well, two inches from the belly, as the child may otherwise lose its life by the loss of blood; and attend to it, and see that this string does not come off. If it does, another string must be tied immediately around it, as serious accidents have frequently occurred by this neglect, or, in other words, the child will bleed to death.

The child, being taken away or separated from the mother, is to be placed in flannel in the arms of the nurse, and put in a moderately warm situation. This being done, the after-birth should next be attended to. Generally, from twenty to thirty minutes elapse between the birth of the child and the expulsion of the placenta or after-birth. In a large proportion of cases, the after-birth will come away generally in the course of ten or twenty minutes after the birth of the child. It will be proper, however, to catch hold of the cord, and carefully ascertain if the placenta can be felt in the passage, and whether the woman is losing more blood than is natural at such a time. If the placenta can be felt in the passage, no great danger need be apprehended, as a few pains, a little gentle motion, and very moderate extension of the cord will deliver it safely. When it remains high up in the abdomen, however, and especially if there be an unusual loss of blood, the action of the womb must be promoted by pressing and kneading the lower part of the bowels, by grasping or squeezing the womb gently in the hand, at the same time pulling very moderately and moving the cord, so as to make the extension in different directions. Great care is necessary in these cases, not to pull so hard as to break or to separate the cord from the placenta, nor to bring down the womb; the latter of which accidents would be fatal, and the former, to say the least of it, very troublesome.

In cases of excessive flooding, whether the after-birth has come away or not, the woman should be placed upon her back, with her head and shoulders low, be kept perfectly still, and cold vinegar, spirits, or water applied freely to the lower part of the bowels and birth-place, by means of cotton or linen cloths, which should be changed every few minutes until the flooding is arrested. If the after-birth be retained in the womb, do not hurry it, but wait for the pains to come on. Great care is necessary, in these cases, not to pull so hard as to break or to separate the cord from the placenta, nor to bring down the womb, which accidents would be very troublesome. The woman should blow in her hands, so as to force the pain downwards; sometimes a change of position assists the after-birth to come away.

If the pains are severe after the after-birth has come away, give twenty or thirty drops of Laudanum, or one or two teaspoonsful of Paregoric. No attempt should be made to move the patient or change her clothes for some minutes. Put a warm application or dry towel to the birth-place. After she has fully revived, change her clothing, and put dry articles beneath her; and the greatest caution should be observed that she does not exert herself so as to cause the loss of blood. For two or three days after delivery, the patient should be kept perfectly quiet, see but little company, and be restricted to a plain diet; and the bowels if necessary, should be moved with some gentle laxative, as castor oil. The milk generally comes on the third or fourth day, but may come on the second or even the fifth. At this time, there is generally some little fever, headache, hot skin, and thirst. The lochia, or discharge which a woman has for several days after the birth of the child, may be stopped; this produces the slight fever; but when the milk comes on, this secretion is established.

27

Should the breasts become hard, they must be softened by the frequent application of lard or oil, gently rubbing them with the hands, and by the application of flannel cloths squeezed out of the lard as hot as can be borne. Draw the breasts frequently with the mouth, as the milk can be seldom drawn out by the child during the first week or ten days. An error is often committed, and much suffering is frequently caused by attempting to force out the milk when the breasts are hard and caked, sore and painful, before they are softened by rubbing them with warm oil, as before described, or with fomentations of hot vinegar or poultices.

In closing these remarks, I can not urge too strongly upon every midwife the great importance, during the time of labor, of keeping the woman calm and composed, and not to hurry Nature; by this course of conduct, a great variety of accidents may be prevented. Besides, women should be aware of the fact that hundreds of deaths and a long train of diseases are produced by too great haste, and not waiting patiently for the operations of Nature; her laws can not be infringed upon with impunity. A long experience in my profession, in the obstetric art, has convinced me that patience is the great remedy in childbirth.

An occurrence transpired in Louisville, Ky., a few years since, when a lady was in labor with her first child for three days, attended by two of the most distinguished physicians of that city. From the intense suffering of the patient, a surgical operation was deemed necessary. With much difficulty, I succeeded in delaying the surgical operation, and on the fourth morning she was delivered of a healthy female child, which is now living. How many lives of women and children might be saved by avoiding active remedies! But the desire to be thought superior has often made men selfishly alive to what the world calls professional skill; and instead of waiting for what God, in his infinite wisdom, has directed in his own due time, they often force Nature, which frequently destroys the general health, and in many cases ends in the most fatal consequences, such as flooding, inflammation of the womb, childbed or puerperal fever, prolapsus uteri, or falling of the womb, etc. Most cases of inverted uterus are produced by improper pulling at the cord.

The pains and suffering connected with birth, or parturition, are often great and protracted; and when the trial comes, the fortitude and resignation with which those sufferings are borne by the most delicate females, should call forth the kindness and sympathy of man. That women generally endure pain and sickness with more fortitude and patience than men, is evident. Looking forward with the pleasing hope of being the mother of a tender offspring, upon which she can lavish her affection and tenderness, sustains her in fulfilling the conditional requirements of Nature. Few men could be induced, for any consideration, to suffer in a similar manner. Then, how imperative the duty of every husband to be kind to the mother of his children, and to sustain her amidst her trials and afflictions; for God has devolved this sacred trust on man, that his smile and tenderness should soothe her in the moments of sorrow. And it is moments like these which give exercise to the noblest charities of our nature, in offices of kindness to our best and most devoted friend on earth, a faithful and affectionate wife. The female constitution being of a more delicate conformation, and having a finer texture of nerves than the other sex, gives women a quickness of sensibility and great promptitude of expression; it softens their manners, refines their ideas, and produces a lively sense of pleasure and pain; but while they enjoy these advantages, the peculiar construction of their frame subjects them to painful and critical vicissitudes, that affect not only their health, but also their temper, which fully entitle them to all possible lenity and indulgence.

In a long practice in my profession, how continually have I witnessed effects from these causes on the functions of the sexual organs! Every part of the animal economy is inflamed by the passions, but none more so than the uterus or womb. Anger, fear, grief, and other affections of the mind, often occasion obstructions of the menstrual discharge and of the womb, which prove very difficult of cure, unless by careful and skilful treatment.

I trust I have now said enough to place you upon your guard, so that you may be circumspect in your conduct at such times, for females have sufferings to endure, from diseases of the womb and other afflictions that will be hereafter mentioned, which should call forth the indulgence and sympathies of every noble and feeling heart.

DIRECTIONS FOR MIDWIVES.

The directions which I shall give, will impart such information as will enable a woman of good common sense and observation to manage almost any ordinary case of labor without the aid of a physician. Most of the accidents that occur during the course of labor, or childbirth, we believe, are produced by not giving Nature time to perform her operations; and a *Midwife* should remember, if she wishes to be considered skilful in her calling, never to force Nature, but to give her time to perform those operations; this will enable her to be successful in almost every case. It fortunately happens, that in those cases which terminate quickly, but little assistance is necessary; and thousands would pass safely through this ordeal of Nature, if they would but exercise patience and fortitude. A careful and experienced accoucher or midwife gives a woman confidence, relieves her of much anxiety, and thereby enhances her chance of going through her confinement safely.

When a woman is expecting to be confined, her arrangements as to her room should be attended to, as frequently, when the morning and evening winds are chilly and damp, a little fire may be necessary; a mattress is preferable to a feather bed, as feathers retain the heat and perspirable matter; and, besides, sink down, rendering the patient uncomfortable in her position during her pains. The bed-clothes and clothing intended for this purpose should be well dried and aired, and if the weather is cold, warmed, and every thing in readiness, that there may be no delay when needed, or a change in them required. The bed should be well protected, to avoid a draught of air, and so arranged that every thing that is wet or soiled be removed, in order to avoid the disagreeable and unwholesome smell that would arise during the woman's confinement in a warm or close apartment.

Labor usually takes place about the end of the ninth month, or the thirty-ninth week of pregnancy; which varies but little from four and a half months after quickening, or the first perceptible motion of the fœtus, or child. It frequently happens that a woman is troubled with irregular pains, more or less severe, for a number of days, or even several weeks, previous to her confinement, which, from their not producing any effect towards expelling the child, are called false pains. These pains are usually most troublesome during the night, and may be entirely absent during the day. They are often sharp and tedious to bear, coming on at irregular intervals, and do not bear down like efficient and expulsive contractions of the womb, nor are they attended with a discharge from the vagina or womb, as is usually the case in real labor. When the pains are suspected to be false, they may be relieved by warm sudorific and anodyne drinks, such as Hop tea, or Sage tea, or Poppy tea, or Camphor and a little hot water; or if these do not give relief, one teaspoonful of Paregoric, or twenty or thirty drops of Laudanum may be given, and repeated, if it becomes necessary, in three or four hours. No danger need be apprehended from giving an anodyne, for if the labor be actually beginning, it will often have a good effect in regulating the pains, and advancing the labor or child-birth.

When the pains come on at regular intervals, gradually increasing, called grinding pains, and attended with a bearing down sensation, looseness of the bowels, frequent desire to make water, and a slight discharge or a moisture from the birth-place, it will be a strong evidence that true labor has commenced, however slow or feeble the pains may be at the commencement. The discharge, which is called a "show,"—(*a medical term,*) is often colored, and sometimes there is an appearance of fresh blood. Should the case prove tedious, or the pains feeble or irregular, bathe the feet in hot water, and give some warm tea, such as Tansy, Sage, Balm, Pennyroyal, etc., which will be followed by beneficial effects; the pains will now become more regular, and stronger. A dose of Castor Oil, or a clyster of salt, lard, and water, by evacuating or purging the lower portion of the bowels, adds greatly to the ease and celerity of the birth of the child, and ought never to be neglected when the bowels are not free. When it is desirable to increase the force of the pains, let the patient drink freely of cold water. I have often produced or increased the pains by this valuable, though perfectly simple and safe remedy.

The Ergot or Spurred Rye is sometimes administered for this purpose, but is an unsafe medicine, except in cases of a peculiar and difficult kind, and should not be given without the advice of a physician, as it will endanger the life of the child, and probably that of the woman.

The location of the pains is different in different cases, though the usual pain in the commencement of labor is in the lower part of the abdomen or belly, and extending round into the back and hips; and in the last stage they are usually confined to the lower part or small of the back. When these pains continue hard in the lower part of the belly for a *long time*, and confined to this part alone, there may be reason to fear the case may be a tedious one, as in some instances the child rests upon the pubic bones for some time before it descends into the pelvis.

When you have ascertained that the actual labor is progressing, which is known by the symptoms I have before described to you, and by introducing your finger up the birth-place, well greased with lard, which should be done with much care and tenderness, your nails being closely and smoothly pared, you will feel the mouth of the womb dilate or open during the time that each pain comes on, and the bladder or bag containing the waters. If the labor be not much advanced, you will feel the mouth of the womb, and its dilation or opening more gradually at every pain.

The patient herself ought to be encouraged by cheerful conversation, and her mind kept occupied, and light nourishment should be given, as a cup of tea or cold water; and, above all things, the pernicious and too prevalent custom of giving stimulants, such as brandy, whisky, etc., is to be avoided, except in cases of great prostration or weakness. The patient should, if convenient and comfortable to her, lie upon her left side. Some, however, prefer a different position. A practice which is often followed by midwives, which can not be too strongly condemned, and which is highly dangerous, is that of delivering a woman kneeling on the floor. As I have before told you, the best position is generally on the left side, the body and head being elevated by pillows. Such a position, however, need not be taken until the labor has advanced, and the pains have become frequent and severe; until which time, I allow my patient to sit up, or walk about, or take her pains in any position which she may prefer. The time for her being put to bed may be known by the change of the pains becoming longer, causing her to hold her breath, and to make bearing down efforts, showing fully that labor is coming to a close. Presently the membranes will break and the water be discharged.

Now place a pillow between her knees, and occasionally elevate or lift her upon her knees at each pain, for the purpose of affording free space. A towel, sheet, or some convenient article, should be tied to the bed post, so that at each pain the woman may pull it, as it assists her very much in her efforts, and she derives great benefit and comfort from the support. The feet must be kept warm, as cold feet will often retard the progress and force of the pains.

When the last strong pains of labor are expelling the head of the child, as it advances, press the right hand steadily and firmly against the part between the fundament and birth-place, called by physicians *perineum*, so as to give that part support, and prevent its rupturing or tearing; at the same time inclining the child's head to the pubes, which means the parts that form the arch in front. It is really important to attend to this matter, as by want of proper care, from hurrying the birth of the child, many serious accidents have occurred, or in plain language, the birth-place and fundament have become one opening from the tearing of the *perineum*.

As soon as the head of the infant is born, the midwife ought to pass her fingers carefully around its neck, to ascertain that it is free, as it sometimes happens, that the navel-string is twisted around the child's neck; and should this be the case, she must endeavor gently to slip it over the head, otherwise the neck may be so strongly compressed as to strangle the infant. At this period, the mouth and nose of the child, if there is any delay in the passage of the body, ought to be kept as free as possible from the discharges from the birth-place, as they may be drawn in by the infant in its efforts to breathe. Neither ought the body, or even the legs, of the infant, to be drawn from the mother too rapidly; their expulsion should be left to the natural efforts of the womb; for if too suddenly pulled away, the natural action of the womb becomes changed, and irregular contraction, accompanied by unnecessary pain and discharge, may be the consequence.

After the child is born or expelled, if it should not cry immediately or breathe freely, it should be chafed upon the stomach with a flannel cloth; a little cold water, or vinegar, or spirits, should be sprinkled upon it, and occasionally a sudden gust of air blown upon it with the mouth or a fan, so as to resuscitate it. These means, if the child be alive, will generally revive it, and the breathing will be established; but should it remain apparently lifeless for some minutes, put it in warm water and rub it well, inflating the lungs by putting a quill or pipe stem into one nostril, and then closing the nose and mouth so as to prevent the escape of the air through them, gently, until the lungs are filled. When the lungs are filled, the mouth and nose should be unclosed, and the air forced out by gentle pressure upon the chest and abdomen, or belly. This application may be done several times, and the motion of breathing be imitated by the pressure of the hand upon the belly of the child, giving it a rising and falling motion. By using this means, I have known many children that were apparently dead, resuscitated, or brought to life.

The new born infant should be allowed to feel its new mode of life before that by which it previously existed is cut off; therefore, too much haste in tying the navel-string should be avoided. It ought to remain a few minutes before separating it from the mother, or until the cord ceases to beat. If separated too soon, it may produce injury to the child. The navel-cord must now be tied with a strong twisted cord, or thread, about as large as a knitting needle, so that it will not break, as a string that is too fine is liable, when tied firmly, to cut the vessels and cause bleeding. It is necessary, particularly if the navel-string be large, to tie it very firmly, as the child may otherwise lose its life by the loss of blood, a circumstance that has frequently occurred within our knowledge. Tie the navel-string three fingers' breadth from the belly of the child. This leaves it long enough to tie again, should it be required. Then tie it again about two inches further from the first string, leaving a space between the two strings, and then divide the navel-cord with a sharp pair of scissors. The infant is now to be placed in flannel until washed and dressed. The child should be frequently examined, to ascertain whether it bleeds at the end of the cord; and if so, another string should be immediately tied to make it perfectly secure. As soon as the infant is separated from the mother, it is proper to ascertain, by the hand placed upon her belly, that there is not a twin child; if there be, the remaining bulk will indicate it in a way that can scarcely be mistaken; and should it prove so, the recurrence of the pain which is to effect the expulsion of the second child must be quietly waited for. In most cases of twin children, the second is quickly and easily born, after the pains set in.

The child being taken away, as before directed, the after-birth should next be attended to. In a large proportion of cases, this will come away without assistance, in the course of ten or twenty minutes after the birth of the child.

It will be proper, however, to take hold of the cord, and carefully ascertain if the placenta or after-birth can be felt in the passage, and whether the woman is losing more blood than is natural at such times.

If the after-birth can be felt in the passage, a few pains or a little gentle motion, or a very moderate pulling of the cord will deliver it safely. When it remains high up in the belly, however, and especially if there be any unusual loss of blood, the action of the womb must be promoted by pressing and kneading the lower part of the bowels; by grasping and squeezing the womb gently in the hand, at the same time pulling very moderately, and moving the cord, so as to make the extension in different directions. Great care, however, is necessary in these cases, not to pull so hard as to break or separate the cord from the after-birth, or to bring down the womb, the latter of which accidents would be fatal, and the former, to say the least, would give a great deal of trouble. I have known cases where midwives and doctors have been in such great haste to get through with their labor as quick as possible, to pull the after-birth away so fast as to produce flooding or serious difficulties, which may eventually, as it does in many instances, produce a falling of the womb.

Every midwife, or physician, has, or ought to have, sufficient experience to be aware of this danger of pulling away the after-birth, instead of waiting the regular efforts of Nature, and assisting her in her operations, and should understand that child-birth is a natural process, and that Nature is fully competent in all *ordinary* cases, and in *extraordinary* ones oftener than is usually imagined.

In cases of excessive flooding, whether the after-birth has come away or not, the woman should be placed upon her back, with her head and shoulders low; should be kept perfectly still, and cold vinegar, or water, or spirits, applied freely to the lower part of the bowels, and to the birth-place, or genital organs, by means of cotton or linen cloths, which must be frequently changed until the flooding is stopped. If, under these circumstances, the after-birth be retained in the womb, the most skilful assistance ought to be obtained to deliver it safely.

The drinks, after the delivery, should be cold water, toast-water, or tea. If the patient should feel weak or faint, a little camphor, or a few drops of hartshorn may be given in a little cold water.

If the pains called after-pains, should continue severe, after the after-birth has come away, twenty or thirty drops of Laudanum or one or two teaspoonsful of Paregoric may be given, or a cup of Hop tea, or Poppy tea, or some other anodyne, which will generally moderate the pain, and afford rest. No attempt should be made to move or change the clothes until the flooding is done, particularly if the woman is faint or nervous; and in changing her clothing and putting her to bed, whether there is flooding or not, be cautious that she does not exert herself so as to bring it on. She ought in no case to be permitted to walk or stand upon her feet without being supported. After her delivery, she ought to be kept perfectly quiet for two or three days, see but little company, and use plain unstimulating diet, and such kinds of food as are the easiest of digestion. If her bowels are costive on the third day after delivery, give a dose of Castor Oil or some gentle purgative; if the bowels are sufficiently open and the woman be comfortable, this will be unnecessary.

Should the bowels be sore or painful, make a liniment of an ounce of Gum Camphor dissolved in half a pint of Sweet Oil; rub the bowels with it, and apply, after rubbing, warm flannels. If the breasts are swelled, hard, and painful, this liniment will likewise be very beneficial in such cases. On the third or fourth day, and sometimes earlier, the milk generally comes; and if delayed to the fifth day, it is usually more or less accompanied with some little fever, headache, hot skin, and thirst; and the countenance has a red appearance Should the breasts become hard or swelled, they must be softened by frequent applications of lard or sweet oil, or some liniment, or by gently rubbing them with the hand with lard or oil, as hot as it can be borne. This usually affords speedy relief. After rubbing them well, apply flannel, dipped in the lard or oil, as hot as it can be borne. The milk should also be drawn occasionally with the mouth, or with a suitable instrument called a breast pump. When this proper course is taken, there is very seldom any trouble with the breasts. A great error is frequently made, and produces much suffering, by attempting to force out the milk when the breasts are hard, sore, and painful, before they are softened by rubbing them well with warm lard or oil, as I have before mentioned, or with hot vinegar or poultices. Be careful then, in the first place, not to allow them to get very full and hard, until they are softened; and never attempt to force out the milk with the breast pump or pipe, until they are soft. It is in vain to make the attempt to force it until they are so, as you will only give your patient a great deal of unnecessary suffering. The pump is a valuable instrument when judiciously used; but it is too powerful in the hands of those who do not understand how to use it properly.

The too common practice of confining a woman to her bed two or three weeks is unnecessary, as it debilitates her system, and frequently, instead of benefiting, does injury to the general health. The length of time, as to her confinement, depends much on habit, as some women can do that which, in similar circumstances, others would suffer much from. If the confinement, or child-birth, has not been very severe, after the third or fourth day, allow her to sit awhile on an easy chair, until her bed is aired and made up, particularly if it is warm weather; and by the fire in a comfortable room in winter. This will, from the change of position, afford her great comfort, and assist her much in regaining her strength. It may not, however, be proper to permit her to stand or walk much for a week or ten days after her confinement, particularly if she is weak, or has lost much blood, by which I mean flooding, during her labor. I have, under ordinary circumstances, permitted women, whose habits have been regulated by an active life, to return, with care, after the tenth day, to their usual mode of domestic affairs, attending particularly at this time to the feet being well protected from cold and dampness, and the clothing sufficiently warm, allowing such food as may assist in affording a sufficiency of wholesome milk for the infant.

In regard to the bandage immediately after the delivery, it will always be found beneficial, if there has been an unusual loss of blood, attended with weakness, relaxation, etc. A moderately tight and well adjusted bandage is proper, and should be worn for a few days. The bandage should always feel comfortable and easy, as a tight one interferes more or less with respiration or breathing, by preventing a free and easy descent of the diaphragm, and by limiting the action of the respiratory muscles, and interrupting, in some degree, the digestion, and the healthy motion of the intestines, by compressing the stomach and bowels. In every case, let the bandage be smooth and comfortable to the feelings of your patient.

The management of a new-born infant is a matter of great importance in the preservation of the life of the child; and a few brief observations on this subject may be necessary. As soon as the infant is separated from the mother, by cutting the cord or navel-string, it should be wrapped in warm, dry flannel, so as to prevent it from taking any cold; then great pains should be taken to wash it clean, gently and tenderly, with warm water and mild soap, and particularly about the eyes, under the arms, and in the groins. If it is thickly covered with the white unctuous matter, which is often very difficult to remove, you will, before washing, find it much easier to remove this by rubbing the infant over with fresh lard or sweet oil. The child should be as little exposed to the air as possible during this process, and wiped dry and wrapped up in a warm flannel cloth, as soon as possible, so as to prevent its taking cold.

In dressing the navel, draw the cord or navel-string through a folded cotton or linen rag, in which a hole is to be made of suitable size to admit the cord to pass through it; grease the hole well with lard or oil, so as to prevent the edges of the rag from excoriating it, or making it sore; then turn the end of the cord up, and place over it another fold of cotton or linen rag; and put a bandage moderately tight around the child, in such a manner that it will keep the rags on the navel-string from slipping or being drawn up. Keep the bandage smooth and in its proper place, not by pins, but by sewing it with needle and thread, and yet not so tight as to interfere with the breathing. It is frequently the case that the infant does not pass any water for several days after it is born; the cause is from a want of a secretion by the kidneys. In such cases, a little flax-seed tea or watermelon-seed tea may be administered; and a thin bag of fine salt made warm and applied to the lower part of its belly, will assist the secretion by the kidneys. This occurrence, however, is not of much importance, as nature generally remedies it in a few days.

OBSTETRICS AND THE WORK OF DOCTORING IN THE MID-NINETEENTH-CENTURY AMERICAN SOUTH*

Steven M. Stowe

"Obstet[rical] call at midnight, last night, at Gen. DuPont's plantation 2½ miles from town," begins an entry in the clinical notebook of a physician practicing in northern Florida in November, 1858. "Negress Lottie —; has had many children—taken suddenly in labor last night—; had been up all day with one of her own daughters who was confined—wasn't looking for her own confinement till Christmas." The physician, Charles Hentz, entered the slave woman's cabin and "found left leg of child hanging from vulva—; face of child to the front—; right leg lying to the right iliac fossa of mother." With great "difficulty," and with "considerable tardiness on the part of nature," Hentz performed a podalic version and delivered the child. "Fine, very large female child," he observes in his notebook. "Asphyxiated —placenta came with it. I inflated its lungs, but to no avail; it was born dead."[1]

When he entered Lottie's home, Hentz also was entering one of General DuPont's slave cabins. Lottie herself was DuPont's property, as Lottie's daughter would have been had she lived, and so as Hentz rode through the darkness toward the birth of a slave woman's child, he also hastened toward the increase of a master's wealth. The white physician's image of himself struggling to breathe life into a "fine" black infant powerfully suggests how physicians' work brought them close to the center of the many-layered social relations of the Old South. What follows is an attempt to illuminate the work of doctoring in that society in light of the social relations that surrounded it and the meaning physicians discovered in it. Throughout, the focus is on obstetrics and on the ways in which physicians used childbirth as an occasion to reflect upon their work. Because obstetrics was relatively unspecialized and was closely tied to key aspects of social life—family and kinship, sexuality, and the experience of pain and death—it provides a particularly valuable perspective on the general practice of ordinary physicians.

*Thanks to my colleagues Ann G. Carmichael, Ellen Dwyer, and M. Jeanne Peterson for their helpful criticism.

[1] Charles A. Hentz, Diary, 7 November 1858, Hentz Papers, Southern Historical Collection, University of North Carolina, Chapel Hill (hereafter SHC).

Surprisingly little has been written about routine medical practice in any region of the United States in the mid-nineteenth century. The voices and the points of view of southern physicians in this paper doubtless were similar in many respects to those of their counterparts in the North or the West. This paper does not seek to make a regional comparison, but rather to plumb one regional context for American medicine. Southern physicians had much to say and many stories to tell, not only about epidemics and anomalies but also about the work they did most of the time. Their day-books, letters, and diaries, together with the more structured but often informal clinical notes, medical theses, and journal articles, form a rich body of relatively untapped source material on the experience of doctoring in that time and place. In certain ways, the "southernness" of these physicians is obvious in their views on race, slavery, medical politics, and plantation life; but other distinctly southern characteristics of their medicine will not be revealed fully until we have close studies of doctoring elsewhere.

The three sections of this paper explore, first, the context of physicians' obstetrical training and the place of obstetrics in their overall practice. The focus turns next to the childbed and the particular social and medical realities revealed in doctors' attempts to describe the experience of birth. The final section expands this view by suggesting that when physicians struggled to communicate the meaning of obstetrical medicine—when they fashioned experience into stories—they discovered much about their work in general and brought into view the wider social relations that helped to shape it. Although the influence of southern physicians on their society was affected both by medicine's sharp limitations and by its emerging professionalism in this era, it was even more deeply rooted in the identity physicians drew from the rural community and its relationships.[2]

I

The training young physicians received suggests how they first came to grips with medicine as both a professional and a social realm. Early in the century, most students of medicine in the South, as elsewhere, obtained their training from mentors who took them on as apprentices. Here the student saw as much or as little of a particular kind of medicine as his mentor's practice or inclination allowed. By the 1830s, however, increasing numbers of young men entered medical schools throughout the South. (Outside the South, women began receiving medical degrees in the mid-1850s—a very few.) The schools themselves were growing in number,

[2] The history of American obstetrics has become a seismograph for the larger jolts and shifts in the historiography of medicine and society in the United States. The history of medicine in the South has tended to cluster around two major themes, the sectional nature of southern medicine and the relation between medicine and slavery, which at least touch upon the work of ordinary doctors. For a brief discussion of the historiography of American obstetrics and southern medicine, see the Appendix, below.

and several larger institutions soon became major centers of southern medical education. Obstetrics emerged as a specialty whose status was dignified with curricular time, faculty appointments, and, by the 1840s, efforts to train students in hospitals or other clinical settings. In short, despite the fact that medical education remained unstructured and piecemeal by modern standards, and despite the fact that many men who ended up practicing medicine still had little or no formal schooling, there existed after the 1830s the beginnings of systematic obstetrical training.[3]

Reading in obstetrics was a part of many apprenticeships and nearly all school curricula. A wide range of textbooks, by authors such as Fleetwood Churchill and Francis Ramsbotham (both British), as well as the Americans Charles Meigs, William Dewees, and Henry Miller, appear in the notes of students and teachers, and in their libraries. These texts and similarly organized faculty lectures clearly placed obstetrics in the domain of medicine, as opposed to the eighteenth-century view that collapsed childbirth into a mechanical, essentially surgical event. Nineteenth-century teachers taught the anatomy and physiology of women, the circumstances of conception and abortion, diagnostic skills, and hygiene. Even so, the focus of such education remained on labor and delivery and on the "natural" and "unnatural" turns they could take. By the 1840s, students with access to hospital clinics learned how to determine the dilation of the cervix and studied the presentation of the fetus, as well as the techniques of ballottement and version, largely by observing and practicing on poor white women. Students learned how to use crochet and forceps, and how to respond to bleeding, convulsions, and "hysteria"; and they learned that the woman's life and health always came first.[4]

[3]There is no single study of medical education in the antebellum South. By 1840, notable schools included Transylvania University Medical Department, the University of Maryland School of Medicine, the Medical College of Louisiana, the Medical College of the State of South Carolina, the University of Louisville School of Medicine, the Medical College of Georgia, and the Medical College of Virginia. For an overview of medical education in the United States as a whole, see Kenneth M. Ludmerer, *Learning to Heal: The Development of American Medical Education* (New York: Basic Books, 1985). State studies include Joseph I. Waring, *A History of Medicine in South Carolina,* 2 vols. (Columbia, South Carolina: R. L. Bryan, 1967); Kenneth M. Lynch, *Medical Schooling in South Carolina* (Columbia, South Carolina: R. L. Bryan, 1970); Phinizy Spalding, *The History of the Medical College of Georgia* (Athens: University of Georgia Press, 1987); John Duffy, ed., *The Rudolph Matas History of Medicine in Louisiana,* 2 vols. (Baton Rouge: Louisiana State University Press, 1958–62); John D. Wright, Jr., *Transylvania: Tutor to the West* (Lexington, Kentucky: Transylvania University Press, 1975); Dwayne Cox, "A History of the University of Louisville" (Ph.D. diss., University of Kentucky, 1984); Lewis J. Moorman, "Influence of Kentucky medical schools on medicine," *Bull. Hist. Med.,* 1950, *24:* 176–86.

[4]For students' comments on their clinical training, see, for example, John M. Richardson, Diary, 29 December 1853, 15 February 1854, SHC; J. S. Blair to Lafayette Strait, 24 May 1857, Gaston-Strait-Wylie-Baskin Papers, South Caroliniana Library, University of South Carolina, Columbia, South Carolina (hereafter SCL); M[armaduke] D. Kimbraugh to Nathan Hunt, 17 December 1859, Manuscripts Department, William R. Perkins Library, Duke University, Durham, North Carolina (hereafter Duke). Frequently mentioned textbooks, in various editions, include William P. Dewees, *A Compendious System of Midwifery, Chiefly Designed to Facilitate the Inquiries of Those Who May Be Pursuing This Branch of Study,* 9th ed. (Philadelphia: Lea & Blanchard, 1839); Henry Miller, *Theoretical and Practical Treatise on Human Parturition* (Louisville, Kentucky: John V. Cowling & George C. Davies, 1849); Charles D. Meigs, *Obstetrics: The Science and the Art* (Philadelphia: Lea & Blanchard, 1849); Fleetwood Churchill, *On the Diseases of Women, Including Those of Pregnancy and*

Some of the advice students recorded made obstetrics seem simple and matter-of-fact. In Jeptha McKinney's notes, Augustus Cenas tells his University of Louisiana class in 1857 to go to the patient as soon as she calls, "& take with you a cathiter [*sic*], tinct[ure] opi[um], tart[ar] emet[ic], lancet, wax pack, thread or silk, & if you have them, a pair of forceps." Yet the accurate diagnosis of a pregnancy was not so simply achieved, in part because it raised complex issues in the physician's relation to the patient and her family. Confirming a pregnancy was difficult for everyone. A family's deepest fears and wishes were involved. Henry Miller cautioned his students that a woman "is liable to imagine she feels [quickening] when she does not." On the other hand, he warned, "you may make a mistake yourself." Changes in a woman's behavior apparent only to her intimates might be the first signs of pregnancy. A physician must come to *know* his patient, and engage himself with her and her family, in order to learn not only the signs of pregnancy but also methods of eliciting and interpreting them in individuals. In the opinion of some teachers, much of obstetrical medicine was "most repugnant" to women, and "if a man possesses not delicacy of feeling and manners he had better not undertake the practice of midwifery." Yet delicacy of feeling did not imply indecisiveness. There was no place for "*do nothing* management," as Miller told his students at Louisville in 1841. The physician must *act:* that was his job. Francis Porcher urged his students to think of themselves as active instruments of that "mercy which has granted medicines" to humanity.[5]

Despite obvious undertones of whistling in the dark, such expressions of professional aspiration were common in medical school obstetrics as early as the 1830s. Although some physicians and their students took note of the "delicate" sexual implications of a male practitioner attending a woman, a man's sex was no longer an appropriate reason to exclude him as a woman's choice of attendant. Far more vexing was the question of whether a woman in labor needed any trained attendant at all. Even many physicians admitted that most births were "natural" and that a woman could do very well with no more assistance than the help of her friends. This issue fed into a larger discussion of the nature of women. Like other male enclaves, medical school was a place where young men speculated about mysterious (and

---

Childbed, new Amer. ed., rev. (Philadelphia: Blanchard & Lea, 1857); Francis H. Ramsbotham, *The Principles and Practice of Obstetric Medicine and Surgery in Reference to the Process of Parturition* (Philadelphia: Blanchard & Lea, 1856). For examples of students' reading and lectures, see Richardson, Diary, 1853–54, SHC; Francis P. Porcher, Commonplace Book, 1842–45, South Carolina Historical Society, Charleston, South Carolina (hereafter SCHS); William Henderson, Student Notebook, 1840–41, Historical Collections, Library of the College of Physicians and Surgeons, Philadelphia, Pennsylvania (hereafter CPS).

[5] Jeptha McKinney, Student Notebook, 22 January 1857, Louisiana and Lower Mississippi Valley Collections, Hill Memorial Library, Louisiana State University, Baton Rouge, Louisiana (hereafter LSU). Miller's observations on quickening and management are in Courtney Clark, Student Notebook, 1841, Duke. The opinion on "repugnant" obstetrics is in Samuel H. Stout, Student Notebook, 1842, CPS; Clark, Student Notebook, 1841, Duke; Francis P. Porcher, "Notes on Anesthetics," n.d., Porcher Papers, Waring Historical Library, Medical University of South Carolina, Charleston, South Carolina (hereafter Waring).

mostly absent) women. The practical problems of obstetrical care thus were inseparable from the larger antebellum context of the separate sexual spheres and men's fascination with women's otherness. Medical school provided a spacious arena for such fascination. M.D. theses on women, for example, ranged from close anatomical discussions in which social views of womanhood did not appear, to tributes to womanly purity in which the medical sciences played almost no part. Lecture and clinic alike lent themselves to generalizations on gender and class. After witnessing a poor white woman undergo a taxing surgical procedure with great courage, a medical student in Kentucky observed that a *lady,* with her finer sensibilities, would likely have died. The crude and hearty outlive the delicate, he concluded, no matter how much delicacy is to be preferred among females. Yet even among ladies, a woman's suffering and vulnerability in childbirth were matched by her impressive powers of reproduction—powers "more important" than a man's, according to the view one student expressed in his M.D. thesis. Indeed, unlike any single aspect of medicine pertaining to men as a sex, the changes wrought by pregnancy and childbirth revealed to at least some medical students "the wise & benevolent order of Providence."[6]

If speculation about woman's nature sometimes lifted obstetrical training out of the practical and into the teleological, medical educators were more grounded on a related point: whatever the nature of woman and childbirth, it increasingly was agreed that southern women and southern childbirth were best off in southern hands. Southern women, responding physiologically to the South's distinct climate and medical topography, were properly within the orbit of native caregivers who were themselves rooted in regional understandings of health. Menstruation, it was generally believed, occurred earlier in southern climates, and conception was more easily achieved. The mechanics of birth were enhanced by the relaxing southern air. Southern sickness almost invariably depended on conditions

[6] The observation on ladies is in Charles A. Hentz, Diary, 6 April 1848, Hentz Papers, SHC. For views on the nature of women, see L. S. Hill, "Woman" (M.D. thesis, Medical College of South Carolina, 1857), pp. 1–17, Waring Library; Larkin G. Jones, "The Peculiarities of the Female," (M.D. thesis, Medical College of South Carolina, 1829), Waring; Wilfred DuPont, "The Peculiarities of the Female" (M.D. thesis, Medical College of the State of South Carolina, 1858), Waring; cf. Thomas Dancer, *The Medical Assistant; or, Jamaica Practice of Physic*... (Kingston, Jamaica: Alexander Aikman, 1801), pp. 14–15; William P. Dewees, *Essays on Various Subjects Connected with Midwifery* (Philadelphia: H. C. Carey & I. Lea, 1823), pp. 24–25. For physicians' notions of the physiology and physical diagnosis of pregnancy, see James K. Chapman, Student Notebook, 1869, SCL; William Henderson, Student Notebook, 24 and 26 November 1840, CPS; Courtney Clark, Student Notebook, 1841, p. 81, Duke; Charles E. Kennon, Student Notebook, 11 and 19 December 1865, Rudolph Matas Medical Library, Tulane University, New Orleans, Louisiana (hereafter Matas); Francis P. Porcher, "Notes on Anesthetics," n.d., Porcher Papers, Waring. On the necessity of having a trained person at the bedside, see the differing opinions of Fleetwood Churchill, *Observations on the Diseases Incident to Pregnancy and Childbed* (Philadelphia: Adam Waldie, 1840), p. 15, and Thomas Denman, *An Introduction to the Practice of Midwifery* (New York: E. Bliss & E. White, 1821), pp. xvi–xvii. See also Frances E. Kobrin "The American midwife controversy: a crisis of professionalization," *Bull. Hist. Med.,* 1966, *40:* 350–63; Judy Barrett Litoff, *American Midwives: 1860 to the Present* (Westport, Connecticut: Greenwood Press, 1978); Jane B. Donegan, *Women and Men Midwives: Medicine, Morality, and Misogyny in Early America* (Westport, Connecticut: Greenwood Press, 1978).

of "relaxation and debility," while in the North it quite differently involved "an excess of strength."[7]

This concern for a medicine founded in the southern environment was significant, too, in assessments of medical care for slave women. By the 1830s, many educators argued that only southern physicians could be depended upon to understand the peculiar nature of slaves and slavery. As mothers, and as property, pregnant slave women were safest under the management of physicians with local training and experience. These men would appreciate how black women were closer to nature than white women and thus were likely to have quicker, easier births. In fact, some educators wrote admiringly of strong slave women who were able to give birth and return to work immediately, and they typically criticized "over-civilized" ladies. Physicians' sense of the social and physiological differences between the races carried over into therapeutics. For example, medical educators typically believed that black women more successfully resisted sickness and the complications of pregnancy but, once weakened, were unable to withstand heroic treatment as well as white women; and medical schools nourished certain world views regarding slaves and slavery, as well as white womanhood, which were often sharply etched in M.D. theses. Typical of one line of medical thought was H. W. Moore's 1856 thesis on slave hygiene, which mirrored an ideal of harmonious race relations within the class structure of slavery, emphasizing the master's duty to supply both good food and humane treatment. Common, too, however, were the views of medical students who, like M. D. McCloud, observed that slaves' complaints of sickness "demand at the hands of the Physician a more careful investigation than those of whites," not for reasons of pathology or physiology but because slaves' complaints almost always involved some degree of deception. A wise southern master would hire a southern-trained physician who could be expected to know the wiles of slaves, and it was a wise southern practitioner who began every diagnosis of a slave patient suspecting a lie.[8]

[7] Quotations are from Dancer, *Medical Assistant,* p. 17; see also pp. 263–64. Popular medical handbooks as well as regular textbooks emphasized the influence of southern climate and topography. See, e.g., John S. Wilson, *Woman's Home Book of Health* (Philadelphia: J. B. Lippincott & Co., 1860), pp. 153–55; Samuel K. Jennings, *A Compendium of Medical Science, or Fifty Years' Experience in the Art of Healing* (Tuscaloosa, Alabama: M. J. Slade, 1847), pp. 134–35; Joseph Brevitt, *The Female Medical Repository* (Baltimore, Maryland: Hunter & Robinson, 1810), p. 39. For practicing physicians' notes on typical diseases and their remedies, see C. V. Barnes, Medical Reference Book, c. 1860, SCL; J. C. Eliason, "Notes in the Practice of Physic," manuscript c. 1830, Waring; [Anonymous Physician], Memorandum Book, c. 1860, SHC; Isabella Porcher, Plantation Prescription Book, manuscript c. 1834, Waring. See also John Harley Warner, "The Idea of Southern Medical Distinctiveness: Medical Knowledge and Practice in the Old South," in *Sickness and Health in America: Readings in the History of Medicine and Public Health,* ed. Ronald Numbers and Judith Leavitt, 2d ed., rev. (Madison: University of Wisconsin Press, 1985); Elizabeth Barnaby Keeney, "Unless Powerful Sick: Domestic Medicine in the Old South," in *Science and Medicine in the Old South,* ed. Ronald L. Numbers and Todd L. Savitt (Baton Rouge: Louisiana State University Press, 1989), pp. 276–94.

[8] See H. W. Moore, "On Plantation Hygiene" (M.D. thesis, Medical College of the State of South Carolina, 1856); the quotation is from M. D. McCloud, "Hints on the Medical Treatment of Negroes" (M.D. thesis, Medical College of the State of South Carolina, 1850), p. 2. See also William P. Hort, "Report of the Committee

It is impossible to say, of course, how the medical school environment specifically shaped all of the young men who learned basic obstetrics there. Differences in quality and seriousness among schools and among students made for various combinations of knowledge, aptitude, and hard work. A student might coast through his requirements, recopying his lecture notes as a thesis, and still receive the M.D. degree at commencement; and even in the best schools, an individual teacher's limited experience often sufficed for the totality of obstetrical knowledge. As student William Cozart shrewdly noted: "The faculties of every institution teaches [*sic*] what they have observed of diseases ... during their own practice. In fact, their practice is their teaching." However, the wide range of what counted as obstetrical knowledge, and its sometimes idiosyncratic organization, is precisely my point. Obstetrical learning ranged from exact technical procedures, many of which were apt, to overblown speculations regarding women and slaves; and it occupied the entire medical landscape from surgery to rhetoric. An attentive student might recall his training in version or in the medications of parturition when he entered the birthing room for the first time; he also might feel mingled fear and respect for womanhood and some mixture of pity and suspicion toward slaves. In all of this, the elevated values of elite physicians, while scarcely mirror images of ordinary medical practice in the South, did carry some weight as ideals. The exemplary young physician realized that obstetrics, and much of medicine in general, was not merely technique but an intense drama of personal and familial meaning. Hugh Hodge, from whom many southern medical students received their first obstetrical training, taught a lesson that doubtless influenced the later practice of his students when he observed that the physician at a childbirth "enters into the privacy of families, and penetrates even to the most retired chamber." The wise physician knows that "the question of a woman's pregnancy is of immense importance" because the "happiness of the female and that of her friends may depend upon the mere expression of your opinion."[9]

---

on Midwifery and the Diseases of Women and Children," *New Orleans Med. Surg. J.*, 1852–53, *9*: 570–71; Hugh L. Hodge, *An Introductory Lecture to the Course on Obstetrics and Diseases of Women and Children* ..., (Philadelphia: William S. Young, 1846), pp. 9–11; Wilson, *Woman's Home Book*, pp. 311–13. For southern physicians' correspondence reflecting their sense of regional identity—often in a negative way—see E. D. Fenner to John Y. Bassett, 24 October 1849, Bassett Papers, SHC; "G" to James M. Gage, March [1837?], Gage Papers, SHC; Isaiah D. Clawson to Charles Clawson, 4 May 1845, Clawson Papers, SCL; L. A. Dugas to J. E. Hawkins, 19 December 1859, Hawkins Papers, LSU. See also John Harley Warner, *The Therapeutic Perspective: Medical Practice, Knowledge, and Identity in America, 1820–1885* (Cambridge, Massachusetts: Harvard University Press, 1986), pp. 76–80; Todd Savitt, "The use of blacks for medical experimentation and demonstration in the Old South," *J. Southern Hist.*, 1982, *48*: 331–48.

[9] William W. Cozart, "The Place Where Southern Students Should Acquire Their Medical Knowledge" (M.D. thesis, Medical College of the State of South Carolina, 1856), p. 4; Hugh L. Hodge, *A Lecture Introductory to the Course on Obstetrics and the Diseases of Women and Children* (Philadelphia: J. G. Auner, 1835), pp. 16–17. On the moral and professional importance of obstetrics, see also Denman, *Practice of Midwifery*, p. 557; S. N. Harris, "Observations on some of the opinions generally received among medical men," *Charles-*

Once in practice, a young physician continued his education in the social and professional realities of childbirth. In their daybooks, physicians denoted their childbirth calls with a variety of terms: "obstetrical visit," "parturition," "accouchement," "delivering wife," and, in one practitioner's standard phrase, "officiating in case of midwifery." Most practitioners kept track of miles traveled, time devoted to the case ("all day, all night"), and weather conditions, largely because they charged higher fees for night travel and bad roads.[10]

Account books also reveal the range of patients typical of country practice. A rural physician might have a "town practice," usually no more than a cluster of people living near a crossroads or mill, in addition to patients scattered around the countryside within a five- or six-hour horseback ride. Plantation practices, often secured by contract with a large slave owner, were particularly prized and meant that in some instances, as in the practice of Arthur B. Flagg of Georgetown County, South Carolina, more than two-thirds of a physician's patients would be slaves. The institution of slavery appears throughout physicians' daybooks. Slaves usually were identified as part of their owner's household, with slaves' names sometimes placed in quotation marks, setting them off as not quite fully persons, as in the typical notation " 'Scilla' (of Josiah Collins)"; or slaves such as Retta and Sally were grouped with their owner's account, as "Mrs. Wilson/Retta, Mrs. Wilson/Sally." Not infrequently, a slave patient was merely noted as "negro woman," with no name to distinguish her as a person with a particular medical history. Practitioners regularly set down certain details of the medical intervention as well. The presentation of the fetus, the use of instruments or medications, and the occurrence of "unnatural" events (protracted labor, difficult version, or convulsions) are among the details most commonly

---

ton Med. J. Rev., 1852, *7:* 751; Samuel H. Stout, Student Notebook, 1842, pp. 42–43, CPS. For antebellum medical school humor with a sharp edge, see Henry Clay Lewis, *Odd Leaves from the Life of a Louisiana Swamp Doctor* (1858; Upper Saddle River, New Jersey: Literature House/Gregg Press, 1969); cf. William Barlow and David O. Powell, "A dedicated medical student: Solomon Mordecai, 1819–1822," *J. Early Republic*, 1987, *7:* 377–97.

[10] For the physician who "officiated" at childbirths, see J. H. Davis, Account Book, 1827–43, SCL. Other account books that give a representative picture of how physicians kept track of their practice include those of Benjamin Robinson and Benjamin West Robinson, 1805–63, SHC; Francis Jones Smith, 1838–42, SHC; James Strudwick Smith, 1823–32, SHC; William L. Jenkins, 1840–52, Waring; Thomas Lee Settle, 1855–1914, Duke. Suggestive works on the practice of ordinary physicians are Thomas B. Jones, "Calvin Jones, M.D.: a case study in the practice of early American medicine," *North Carolina Hist. Rev.*, 1972, *49:* 56–71; Irvine Loudon, "The concept of the family doctor," *Bull. Hist. Med.*, 1984, *58:* 347–62; Ian Inkster, "Marginal Men: Aspects of the Social Role of the Medical Community in Sheffield, 1790–1850," in *Health Care and Popular Medicine in Nineteenth-Century England: Essays in the Social History of Medicine*, ed. John Woodward and David Richards (New York: Holmes & Meier, 1977); Judith S. Lewis, *In the Family Way: Childbearing in the British Aristocracy, 1760–1860* (New Brunswick, New Jersey: Rutgers University Press, 1985), esp. chap. 4; Sylvia D. Hoffert, *Private Matters: American Attitudes toward Childbearing and Infant Nurture in the Urban North, 1800–1860* (Urbana: University of Illinois Press, 1989), esp. chap. 2; Judith Walzer Leavitt, *Brought to Bed: Childbearing in America, 1750–1950* (New York: Oxford University Press, 1986), esp. chap. 4. See also Barrington Moore, Jr., "Historical notes on the doctors' work ethic," *J. Soc. Hist.*, 1984, *17:* 547–71.

logged. Frequently, too, the physician recorded the sex of the child and its apparent health.[11]

This outline of obstetrical work, however, must be seen in the context of a physician's general practice. Most practitioners appear to have done relatively little obstetrical work. Account books suggest that country doctors typically attended only five to eight obstetrical cases a year. Other types of surgical work, such as reducing bone fractures or suturing wounds, were far more common, occurring scores of times within a given year. Pulling teeth was almost a daily surgical task; and compared to this and to the constant routine of prescribing drugs for various ills, childbirth was an uncommon event in a physician's work. Nor was obstetrical practice particularly lucrative. Throughout the years covered here, physicians' fees remained fairly constant: typically, one to three dollars for a routine medical call and prescription. The fees for childbirth cases—typically eight to fifteen dollars—reflect the time devoted to the case, not a higher rate. Thus, in the context of a physician's general practice, obstetrical work typically yielded neither a substantial portion of the practitioner's total income nor sufficient clinical experience to sharpen his technique and develop his professional self-confidence.[12]

II

At the heart of obstetrical work was the bedside itself. Here the physician's technical abilities, his perceptions of women and of race, and his attitudes toward the scope of his general practice and toward the promise of his profession ran up against the realities of daily work: doctoring at its most basic level. No medical school lecture or clinic, no matter how careful or cautionary, came close to the drama of the bedside encounter between patient and practitioner, the call for help and the awkward, hopeful joining of needs and effort. Thus a view of obstetrics as a matter of professional training, ideals, and income, while importantly establishing the place of

[11] For the names of slaves and owners, see [Anonymous Physician], Account Book, 1849, ms. no. 598, SHC; James A. S. Milligan, Account Book, 1857, Milligan Papers, SHC. For Arthur B. Flagg's practice, see his account book, 1850–55, SCHS.

[12] In addition to the account books cited in n. 10, see the obstetrical work noted in Alexander R. Hendry, Account Books, 1839–42 and 1849, LSU; J. W. Brigham, Account Books, 1842–75, SHC; Thomas Edward Cox, Account Books, 1815–1855, SHC; C. V. Barnes, Account Books, 1858–59 and 1867, SCL; George Dunlap, Account Book, 1829–39, SCL. The small number of obstetrical cases typical of most physicians' practices should not obscure the fact that for a few practitioners, the childbed became a big business. See, for example, the daybooks of William Jenkins, 1840–52, Waring, which show him attending only two births between 1840 and 1847 but called to forty-seven births between 1848 and 1852. See also Charles G. Roland, "Diary of a Canadian country physician: Jonathan Woolverton (1811–83)," *Med. Hist.*, 1971, *15:* 168–80; Jacalyn Duffin, "A rural practice in nineteenth-century Ontario: the continuing medical education of James Miles Langstaff," *Canadian Bull. Med. Hist.*, 1988, *5:* 3–28; Joan Lane, "A provincial surgeon and his obstetric practice: Thomas W. Jones of Henley-in-Arden, 1764–1846," *Med. Hist.*, 1987, *31:* 333–48; Hoffert, *Private Matters*, pp. 68–75. On fees, see George Rosen, *Fees and Fee Bills: Some Economic Aspects of Medical Practice in Nineteenth-Century America* (Baltimore, Maryland: Johns Hopkins Press, 1946); Irvine Loudon, *Medical Care and the General Practitioner, 1750–1850* (Oxford: Clarendon Press, 1986), pp. 94–97; Hoffert, *Private Matters*, p. 62.

obstetrics in a southern physician's practice, now requires something more: a closely textured view of the physician at work. This latter view may be somewhat reconstructed by piecing together the fragmentary writings of physicians as they went about their work. At the bedside, the ironies of doctoring—the fortuitous successes, the sudden failures, and the great, dreary expanse of suffering—revealed what it was to make good one's professing medicine. When a practitioner made his case notes, logged a daybook entry, wrote to a colleague, or reflected in his diary, he in effect composed a running account of his practice, closely tied to the rhythm of bedside experience. Over time the accumulated writings of many physicians form an unselfconscious narrative of doctoring: complicated, immediate, terse, and therefore historically enormously valuable.[13]

Several distinct circumstances or moments in obstetrical work were recurrent themes in physicians' writings. The first of these was the summons to the bedside, the timing of which is crucial to an understanding of the nature of obstetrics in a practitioner's experience. Routine prenatal examinations were rare in nineteenth-century practice. Most childbirth cases first appear in physicians' daybooks as patients' requests for help, usually medication. After describing his wife's cough and pain, for example, H. R. Kingly wrote to the physician John Lyle, "Please send her such prescription & medicine as you think her condition requires." Hunter McCullock wrote his physician for "a vial of those kind of drops that promote labour," and implied that the family would take it from there. "Dier Sir," a woman's husband wrote to the physician James S. Smith, "my wife had hur baby last night the child was dead.... please to send hur medicine & directions how to use it & what diet she must have & hur drink likewise. She is as well as cood be expected." Physicians often simply sent the medicine requested. This mode of practice prevailed even when the patient was a slave woman. As one physician observed, many slave owners acceded to their slaves' wish to "do their own doctoring as long as they can. When they can't go further they call for help."[14]

[13] The emphasis here on informal discourse does not imply that published works are not revealing of physicians' feelings and attitudes about their work. In fact, papers and clinical notes published in antebellum medical journals frequently display the dynamics of doctoring as work; but the argument here is that informal writing has been given far less attention by historians and others interested in mid-nineteenth-century medical practice. In thinking about the stories physicians told, I have been aided by work on modern medicine. See, for example. Howard Brody, *Stories of Sickness* (New Haven, Connecticut: Yale University Press, 1987); Arthur Kleinman, *The Illness Narratives: Suffering, Healing, and the Human Condition* (New York: Basic Books, 1988); Jay Katz, *The Silent World of Doctor and Patient* (New York: Free Press; London: Collier Macmillan, 1984); John Pekkanen, ed., *Doctors Talk about Themselves* (New York: Delacorte Press, 1988); Larry Churchill and Sandra W. Churchill, "Storytelling in medical arenas: the art of self-determination," *Lit. Med.*, 1982, *1*: 73–79.

[14] H. R. Kingly to John Lyle, 19 April 1856, Lyle-Siler Papers, SHC; Hunter McCullock to James S. Smith, 9 December 1831 [letter attached to p. 117 of Smith, Account Book, 1828–32], and William N. Jordan to James S. Smith, 3 November 1831 [attached to p. 183 of Smith, Account Book, 1823–28]—both SHC; R. A. Edmonston to John Lyle, 2 November 1851, Lyle-Siler Papers, SHC. For similar notes on patients, see Edward Campbell to Josiah Hawkins, 11 August 1879, Josiah Hawkins Papers, LSU; M[argaret] A[nn] Ott to George Colmer, 26 May 1866, Colmer Papers, Manuscripts Department, Howard-Tilton Memorial Library, Tulane University, New

Thus the summons to the bedside did not depend on or follow from the physician's knowledge and advice. Instead, the summons was based on the decision of the patient and her attendants. It usually occurred only after they had exhausted their other resources. Accepting the diagnosis made by the woman and her attendants, and responding to their requests from afar, the physician found the call to the bedside synonymous with emergency. While emergency situations were common in nineteenth-century medicine, the highly charged expectations surrounding childbirth amplified the urgency. Letters from women's families have the heaviness of near panic that gave the obstetrical call its peculiar weight. "Dr. Sir—I send for you in a case of emergency. I want you to come as soon as you can get here. My wife cannot live 24 hours without relief.... I do not know anything I can do. I now call upon your skill." Physicians' personal accounts show them literally racing against the progress of a woman's labor. Although it is debatable whether a physician could have given a woman better care had he known her habits and physical condition ahead of time, his haste in the last, urgent stages of a pregnancy certainly did not increase his chances of applying his knowledge and skills most effectively. Experienced practitioners advised apprentices to seek consultations with other physicians but added almost in the same breath that most obstetrical calls would be too closely timed to permit this. Similarly, young physicians were advised to be calm when entering the birthing room, to "avoid all [excessive] decorum, be mild ... speak kindly." In reality, however, practitioners found themselves "embarrass[ed]" when, arriving late, they had to balance mildness and kindness with a pressing need to assess the labor and its complications. In more than a few instances, they simply arrived too late. After a frenzied horseback ride of seven miles, Charles Hentz noted in his record book, "When I arrived I found a neighbor holding the baby in her hands, the afterbirth lying on the floor. Patient back in bed. It all dropped out as she got up to [go to] the chamber." These accounts became, at best, wry little stories. Samuel Leland wrote in his diary in 1852, "A message was awaiting me to repair *instanter* to Mrs. Wm. Myers' plantation to the relief of a woman in labour. Six miles of rough, dark road; but I did not hesitate, and arrived there in time to congratulate the mother (Mrs. Gallegan) on the birth of a daughter."[15]

Orleans, Louisiana (hereafter Howard-Tilton). On the patient's perspective in the history of medical practice, see Roy Porter, "The patient's view: doing medical history from below," *Theory Soc.*, 1985, *14:* 175–98; Elliott J. Gorn, "Black Magic: Folk Beliefs of the Slave Community," in Numbers and Savitt, eds., *Science and Medicine in the Old South*, pp. 295–326; Martia G. Goodson, "Medical-botanical contributions of African slave women to American medicine," *Western J. Black Studies*, 1987, *11:* 198–203; Samuel J. Touchstone, *Herbal and Folk Medicine of Louisiana and Adjacent States: Arkansas, Mississippi, Oklahoma, Texas* (Princeton, Louisiana: Folk-Life Books, 1983); Leavitt, *Brought to Bed*, chap. 3; Hoffert, *Private Matters*, chap. 1.

[15] A. B. Johnston to Beverley Jones, 17 August 1854, Jones Papers, SHC; William Henderson, Student Notebook, 4 January 1841, CPS; Joseph Milligan to Joseph A. S. Milligan, 9 July 1846, Milligan Papers, SHC; Charles A. Hentz, Obstetrical Records, 2 February 1859, Hentz Papers, SHC; Samuel Leland, Diary, 5 September 1852, SCL. See also Samuel I. Fowler to Calvin Taylor, 21 July 1847, Taylor Papers, LSU; Benjamin F. Fessenden, "A case of puerperal apoplectic convulsions, with spontaneous expulsion of the foetus after death," *Med. J. North Carolina*, 1858–59, *1–2:* 16.

At other times, the convulsions, bleeding, or other alarming condition that had prompted the patient and her family to summon the physician suddenly eased, leaving everyone more hopeful—and the child still unborn. Tedious periods of waiting are the obverse side of emergency, and physicians' writings also play upon this theme. Walking the floor or seated at the family table, the physician in the quiet backwash of an emergency had time to review his childbirth training and experience. What becomes clear in these instances is the mixed message purveyed by the practitioner's best authorities. Many physicians, especially those new to practice, regularly consulted their obstetrics texts or lecture notes only to find them disturbingly ambiguous about when a doctor should act in a labor and delivery, and even about the very nature of pregnancy itself. Some authorities, Fleetwood Churchill and Thomas Denman among them, counseled patience, observing that the powers of nature were almost always adequate to the labor of birth. Others, however, while not advocating the aggressive behavior imputed to midwives, strongly suggested that the doctor who practiced "expectant medicine," to use Augustus Gardner's sarcastic term, was remiss in his duty to act, to doctor. As Henry Miller wrote, "A meddlesome midwifery is bad, but a shilly-shallying midwifery is worse." This debate, which appears not only in textbooks but also in the case notes published in medical journals and in physicians' correspondence, was tied to the even larger question of whether or not a woman was ill when she was in labor. Unlike many eighteenth-century authorities who, like Benjamin Rush, concluded that pregnancy was a disease, and unlike most popular midwifery books that simply avoided the question altogether, physicians in the mid-nineteenth century typically maintained that pregnancy was "natural" and therefore not a disease. At the same time, however, it was a disturbingly "altered" state in which the woman, although fulfilling her natural destiny, was terribly vulnerable.[16]

Thumbing through their old lecture notes, consulting Dewees or Churchill, waiting at the now-quiet bedside, country physicians puzzled over the "natural" condition that was marked nonetheless by "the disordered state of the stomach and bowells, the stitches and pains in the back and sides, the vertigo and headach[e], the heavy, unwieldy, listless sensations." Attending one of his first childbirths, young Larkin Jones reviewed what he had learned about the naturalness of pregnancy but wondered, "What other organ than the womb can, when in good health, produce vomiting, heart-burn, salivation, sleeplessness, palpitation of the Heart, laughing,

[16] Augustus K. Gardner, *A History of the Art of Midwifery: A Lecture* (New York: Stringer and Townsend, 1852), p. 29; Miller, *Theoretical and Practical Treatise*, p. 161. For Benjamin Rush, see his "On the means of lessening the pains and danger of child bearing and of preventing its consequent diseases," *Med. Repository*, 1803, 6: 26–31; cf. Fielding Ould, *A Treatise of Midwifery; in Three Parts* (Dublin: Printed by and for Oli. Nelson, 1742). For authors generally wary of aggressive action, see Churchill, *Observations on the Diseases*, p. 15; Denman, *Practice of Midwifery*, pp. 306–7. For an author emphasizing intervention, in addition to Henry Miller, see Hodge, *A Lecture Introductory*, pp. 12–13. William Dewees played both sides of the debate; see his *Compendious System*, pp. x–xii, 184–85.

crying, and a variety of other [symptoms]?" Observing his own wife's pregnancy, a Georgia doctor remarked in 1859 that in her eight months of carrying the child his wife had experienced "only one week of *any* thing like health." Only at the bedside was the ambiguity of calling childbirth both healthy and altered, natural and extreme, made so plain.[17]

This ambiguity in physicians' accounts is important for a number of reasons. First, despite the orthodox view that pregnancy was a state of health, doctors at the bedside regularly dispensed far more medicine than orthodoxy would allow. Sometimes at the request of the patient, sometimes on their own initiative, doctors routinely purged or sedated once they arrived on the scene. In a larger sense, the definitional fuzziness of the "pregnant state" makes it clear that if the conscientious physician had any time to pause and take stock of the situation, he might well find himself confronting very large issues—issues involving the definition of disease and the appropriateness of his action—as well as smaller technical questions. He might find himself uncertain at a time when certainty was much desired by the patient and her family. Should he follow this authority or that one? Should he act now or later? The medicalization of childbirth in this period has been discussed in terms of the physician taking control of the childbed. While southern country doctors were part of this general trend, their writings about the troublesome definition of pregnancy and labor nevertheless suggest that this control was undercut by their own doubts as well as whatever misgivings their patients had. Finally, the physician's uneasiness or, at least, puzzlement, helped to shape the social relations at the bedside when the waiting was over and the time came to act.[18]

Doctors' accounts of taking action often feature the patient herself as the center of attention, a theme that scarcely seems to need mention were it not for the fact that formal obstetrics texts rarely depicted the patient at all. Even when interesting or unusual births led physicians to make extensive notes of clinical details as such, many physicians included at least some picture of the patient: a physical description, some praise of her courage, a sense of her personality. Moreover, while textbooks continued to advise students not

[17] John Esten Cooke, "Essay on the use of cathartic medicines in the diseases of women," *Transylvania J. Med. Associate Sci.*, 1829, *2*: 399; Jones, "Peculiarities of the Female," p. 19; R. A. Edmonston to John Lyle, 10 February 1859, Lyle-Siler Papers, SHC. See also Daniel Drake, Medical Diary (typed copy), 1805, pp. 10–11, Matas, for a young doctor working out a definition of "natural" labor.

[18] On the ambiguity of defining the "pregnant state," see Hoffert, *Private Matters*, pp. 20–21; Leavitt, *Brought to Bed*, pp. 142–47. Many physicians seem to have felt a kind of relief when confronted with the risk of an unquestionably "diseased" state, such as puerperal convulsions. Convulsions, more than any other obstetrical topic, generated articles in medical periodicals during the mid-nineteenth century. See, e.g., Milton Antony, "Puerperal convulsions," *Southern Med. Surg. J.*, 1836–37, *1*: 24–26; James Conquest Cross, "Cases and observations on convulsive affections," *Transylvania J. Med. Associate Sci.*, 1828, *1*: 153–74; Horatio Hollifield, "Convulsions in the early stages of pregnancy," *Georgia Med. Surg. Encyclopedia*, 1860, *1*: 107–11. See also John Bernard Vandergriff, "Southern Practice of Medicine, Surgery, Obstetrics, Therapeutics, Toxicology, and Useful Notes of Various Methods of Treatment ..." (New Orleans, n.d.), manuscript volume, p. 85, Matas; William Holcombe, Diary, 10 March 1855, SHC; Lunsford Yandell, Diary, 1830, Manuscript Department, The Filson Club, Louisville, Kentucky.

to speak to the patient while conducting a physical examination, for fear of mutual embarrassment, informal bedside accounts make it plain that such modesty did not survive the realities of labor and delivery. Practitioners typically accepted without question that the patient should determine who talked to whom and what position was most comfortable. The physician should help prevent patients from "losing their self-possession" and stop others from acting inappropriately, especially in ways that would unduly hasten the birth. Otherwise, he should do as the patient wished.[19]

This is not to say, however, that the picture of the birthing woman that emerges from physicians' writings is simple or simply approving. The drama of birth was such that the woman frequently was seen as someone who needed and deserved the physician's combined kindness and skill—his "ministering," as it frequently was termed—yet who might also become his adversary because she hoped for too much from him. Arriving late in the midst of a crisis, the experienced physician did not forget that although he had been called for his particular abilities, whatever skills he possessed might be overvalued in the heat of the moment. He had to calm the patient or inspire her in her efforts without letting himself be pushed into making "any positive declaration" as to the outcome. As the last person on the scene and the one reserved for crisis, the physician had to hold himself somewhat aloof from the family's hopes. "The business of the practitioner is at the bedside," one young apprentice affirmed to himself in his journal. "He should not leave it. He must encourage his patient, keep up her spirits, give as favorable a prognosis as is consistent with truth, but make no promises." Keeping this distance from his patient was made all the more difficult by the intensity of labor and delivery. Physicians' writings often take note of women utterly changed by labor. The change would doubtless impress young men who believed that a white, well-to-do woman was especially modest in her self-expression and guarded in her relations with males. Young physicians were impressed with the power of a woman's transformation. "You will always know the pains [i.e., contractions] I speak of, by [her] irresistable desire to catch hold of everything in reach," one teacher wrote. The woman's "face and neck becomes [*sic*] red or livid," medical student William Henderson recorded in 1840, "the jugular veins acquire an enormous size"; at this point he had learned to "be careful and keep out of her way."[20]

Most women were not alone at this time, of course, and physicians' writings depict the social scene at the bedside as well as the encounter with the patient. Some physicians complained about the "officiousness of nurse

[19] Denman, *Introduction to the Practice of Midwifery*, pp. 274–75. See also Judith Walzer Leavitt, "Under the shadow of maternity: American women's responses to death and debility fears in nineteenth-century childbirth," *Feminist Studies*, 1986, *12*: 129–54.

[20] Dewees, *Compendious System*, p. 187; Samuel H. Stout, Student Notebook, 1842, p. 77, CPS; A. G. Goodlett, *The Family Physician; or, Every Man's Companion*... (Nashville, Tennessee: Printed at Smith and Nesbit's Steam Press, 1838), p. 519; William Henderson, Student Notebook, 18 December 1840, CPS.

and friends," and one popular author advised patients to "avoid all foolish *babbling* which only worry [*sic*] attendants." However, many practitioners seem to have accepted the crowded bedside, even welcoming it. A South Carolina physician recorded the birth of a "fine little girl" to one of his slave women in the happy company of her husband in the physician's own house. George Colmer, a Louisiana physician, frequently recorded the bedside group in his daybook ("Present Mrs. Robinson the midwife, Wash King & wife, J. Thomas King & wife, old Mr. George Efferson") and accepted the help of experienced attendants, black and white. Physicians stayed for many hours in some cases, rubbing the patient's abdomen with oil or butter or, sometimes, alcohol; giving her water or ice if possible; alternately watching the labor, talking with the patient and her friends, and eating or otherwise joining in the rhythms of family life in the household. At the end of a long labor, Charles Hentz noted with satisfaction a happy outcome as the patient, to her great relief, "delivered sitting in her husband's lap." And Colmer was delighted after one lengthy birth to prescribe a celebratory bottle of "medicinal whisky" for the woman's husband.[21]

Because so many physicians arrived only after complications had developed, rivals and colleagues as well as families and friends populate their accounts of bedside scenes. Practitioners often accepted the help of female midwives they considered competent, but could also single out for blame, as did one physician, "an ignorant midwife" who had struggled unsuccessfully for thirty hours before calling him. Protracted labors and, especially, difficult versions prompted physicians themselves to summon assistance. Courtney Clark, called by another physician when the patient "would not suffer him to introduce his hand to make an examination" in a breech presentation, managed to talk the patient into an examination and delivered the child with help from his colleague. In another case, three physicians worked to achieve a version, and at the next contraction the child was thrown "far out on the bed—like a melon seed popped between the fingers." Calling in another physician was a difficult decision to make, for not only was it an admission of a practitioner's own limits, but it also introduced someone else into the social scene at the bedside who might prove more of a distraction than a help. Yet successful consultations took place under the most difficult of emergency conditions, and happy outcomes left families satisfied and physicians with new collegial bonds.[22]

[21] Edward H. Dixon, quoted in Leavitt, "Under the shadow," p. 146; Isaac Wright, *Wright's Family Medicine, or System of Domestic Practice*... (Madisonville, Tennessee: Printed at the office of Henderson, Johnson & Co., 1833), p. 209; Samuel Leland, Diary, 30 January 1851, SCL; George Colmer, Journal, 1871, p. 55, Howard-Tilton; Charles A. Hentz, Obstetrical Record, 8 October 1853 (see also entries for 16 January 1853, 2 December 1858, 11 November 1858), Hentz Papers, SHC; George Colmer, Journal, 27–28 November 1873, p. 55 (see also entry for 10 February [1870], p. 61), Howard-Tilton.

[22] Charles A. Hentz, Obstetrical Record, 3 August 1858, Hentz Papers, SHC; Courtney Clark, Clinical Notebook, n.d. [1844–77], p. 14, Duke; Hentz, Obstetrical Record, 20 October 1858, Hentz Papers, SHC. For other examples of physicians' consultations, see F.J. Smith, Daybook, 18 August 1863, Mary Ruffin Smith Papers, SHC; William Holcombe, Diary, 9 February 1855, SHC; Frederick Bates to William Stump Forwood, 19 January 1854, Forwood Papers, SHC.

At other times, conflict twisted birth into disaster. Although in most instances physicians reacted with resourcefulness or at least calm, fear and defensiveness sometimes crowded in. In a difficult labor, warned one medical adviser, a physician had to watch out for the erosion of a family's confidence from within the bedside circle. "Some one of the woman's friends . . . whispers first to one, and then to another—takes the mother, or sister, or husband aside, makes oblique insinuations. . . . The alarm begins to spread, and a messenger is dispatched post haste, to call in the assistance of some distant dame, or doctor." An Alabama physician, A. Armstrong, disputed with a colleague in 1839 whether or not to remove a patient's placenta immediately after the birth; the colleague was set against it and Armstrong was equally adamant for removal. "At the request of the patient," Armstrong wrote, "I removed the placenta" only after the two doctors angrily argued in front of the patient that her life depended upon the decision. Their conflict spread, growing to include the woman's brother-in-law, a folk practitioner (a "Steam Dr. as he called himself"), until in the intensity of argument over medications and other treatment Armstrong denounced the others, leaving the house shouting, "I shake my hands free of this Lady's blood." Though the woman survived the scene, she died a few days later, leading Armstrong to write his account in anger and guilt.[23]

III

Sometimes after disaster, sometimes in triumph, often in the terse language of routine work, southern physicians thus recorded the bedside circumstances of childbirth as clustering around certain key themes: the summons to the bedside, the decision whether or not to act *now*, the encounter with the woman, and the surrounding web of social relations that might impede or advance this most intense drama. These themes were the elements of the doctor's bedside story, which significantly took place in surges of activity more akin to modern emergency medicine than to modern obstetrics. The shifts between hurrying and waiting, patience and urgency, between taking dinner with the family while the patient rested and then, suddenly, finding a footling presentation and blood all over, are the crucial dynamic of these accounts. In this way, physicians' writings reveal not simply how marginal the physicians were to childbirth—everyone but the woman in labor was marginal in some way—but how specific was their work. They were called at a particular time in the drama, they joined an already established social scene with its own pattern of emotions and relationships, and they were expected to do something terribly important but also quite exact: stop the convulsions, give an opiate, get the baby out. In no

[23] Thomas Hersey, *The Midwife's Practical Directory* . . . , 2d ed. (Baltimore: "published by the author," 1836), p. 65; A. Armstrong, Notebook, 1839, pp. 24–29, LSU. See also William Blanding, Diary, 14–21 November and 6 December 1810, Blanding Papers, SCL.

way do country doctors emerge from their own writings as the dominant men whom textbooks depicted as possessing all the answers—or even as dictating the questions—that mattered in childbirth.[24]

What does emerge, however, is a clear sense that many physicians took the themes of childbirth and fashioned them into brief narratives of rural life and professional ambition, personal hope and a community's values, that reveal something important about the meaning physicians found in their work. When they permitted themselves a fuller narrative, whole stories appear that in some way sustained them and also placed them in their culture. Even the most routine clinical notes concealed such stories, some of which were told almost by accident through some need or whim on the physician's part: because certain calls had periods of time when nothing was stirring but the physician's mind; because individual cases for one reason or another were pulled from notebooks and embellished in letters to families or friends; because there was something in a particular case that illuminated ordinary routine, flooding it with the light that narrative sheds. Whatever the reasons for their creation, these stories reflected the physician's attempt to say something more about the meaning of his obstetrical work. Charles Hentz's notebook contains an entry briefer than most but typical of doctors' record keeping: "[case] 55. Negress Mary, belonging to F. R. Ely, flooding, ergot inefficient—Recovery. Nov. 5 1857." Yet for some reason when Hentz sat down with his diary later the same day, Mary's case unfolded into a story rich with the images and sense of doctoring:

> At about midnight last night, after I had undressed, & was sitting in my gown about retiring, I rec'd a call from Frank Ely's plantation. Dr. Wilson's practice, but he being threatened with a chill, sent the call to me—an obstetrical call—negro woman—placenta praevia—flooding; I went out; found my patient in above condition on floor of her cabin, dangerous symptoms abating. I sat by the fire (after ministering to her) till 5 o'clock, reading a copy of Harper's Magazine rec'd yesterday, by the pine knot blaze. I laid down on a negro bed at 5; slept till 7. Took breakfast with Mr. & Mrs. Woodward—overseer and wife; continued near my patient; took an old rocking chair by a tree near the cabin door cocked my feet against tree, smoked cigar & read Harper's Weekly—(1st of Dicken's tale 'Lazy Tour of two ideal apprentices' in it)—& hymns from my hymn book & looked up into the bright blue sky & white drifting clouds; had dreams & reveries about the great drama of life that is going on over this great world, the great destinies of life. I laid down again & slept from 10 till 12. Continued

[24] Two other childbirth scenes found in many practitioners' daybooks deserve more study: the prolonged postpartum care necessitated by infection or other complications, and the use of instruments, especially forceps. For postpartum accounts of physicians' efforts to stave off women's illnesses, see, e.g., Joseph Milligan, Medical Daybook, 8–29 April 1857, Milligan Papers, SHC; John Calhoun McDowell, Account Book, 23–26 April [1848], SHC; Beverley Jones, Diary, 12–24 January 1862, Jones Papers, SHC; Daniel McPhaul, Account Book, 27 February–7 April 1826, Harnett County Papers, SHC. On physicians' attitudes toward instruments, see Benjamin Robinson and Benjamin West Robinson, Account Book, 10 November 1829, SHC; Daniel Drake, Diary (typed copy), 1805, pp. 23–24, Matas; Alfred Boyd, "Finale of Twenty Years of Study and Practice of Medicine" (M.D. thesis, Medical College of the State of South Carolina, 1856), p. 7, Waring.

> ministering to my patient. Dined with Mrs. Woodward; Mr. W being away; found that they are good Methodists. Finding that things were progressing slowly, & that I must return to town, I sent for Dr. Wilson, to see if he could take my place. He was just starting to the Sullivan settlement taking Toney with him. He came out directly, bringing Toney with him. He agreed that I c[oul]d leave the patient awhile. So we all returned together, Dr. W riding in my buggy & Toney riding in Dr. W's & eating sugar cane all the way. I went out again after supper; found all over, patient doing well; pains had come at dusk, & she quickly got through. So I returned home. 12 miles.[25]

Hentz's burst of story-telling includes several of the typical motifs of country obstetrical practice: the urgent call; the journey alone to the patient's home; the subsequent abatement of the emergency and the long "ministering" wait; the apparently easy yet necessarily choreographed relations with his colleague; the social scene and its conversation; the irony of having a patient who "quickly got through" without him; the closure of logging in the mileage. In this passage the very familiarity of his work, a routine summed up in just two lines in his clinical account, gave rise to something approaching autobiography. The story is *his* story. He wants to tell about the pine-knot fire and the blue sky, Dickens and the "negro bed," because, it seems, these things were the essence of his work. It was work shaped by conditions not of his own choosing but made his own by telling a story. This passage, like others in Hentz's journal, has an edge of self-conscious literary effort; one senses he is striving for a rustic epiphany amid the ordinariness of his life. Yet this is exactly the point: there were times when the routine of obstetrical work revealed meanings beyond the immediate tasks at hand. The way Hentz chooses to tell his story suggests how the social scope of doctoring went well beyond the specifically medical and professional realm. He ministers to his patient, but he describes none of his techniques; he neglects to mention administering ergot. His one Latin term seems oddly out of place in a story so deeply embedded in ordinary language. This is not a story about medical knowledge being applied to a woman in labor but rather a story of the physician's craft taking its shape from persons and conditions beyond his control, professional or otherwise. Hentz's notebooks show him to have been a well-trained, conscientious physician. Yet here the story that is told does not touch on his professional aims but is borne along on nothing more or less profound than the timing of a woman's labor.

Hentz's story, what the doctor said and left unsaid, begins to suggest questions about medicine and the wider society. Though seemingly alone with his reading and his reveries for much of the time, he nevertheless

[25] Charles A. Hentz, Obstetrical Notebook, 5 November 1857, and *idem*, Diary, 5 November 1857—both in Hentz Papers, SHC.

breaks bread with the "overseer and wife" and learns that they are good Methodists, Christian people in an occupation renowned for its brutality. Having these white people placed in the social landscape only emphasizes the absence and thus the silence of the slave owner Ely. Silent, too, is the woman Mary herself, raising questions of race and class once again. She is given neither name nor voice in Hentz's story; for him, she is a "negro woman" who becomes in the telling "my patient." Was she as alone in her labor as she seems, or did Hentz not bother to mention other black people in the house? He must have conversed with her; does he tell us nothing of their conversation because it was unremarkable, or was it unremarkable because she was a slave? He is careful to say he ministered to her; yet is it likely that he would have waited on the lawn of a wealthy white woman's house, feet cocked against a tree, reading *Harper's?* Perhaps Hentz left the cabin oppressed by the woman's race and her enslavement, or perhaps he left because he wanted to be outdoors on a nice day; in any event, the yard of a slave cabin put him no farther away from his patient than he would have been in the parlor of a wealthy white woman's home.

Thus the physician's bedside narrative inevitably gives way to questions about the wider world of southern social relations: between slaves and free, black and white, Methodists and others. Physicians' stories, like their work, move back and forth from the bedside to the social world beyond. Though the glimpses of this world are not in every case unique to the South, the social themes in southern doctors' writings—like the bedside themes—reflect a crucial dimension of doctoring in this time and place. The social themes define medical practice in its largest scope, suggesting how physicians acted as sharp social observers, chroniclers of the social patterns of class and race thriving in the rural distances of the South, and witnesses to a community's character.

Perhaps the most prominent of these social themes is the overarching importance rural distances had for the quality of medical practice. The weather and horses, and the need to have both on one's side, is a constant motif in physicians' writing. Sometimes this aspect of work was delightful: "Give me a bright, bracing day, a fine road and a lonely one, an active, stirring horse—and a certain mood of mind, and nothing is more pleasant than these long solitary rides," wrote a Mississippi practitioner. Yet this same man also recorded the missed turns, the soaking rain, the endless conversations with himself, all leading up to a late arrival at the patient's home, and his sense that "this practice of medicine is next to the soldier's life for vexation and discomfort. Can an equivalent for its toils and anxieties ever be rendered in money? I doubt it." Rural distances affected not only a practitioner's degree of comfort or the timeliness of his arrival but also his opportunities for meaningful consultations, for acquiring new knowledge and colleagues at regional medical association meetings, and for spending time with his family. Thus rural distances may be seen as structuring the very substance of medical knowledge, caring, and ethics.[26]

Not only the rural distances but the close-knit communities that sparsely populated them loom large in physicians' accounts of their work. The doctor worked among relationships that were multiple and intimate. A patient might be new to an established practitioner's care, but there was little chance that she would be completely unknown to him. If she were a slave woman, this in itself would give her an identity in the white community of which the physician was a part. In this way, the country doctor did not know that nemesis of modern medicine, the anonymous patient. He treated people he knew: his landlord, his neighbor's wife, his creditors and debtors. As physicians stepped back from the bedside a bit, they gave accounts that revealed complex community values. Inadvertently or not, they became the bearers of secrets. Because they missed church to treat a patient, they found out who else was not there, and why not; they learned who drank too much and who was dependent on drugs; they learned whose children were healthy and whose abused; they knew who suffered in silence and who was hypochondriacal; they saw the variety in how much slave owners were willing to spend on the health of their slaves.[27]

In terms of obstetrics, the doctor's role as bearer of secrets is nowhere better seen than when he learned something of the sexual lives and attitudes of his neighbors, as of course he inevitably did, and thus encountered a community's values at an especially sensitive point. "*Miss* Ella Grubb," Charles Hentz pointed out in his clinical notes on the woman's pregnancy. "Mrs. Jenkins was delivered of a fine boy about 2 hours since," recorded Thomas O'Dwyer in his 1825 journal, adding, "Conjecture is busy ascertaining its Pa." Such comments, perhaps kept to the confines of clinical notebooks, perhaps not, usually were on the side of "proper" conduct. Yet some physicians, such as Joseph Milligan, became frustrated by standards of sexual modesty or vanity when they were seen as counter to good health. He shared the exasperation of his physician son, whose patient was "so very modest that she will not permit you to cup her on the loins" and whose mother opposed therapeutic blistering because it made an "unpleasant appearance." Less frequently recorded but doubtless common were times when the doctor was sexually tempted. Young Richard Arnold attended a young, poor, white woman who, in the course of the physical examination, "imprinted on my lips a voluptuous kiss which shot through my system like electricity." Indeed, physicians' writing about encounters with sexuality in their obstetrical work show them struggling with their society's widespread fascination with sex yet equally widespread propensity to choke on the

[26] William Holcombe, Diary, 28 January 1855, 20 January 1855, SHC.

[27] For an insightful look at current rural practice, see Ruth Purtilo and James Sorrell, "The ethical dilemmas of a rural physician," *Hastings Center Rep.*, August 1986, *16(4)*: 24–28.

words that gave it expression. The sexual nature of obstetrics thus intruded in personally vexing ways. In the midst of an anxious courtship, and on his way to an obstetrical call in town, Arnold passed by his sweetheart's window, caught a brief glimpse of her, "& ten minutes after I saw her I was engaged professionally viewing a beautiful woman's pudenda. Good heavens thought I at the time, would she [i.e., his sweetheart], be all sorts of scared if she could take a coup d'oil [*sic*] of the whole scene." Her reaction was hypothetical; *he* seems to have been "all sorts of scared." Another young doctor, typically, could only utter euphemisms when talking of his obstetrical work to his fianceé, referring to a woman's "certain condition." It was embarrassing, his father wrote, when the young woman accidentally saw a letter which "contained some comments about obs[tetrical] forceps, &c."[28]

Working with pregnant women and expectant families in close-knit communities led doctors to become not only the bearers of sexual and other secrets but also the tellers of certain kinds of tales. Many physicians wrote down stories that illuminated something about the quality of the community's life as it was revealed to them in their rounds. Samuel Leland, for example, set down in his 1852 diary the story of a pregnant woman who went into labor while alone on her farm, shouted for help, and when no help came, delivered the baby herself. Leland wrote (and, we may imagine, told others) the story with a clear sense of amazement at the woman's strength. "Her child being born, she gets out of bed, goes out, and cuts up wood, then gets water from the spring, makes up a fire, and warms the water to wash the child." Step by step, Leland follows the woman in her self-sufficient work until a neighbor finally comes by to see if she needs help—but none is needed, least of all a physician's. This woman was, Leland wrote, *"some pumpkins."* Charles Hentz told of treating a woman with a prolapsed uterus, finally suggesting (to her husband) that the couple have a child and thereby, he hoped, cure the prolapse. Though everyone was embarrassed by the boldness of Hentz's suggestion, he was delighted when some weeks later the husband confided to him that his wife was in "that interesting condition." The child was born with Hentz's assistance and the prolapse was cured, providing the doctor with so happy a story of medicine's success that he retold it many years later in his autobiography. A medical student at the Kentucky School of Medicine related a quite different story, an appalling tale in which a number of students and a professor of obstetrics took the stillborn infant of a woman "of good family" and gave it to another student for dissection instead of burying it as they had promised. The professor, "as a joke," then pretended that a local newspaper had

[28] Charles A. Hentz, Obstetrical Notebook, 22 September 1867, Hentz Papers, SHC; Thomas O'Dwyer, Diary, 21 March 1825, Wheeler Papers, SHC; Joseph Milligan to Joseph A. S. Milligan, 9 July 1846, Milligan Papers, SHC; Richard Arnold, Diary, 4 and 7 December 1832, Arnold Papers, Duke; Joseph A. S. Milligan to Octavia Camfield, 17 July 1848, Milligan Papers, SHC; Joseph Milligan to Joseph A. S. Milligan, 11 June 1846, Milligan Papers, SHC. See also Beverley Jones, Account Book, 13 January 1851, Jones Papers, SHC.

learned of the disposition of the corpse. The student, terrified of being held responsible for graverobbing, confessed to the surprised newspaper editor, setting off an investigation in which the grieving parents pressed charges against the school. The professor was fined and the students reprimanded; but the story remained, testimony to the physician's power to touch the deepest nerves of family and community, or, in the unrepentant observation of the professor involved, to create "a Hurricane from such a little breeze!!!"[29]

These stories, taken from many that can be found in physicians' notebooks and letters, are obviously different in many respects; but they share features that indicate why physicians wrote them down. Each of these tales suggests how everyday doctoring exposed the grain of people's desires and capacities; each of these tales shocked, amused, or astonished, and on one level is told in the spirit of "here's what happened to me today." Yet such stories, appearing at random amid the routine accounts of visiting patients and prescribing drugs, also imply lessons to be learned, and conclusions to be drawn about people's potential for coming to terms with the shared mysteries of death, birth, and sexuality. In this sense, they are stories not only about "what happened today" but "what happens in my work." They gave shape to something important about obstetrical practice itself, about how it provided the doctor with a unique vantage point in a community's life, as a witness to the ways people submitted to or triumphed over the biological—the "animal nature" so fearful and so fascinating in the mid-nineteenth century.

The extent to which many physicians drew upon the personal and moral lessons of childbirth may be seen, too, in the brief narratives they frequently wrote of the birth of their own children. "Dear Wm.," wrote a Mississippi doctor to his friend in 1853, "Your letter of 25th Dec. reached me last evening . . . whilst I was in attendance on my wife in her trouble. In an hour after she was safely delivered of *another daughter!!* This morning both are doing fairly well, though [my wife] is not *so* well as I could wish. I *hope* for the best, however." Similar in its brief story of birth, though entirely different in outcome, is physician Samuel Leland's diary entry for 8 August 1851: "After great pain and suffering my wife was delivered of a dead infant. The little girl though dead, was so well formed, plump, and beautiful, it grieved me, to commit it to the earth. But—God's will be done. My wife bears up against the loss, but her disappointment is keen." In both stories, hope is the motif. Neither event—a new girl child in the family, the loss of a daughter—calls for the least bit of medical or technical language. In both, the narrator speaks most compellingly as father and husband. The timing of

[29] Samuel Leland, Diary, 1 March 1852, SCL; Charles A. Hentz, autobiographical manuscript, pp. 277–78, and *idem*, Obstetrical Notebook, 25 June 1856 (the couple named the child Bettie Hentz McNealy, after Hentz's wife) Hentz Papers, SHC; J. H. Parker to Josiah Hawkins, 11 March 1883, Hawkins Papers, LSU.

the birth and the condition of the woman are duly recorded; but the medicine they practiced is not as important to recount as is their reliance on God and on the habit of hoping for the best.[30]

In addition to containing secrets, and telling stories instructive and moral, country doctors' informal writings comment on a further plane of a community's life, the broad one defined by race and class. The full extent of this area of social experience appears only obliquely within the tight confines of the clinical notes and correspondence looked at here. Many more questions than answers are suggested about the particular perspective on race and class forged by physicians' daily work, but these issues obviously are crucial. For example, the explicit ideological racism in medical thought (typified by the ideas of Josiah Nott and Samuel Cartwright), which grew in stature during the antebellum years, was available to practitioners in their obstetric work as well as in their general medicine. However, the ideas of Nott and Cartwright rarely appear in terse clinical accounts, or even in physicians' stories, so the extent to which bedside work was influenced by the finer points of ideological racism is difficult to determine.[31]

It can be recalled that country doctors learned certain general "truths" about black patients which they most likely applied to pregnant black women. For example, the conviction that black patients were by nature deceptive could easily have made a white physician harshly skeptical of a black woman's pain or her assessment of her own pregnancy. On the other hand, the notion that black patients required less heroic dosages of drugs could have resulted in a practitioner prescribing medicines that were less unpleasant or harmful than the ones he prescribed for white patients. It is difficult to find direct comments on these racial discriminations in physicians' bedside accounts. However, it is clear that certain "structural," as opposed to ideological, racist distinctions obtained in the practical routines of obstetrical work: the slave owner, not the patient or her friends, issued the call to the doctor; black patients were experimented upon with greater freedom than were white patients; still-born black infants were more easily acquired as subjects for dissection; black women and the children born to them were more easily made anonymous in clinical records. White practitioners could write of a black woman, as they never could of a white woman, "Delivered Mr. Lee's Wench, the child dead."[32]

However, in other ways that require more study, the immediacy of labor and delivery probably provided for a kind of racial blindness; in the

[30] J. Warren Brickell to William G. Conner, 1 January 1853, Conner Papers, LSU; Samuel Leland, Diary, 8 August 1851, SCL.

[31] See Mary Louise Marshall, "Samuel A. Cartwright and states' rights medicine," *New Orleans Med. Surg. J.*, 1940, *93:* 74–78; and Reginald Horsman, *Josiah Nott of Mobile: Southerner, Physician, and Racial Theorist* (Baton Rouge: Louisiana State University Press, 1987).

[32] William Blanding, Diary, 20 October 1810, Blanding Papers, SCL. See John Campbell, "Work, pregnancy, and infant mortality among southern slaves," *J. Interdisciplinary Hist.*, 1984, *14:* 793–812.

urgency of much obstetrical work, race was pushed into the background. The time a physician spent on a case, the decision to use instruments, the sense of success measured by the delivery of a live, healthy infant, the relief of "getting through" a delivery, all appear to have depended little on the physician's perception of race as such. To be sure, a slave woman's value to her owner might inspire a physician's best efforts. Yet just how the economics of race relations affected the clinical efforts of practitioners cannot be reduced to a simple equation. The racism of everyday life in a slave society was protean and only partly acknowledged. When Courtney Clark, attending a slave woman, decided that he should "take the trouble" to time the intervals between her contractions "by the watch," was he performing an experiment he would not have made if he had been attending a white patient? Did he believe that whatever he learned would be applicable only to black women, or would it tell him something about his white patients as well? He did not say; but in such small actions did clinical racism take its shape.[33]

In striking contrast is the explicit contempt expressed by many physicians toward patients referred to as "pauper cases"—the white people too poor to pay for a physician's care. As a class, poorer whites appear to have been the scapegoats for much that physicians hated about the practice of medicine. The sense of encouragement that Charles Hentz usually brought to his work, and to his diary, was marred by his prejudice against poorer whites; he almost invariably expected to be manipulated, lied to, and not paid. William Holcombe, a homeopathic practitioner who saw his "new medicine" as part of his benign Swedenborgian optimism, nevertheless sneered at his Irish patients ("ignorant, vacillating, prejudiced"), as did many other practitioners. Poorer white women were singled out for a kind of disparagement that betrayed some doctors' intense uneasiness—even hatred—directed toward the sexual implications of childbirth. "Poverty, Filth, & Vice in its lowest degree," wrote young Richard Arnold of one visit to a poor woman, adding that he found it repellent to imagine how she "had shared a venereal orgasm with some brute whose passions were on the same scale with hers." Degraded sexually, despised for being Irish, "upcountry," or "sand hills," poor whites could not have received the same kind of medical care as the well-to-do or, for that matter, as well-treated slaves. Poor patients must have felt the physician's prejudice even if the care he gave was technically appropriate. The openly expressed class contempt in physicians' informal writing thus reverses the thrust of the published medical literature, with its emphasis on the nature of race. This suggests that African Americans under slavery were perhaps free from a certain bedside

[33] Courtney Clark, Notebook, n.d. [1844–74], p. 82, Duke. See also Charles A. Hentz, Obstetrical Record, 1849–91, and *idem,* Medical Diary, 19 September 1860—both in Hentz Papers, SHC; Joseph A. S. Milligan, Case Notes, 1859, manuscript vol., Milligan Papers, SHC.

disparagement reserved for poor whites, making blacks comparatively "invisible" both as objects of contempt and as persons.[34]

Class, race, the bonds of community, and the rural conditions that embraced all of these aspects of southern life in the mid-nineteenth century thus emerge as unifying themes in the informal writings of physicians. It is significant how the broad social context of medical practice shaped rural practitioners' values and routines far more extensively than did elite physicians' calls for professional dominance. Seen from the perspective of the ordinary physician, even an ideologically charged area of medicine such as obstetrics was more profoundly shaped by the craftlike work of doctoring than by professional ideology. Trained in various, conflicting ways, having little obstetrical work, called mostly in emergencies, and then performing sharply circumscribed tasks, southern country doctors had few opportunities to reinterpret their community's values. For some physicians, the profoundly local scale of their work was a mark of failure. They felt stifled by the rural isolation and its intellectual emptiness.[35] However, other practitioners drew an identity and a kind of power from their routine care-giving and the stories that flowed from it. Their work countered the restrictiveness of medical practice by placing them squarely into the midst of their society; while not always effective, neither were they alienated. Rediscovering the informal voices of doctors, it is possible to recover something of this crucial local scale, to see how a physician's sense of his own work was affected by the vagaries of his training, the intense bedside drama, and the wider circle of class and race relations in a rural society. Physicians' words mirrored the plain, fragmentary, often intense nature of their work. Their stories suggest how much there is to be learned about southern country medicine's complexity and surprise.

APPENDIX

Older works that have been largely supplanted but remain useful include J. Whitridge Williams, *A Sketch of the History of Obstetrics in the United States Up to 1860* (n.p., 1903); and Irving S. Cutter and Henry R. Viets, *A Short History of Midwifery* (Philadelphia: W.B. Saunders, 1964). Beginning in the 1970s, feminist historians and other scholars challenged the tight clinical focus and professional complacence that marked much of traditional obstetrical history. Important critiques include Barbara Ehrenreich and Deirdre English, *Complaints and Disorders: The Sexual Politics of Sickness* (Old

[34] William Holcombe, Diary, 3 February 1855 (see also 1 February 1855), SHC; Richard Arnold, Diary, 17 August 1832, Arnold Papers, Duke. See also Charles A. Hentz, Diary, 6 April 1848, Hentz Papers, SHC; M[armaduke] D. Kimbraugh to Nathan Hunt, 17 December 1859, Hunt Papers, Duke.

[35] For a look at this aspect of doctoring, see Steven M. Stowe, "Religion, Science, and the Culture of Medical Practice in the American South, 1800–1870," in James Gilbert, Amy Gilman et al., eds., *Essays in American Culture and Society* (Los Angeles: Wadsworth Publishers, forthcoming, 1991).

Westbury, New York: Feminist Press, 1973); G.J. Barker-Benfield, *The Horrors of the Half-Known Life: Male Attitudes toward Women and Sexuality in Nineteenth-Century America* (New York: Harper & Row, 1976); Linda Gordon, *Woman's Body, Woman's Right: A Social History of Birth Control in America* (New York: Grossman, 1976); Ann Oakley, *The Captured Womb: A History of the Medical Care of Pregnant Women* (Oxford and New York: Basil Blackwell, 1984); Flavia Alaya, "Victorian science and the 'genius' of women," *J. Hist. Ideas,* 1977, *38:* 261–80; Janet Bogdan, "Care or cure? childbirth practices in nineteenth-century America," *Feminist Studies,* 1978, *4:* 92–99; Martha Vicinus, "Sexuality and power: a review of current work in the history of sexuality," *Feminist Studies,* 1982, *8:* 133–56; Mary Poovey, " 'Scenes of an indelicate character' in the Victorian delivery room," *Representations,* 1986, *14:* 137–68. The feminist critique has helped to shape broader social histories of women and childbirth in recent years. A watershed work is Judith Walzer Leavitt, *Brought to Bed: Childbearing in America, 1750 to 1950* (New York: Oxford University Press, 1986). Also important are Sally G. McMillen, *Motherhood in the Old South: Pregnancy, Childbirth, and Infant Rearing* (Baton Rouge: Louisiana State University Press, 1990); Sylvia D. Hoffert, *Private Matters: American Attitudes toward Childbearing and Infant Nurture in the Urban North, 1800–1860* (Urbana: University of Illinois Press, 1989); Judith Walzer Leavitt, "The growth of medical authority: technology and morals in turn-of-the-century obstetrics," *Med. Anthropol. Quart.,* 1987, n.s., *1:* 230–55. These studies build on work in the social history of childbirth which includes Richard W. Wertz and Dorothy C. Wertz, *Lying-in: A History of Childbirth in America* (New York: Free Press, 1977); Catherine M. Scholten, *Childbearing in American Society, 1650–1850* (New York: New York University Press, 1985); Charles Rosenberg and Carroll Smith-Rosenberg, "The female animal: medical and biological views of woman and her role in nineteenth-century America," *J. Amer. Hist.,* 1973, *60:* 332–56; Martha H. Verbrugge, "Women and medicine in nineteenth-century America," *Signs,* 1976, *1:* 957–72. Useful comparative works are Judith S. Lewis, *In the Family Way: Childbearing in the British Aristocracy, 1760–1860* (New Brunswick, N.J.: Rutgers University Press, 1986); Claire E. Fox, "Pregnancy, Childbirth, and Early Infancy in Anglo-American Culture, 1675–1830" (Ph.D. diss., University of Pennsylvania, 1966); Edward Shorter, *A History of Women's Bodies* (New York: Basic Books, 1982).

For sectional influences, see Joseph I. Waring, *A History of Medicine in South Carolina, 1620–1825,* 2 vols. (Columbia, South Carolina: R. L. Bryan, 1967); Reginald Horsman, *Josiah Nott of Mobile: Southerner, Physician, and Racial Theorist* (Baton Rouge: Louisiana State University Press, 1987); Richard H. Shryock, "Medical Practice in the Old South," in his *Medicine in America: Historical Essays* (Baltimore, Maryland: Johns Hopkins Press, 1966), pp. 49–70; John Duffy, "A note on ante-bellum southern nationalism and medical practice," *J. Southern Hist.,* 1968, *34:* 274–76; John Duffy, "Medical practice in the ante bellum South," *J. Southern Hist.*, 1959, *25:* 53–72; John Harley Warner, "The Idea of Southern Medical Distinctiveness: Medical Knowledge and Practice in the Old South," in *Sickness and Health in America: Readings in the History of Medicine and Public Health,* ed. Ronald Numbers and Judith Leavitt, 2d ed., rev. (Madison: University of Wisconsin Press, 1985); John Harley Warner, "A southern medical reform: the meaning of the antebellum argument for southern medical education," *Bull. Hist. Med.,* 1983, *57:* 364–81; James O. Breeden, "States-rights medicine in the Old South," *Bull. New York Acad. Med.,* 1976, *52:* 348–72; Mary Louise Marshall, "Samuel A. Cartwright and states' rights medicine," *New Orleans Med. Surg. J.,* 1940, *93:* 74–78; James Denny Guillory, "Southern Nationalism and the Louisiana Medical Profession, 1840–1860" (M.A. thesis, Louisiana State University, 1965). On slavery see Todd L. Savitt, *Medicine and Slavery: The Diseases and Health Care of Blacks in Antebellum Virginia* (Urbana: University of Illinois Press, 1978); Richard B. Sheridan, *Doctors and Slaves: A Medical and Demographic History of Slavery in the British West Indies, 1680–1834* (New York: Cambridge University Press, 1985); David O. Whitten, "Medical care of slaves: Louisiana sugar region and South Carolina rice district," *Southern Studies,* 1977, *16:* 153–80; Weymouth T. Jordan, "Plantation medicine in the Old South," *Alabama Rev.,* 1950, *3:* 83–107; John S. Haller, Jr., "The Negro and the southern physician: a study of medical and racial attitudes, 1800–1860," *Med. Hist.,* 1972, *16:* 238–53; Walter Fisher, "Physicians and slavery in the antebellum southern medical journal," *J. Hist. Med. Allied Sci.,* 1968, *23:* 36–49. Suggestive of the broader medical environment are Kenneth F. Kiple and Virginia H. King, *Another Dimension to the Black Diaspora: Diet, Disease, and Racism* (Cambridge: Cambridge University Press, 1981); Kenneth F. Kiple, "A survey of recent literature on the biological past of the black," *Soc. Sci. Hist.,* 1986, *10:* 343–67; Richard H. Steckel, "A peculiar population: the nutrition, health, and mortality of American slaves from childhood to maturity," *J. Economic Hist.,* 1986, *46:* 721–41; K. David Patterson, "Disease Environments of the Antebellum South," in *Science and Medicine in the Old South,* ed. Ronald L. Numbers and Todd L. Savitt (Baton Rouge: Louisiana State University Press, 1989), pp. 152–65; James H. Cassedy, "Medical Men and the Ecology of the Old South," *ibid.,* pp. 166–78.

The Woman's Experience of Childbirth on the Western Frontier

Charles R. King

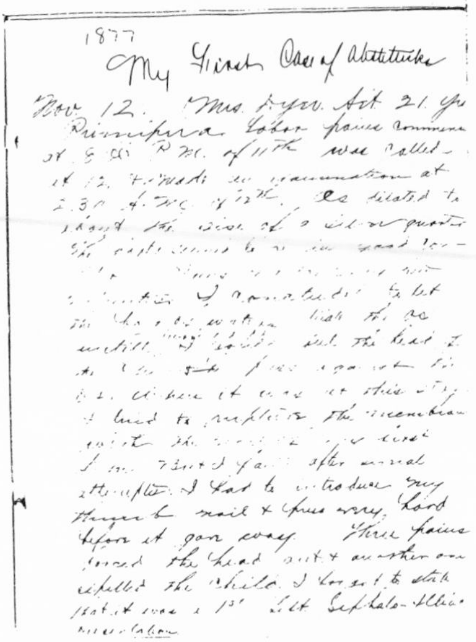

"My First Case of Obstericks" from the case book of Dr. J. C. Parrish, who began his rural Missouri medical practice in the summer of 1877.

WOMEN have been and continue to be controlled by the biological clock of menstruation and birth. Modern woman is a sister of both her prehistoric and her frontier counterparts, and in the process is clearly separated from man, who does not physically experience childbearing. Urban women, rural women, educated women, illiterate women, all women individually confront this process. To this biological framework frontier women added the problems of simply surviving, and these circumstances multiplied the complications of life and the price of death. These limitations were graphically recorded on a nineteenth-century Ohio tombstone:

> Thirteen years I was a virgin
> Two years I was a wife,
> One year I was a mother
> the next year took my life.

Childbirth on the nineteenth-century American frontier dramatically illustrates the difficulties that the natural function of childbearing created for frontier women.

Frontier Women

The nineteenth-century American woman was increasingly confined by the growing "cult of domesticity" to the expected biological and growing social roles of wife and mother. This "cult of true womanhood by combining piety and domesticity with submissiveness and passivity, controlled women and narrowed their options."[1] At the same time, Westward expansion placed additional burdens on women and provided new opportunities for women to undertake positions, responsibilities, and roles not generally open in the structured, developed Eastern society. The important role that women played in the Westward migration and the development of the Western frontier were not always acknowledged by men, but Andrew McClure in his diary from the Oregon Territory noted: "Much praise for our success of this enterprise is due to our women."[2]

Women were expected not only to perform typical women's chores — the "drudgery of housework and homemaking"[3] (if the toil of day-to-day work on the frontier or on the trail can be equated with homemaking in a more refined setting), but as Helen Carpenter, the daughter of a Kansas homeopathic physician traveling West, noted in June 1857:

> The days are all very much the same except when we stop for a day. Then there is washing to be done and light bread to make and all kinds of odd jobs. Some women have very little help about being obliged to get the wood and water, make camp fires, unpack at night and pack in the morning.[4]

Ultimately, "it was a human experience in which all — men, women and children — participated as individuals,"[5] and each person responded as an individual, positively or negatively, to the hardships and the rewards of the Westward movement.

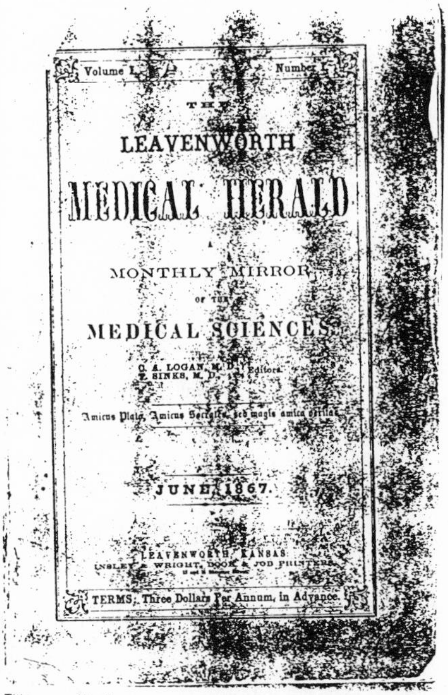

Volume I. Number I.

THE

LEAVENWORTH

MEDICAL HERALD:

A

MONTHLY MIRROR

OF THE

MEDICAL SCIENCES.

C. A. LOGAN, M. D., T. SINKS, M. D., Editors.

Amicus Plato, Amicus Socrates, sed magis amica veritas.

JUNE, 1867.

LEAVENWORTH, KANSAS:

INSLEY & WRIGHT, BOOK & JOB PRINTERS.

TERMS: Three Dollars Per Annum, in Advance.

Title page of the *Leavenworth Medical Herald*, the first medical journal published in Kansas. The first article in the premier issue concerned infant feeding.

Frontier Pregnancy

The details of nineteenth-century urban American childbirth have been previously recorded, and will not be repeated here.[6] Reports of frontier pregnancy and childbirth have rarely survived, probably because such topics were a part of women's oral tradition and knowledge rather than recorded history and, when recorded, the temporary nature of frontier life often prevented their survival. Nonetheless symptoms of pregnancy, such as morning sickness, were alluded to by frontier women. For example, a report of "sick headaches," "I must make some tea; that will help it if not cure it."[7] is one such reference. Frontier women, along with women in general, were also well aware of the symptoms of pregnancy. Martha Shaw, the wife of a Topeka, Kansas, mail carrier, noted "I get so deathly sick of mornings. . . . I know its symptoms."[8] At the same time the accurate diagnosis of pregnancy and its duration was difficult for both physicians and patients. Mary Kincaid, in a letter from Dakota Territory in December 1896, noted she had not had a menstrual period for three months and wondered if she were pregnant. "I don't know if I am or not, I don't feel so I was."[9]

Even with the difficulties of frontier pregnancy, many women expressed disappointment following a miscarriage or even an unexpected menstrual period. Martha Shaw, in October 1890, reported the onset of her menstrual flow: "It means I might have been a mother, but not this time. I'm sorry." Five months later she recorded in her diary, "I was two months, or more, in maternity and had so much hoped, and was so happy in the hope of being a mother; only to meet with bitter disappointment this morning, so have been in bed all day. O it is dreadful."[10]

When pregnancy continued, women were often uncertain about the outcome. Millicent Hunt reported with her fifth pregnancy.

> Yes even now the frail foetus within me is the abode of an immortal spirit, and this has caused thoughts of discontent. I would it were not thus, I love my liberty, my ease, my comfort and to not willingly endure the inconvenience and sufferings of pregnancy and childbirth.[11]

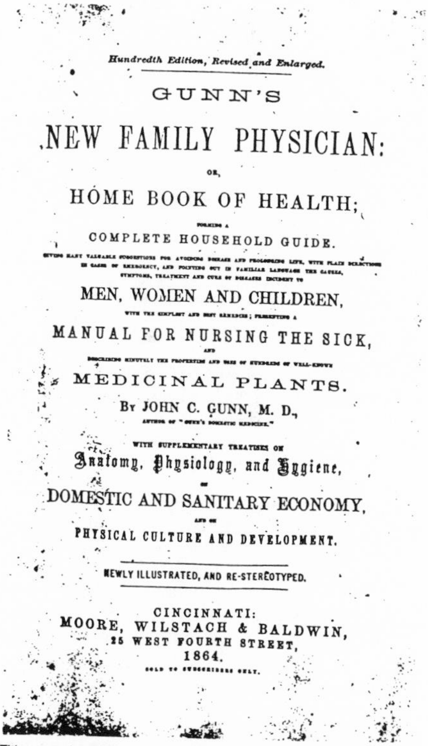

Hundredth Edition, Revised and Enlarged.

GUNN'S

NEW FAMILY PHYSICIAN:

OR,

HOME BOOK OF HEALTH;

FORMING A

COMPLETE HOUSEHOLD GUIDE.

MEN, WOMEN AND CHILDREN,

MANUAL FOR NURSING THE SICK,

AND

MEDICINAL PLANTS.

BY JOHN C. GUNN, M. D.,

WITH SUPPLEMENTARY TREATISES ON

Anatomy, Physiology, and Hygiene,

DOMESTIC AND SANITARY ECONOMY,

PHYSICAL CULTURE AND DEVELOPMENT.

NEWLY ILLUSTRATED, AND RE-STEREOTYPED.

CINCINNATI:
MOORE, WILSTACH & BALDWIN,
25 WEST FOURTH STREET,
1864.

Title page from *Gunn's New Family Physician* . . . the Hundredth Edition, Revised and Enlarged. Aside from the Bible this book was one of the most popular advice books of the nineteenth century.

Portrait of Dr. J. C. Gunn, author of one of the most famous nineteenth-century manuals of domestic medicine *Gunn's New Family Physician.*

Such feelings are understandable, when the difficult circumstances of frontier childbirth are considered.

Even with obvious hardships pregnancy and childbirth were not considered reasons to excuse women from either daily chores or the new burdens of the frontier. "Women were expected to be strong enough to serve the common needs of the day, and strong enough to meet the uncommon demands as well."[12] Motherhood was not recognized as an impediment to either Westward emigration or to work once established on the frontier. To even consider reducing the expected contribution of women would have been difficult, since nearly one woman in five was at some stage of pregnancy and "virtually every married woman travelled with children"[13] on the Westward trail. Limitations on women's workload would dramatically reduce their contribution to the success of the journey and possibly lead to failure. Delays in the journey, whether to celebrate the Sabbath or the birth of a child, were rare, and only physical incapacity created by sickness, injury, or the environment were recognized as reasons to delay the journey. "Delay for the dying," or the sick, or the woman laboring in childbirth, "threatened the life of the living."[14]

The toll of childbearing on the frontier is clearly evident in the high rates of perinatal and infant mortality. At least one in five children died before age five, and probably an equal number of prenatal losses occurred as well. The need for large families to complete work on the frontier, and frequent unsuccessful pregnancies, created a 2 1/2 year reproductive cycle for the frontier woman, for 19 or 20 months of which she was either pregnant, nursing a newborn baby, or providing infant care. This cycle was likely to repeat itself from the mid- or late teens through at least the mid-thirties. The harsh "physical and emotional wear and tear of nearly constant pregnancy and suckling"[15] on the frontier woman is obvious. The stark reality of spontaneous abortion, stillbirth, childhood death, maternal mortality, and not infrequent paternal deaths led Harever to the conclusion that "before 1900 only about 40% of the female population" was able to achieve "an ideal family cycle" with both surviving children and parents.[16] Thus, there is little surprise that the "overland journey and the frontier experiences were primarily for the very young."[17]

Frontier Emigration

Despite hardships for both men and women Westward emigration continued. An estimated 350,000 persons migrated to California and Oregon during the 26 years between 1841 and 1867. By 1880, 20 percent of the population lived in the Western half of the continent. Women participated in this migration for obvious reasons. Most came West because the family, generally the husband, had already moved there, and maintenance of family structure was desired.[18] Others came to promote health, "in the hope that that far-fair atmosphere would drive away all tendency to intermittent fever and invigorate our shattered conditions."[19] Others came for matrimonial possibilities like the young woman who advertised in the Waterloo *Courier* in 1860, and desired "correspondence with some young man in the West."[20] Family continuity, better health, and matrimonial prospects were all circumstances likely to result in childbearing, "one of the most important economic functions of frontier women."[21]

Specific advice for women to prepare for the journey West was limited. In fact, written records were "typically silent about illness, depression, menstruation, pregnancy, childbirth, birth control, abortion, and sexual relationships" during the mid-nineteenth century.[22] The surviving advice-books for overland travelers generally failed to mention women, and even though they advised emigrants to carry flour, sugar, and other provisions they rarely recorded useful medications.[23] One journal even advised that medications are not needed since "it is seldom, that emigrants are sick."[24] The experiences of the emigrants generally did not confirm this advice.

Frontier Medicine

Although the overland guides did not advise specific medications for the frontier journey, the best medications of the day were often included by emigrants making the journey. For example, Catherine Haun included in her trunk "a few treasures: a Bible, medicines, such as quinine, bluemass (a quinine derivative), opium, whiskey, and hartshorn for snake bites, and citric acid — antidote for scurvy."[25] Most of the therapeutically effective drugs of the mid-nineteenth century are found among these "treasures" for the journey.

Few medications were specifically effective for childbearing women, yet women's pills and remedies were widely available. These ranged from "Moffat's vegetable life pills; and Dr. Duponco's golden periodical pills for females," especially useful for "female weaknesses," to "Dr. Baker's specific for seminal weakness and venereal disease, may be used with entire secrecy for either sex." "Merchant's Celebrated Gargling Oil" was useful not only for "caked breasts" and "sore nipples" but also for "chapped hands" and even "spavins" and "sweeney" in horses. Liniments, especially "The Mother's Companion," were reported to aid "various problems associated with childbirth and the nursing of a newborn infant.[26]

In spite of the good health suggested by the trail guides, illness was the most frequent cause of death on the frontier. The most common cause of death, accounting for 40 to 70 percent of reported deaths, was infectious diseases — tuberculosis, malaria, dysentery, typhus, typhoid fever, pneumonia, measles, and epidemics of smallpox, yellow fever, and cholera.[27] Even though limited effective treatments, were available, *i.e.*, cinchona bark for malaria, opium for pain, digitalis for heart failure, vaccine for smallpox, and surgical corrections for fractures, trauma, and dental problems, self-medication was the first line of defense. During a typical four months at mid-century, 469 deaths in children under the age of five were reported in St. Louis. Only 30 percent of these infants were treated by a physician and another 40 percent of them had no care.[28] A quarter-century later, during the decade of the 1870s, in Lawrence, Kansas, only two childhood deaths were reported from smallpox, but "teething" accounted for 21 deaths, if medical records are to be believed.[29]

Disease and childbirth could coincide with each other, and either compound the difficulty or prove fatal. The wife of the post surgeon at Fort Harker died from this combination.[30] Lavincia Porter, traveling to Colorado with her husband, became ill with dysentery during her pregnancy. In her own words she became

> . . . so weakened that I could no longer climb in and out of the wagon, I was compelled to keep my bed . . . the jolting motion of the wagon soon became a perfect torture to me, and at last became so unendurable, that I implored my husband to take me out, make my bed on the sand and let me die in peace.[31]

She survived after consuming a dose of hair tonic that

Breast feeding was essential for infant survival in the nineteenth century. Artificial feeding rarely permitted infant survival until the end of the century.

ostensibly "acted like a charm."

Women were recognized as naturally useful and knowledgeable about health and the treatment of sickness. These "angels of mercy" functioned both with and without "professional medical help."[32] Many women became "surprisingly skilled in their mastery of herbs, bone setting and delivering children.[33] These "skillful women" were always in demand for the care of the sick and the burial of the dead.[34] Although some medical knowledge may have been learned from physicians, most of the expertise of these "wise women" was derived from the folklore of medicine, available volumes on domestic medicine, and popular periodicals and newspapers.[35]

By mid-century, the public view of the physician was often similar to Ruth Hoppin's description in her *Personal Recollections.* "He came every day, he purged, he bled, he blistered, he puked, he salivated his patient, he never cured him."[36] This negative view of the physician was not improved by the poor training of many physicians. For example, in Wichita, Kansas, the city directory indicated that only a minority of doctors actually had a medical degree.[37] Irregular physicians, including hydrotherapists, animal magnetizers, phreno-magnetizers, urine and poudrette doctors, as well as electics and homeopaths, continued to practice medicine actively on the frontier.[38] In fact, shortly after mid-century the Michigan Supreme Court

Infant death was all too common on the Western frontier. Few families did not experience an empty chair at the family table. (Photograph from the Pennell Collection, courtesy of the Kansas Collection, University of Kansas Libraries, Lawrence, Kansas)

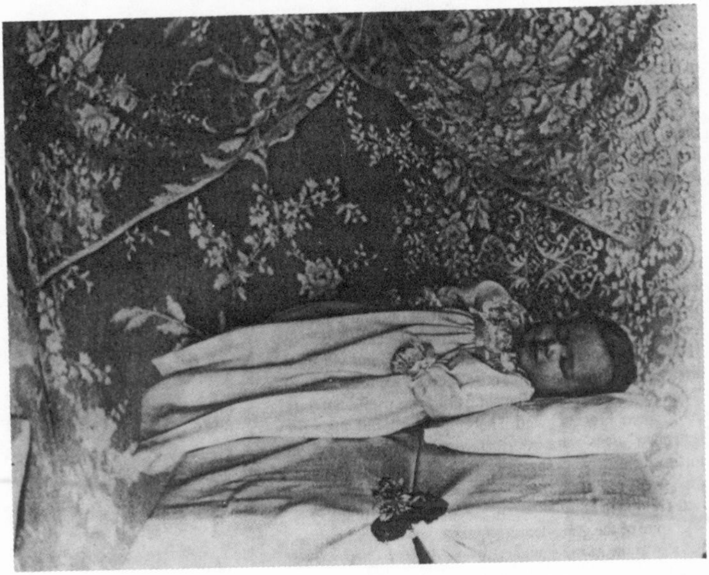

ruled that "a doctor is any person calling himself such,"[39] a status which continued until formal medical licensure developed later in the century. With this definition of a doctor it is easy to understand Helen Carpenter's report in June 1884 of a filthy, "most repulsive looking person" who appeared as bacon was frying at a trail camp and claimed to be a doctor.[40] Apparently he was accepted as such, because he soon went to another train to provide medical services. Albert Britt, of frontier Illinois, summarized the layman's view of the physicians, "the learned profession was not so very learned" and doctors were "to be sent for only as a last resort."[41]

Frontier Midwifery

Texts of domestic medicine, *e.g.*, *Buchan's Domestic Medicine*,[42] or Gunn's *New Family Physician*, were really home health guides designed to provide medical care in the physician's absence. Gunn advised the would-be accoucheur to "wait patiently for the operations of Nature." Further, "the directions which I shall give, will impart such information as will enable a woman of good common sense and observation to manage almost any ordinary case of labor without the aid of a physician."[43] Recommendations to position the woman in bed with "multiple pillows," to tie the umbilical cord with "strong twisted cord or thread," and the dangers of using "ergot or spurred rye," to strengthen uterine contractions, were included. At delivery if the infant did not immediately cry, "chafe with a flannel cloth," sprinkle with "cold water, vinegar, or spirits," or a "sudden gust of blown air" was reportedly helpful for resuscitation.[44] Remedies for false labor and even morning sickness were also included.

Further advice was found in *The American Lady's Medical Pocket Book*. According to this popular text, health during pregnancy was promoted by exercise, fresh air, good diet, and proper dress, all of which, except perhaps diet, were readily available on the frontier. Unfortunately, the physical birthing requirements were conditions hard to meet on the frontier. The lying-in chamber should be "large and airy" and guarded from "drafts of air," and following delivery the woman should receive "the most perfect tranquility."[45] "Under prudent management" according to one optimistic view, pregnancy became a state of "great suffering or danger" for only a few women, and "all danger may very generally be avoided, and the whole period of pregnancy be entirely divested of suffering."[46]

Frontier women continued the centuries-old role of midwife. Since the frontier population was rapidly increasing, these women were extremely active. Jim Marshall of Jefferson County, Kansas, recalls that his mother delivered scores of babies, which left "little time for our own family.[47] Patty Sessions, often described as the mother of Mormon midwifery, reportedly delivered 3,977 infants. On the way to the Great Salt Lake, while camped at winter quarters in Nebraska during 1847, she delivered 14 babies. She even delivered the children of her husband's polygamous second wife, Harriet.[48] During her 52nd birthday party on 4 April 1847 Patty Session described her obstetrical duties. "I then went and put James Bullock's

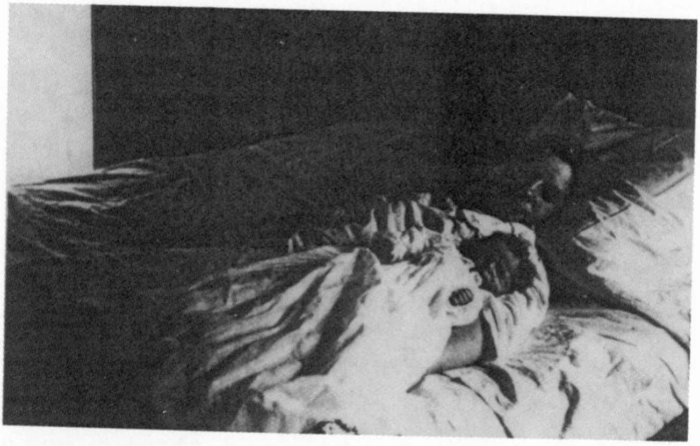

Photographic records of birth, even late in the nineteenth century are rare. The Kansas mother and son were portrayed within hours of birth. The calm, fatigue, and importance of the event were recorded by the father. (Photograph from the Bourquin Collection, courtesy of the Kansas Collection, University of Kansas Libraries, Lawrence, Kansas)

wife to bed, then to the party, had a good time singing and praying. I was called away to Sister Morse, then to Sister Whitney then back to Sister Morse; put her to bed at 2 o'clock.[49]

Childbirth was commonly attended by a midwife and women from the community. As Clarinda Tate of Calloway County, Missouri, recorded in a letter to her mother in Virginia in April 1823:

> On the 13 of the month I was delivered of a seven month babe, a *son* it lived but one hour. and in great misery it was choaked [*sic*] with phlegm, though breathed its last without a groan. I felt mutch [*sic*] resigned to the will of *God*. Blessed be his name he only took but what he gave — I suffered very mutch [*sic*], but was mercifully supported (by) Mrs. Henderson and 2 other of our neighbor women was all that was with me. the woman that delivered me was not skilld [*sic*] but had done in plane [*sic*] cases. . . . Mis[s] Henderson staid [*sic*] with me 3 weeks almost constent [*sic*] she attended me night and day as tenderly as she could have done her own child. She scarcely closed her eyes to sleep for two nights — James and her nursed me with so mutch [*sic*] tendernes[s] and care. . . . our neighbours were all as friendly as they could be[.] they came to see me often and done evry [*sic*] thing for me being in their power[.] some came to see me that had not been acquainted with us before. I fell very mutch [*sic*] attached to our acquaintences here[.] they appear moore [*sic*] like relations than strangers.[50]

Thus, the community of women was maintained as a support network for the childbearing woman. This community provided physical and psychological support during both pregnancy and childbirth, and maintained an oral tradition of knowledge about the care of women in health, disease, pregnancy, and childbirth.

Frontier Obstetricians

By mid-century most doctors carried obstetrical instruments, but many were poorly trained.[51] Francis Long, a graduate of the University of Iowa in 1880, had seen no obstetrical cases while in school.[52] Samuel Crumbine of Dodge City, who was later to become State Health Officer of Kansas, described his first obstetrical case:

> Mothers are often less nervous than the young doctor when he must make his first delivery, but once the case is happily terminated, the inexperienced physician, without meaning to, finds himself assuming unconsciously the attitude of a gifted obstetrician, as if the whole proceedings had been no more than an incident in his busy professional life.[53]

Dr. Long also reported his hesitance at a first delivery, an event that he had "schooled himself" to handle.[54]

A University of Pennsylvania graduate, Bernard James Byrne, practicing as an Army surgeon at Ft. Lewis, Colorado, in the 1880s, attended Mrs. Norris at her delivery, which occurred in a lean-to with "large cracks in the boards that let in a plentiful supply of air" and snow. The space was not heated. There were no bed springs, no sheets, only one pillow, and no female birth attendants for the delivery. The husband apparently classed his wife "among the cows and mares" and treated her as such. The infant died on the third day, a fate met by the previous children born into the family.[55]

Frontier Birth

Typically, childbirth was conducted by women, and the physician was only involved with complicated cases.

> There was of course no doctor. No one thought of calling a doctor for so casual a matter. Mary Bartsch came down to help, and afterward one of the Bartsch girls came and stayed for a week. Then Rosie re-

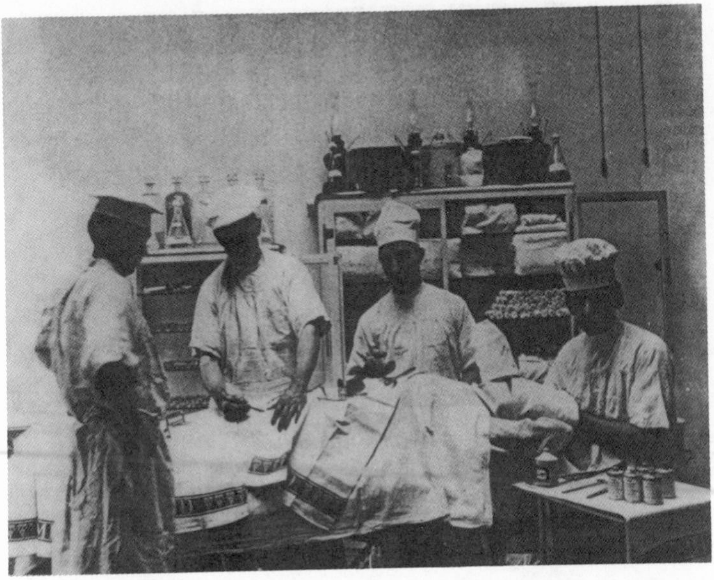

Near the start of the twentieth century many features of the contemporary operating room were in use, such as either anesthesia and rudimentary aseptic technique, but note the surgeons wear neither surgical masks nor surgical gloves.

sumed her work, with the added task of caring for her bright-eyed baby.[56]

In 1844, Thomas Steel, an 1833 graduate of the University of Glasgow, who had been called to attend few obstetrical cases, reported that he had performed "every exertion to save the life of a woman in childbirth during eight days, in which time I never had my clothes of(f)."[57] Finally, the delivery of the dead infant in pieces was completed "in the presence of all the old women of the neighborhood." The mother also died.[58]

A better outcome in the absence of a doctor was reported by Dr. Long. "No one was available but the 12-year-old daughter. The labor was precipitate. The mother called the girl and told her what to do and together they managed to tie the cord, get the baby dressed and the mother's toilet arranged."[59] Caroline Findley recorded the following delivery scene in her overland journal. The birth occurred "within a tent, during the storm, nurses wading around a bedside placed upon chairs set in a wagon leaking rain, with the floor deep in water."[60] Little wonder that childbirth raised concern for frontier women.

Mary Larrabee, the first white woman to give birth in Foster County, Dakota Territory, reported in a letter to her friend Patty: "We have a new baby at our house, a boy . . . whose arrival on the 12th was very much dreaded by me because we had no doctors or nurses." Two years later she delivered the second white child in the county and entered the birth in the family Bible because "there are no county organizations." Martha Morrison Minto also gave birth without the aid of other women. "When my second child was born my husband and I were alone three miles from any woman or doctor."[61]

Complications of childbirth were not uncommon. Labors were often slow and protracted, and older mothers could have problems, as Emily French noted, "she 30 past; no wonder she has a hard time."[62] Twin pregnancies occurred and were often complicated by premature delivery. Rose Williams reported "a young married girl," who gave birth to twins at seven months. "They were the littlest things I ever saw, did not think both together would weigh more than 4 pounds."[63] Infection or hemorrhage could follow birth and lead to maternal death: "The poor mother lived one day and night after the baby came."[64] Some women did recover from the grasp of maternal death. Emily McCorkle Fitzgerald, who accompanied her husband, an Army surgeon, to Sitka, Alaska, was one.

> The attack that sent me back to bed and nearly put an end to me. It makes me about sick now to think of my poor little babies if I had not gotten well. I can imagine the sorrow it would be to Doctor to lose me because I know what it would be to me to lose him. But what would my poor little babies do.[65]

With her second child, although she "suffered dreadfully" and "begged doctor to give me chloroform," she received little assistance for "he was afraid to have me under the influence of it long."[66]

The birth itself could be difficult, and both physicians

and midwives might intervene traumatically. Martha Shaw reported her own difficult delivery in 1892:

> But what a time we had, to get her, motherhood was all but impossible, by stopping the pains, . . . and part of the time I lay exhausted and unconscious, until in sheer desperation; Mrs. Gordon, tore me, with her fingers and pulled baby away, scratching the little forehead and almost crushing the head until [there] was a great ridge across the top, big as my finger and the darling was black with strangulation — the cord twice around the neck.[67]

In spite of these difficulties most women agreed with Mary Walker, who following the birth of her child in 1838 noted, "Just as I supposed the worst was at hand, my ears were salvated [*sic*] with the cry of my child."[68]

Frontier women worked up until delivery and often did not stop until forced to do so by the progress of labor. Amie Greenwood, an Idaho farmer's wife, reported:

> The week Rhoda was born I cooked for 15 men who had come to help stack hay. And in the intervals of serving them I would creep into my bedroom to sink across the bed. I was so tired. Through the bedroom window I could see the mare and the cow turned out to pasture for weeks because they were going to have their young.[69]

This could be carried to the extreme, as for example the wife of Charles W. Penrose, who "left his wife behind on the trail (to Salt Lake) to bear a child and then to catch up with the ox team by forced march."[70] She later bore her husband 18 children.

Conclusions

Childbirth was an expected biological and social function for nineteenth-century frontier women. Pregnancy and childbirth were complicated by the daily tasks of frontier life, which only made women's sense of loss from a failed pregnancy more acute. Physicians and the nineteenth-century practice of medicine often provided little comfort or benefit to the pregnant woman, and they possessed few abilities not available to the skilled midwife. More important than available professionals was a close circle of other women and female friends, who provided a social and personal support system for the pregnant woman and a ready source of medical advice handed down from the oral tradition and continued in manuals of domestic medicine.

NOTES

1. C. N. Degler, *At Odds, Women and the Family in America from the Revolution to the Present* (New York: Oxford, 1980), 26.
2. A. McClure, *The Diary* (Eugene, OR: Lane County Historical Society, 1959), 69.
3. J. L. Stratton, *Pioneer Women: Voices from the Kansas Frontier* (New York: Simon and Schuster, 1981), 57.
4. S. L. Myres, *Ho for California: Women's Overland Diaries from the Huntington Library* (San Marino: Huntington Library, 1980), 114.
5. *Ibid.*, x.
6. J. Leavitt, *Brought to Bed: Childbearing in America, 1750-1950* (New York: Oxford, 1986).
7. L. Schlissel, *Women's Diaries of the Westward Journey* (New York: Schocken, 1982), 202.
8. P. Franklin, *Private Pages: Diaries of American Women, 1830s-1970s* (New York: Ballantine, 1986), 328-329.
9. E. Hampsten, *To All Inquiring Friends: Letters, Diaries, and Essays in North Dakota, 1880-1910*, 2nd ed. (Grand Forks, ND: 1980), 20.
10. Franklin, *Private Pages*, 328-329.
11. H. Adams, "A Puritan Wife on the Frontier," *Mississippi Valley Historical Review*, 27 (1940): 81.
12. Schlissel, *Women's Diaries*, 35.
13. *Ibid.*, 14.
14. *Ibid.*, 131.
15. J. M. Faragher, *Women and Men on the Overland Trail* (New Haven: Yale, 1979), 58.
16. T. K. Harever, "Family Time and Historical Time," *Daedaleus*, 10 (1977): 62.
17. Schlissel, *Women's Diaries*, 51.
18. J. R. Jeffrey, *Frontier Women: The Trans-Mississippi West, 1840-1880* (New York: Hill and Wang, 1979), xi.
19. C. O. Van Cleve, *Three Score Years and Ten: Life-Long Memories of Fort Snelling, Minnesota, and other Parts of the West* (Minneapolis: Harrison and Smith, 1888), 130.
20. G. Riley, "Women Pioneers of Iowa," *Palimpsest*, 57, 2 (1976): 36.
21. Riley, *Frontierswomen: The Iowa Experience* (Ames: Iowa State, 1981), 82.
22. C. Fairbanks, *From Women on the Prairie Frontier* (Metuchen, NJ: Scarecrow Press, 1983), 55.
23. L. Hastings, *The Emigrant's Guide to Oregon and California* (Princeton: Princeton University Press, 1932), 143; A. Child, *Overland Route to California* (Los Angeles: N. A. Kovack, 1946).
24. J. Palmer, *Journal of Travels over the Rocky Mountains to the Mouth of the Columbia River; Made during the Years 1845 and 1846* (Cleveland: A. H. Clark, 1906), 260.
25. Schlissel, *Women's Diaries*, 168.
26. Riley, *Frontierswomen*, 283-285; 81-82.
27. T. B. Hall, *Medicine on the Santa Fe Trail* (Dayton, OH: Morningside Bookshop, 1971); W. G. Rothstein, *American Medical Schools and the Practice of Medicine* (New York: Oxford, 1987), 18.
28. J. Duffy, "Medicine in the West: An Historical Overview," *Journal of the West*, 21, 3 (1982):7.
29. T. N. Bonner, *The Kansas Doctor: A Century of Pioneering* (Lawrence: University Kansas Press, 1959), 36-37.
30. *Ibid.*
31. Schlissel, *Women's Diaries*, 129.
32. M. DeMund, *Women Physicians of Colorado* (Denver: Range Press, 1976), 1.
33. C. Luchetti, *Women of the West* (St. George, UT: Antelope Island Press, 1982), 27.
34. Bonner, *The Kansas Doctor*, 12.
35. M. E. Pickard and R. C. Buley, *The Midwest Pioneer: His Ills, Cures, and Doctors* (New York: Schuman, 1946), 154.
36. R. Hoppin, "Personal Recollections," *Michigan Pioneer & Historical Collections, 1912*, 38:414.
37. J. Duffy, "Medicine in the West," 11.
38. R. L. Numbers and J. W. Leavitt, *Wisconsin Medicine: Historical Perspectives* (Madison, WI: 1981), 17.
39. R. C. Buley, "Pioneer Health Prior to 1840," *Mississippi Valley Historical Review*, 20 (1933-1934): 510.
40. Myres, *Ho for California*,145.
41. A. Britt, *An America That Was: What Life was Like on an Illinois Farm Seventy Years Ago* (Barre, MA: Barre Pub., 1964), 135.
42. William Buchan, *Domestic Medicine, or, A treatise on the Prevention and Cure of Diseases by Regimen and Simple Medicines with an Appendix, Containing a Dispensatory for the Use of Private Practitioners*, 5th ed. corrected (London: W. Strahan, T. Cadell, J. Balfour, and W. Creech, 1776).
43. J. C. Gunn, *New Family Physician*, 100th rev. ed. (Cincinnati: Moore, Wilstack and Baldwin, 1864), 529-533.
44. *Ibid.*
45. *The American Lady's Medical Pocket Book, and Nursery Adviser*

(Philadelphia: Hayes and Zell, 1854), 129-132.
46. *Ibid.*, 113.
47. Bonner, *The Kansas Doctor*, 12.
48. K. L. Holmes, *Covered Wagon Women: Diaries and Letters from the Western Trails 1840-1890*, vol. 1 (Glendale CA: Arthur Clark, 1983), 158.
49. C. L. Bushman, *Mormon Sisters: Women in Early Utah* (Cambridge: Emmeline Press, 1976), 52.
50. Clarinda Tate letter to her mother Mrs. James Tate, 24 April 1823 in the Tate Papers, Missouri State Historical Society, St. Louis, MO.
51. Pickard and Buley, *The Midwest Pioneer*, 101.
52. F. A. Long, *A Prairie Doctor of the Eighties* (Norfolk, NE: Huse Pub., 1937), 23-29.
53. S. Crumbine, *Frontier Doctor* (Philadelphia: Dorrance, 1948), 39.
54. Long, *A Prairie Doctor of the Eighties*, 23-29.
55. B. J. Byrne, *A Frontier Army Surgeon, Life in Colorado in the Eighties* (New York: Exposition Press, 1935), 63.
56. J. Ise, *Sod and Stubble, The Story of a Kansas Homestead* (Lincoln: University Nebraska Press, 1936), 86.
57. Numbers and Leavitt, *Wisconsin Medicine*, 20.
58. *Ibid.*
59. Long, *A Prairie Doctor of the Eighties*, 68.
60. Schlissel, *Women's Diaries*, 57.
61. Hampsten, *To All Inquiring Friends*, 57-58.
62. Schlissel, *Women's Diaries*, 45.
63. E. French, *Emily, the Diary of a Hard-working Woman* (Lincoln: University of Nebraska Press, 1987), 92.
64. Hampsten, *To All Inquiring Friends*, 121.
65. E. P. Stewart, *Letters of a Woman Homesteader* (Lincoln, NE: 1961), 47.
66. A. Laufe, *An Army Doctor's Wife on the Frontier: Letters from Alaska to the Far West, 1874-1878* (Pittsburgh: University of Pittsburgh, 1962), 68-70.
67. *Ibid.*
68. Franklin, *Private Pages*, 337-338.
69. Degler, *At Odds*, 62.
70. M. S. Wyoski, "Women and Mining in the Old West," *Journal of the West*, 20 (1981):44.

Charles R. King, M.D., is Professor of Reproductive Genetics at the University of Kansas Medical Center in Kansas City and a specialist in gynecology and obstetrics. He also has recently obtained an M.A. in history from the University of Kansas (Lawrence). He has a grant currently for maternal mortality surveillance and another to study the social explanations for the decline in maternal mortality. He has published widely in the medical professional journals and is now also publishing books in the history of medicine.

BLOODLETTING IN AMERICAN OBSTETRIC PRACTICE, 1800-1945*

A. CLAIR SIDDALL**

Prior to the twentieth century, bloodletting was considered by orthodox practitioners of medicine to be the most powerful remedy in the treatment of disease.[1] This therapy perhaps originated in folk lore[2] and was elevated to respectability when Hippocrates incorporated it into regular medical practice.[3] In 1847, it was said, that since his time, very little had been added to the stock of practical medical knowledge.[4]

In the United States and Europe, this ancient and popular cure-all was not employed in pregnancy as long as the midwives maintained their traditional monopoly over childbirth.[5] But the eighteenth and nineteenth centuries brought dramatic social changes for women in their maternal care. This was accomplished by well-trained male physicians who were willing to face popular prejudice and hostility and inaugurate scientific studies concerning this neglected area in medicine.[6]

Leadership in the development of obstetrics took place first in Great Britain and Europe. William Smellie (1697-1763) and William Hunter (1718-1833) and others, carried out careful investigations on the anatomy of the pregnant woman, on the mechanism of labor, and on the effective use of the obstetric forceps.[7] They introduced and extended the therapy of bloodletting for almost every complaint in pregnancy.

* Revised version of a paper read at the 52nd annual meeting of the American Association for the History of Medicine, Pittsburgh, Pa., May 4, 1979.

** The author acknowledges with gratitude the assistance of archivist Ms. Glen Jenkins and her staff at the Allen Memorial Library in Cleveland.

[1] James Wardrop, *Curative Effects of the Abstraction of Blood* (Philadelphia: A. Waldie, 1837), p. 1. See also, Henry Clutterbuck, "Lectures on bloodletting," *London Med. Gaz.*, 1838, *2:* 9 and Marshall Hall, *Cyclopaedia of Practical Medicine*, ed. by John Forbes, Alexander Tweedie, John Conolly, 4 vol. (Philadelphia: Lea & Blanchard, 1845), 1: 296.

[2] Henry E. Sigerist, *A History of Medicine*, 2 vols. (New York: Oxford Univ. Press, 1951), 1: 116, 202.

[3] Francis Adams, *The Genuine Works of Hippocrates* (Baltimore: Williams & Wilkins, 1939), p. 80. "Bleed in acute infections if the disease appears strong and the patient in the vigor of health."

[4] Clutterbuck, "Bloodletting," p. 89.

[5] Alex Lee, *Celsus on Medicine*, 2 vols. (London: E. Cox, 1831), 1: 99. "It is not a new practice to let blood by the incision of a vein: but it is new to embrace this remedy in almost every disease."

[6] William F. Mengert, "The origin of the male midwife," *Annals Med. Hist.*, 1932, *4*, n. 5: 453-465.

[7] J. W. Johnson, *William Smellie: The Master of British Midwifery* (Edinburgh & London: E. and S.

BULLETIN OF THE HISTORY OF MEDICINE **Vol. 54** Pp. 101-110
0007-5140/80/0541-0101 $01.00

Bloodletting in American obstetric practice developed in two waves; the first larger swell is visible from 1800 to 1850;[8] the second smaller rise lasted from mid-century to 1945 with such treatment being confined to toxemia. Due to the scientific advances in physiology and pathology and due to an increasing popular resistance to all venesection therapy, the long accepted remedy of bloodletting in pregnancy was finally abandoned in the twentieth century.

It is not surprising that the practice of bloodletting in the treatment of disease had a direct influence on its use in pregnancy. In the gradual growth of an American obstetrical speciality after 1800, the leaders of greatest influence were located in Philadelphia. There they came under the shadow of Benjamin Rush (1746-1813) who had gained the reputation as the most renowned proponent of bloodletting in medical practice. He professed to have no faith in the healing power of nature. He believed and he taught his students at the University of Pennsylvania that the blood volume of the human body was 25 to 28 pounds[9] (12,500-14,500 cc.), instead of the true value of less than half that amount.[10]

Rush's observations on the immediate effects of bloodletting seemed to furnish evidence for multiple benefits which the obstetricians later found to be most desirable in pregnancy. Rush noted that bloodletting relieved pain, promoted relaxation and sleep, subdued fever and the force of the circulation, checked vomiting and prevented hemorrhage.[11] Finally, when

Livingston, 1952); Fielding H. Garrison, *History of Medicine*, 4th ed. (Philadelphia: W. B. Saunders, 1960), pp. 339, 340.

[8] B. Fordyce Barker, "Bloodletting as a therapeutic resource in obstetric medicine," *Med. and Surg. Reporter*, 1871, *24*, n. 1: 1: "In all the consultations in obstetric practice . . . during the last fifteen years, I do not recollect of an instance where bloodletting has been resorted to except a few cases of puerperal convulsions."

[9] Carl Binger, *Revolutionary Doctor: Benjamin Rush* (New York: W. W. Norton, 1966), pp. 215, 227, 229. See also, Albrecht von Haller, *First Lines of Physiology*, First Am. ed. Trans. 3rd Latin ed. (Troy: Obadiah Penniman & Co., 1803), ch. 5, p. 66. True blood volume is 28 pounds.

[10] F. A. Bainbridge & J. Acworth Menzies, *Essentials of Physiology*, 3rd ed. (London: Longmans, Green, 1919), p. 191: "Results indicate the blood forms one-thirteenth of the body weight." Also, W. W. Wintrobe, *Clinical Hematology*, 7th ed. (Philadelphia: Lea & Febiger, 1975), p. 696. Average blood volume is 5000 cc.

[11] Binger, *Benjamin Rush*, p. 228. Note: In addition to the obvious immediate physical effects of bloodletting, the patient, no doubt, experienced a strong psychological impact. The shedding of one's own blood could well have carried a feeling of purification, mystery and magic. Because of the traditional prestige of bloodletting therapy, the patient likely felt a sense of security when he submitted to venesection. Other aspects of bloodletting must have been intriguing. It involved total cooperation between the patient and the physician. It was easy to carry out. It was an economical remedy since the cure of disease lay inside the body of each person. Bloodletting was quickly available to rich and poor alike, whether in the city or in more remote regions. In moderation it caused no harm but on the other hand, it was easily abused.

bloodletting was carried to the point of syncope, a temporary total anesthesia of the patient resulted.

In 1803 Rush did not hesitate to advise bloodletting in pregnancy. In answering a query from Dr. Edward Miller, he asserted that no woman need suffer in parturition. Bloodletting, he wrote, would not only relieve pain and accelerate labor, but would also prevent many complaints, such as, "after pains, hemorrhage, obstructed lochia, retained placenta, prolapse of the uterus, inflamed breasts, colics, madness, and fever." Furthermore, he said the patient's friends would be pleased to see these good effects.[12] Therefore, at a time when no specific medications were available, bloodletting held top priority as long as it elicited the support of both the patient and the physician because, in the case of the patient, it was she who in the long term dictated her own treatment.

Bloodletting therapy became the fashion in American obstetric practice when the patients received little or no prenatal care and often endured ten or twelve pregnancies. Inevitably they must have suffered from a greater or lesser iron deficiency anemia with the demands of these multiple gestations.

The man who exerted the most influence on American obstetrics was William P. Dewees (1768-1841).[13] He was twenty-two years younger than Rush under whose patronage he came to Philadelphia in 1793. He was a forceful character and an independent thinker. He became Adjutant Professor of Midwifery at the University of Pennsylvania in 1825 and was noted as a fearless bloodletter in his speciality. Early in his career, Dewees wrote that, "few women object to being bled, nay they, for the most part, think it indispensable, when labor has been protracted."[14] As late as 1864, Hugh L. Hodge (1796-1873) who succeeded Dewees on the faculty, also testified that patients in labor expressed in the strongest terms, their sense of relief, mentally and physically, after the loss of 15 to 20 ounces (425-600 cc.) of blood.[15]

Like Rush, Dewees believed that excessive pain was not necessarily an accompaniment of labor.[16] Therefore, in a case of protracted labor with rigid uterine cervix, he proceeded as follows: While the patient stood

[12] Benjamin Rush, "Letter to Dr. Edward Miller," *Medical Repository*, 1803, *6:* 26-29.

[13] Harold Speert, *Obstetric and Gynecologic Milestones*.(New York: Macmillan, 1958), p. 605.

[14] William P. Dewees, *Essays on Various Subjects Connected with Midwifery* (Philadelphia: H. C. Carey & I. Lea, 1823), p. 59.

[15] Hugh L. Hodge, *Principles and Practice of Obstetrics* (New York: Blanchard and Lea, 1864), p. 424.

[16] William P. Dewees, *An Essay of the Means of Lessening Pain and Facilitating Certain Cases of Difficult Parturition* (Philadelphia: J. H. Oswald, 1806), p. 95.

erect, 30 or 40 ounces (900-1200 cc.) of blood were withdrawn from the arm vein to the point where the patient fainted. After being replaced in bed, she was often totally unaware when the infant was delivered.[17] He found that this method required less blood loss than if venesection was done with the patient recumbent. Dewees reported two cases in this category; one of which required the withdrawal of two quarts (2000 cc.) of blood for satisfactory results and another requiring 50 ounces (1500 cc.).[18] Such therapy was employed only if the patient was plethoric and strong. Prior to the discovery of the general anesthetics at mid-century, nothing could surpass bloodletting in its prompt lessening of pain for the laboring patient.

During this time new theories were added to the old in order to justify the wide utilization of bloodletting in pregnancy. Thomas Denman (1733-1815) of London was the clearest spokesman for these generally accepted concepts. He stood next to Dewees in influence on American obstetric practice.[19] The fifth edition of his comprehensive and well written text was published in the United States in 1821. Its value was enhanced by the inclusion of numerous notes and comments by New York's outstanding obstetrician, John W. Francis (1789-1861).[20]

Denman's writings reflected the serious lack of progress in physiology and pathology when he stated that pregnancy was accompanied by a plethora which was due to the retention of the menses.[21] The term plethora had survived from early times and connoted an excess of blood in the system of the patient who exhibited a robustness with flushed face and a strong pulse.[22] Thus, plethora was a morbid condition fraught with

[17] William P. Dewees, *Compendious System of Midwifery* (Philadelphia: H. C. Carey & I. Lea, 1824), pp. 370-371. Also, William Smellie, *Cases and Observations in Midwifery* (London: T. Wilson & T. Durham, 1758), p. 251. In contrast to Dewees, Smellie withdrew only eight ounces of blood in cases of protracted labor.

[18] Dewees, *Compendious System*, pp. 373, 377. See also, Daniel Drake, "Comment," *Med. and Physiol. J.*, 1827-8, *10:* 162, "In country practice 19 out of 20 cases, where females are generally of a strong plethoric habit, bleeding is required for protracted labor from rigid cervix" and Henry Miller, *Principles and Practice of Obstetrics* (Louisville: John V. Cowling & Geo. C. Davies, 1849), p. 158. He agreed with the practice of the day—for the treatment of a rigid cervix—"the most successful remedy is bloodletting."

[19] Speert, *Milestones*, p. 605.

[20] Thomas Denman, *An Introduction to the Practice of Midwifery*. From the last London ed., with notes and emendations by John W. Francis (New York: E. Bliss & E. White, 1821).

[21] *Ibid.*, p. 232. and footnote by John Francis, same page. Also see John Vaughan, "An inquiry into the utility of occasional bloodletting in the pregnant state of disease." *Medical Repository*, 1803, *6:* 152. This author believes that during the ten lunar months of pregnancy, the patient retains in her circulation five to ten pounds of unshed menstrual blood.

[22] Robert M. Green, *Galen's Hygiene* (Springfield: Charles C Thomas, 1951), p. 251.

danger which could be removed by bloodletting. Therefore, during the first half of the nineteenth century, the obstetricians bled their plethoric patients for almost all of their minor antenatal complaints, such as headaches, vertigo, nausea and vomiting, insomnia, apprehension, palpitation, muscle cramps, diarrhea, hemorrhoids, varicose veins, and swelling of the lower extremities.[23] Almost without fail, the patients expressed relief after venesection of a small amount of blood.

It is difficult to see how any patient under an obstetrician's care could escape venesection since it was considered an effective preventive measure as well as a curative one. For example, if a woman had a history of a previous abortion, it was considered beneficial during a subsequent gestation to withdraw a few ounces of blood soon after the first missed period, and then two or three times thereafter.[24] It was also considered a wise precaution to take 8 or 10 ounces of blood from normal prenatal patients in order to relax a rigid cervix and prevent post partum hemorrhage, infection and convulsions.[25] Prior to performing a podalic version, most practitioners employed moderate bloodletting.[26]

Denman explained further the generally accepted concepts, that in addition to plethora during pregnancy, the woman manifested a feverish disposition with increased body heat,[27] and an extreme irritability of the uterus.[28] Finally, he referred to the constant finding of a "buffy coat" on the surface of the patient's clotted blood.[29] John Hunter (1728-1793) discovered that, "women who are breeding, and are in perfect health, always

[23] Denman, *Practice of Midwifery*, pp. 227-8, 234, 248. See also, John Burns, *The Principles of Midwifery*, with notes by N. Chapman (Philadelphia: Hopkins & Earle, 1810), pp. 172-185 and Charles White, *A Treatise on the Management of Pregnant and Lying-in Women* (London: Edward and Charles Dilly, 1777), p. 221. White did not do bloodletting for minor complaints in pregnancy, "I never found bleeding necessary except when inflammation of the womb has been brought on by violence used in the extraction of the child or of the secundines."

[24] Denman, *Practice of Midwifery*, p. 642. Also, Samuel Bard, *A Compendium of the Theory and Practice of Midwifery* (New York: Collins & Perkins, 1808), p. 96.

[25] Charles Severn, *Practice of Midwifery* (London: S. Highley, 1831), p. 51.

[26] Denman, *Practice of Midwifery*, p. 633. Also see p. 651 under "Aphorisms" for Denman's more conservative attitude: "The less the quantity of blood discharged postpartum, the better women in general recover." Also, Charles D. Meigs, *The Philadelphia Practice of Midwifery, Lectures on Midwifery* (Philadelphia: James K. Jun & Brother, 1838), p. 254: "A patient bled to syncope will be more capable of undergoing safely the operation of turning"; Edward Rigby, *A System of Midwifery* (Philadelphia: Lea & Blanchard, 1841), p. 236.

[27] Denman, *Practice of Midwifery*, p. 231.

[28] *Ibid.*, p. 226.

[29] *Ibid.*, p. 231. See also Wardrop, *Curative Effects*, p. 14. and Hodge, *Principles of Obstetrics*, p. 94: "The buffy coat of coagulated blood is seen in cases of acute inflammation. The same is seen in pregnancy."

have sizy blood (buffy coat) . . . this blood should not be called inflammatory blood, but blood whose powers of life are increased.''[30] As blood from medical cases of severe inflammation manifested a similar buffy coat and they were traditionally treated by bloodletting, many obstetricians felt justified in using this sign for bloodletting in pregnancy. Many patients suffered needless loss of blood.[31] Denman hastens to reassure the reader that we have no reason to be solicitous about the investigation of the cause of all these changes in pregnancy because bloodletting would relieve them.[32]

During the first half of the nineteenth century, the American obstetricians, like their European colleagues, adopted bloodletting not only for minor complaints but also for the major complications in pregnancy, such as hemorrhage, post partum infection and puerperal convulsions.

The paradoxical treatment of hemorrhage in pregnancy by bloodletting was based on the logic that in producing a faint by venesection, the force of the patient's circulation was so reduced that clotting could take place, which in turn checked the hemorrhage.[33] With this reasoning, Dewees proceeded in one case of threatened abortion at four months, complicated by hemorrhage, to bleed the patient 17 times in seven days. He withdrew a total of 110 ounces (3300 cc.) of blood. The hemorrhage was controlled and later a safe delivery achieved.[34] Francis reported a case who, in the last month of her eleventh pregnancy, suffered three gross hemorrhages. Each episode was treated by bloodletting of 8 ounces (240 cc.), 10 ounces (300 cc.) and 6 ounces (180 cc.) respectively. The patient fainted several times before the delivery of a stillborn infant. She suffered blindness for a period of six weeks post partum. Francis later asserted with obvious pride that it was incredible how much blood could be lost and yet with careful management, the patient would recover.[35] It is obvious that his only criterion for success was the survival of the patient.

[30] John Hunter, *A Treatise on the Blood, Inflammation and Gunshot Wounds* (London: John Richardson, 1794), p. 316, footnote. Also, Louis M. Hellman & Jack A. Prichard, editors, *Williams Obstetrics*, 14th ed. (New York: Appleton, Century, Crofts, 1971), p. 762: "The rapid sedimentation rate of erythrocytes is induced by the hyperfibrinogenemia of pregnancy."

[31] Hodge, *Principles of Obstetrics*, p. 94. "The argument therefore is strong, that as similar conditions of the blood exist in pregnancy and in inflammatory complaints, there is an analogous condition of the general circulation." Also, Gunning Bedford, *System of Obstetrics* (New York: Samuels & W. Wood, 1861), p. 134: "The general belief, so far as practice is concerned, is that wherever the buffy coat is recognized, it is an urgent indication for further depletion."

[32] Denman, *Practice of Midwifery*, p. 232.

[33] Denman, *Practice of Midwifery*, p. 473; Bard, *Compendium of Midwifery*, p. 156; Wardrop, *Curative Effects*, p. 5.

[34] Dewees, *Compendious System*, p. 408.

[35] John W. Francis, Footnote in Denman, *Practice of Midwifery*, pp. 507-510. See also, Charles D.

Toward mid-century voices were heard opposing this practice of treating hemorrhage by bloodletting,[36] the most cogent being the report of Gabriel Andral (1797-1876) of France. His laboratory studies revealed that during gestation there was a diminished number of red blood cells in the circulation of the pregnant woman and not an excess as long believed.[37] This proved to be the death blow to the old theory of plethora in pregnancy, and in time, bloodletting for hemorrhage became obsolete.[38] Gradually, substitutes for bloodletting were employed, such as wider use of ergot post partum, pouring cold water on the abdomen,[39] placing a piece of ice into the uterine cavity,[40] compression of the aorta,[41] intravenous saline solution, binding the lower extremities, routine use of iron medication,[42] and careful suturing of perineal and cervical lacerations.[43] Finally, this problem of hemorrhage was most successfully met when blood transfusions became practical in 1915.[44]

No complication of pregnancy presented a more distressing challenge to the nineteenth-century practitioner than puerperal fever. It is significant that no epidemic of this disease seems to have occurred as long as the midwives held their monopoly over childbirth.[45] The medical profession seemed mystified when early in the century, this disease spread rapidly

---

Meigs, *Puerperal Fever* (Philadelphia: Ed Barrington & Geo. D. Haswell, 1842), p. 233: "I think it a great mistake to say that loss of blood, even an enormous loss, is in any degree injurious to the constitution of the individual." Also, Francis H. Ramsbotham, *Principles and Practice of Obstetric Medicine and Surgery*, 1st. Am. ed. (Philadelphia: Lea & Blanchard, 1842), p. 394. While most authors during this time paid close attention to the immediate effects of bloodletting and ignored the long-term ill effects, this writer was aware of the injuries which resulted. "Some effects are evanescent, others do not occur for some time, but are more permanent. Of the latter kind are cachexia, wasting, purgings, dyspepsia, dropsies, especially an edematous state of the legs and feet."

[36] Ramsbotham, *Obstetric Medicine*, p. 397: "All evacuant remedies must be avoided, particularly bleeding (for hemorrhage), although I have known venesection repeatedly resorted to." William H. Byford, *Theory and Practice of Obstetrics* (New York: Wm. Wood, 1873), p. 389: "I cannot recommend bloodletting (for hemorrhage), it increases the effects of loss of blood . . . and is productive of great evil."

[37] G. Andral, *Essai d'hématologie pathologique* (Brussels: Societé Encyclopédique des Sciences Médicales, 1843), p. 153.

[38] Charles Jewett, ed., *Obstetrics* (New York: Lea Brothers, 1899), p. 124. "Thirty-five years ago it was common to bleed the obstetric patient. Now that the blood conditions are better understood such practices have very properly become obsolete."

[39] Bedford, *System of Obstetrics*, p. 392.

[40] William Thompson Lusk, *The Science and Art of Midwifery* (New York: D. Appleton, 1882), p. 543.

[41] *Ibid.*, p. 539; P. Cazeaux, *Obstetrics*, 6th Am. ed. (Philadelphia: Lindsay & Blakiston, 1877), p. 892. For postpartum hemorrhage, "I use compression of the aorta up to four hours."

[42] Barton Cooke Hirst, *A Text Book of Obstetrics* (Philadelphia: W. B. Saunders, 1898), p. 542.

[43] Edward P. Davis, *Practical Obstetrics* (Philadelphia: P. Blakiston, 1898), p. 250.

[44] Arthur Hale Curtis, *Obstetrics and Gynecology*, 3 vols. (Philadelphia: W. B. Saunders, 1933), 3: 951.

[45] Bard, *Compendium of Midwifery*, p. 308. "I have been so fortunate as never to see this fatal disease (puerperal fever) epidemic in the city of New York."

and was accompanied by a high mortality among those women admitted to the lying-in wards of the newly-formed hospitals across the country.[46] Controversy continued for many years as to its cause and treatment. Most practitioners believed that sporadic cases were inflammatory in nature and called for prompt bloodletting. On the other hand, the epidemic form was thought of as a fever peculiar to pregnancy and characterized by exhaustion and severe debility which contraindicated any general bloodletting.[47]

Denman admitted that early in his career he doubted the propriety of bloodletting in puerperal fever but later became convinced that it was absolutely necessary.[48] John Armstrong (1784-1829) of London presented other reasons for his early withdrawal of 24 ounces (720 cc.) of blood in such cases. He wrote that it was dangerous not to deplete these patients, "dangerous to one's reputation."[49] Did this not mean that he was bowing to the demands of his patients? Charles D. Meigs was convinced that all cases of post partum infection were peritonitis and therefore required bloodletting. He never believed that puerperal fever was contagious[50] even after the publication in 1843 of the convincing essay by Oliver Wendell Holmes (1809-1894)[51] and the life-saving experiments of Ignaz P. Semmelweis (1818-1865) three years later.[52] In one successful case, Meigs withdrew 52 ounces (1560 cc.) of blood over a span of 12 hours and followed this with calomel and opium by mouth.[53]

[46] T. G. Morton, *The History of the Pennsylvania Hospital* (Philadelphia: Times Printing House, 1895), p. 238. The Lying-in ward was opened in 1803 but closed in 1830 due to puerperal fever. It was reopened in 1835 but abandoned in 1851 for the same reason. See also, B. Fordyce Barker, *Puerperal Diseases* (New York: D. Appleton, 1874), p. 439. From all sources: "More than 200 epidemics of this disease (puerperal fever) have been described by different authors since 1740 . . . a terribly fatal disease in lying-in hospitals in all great cities where such hospitals exist."

[47] Denman, *Practice of Midwifery*, p. 565. See also footnote on p. 590 by John Francis: "It must be remembered that the typhoid form of this complaint is rarely to be seen in this country." John Armstrong, *Facts and Observations and Practical Illustrations Relative to Puerperal Fever*, 1st Am. Ed. from 2nd London ed. (Hartford: Oliver Cooke & Sons, 1823), p. 8. He treated both types of puerperal fever by bloodletting. Robert Collins, *Midwifery*, (London: Longmans, 1826), p. 180. For puerperal fever he applied 48 leeches to the abdomen.

[48] Denman, *Practice of Midwifery*, pp. 575, 576. Another practitioner with similar opinions was Charles White, *Treatise on Pregnant Women*, p. 72. "I am convinced that bleeding is too indiscriminately used and too often."

[49] Armstrong, *Puerperal Fever*, pp. 42, 44.

[50] Meigs, *Midwifery*, p. 352.

[51] Oliver Wendell Holmes, "On the contagiousness of puerperal fever," *N. Engl. Quart. J. Med. and Surg.*, 1842-43, *1*: 503-530.

[52] Ignaz Philipp Semmelweis, "Höchst wichtige Erfahrungen über die Aetiologie der in Gebaranstalten epidemischen Puerperalfieber," *Zeit. Gesellsch. d. Aerzte in Wien*, 1847, *2*: 242; 1849, *5*: 64.

[53] Meigs, *Puerperal Fever*, p. 27.

Bloodletting for puerperal fever lost its prestige as practitioners accepted the increasing evidence of the contagious nature of the disease. It must have been a sad day when the obstetricians admitted their part in the spread of this dread infection. Cleanliness soon became the guideline and prevention the goal. Benjamin Fordyce Barker (1818-1891) in 1847 used Veratrum viride to serve as a substitute for bloodletting in treating this disease and for a time he followed the Alonzo Clark (1807-1887) method of utilizing only repeated doses of morphine.[54] In the 1870s bacteriologic research revealed the cause of this disease[55] and vaginal douches of 2 per cent carbolic acid solution became an accepted preventive measure. Finally, in 1935, sulfanilamide came on the market, followed in a few years by the antibiotics, which reduced maternal death from infection to its lowest level.

One of the most frightening complications of pregnancy is a convulsive disorder called eclampsia which is cured by the delivery of the infant. Dewees faced this syndrome with characteristic courage. In one case he drew a total of 120 ounces (3600 cc.) of blood from the vein over a period of seven hours and the patient survived.[56] He believed there was some advantage in releasing blood near the patient's head, therefore he often opened the jugular vein. Otherwise he used both arm veins at one time. Meigs found that a venesection of 30 to 60 ounces (900-1800 cc.) was adequate in controlling the convulsions if he then applied cups to the temples and nape of the neck.[57] In addition, he directed that the patient's head should be shaved so leeches could be applied to the scalp.

By mid-century professional and popular disapproval became significant factors in restraining all bloodletting therapy.[58] About this time ether and chloroform were employed, sometimes continuously for six hours, in order to control the eclamptic convulsions.[59] But the most effective substitute for bloodletting in this affection was Veratrum viride which came

[54] Garrison, *History of Medicine*, p. 577; also, Barker, *Puerperal Diseases*, pp. 344, 492.

[55] Louis Pasteur, "Septicémie puerpérale," *Bull. Acad. de Méd.*, 2ᵉ Série, 1879, *8:* 260-271.

[56] Dewees, *Essay on Midwifery*, p. 167. He did not force labor, p. 169.

[57] Meigs, *Midwifery*, pp. 280, 282. Also, Hodge, *Principles of Obstetrics*, p. 449: "The introduction of free bleeding (for puerperal convulsions) by Denman, Ramsbotham Sr., and Dewees has greatly relieved the mortality which is now twenty-five per cent." Also, P. Cazeaux, *Obstetrics*, p. 817. As a substitute for bloodletting in eclampsia, Cazeaux used Junod's boots or cups.

[58] Hirst, *Obstetrics*, p. 579: "Venesection at present is somewhat in disfavor. The reaction against the indiscriminate use of the lancet has gone too far." Also, Leon S. Bryan, Jr., "Bloodletting in American medicine," *Bull. Hist. Med.*, 1964, *38:* 517, 528.

[59] Barker, *Puerperal Diseases*, p. 119. Also, Davis, *Obstetrics*, p. 172.

into use in the 1870s.[60] It reduced the excitement of the circulation, lowered the pulse rate to 30 or 40 beats per minute, as well as producing diaphoresis and diuresis.[61] Most important of all, it seemed to prevent any subsequent convulsions.

About the time when the obstetricians were rejecting bloodletting,[62] a new theory was proposed by Gunning Bedford (1806-1870) that puerperal convulsions were caused by a poison or a toxemia in the system.[63] Once again it became logical to withdraw blood in an attempt to remove at least a part of this poison. This new concept accounts for the second rise in bloodletting therapy in pregnancy. J. Whitridge Williams (1866-1931) of the Johns Hopkins University urged in his textbook of 1920, that 1000 cc. of blood should be taken from the eclamptic patient promptly, irrespective of the condition of the pulse.[64] Twenty-five years later the ill effects of such treatment were admitted and the modern conservative management of this still unsolved problem was summarized by Paul Titus (1885-1951) of Pittsburgh. He wrote, "omit venesection, colonic irrigations, gastric lavage, purging, and sweating." Instead he recommended the use of sedatives, hypertonic glucose solution intravenously, magnesium sulfate intramuscularly, and Veratrum viride.[65] Cesarean section was advised if the patient did not improve under medical management, otherwise labor was not forced.

In conclusion, bloodletting, which was uncritically accepted as the fashion in early American obstetric practice, finally was abandoned after one hundred and fifty years. Scientific advances gradually revealed the false premises on which bloodletting rested. In addition, the patients, who alone suffered the long-term ill effects, ultimately rebelled against its use.

[60] Hirst, *Obstetrics*, p. 579. For the past twenty-five years Veratrum viride has been the favored American treatment for eclampsia.

[61] Jewett, *Obstetrics*, p. 527.

[62] B. Fordyce Barker, "Bloodletting as a therapeutic resource in obstetric medicine," *Med. and Surg. Reporter*, 1871, *24*, n. 1: 1: "In all the consultations in obstetric practice with members of the profession during the last fifteen years, I cannot recollect a single instance where bloodletting has been resorted to, or even alluded to as a therapeutic measure to be discussed, except in a few cases of puerperal convulsions."

[63] J. Whitridge Williams, "A sketch of the history of obstetrics in the United States up to 1860," *Am. Gynec.*, 1903, *3*, n.1, 360.

[64] J. Whitridge Williams, *Obstetrics*, 4th ed. (New York: D. Appleton, 1920), p. 587. Also, Joseph B. DeLee, *Principles and Practice of Obstetrics* (Philadelphia: W. B. Saunders Co., 1913), p. 367: "Venesection should be reserved until after delivery."

[65] Paul Titus, *The Management of Obstetric Difficulties* (St. Louis: C. V. Mosby, 1945), p. 432 and Hellman, *Obstetrics*, p. 727.

THE TRAINING AND PRACTICE OF MIDWIVES: A WISCONSIN STUDY*

Charlotte G. Borst

On 1 February 1887, Caroline Kueny, thirty-five years old and the mother of five children, began her six-month course at the Milwaukee School of Midwifery. Attending daily lectures and sixteen confinements during the course of her instruction, Kueny graduated after passing her final examinations to the satisfaction of the school's two physician examiners. She immediately registered with the Milwaukee Health Department and began attending births. Over the next fifteen years she worked as a midwife in Milwaukee and had several more children of her own.[1]

While Caroline Kueny was obtaining her midwifery skills at school, Dora Larson, living in the small, rural village of Woodford, Wisconsin, became a midwife in a more traditional way. Thirty-seven years old and the mother of eight children, Larson learned midwifery techniques from assisting her midwife mother, by helping local physicians attending confinement cases, and from her own childbirth experiences.[2] She noted on her 1910 state license application that her midwife credentials included being the mother of eight children and the grandmother of four.[3]

In 1910, when Kueny and Larson applied for state licenses from the Wisconsin Board of Medical Examiners, many other applicants had experience similar to theirs. Although the women who applied for licenses had become midwives in many ways—demonstrating the problem of defining the "American midwife"—patterns in the manner of their training and in the tempos of their practice are discernible. In addition, Kueny's and Larson's status as mothers and grandmothers typified most of Wisconsin's midwives:

* Revised version of a paper read at the fifty-ninth annual meeting of the American Association for the History of Medicine, Rochester, New York, 1 May 1986. I wish to thank Judith Walzer Leavitt for her patient reading of and generous comments on several drafts of this paper. The Alice E. Smith Fellowship from the Wisconsin State Historical Society, and grants from the University of Wisconsin Graduate School and the Maurice Richardson Fund for the History of Medicine, helped defray the cost of the data entry and computer analysis.

[1] Information about Kueny's schooling was found in her midwife license application, Midwife File, State Board of Medical Examiners, Archives Division, Wisconsin State Historical Society, Madison, Wisconsin (hereafter Midwife File). Information concerning her family life came from the 1900 federal manuscript census for Milwaukee.

[2] License application of Dora Larson, Midwife File.

[3] *Ibid.*

almost every midwife practicing in the state had children. Indeed, it seemed that motherhood was almost a prerequisite for the job.

Historical studies of American midwives generally have employed the term *midwife* to mean any non-physician, usually female, attending a childbirth. The focus of most scholarly studies—the elimination of midwives by a hostile medical profession—explains why many historians have not analyzed the differences among these female practitioners.[4] By evaluating midwives only within the scope of the "midwife question" of the early twentieth century, historians have been forced to see these female practitioners through the eyes of the doctors engaged in the debate. That is, most of their sources of information about midwives are articles from medical journals or papers delivered at conferences controlled by physicians.

Frances E. Kobrin's pioneering article suggested that midwifery was the victim of a crisis of professionalism within the medical profession over the role of the new specialty of obstetrics.[5] However, I will argue in this article that American midwifery also faced its *own* professional crisis. Like physicians and nurses, schooled midwives began to replace their untrained or apprentice-trained counterparts at the turn of the twentieth century. Unlike these other practitioners, however, midwives did not become professionalized as they became more educated; instead, they became increasingly subservient to physicians, their autonomy decreased, and, ultimately, they ceased functioning in the United States.

The evolution of midwifery training is not the sole explanation for midwives' professional problems at the beginning of the twentieth century. Most midwives, however they learned their trade, were married women who practiced within the limited confines of a discrete ethnic and class-based community. Thus, any study of the demise of midwives in the United States must incorporate an analysis of the geographic, class, and ethnic dimensions of the midwives' practices, a consideration of their gender and marital status, and an examination of the results of changes in their training.

The best way to understand this crisis of professionalization is to exam-

[4] For example, Frances E. Kobrin carefully delineates the characteristics of the various physician groups who were debating the "midwife problem" at the beginning of the twentieth century. However, she only identifies the midwives as sharing "race, nationality, and language with their customers" (p. 351). See Kobrin, "The American midwife controversy: a crisis of professionalization," *Bull. Hist. Med.*, 1966, *40*: 350–63.

Judy Barrett Litoff's sweeping study of American midwives also focuses primarily on the physician debates of the early twentieth century. While her evidence demonstrates that many varieties of midwives practiced in different areas of the country, she allows no regional distinctions for the "American midwife," grouping together such diverse women as Southern black grannies, Northern, native-born white neighbor-women, and immigrant midwives trained in the European hospital schools (Litoff, *American Midwives, 1860 to the Present* [Westport, Connecticut: Greenwood Press, 1978], pp. 53, 60). Despite the disparate nature of these groups of female practitioners, Litoff argues that there was no clear distinction between the official midwife and the neighbor (p. 27). Litoff acknowledges the difficulties in drawing a composite picture of early twentieth-century midwives, but this problem relates directly to her sources. Because she did not use data directly focusing on the midwives themselves, such as midwife licenses or birth registration certificates, she could not provide detailed information about individual midwives and their practices.

[5] Kobrin, "American midwife controversy," p. 350.

ine the cultural context of these midwives' practice by focusing on a particular geographic location. This article examines midwifery in Wisconsin.

Wisconsin was fairly typical in its political stance toward midwives. Beginning informally in Milwaukee in the 1870s and culminating with the 1909 Midwife Registration Act administered by the State Medical Practice Board, physicians in the state helped to determine which midwives could deliver babies and the conditions under which they could work.[6] Thus, Wisconsin did not ban these female attendants outright when there were still women who wished to practice, but neither did the state leave the practice of midwifery unregulated.[7]

Sociologists and historians have identified an occupation's movement toward professionalization with the "standardization allowed by a common and clearly defined basis of training.... [This training] is, in fact, the main support of a professional subculture."[8] Instruction, Magali Sarfatti Larson points out, historically proceeds from the relative informality of apprenticeships to formal education based on the standardization and the codification of knowledge.[9] While the rise of a profession is tied intimately to the development of formal training, a profession also needs a distinct body of knowledge to claim exclusively. To then achieve a monopoly of practice, the profession needs to control both the "production of knowledge and the production of producers," preferably within one institution.[10] In summary, the professional practitioner must have mastered a body of knowledge unique to the field within a formal setting and then have the autonomy to decide when and under what circumstances to apply this knowledge.

By the first decades of the twentieth century, leaders in both medicine and nursing recognized clearly that better training fulfilled their professions' objective need for more knowledge and their subjective desire to

[6] The Milwaukee Health Department began requiring midwife registration in the 1870s (see "Dr. Wight's wisdom," *Milwaukee (Wisconsin) Sentinel*, 20 July 1878). In 1909, the state of Wisconsin began requiring all midwives to have licenses. Women already in practice were allowed to continue, regardless of their training. But new practitioners were required to have a diploma from a "reputable school of midwifery," defined as being "connected to a reputable hospital or sanitarium, [offering] a twelve month course in the science of midwifery and experience in twelve confinements." Candidates were also required to pass an exam relating to the clinical practice of midwifery. *Wisconsin Statutes*, 1909, chap. 528, s 1435f-12.

[7] In 1894, Massachusetts was the first state to forbid midwifery practice by refusing to license midwives and by recognizing attendance at childbirth as medical practice. Thus, midwives delivering babies were subject to prosecution for practicing medicine without a license. Midwives continued to deliver babies in Massachusetts, however, often under the protection of a local physician who would sign the birth certificate. Contemporary critics of this approach to midwife regulation noted this discrepancy. See J. M. Baldy, "Is the midwife a necessity?" *Amer. J. Obstet.*, 1916, *73*: 399–400. A recent historical study analyzed this phenomenon: Eugene Declercq, "The nature and style of practice of immigrant midwives in early twentieth-century Massachusetts," *J. Soc. Hist.*, 1985, *19*: 113–29. Many Southern states completely ignored their midwife practitioners in the early twentieth century. Midwives in these states, mainly black "grannies," did not have to hide their practice, but the state had no records about them.

[8] Magali Sarfatti Larson, *The Rise of Professionalism: A Sociological Analysis* (Berkeley and Los Angeles: University of California Press, 1977), p. 45.

[9] *Ibid.*, pp. 44–45. Also Paul Starr, *The Social Transformation of American Medicine* (New York: Basic Books, 1982), p. 81.

[10] Larson, *Rise of Professionalism*, p. 17.

improve the status of practitioners. As one historian has written, "The aim and end of reform ... [was] professionalization."[11] Physician and nursing leaders understood professionalization to mean credentials based on education and licensing standards that would enable practitioners to assert the value to human life of their specialized knowledge and their special skills. At the same time, professionalization was a movement to strengthen status and to consolidate authority, which allowed professionals to seek broad social prerogatives.[12]

This paper asks the same questions about midwives that medical historians have posed for physicians and nurses: Did midwife training follow the classic path toward professionalization—from casual learning to apprenticeships to school training? Were midwives able to claim an exclusive body of knowledge? Were these practitioners then able to control their own institutions and to determine who could practice midwifery?

From state and county records, I identified 893 midwives practicing in Wisconsin between 1870 and 1920. Defined by the sources of their training, these practitioners may be divided into three groups: (1) the neighbor-women, usually native-born women who learned their techniques from giving birth themselves and from assisting at the deliveries of relatives and a few close friends; (2) the apprentice-trained midwives, typically native-born, who learned their craft from other, older, midwives or from physicians; (3) school-educated midwives, usually first- or second-generation immigrants, instructed in formal settings through both didactic and practical work. Though all three kinds of midwives could be found practicing in Wisconsin during much of the fifty-year period of my study, there was a distinct evolution from neighbor-women to apprentice-trained practitioners to school-trained practitioners.

NEIGHBOR-WOMEN

Distinguished from the other midwives by their completely informal training, neighbor-women were the most "traditional" practitioners in my study.[13] Their experiences are exemplified by Susan Washburn, a sixty-year-old Millston, Wisconsin, woman who wrote in 1909: "I would state that my

[11] Barbara Melosh, *"The Physician's Hand": Work, Culture, and Conflict in American Nursing* (Philadelphia, Pennsylvania: Temple University Press, 1982), p. 4.

[12] *Ibid.*, p. 17. For a longer discussion, see Eliot Freidson, *Profession of Medicine: A Study of the Sociology of Applied Knowledge* (New York: Dodd, Mead, 1970).

[13] Judith Walzer Leavitt has defined *traditional* childbirth as a birth that remained within a female-centered network. The attendant was a midwife, trained or not. (Leavitt, *Brought to Bed: Childbearing in America, 1750–1850* [Oxford and New York: Oxford University Press, 1986], pp. 73–74). As I try to show in this article, the definition of who was a midwife in the nineteenth and early twentieth centuries did not remain static. However, despite this occupation's move toward school training, its practitioners and its philosophy toward birth remained tied to gender and cultural norms that were not part of a "modern" professional outlook. For the best interpretation of the historical meaning of modernization, see Thomas Bender, *Community and Social Change in America* (New Brunswick, New Jersey: Rutgers University Press, 1978).

[midwifery] education has been limited; has more in careing [*sic*] for those that are going to be confined in my own neighborhood. I have followed this about 30 years with good results my experience has been much in midwifery: I never lost a case."[14]

Some neighbor-women did not even consider themselves "midwives," for besides their own confinements, their birthing experiences were limited to a few births in their family or among their friends. A 1919 Children's Bureau report of rural problems illustrated how blurred the distinction was between merely helping out one's neighbor and assuming a title conveying authority. Reviewing maternity care in rural America, the report revealed that "many country women are cared for entirely by neighbors who may or may not have acquired some skill from experience." Some women, the report added, were seen "by midwives."[15]

Some neighbor-women midwives developed more than the basic skills out of sheer necessity—many lived in isolated settlements where doctors or trained midwives were scarce and where even rudimentary skills were better than none. A study of childbearing conditions by the Children's Bureau in Marathon County, Wisconsin, portrayed the extreme isolation farm women faced in many rural areas. Only 5 percent of the county's roads were paved; the rest were dirt roads, impassable during much of the year. Families in this area, including childbearing women, were often forced to rely totally on nearby neighbors.[16]

The neighbor-women's skills were probably the most heterogeneous of any of the three midwife groups. However, these women did share several characteristics: their training, which I have described as informal or self-taught, their age, and where they lived.

Thirty-nine of 396 women who registered with the state of Wisconsin after 1909 reported training like Susan Washburn's. These 39 women shared two other characteristics: their mean age of fifty-four was older than the mean ages of other types of midwives,[17] and they were overwhelmingly rural residents—only one neighbor-woman who registered with the state lived in Milwaukee, the state's most urban area.[18]

[14] License application of Susan Washburn, Midwife File.

[15] Elizabeth G. Fox, "Rural Problems," *Children's Bureau Conference,* Children's Bureau Publication no. 60 (Washington, D.C.: Government Printing Office, 1919), p. 187.

[16] Florence B. Sherbon and Elizabeth Moore, *Maternity and Infant Care in Two Rural Counties in Wisconsin,* Children's Bureau Publication no. 46 (Washington, D.C.: Government Printing Office, 1919), pp. 20–22. Many areas of this county were served by minor roads, which the study noted were impassable by automobile and difficult even for wagons. Many parts of the county had no mail delivery, which required that some people send twelve miles for their mail.

[17] Compared with an overall mean age of 47 for all of the registrants and a mean age of 45 for the school-trained group.

[18] All of the licenses (396) were examined and coded for computer analysis. Using SPSS to analyze the data, t-tests to evaluate differences in mean age were run between the different groups. The t-test comparing the mean age of the school-trained versus no training was highly significant, with a t-value of (−4.56) and a significance level (2-tailed) = .00001. To investigate the link between region and type of midwife, the 52

An analysis of the 588 women who registered with the Milwaukee Health Department between 1877 and 1907 reinforced the state license data correlating neighbor-women midwives with rural residence: over the thirty-year period of the Milwaukee registry, only 21 women reported having no training.[19]

Characterized by the female-centered networks within which they operated and by their self-acquired knowledge, neighbor-women midwives were also the most autonomous midwives—the conditions under which they practiced permitted them complete freedom. But even in isolated rural areas, families turned to these practitioners only when they could not find a trained midwife. It is not surprising, therefore, that as frontier areas of Wisconsin became more settled and more accessible, trained midwives replaced neighbor-women. By the beginning of the twentieth century, this family-based practitioner was found only in limited numbers, and in the most rural parts of the state.

APPRENTICE-TRAINED MIDWIVES

Throughout American history, many midwives, like many physicians, received their training by apprenticing themselves to older and more experienced practitioners. Indeed, in the United States until the last decades of the nineteenth century, almost all trained midwives learned their trade in this way.

Anne Nowak of Lublin (in north-central Wisconsin) reported her training in terms that would be familiar to many medical historians: "I have not taken up the course in any school but have practiced with a professional midwife until I was fully-capable of doing the work alone."[20]

---

counties represented by the registrants were grouped into six regions. When these six regions were cross-tabulated with kind of training, the result was highly significant: $X^2 = -102.33$, with 15 df, the significance level = .00001. Particular individual cells contributed to this finding: the western and central regions had three times more than the expected number of midwives reporting "no training," and the city of Milwaukee, where twenty untrained midwives would be expected, reported only one.

19 "Physicians' Register," City of Milwaukee Health Department, Archives Section, Milwaukee Public Library. (The title of this registry is a misnomer, for the registry contains registration data on both physicians and midwives.) There were 658 entries between June 1877 and October 1907, but many of the women who registered in 1906 and 1907 had previously registered. The registry, then, provided information on 588 different women.

A more detailed check of the data suggests that the 21 women who reported "no training" did not represent a larger contingent of unregistered, infrequent practitioners, perhaps singled out for prosecution. Dividing the registry into five-year periods, I found one or two neighbor-women midwives in Milwaukee signing up with the health department every year.

The figures (from the "Physicians' Register") were as follows:

	1877–79	1880–89	1890–99	1900–1907
Total registered:	46	171	227	144
"No training"	6	0	3	12

20 License application of Anne Nowak, Midwife File.

Education at the side of a preceptor midwife was not the only form of apprenticeship. Many nineteenth-century midwives did learn their craft from older, more experienced midwives, but many more female birth attendants apprenticed themselves with local physicians. Indeed, the most striking feature of apprentice-trained midwives in Milwaukee was that virtually all of them were trained by physicians.

Historians who have analyzed physician debates over the "midwife question" of the late nineteenth and early twentieth centuries have emphasized the hostility or at least the indifference most doctors felt toward these female practitioners. My analysis of Wisconsin data, however, indicates that midwives and physicians sometimes worked together in America's birthing rooms.[21]

Of the forty-nine apprentice-trained women who registered in 1909, thirty-five (or 71 percent) reported that their instruction came from physicians. Mary Greeley, a practitioner in western Wisconsin, was typical, declaring, "[I] have been practicing midwifery for 30 years. Usually under the direction of physicians of River Falls, Wis. and Ellsworth, Wis."[22]

The timing of this kind of instruction in the midwife's career varied. Some women received their initial training from physicians and then went out to deliver babies, as Ellen Stiefvator, practicing in northern Wisconsin, related: "I practiced with Dr. Rauls Gadsden [identified by her as an Alabama physician] 30 yrs. ago and have practiced ever since: [the] last 18 years in Merrill, Wis."[23]

This cooperation between midwives and physicians in rural areas is not entirely surprising, as the 1919 Children's Bureau survey of childbirth attendants in northern Wisconsin pointed out. In one village, for example, the authors of the survey found that local doctors had ceded all of the obstetrical work to Mrs. M. (a local midwife). The doctors referred their maternity cases willingly, telling the survey authors, "We don't like that kind of work and have always more or less turned it over to [her]."[24]

A close interaction between midwives and doctors was not limited to rural maternity practice, however. Although historian Jane Pacht Brickman has concluded that "medical animus against the midwife focused on the urban practitioner," in Milwaukee, many physicians trained midwives.[25] While some of these physicians were European-educated émigrés and may have been continuing a traditional association common to their ethnic groups, other Milwaukee doctors who worked closely with midwives had

[21] Leavitt finds that physicians called to a birth in the nineteenth century were forced to work within the limits imposed by "women's domain of the birthing room," which often included a midwife and the parturient woman's friends (*Brought to Bed*, p. 100).

[22] License application of Mary Greeley, Midwife File.

[23] License application of Ellen Stiefvator, Midwife File.

[24] Sherbon and Moore, *Maternity and Infant Care*, pp. 32–33.

[25] Jane Pacht Brickman, "Public Health, Midwives, and Nurses, 1880–1930," in *Nursing History: New Perspectives, New Possibilities*, ed. Ellen Condliffe Lagemann (New York: Teachers College Press, 1983), p. 70.

trained at American medical schools. Furthermore, most of these American physicians were well regarded; they were not marginal practitioners at the fringes of the Milwaukee medical community.[26]

Apprenticeships for these women lasted from several months to a year, and ranged from what seemed to be private lectures to a combination of didactic and practical work. Frances Jahnz, for example, had "Studied with Dr. F. S. Wasielewski for (2) two months, [in 1901] then took up ten (10) practical cases of confinement in his presence according to his instructions."[27] Her entire course of instruction, she stated, lasted one year. Mary Browikowski had "received instructions in the art of midwifery for fully five months from Dr. F. M. Hinz, M.D."[28] Dr. Hinz, a homeopathic physician trained in Illinois, offered seven months of "private instruction" to another Milwaukee midwife.[29] Like Dr. Hinz, other Milwaukee physicians lectured to midwives, offering courses ranging in length from six to twelve months.[30] One physician continued his mentor relationship with his pupil midwife after training her. Mary Holub received "practical training under Dr. Andrew Munro. attended confinement cases with him and studied under him. Have certificate from him and recommendation to practice. Attended cases with him for a period of seven years."[31]

Apprentice training, however, was not a major route to the profession for Milwaukee midwives. Overall, of the 588 women who registered with the Milwaukee Health Department between 1877 and 1907, only 51 (8.6 percent) reported an apprentice education. Analyzed year by year, the Health Department data also reveal the declining popularity of apprenticeships, from 15.7 percent of the registrants in the early years to 6.6 percent in the last years of the registry. The state licenses, covering a somewhat later period (1909–15), also demonstrate the decline in the twentieth century in apprentice training, especially for urban midwives: of 189 Milwaukee midwives who registered with the state, only 12 (6.3 percent) were apprentice educated.

SCHOOL-TRAINED MIDWIVES

Though many European school-educated midwives were among the large numbers of immigrants flooding Milwaukee and the rest of the state in the

[26] License applications for both the city of Milwaukee and the state of Wisconsin required two physician references. Each woman's references were coded, and the names of physicians in selected counties were checked against their own license applications (Physician Licenses, Board of Medical Examiners, Archives Division, Wisconsin State Historical Society). For physicians' activities within the medical community, I consulted Louis F. Frank, *The Medical History of Milwaukee, 1834–1914* (Milwaukee: Germania Publishing, 1915).

[27] License application of Frances Jahnz, Midwife File.

[28] License application of Mary Browikowski, Midwife File.

[29] Frank, *Medical History of Milwaukee*, p. 255. Hinz practiced in Milwaukee from 1875 to 1903.

[30] State license applicants reported four other Milwaukee physicians who offered lectures. Some of these doctors were instrumental later in the several midwifery schools in Milwaukee. (License applications of Kondaneygli Sythowski, Helena F. Mueller, Fredericka Wirth, and Mary Dudek, Midwife File.)

[31] License application of Mary Holub, Midwife File.

late nineteenth century, the most compelling explanation for the decline in apprentice training for midwives in Wisconsin lies with the establishment of two schools for midwives in Milwaukee.

Organized by physicians together with some local midwives, these schools offered Milwaukee and many other Wisconsin women a potentially more rationalized course of instruction in the theory and the practice of midwifery than they could obtain by themselves or in tandem with a physician or experienced midwife. Many Milwaukee midwives received their training at these schools: between 1890 and 1907, over half of the Milwaukee Health Department registrants reported having obtained a local school education.

But did the Milwaukee and other American midwifery schools organized at the end of the nineteenth century offer formal training that resulted from a "codification of knowledge" in midwifery? Did this standardized training become "the main support of a professional subculture" for midwives? A brief examination of these schools' curricula, personnel, and philosophy demonstrates that regardless of the new skills they offered their students, the schools were not, nor were they intended to be, a step in promoting the professionalization of midwifery.

American midwifery schools developed from and found their roots in the immigrant community. It was not surprising, therefore, to find two schools based on the European model in Milwaukee. They were founded in the period of the heaviest immigration from Europe to America. Milwaukee in particular experienced an explosive growth in its population and much of its growth was related to immigration from the Old World: between 1880 and 1910, the city was tied for first place among all American cities in its percentage of foreign stock. By 1910, the total foreign stock—total foreign-born plus natives with at least one foreign-born parent—was 78.6 percent of the total population. Of the foreign-born, almost half of them came from Germany.[32] Partly as a consequence of this immense immigrant population, the city had a large number of practicing midwives. As a *Milwaukee (Wisconsin) Sentinel* article pointed out in 1889: "No city in the country has as many practicing midwives as Milwaukee, it is said, at least in proportion to population. There are only a comparatively few more physicians than midwives, there being 171 of the former and 115 of the latter."[33]

European midwives and physicians emigrating to the United States brought their ideas about training with them. They were proud of their education and sometimes quite critical of unlicensed, unskilled American

[32] Roger Simon, "The Expansion of an Industrial City, Milwaukee: 1880–1910" (Ph.D. diss., University of Wisconsin-Madison, 1971), p. 82.

[33] "Figures that don't lie," *Milwaukee (Wisconsin) Sentinel,* 23 June 1889.

midwives. As Agnes Pradzinska, a Milwaukee midwife, wrote to the *American Midwife:*

> I wish to express my concern that the so-called butcher-women be prohibited. I have already complained many times to the local health department, but I always receive the same answer: "We will look into the matter." But they are [still] looking into it. The births should all be registered in the state of Wisconsin, such female bunglers don't do it though. Among [female bunglers] is one, who calls herself a midwife and has hung out a shield with the inscription "midwife and birth assistant." But she can neither read nor write.... [She] also knows fortune-telling, because when she performs a delivery, she tells the mother how many more children she will have; even better yet is that she also knows how many boys and girls she will have.[34]

The Milwaukee School of Midwifery, founded in 1879 by a German-born and -trained midwife, and the Wisconsin College of Midwifery, established in 1885 by a German immigrant woman trained in America, shared many of the characteristics of other American midwifery schools at the turn of the century. Established by immigrant midwives in conjunction with immigrant physicians and their sympathetic American colleagues, both schools drew their students from Milwaukee's foreign-born community. Like many American midwifery schools, the Milwaukee School of Midwifery offered its course in German and in English. Both schools, like their American and European counterparts, relied heavily on the prestige of local, usually young, physicians for some instruction and for the approval of the final examination.[35]

Finally, both institutions, established as profit-making ventures, encountered many of the problems that plagued proprietary medical schools of the era: poor facilities, a dearth of instructional materials, and course work consisting of didactic lectures with limited clinical instruction. For example, a graduate of the Milwaukee School of Midwifery described her training as a course of "6 months—about 6 lectures per month.... [I] assisted Precepteress Mrs. Wilhelmine Stein at her hospital as well as at [her] private confinement cases."[36] Mrs. Stein's hospital was in her home, having probably fewer than ten beds. The Wisconsin College of Midwifery offered a seemingly more rigorous course. Two graduates reported having taken a five-month course, with lectures in anatomy, biology, and physiology daily for two hours.[37]

[34] Agnes Pradzinska, "Letter to the editor" in the German-language section of the *American Midwife* [St. Louis, 1896], *2*(*6*): 40–41. My thanks to Rebecca Bohling for her expert help in translating this letter for me.

[35] These schools exerted a statewide influence: the Milwaukee registry listed 84 graduates of these schools who gave addresses outside the city, perhaps implying that some students came to Milwaukee specifically for their education. The State License file also showed the wide influence of the schools on Wisconsin's midwives: of the 121 women in the state file who reported their school was based in Milwaukee, 30 gave addresses outside of the city.

[36] License application of Sophia Kegel, Midwife File.

[37] Licenses of Caroline Fuss and Maria Timm, Midwife File.

Students at this college gained practical experience by attending obstetrical clinics at least twice weekly, though many graduates reported that much of their entire four- to six-month term was spent working at the hospital. By 1912, the school's term had expanded to eight months, and students attended at least twenty-five confinements.[38]

These midwifery schools, despite their faults, probably offered more clinical instruction in obstetrics than many of the medical schools surveyed by J. Whitridge Williams in 1912.[39] However, both institutions suffered from a problem unique to the field of midwifery. Unlike nursing or medicine, midwifery could claim neither a distinct body of knowledge nor control over the production of its own practitioners. Milwaukee's two schools, like other midwifery schools in America and in Europe, needed physician support to establish and maintain respectability.

One way to maintain physician support was to downplay midwives' roles as teachers. For example, advertisements for the Wisconsin College of Midwifery in Milwaukee's city directory show how Mary Klaes's central role as founder and teacher was subsumed by her need to attract doctors to her school. Listing the physician president's name first and in bold print, the text of the advertisement urged other doctors to bring their cases to the school's hospital, with the promise that "trained nurses" were "always in attendance."[40] By 1901, the school referred to itself as "St. Mary's Sanitarium," having the "best physicians in charge." Though midwives were still being trained in the hospital at that date, the city directory advertisement made no mention of the fact..[41]

The problem the Milwaukee schools experienced in holding on to their "best physicians" was not unusual, and reflected a lack of autonomy by midwives at even the most prestigious institutions. The Bellevue Hospital School in New York City, for example, generally considered the best American midwifery school, denied any teaching role to midwives. Local doctors taught all of the courses.

The midwifery schools' curriculum reinforced and institutionalized their students' dependent role by emphasizing that good midwives should attend only normal, uncomplicated births. A physician associated with the Bellevue School explained this philosophy:

> Not the least advantage of our primitive attempt to educate the midwife at the Bellevue School is the thorough teaching of each candidate for graduation her

[38] Two 1913 graduates of the school, Stefonia Staigwillo and Gertrude Turzynski, noted on their state license applications that they had attended the "full course" of the Wisconsin College of Midwifery. The full course was described as an eight-month term from September to May with hospital time and attendance at twenty-five cases. (License applications, Midwife File.)

[39] J. Whitridge Williams, "Medical education and the midwife problem in the United States," *JAMA*, 6 January 1912, *58*: 1–7. Williams found the general attitude toward obstetric teaching to be "a very dark spot in our system of medical education" (p. 1).

[40] *Wright's Milwaukee City Directory*, 1893, *26*: 1178, advertisement under "Lying-In Hospitals." Midwives were never mentioned.

[41] *Ibid.*, 1901, *34*: 1262, advertisement under "Hospitals."

> limitations.... One important fact is instilled into the brain of each midwife, and that is the knowledge of her own limitations—the knowledge of what not to do, and when to seek the aid of a practising physician.[42]

In another study by American physicians of midwifery schools in Scandinavia, the writer noted approvingly that the students "are taught to revere the physician, and they are distinctly shown their limitations."[43] Although Milwaukee physicians left no written records of their feelings about the midwifery schools they were associated with, it is reasonable to assume, given their standing in the medical community, that the physicians instructed their students the same way the instructors at the Bellevue and Scandinavian schools did theirs.

Midwifery in the United States in the late nineteenth century faced a "crisis of professionalism" that related directly to the natural evolution of the occupation. At each stage of the developmental process, midwives' practice became less autonomous and more circumscribed. But a limitation on midwife practice that did not diminish as the period unfolded related directly to the nature of the practitioner herself: midwives in Wisconsin, as midwives everywhere else, were almost always mature married women who had children themselves. Most midwives, therefore, were limited by family responsibilities to practicing part time.

Like many other working women in their community, midwives left their homes in order to earn money to help support their families. Their work force participation as married women placed them in a unique category of working women, however. Historians studying women's work have emphasized the rarity of married women participating in the work force in the late nineteenth and early twentieth centuries. Leslie Woodcock Tentler found that only 5.6 percent of all married women were reported at work in 1900, though she admits that the census data probably underenumerated wives working part time in the home.[44] Married women who did work outside the home, she points out, chose jobs with relatively short or flexible hours or part-time employment, and they often tried to stay within the immediate neighborhood.[45]

Eugene Declercq and Neal Devitt have shown that early twentieth-century midwives were predominately mature women of diverse ethnic backgrounds.[46] However, neither of these historians has considered how these women's demographic characteristics might have influenced the prac-

[42] J. Clifton Edgar, "Why the midwife?" *Amer. J. Obstet.*, 1918, *78*: 249.

[43] George W. Kosmak, "Results of supervised midwife practice in certain European countries," *JAMA*, 10 December 1927, *89*: 2009.

[44] Leslie Woodcock Tentler, *Wage-Earning Women: Industrial Work and Family Life in the United States, 1900–1930* (Oxford and New York: Oxford University Press, 1979), p. 137.

[45] *Ibid.*, p. 142.

[46] Declercq, "Nature and style," p. 118, and Neal Devitt, "The statistical case for elimination of the midwife: fact versus prejudice, 1890–1935" (pt. 1), *Women and Health*, 1979, *41*: 82.

tice of midwifery. In my investigation of some of Wisconsin's midwives, I analyzed the effects of age, marital status, immigrant background, and social class on the occupation and professionalization of midwifery in America.

Women training for or practicing midwifery in Wisconsin were usually immigrant, working-class, married women. Midwives' age, marital status, ethnicity, and social class, and the number and the ages of their children, conceivably influenced the patterns of their practice. In order systematically to study midwives as working women, I linked practitioners who identified themselves as midwives to the descriptions of their families in the federal or state manuscript census schedules.[47] From four Wisconsin counties that were chosen to sample an urban-to-rural-to-frontier continuum in a fifty-year period (1870 to 1920), 893 different midwives were identified.[48] Of these practitioners, 398 were then linked to the 1870, 1880, 1900, or 1910 federal census, or to the 1905 state census (see table 1). The census data was then analyzed to determine basic information about midwife families and about the practitioners themselves.[49]

Table 1. Distribution by County and by Census of 398 Wisconsin Midwives

County of Residence	Number	Year of Census	Number
Dane	33	1870 federal	6
Milwaukee	301	1880 federal	77
Trempeleau	26	1900 federal	191
Price	38	1905 state	38
		1910 federal	86

SOURCES: U.S. Census Office, 9th, 10th and 12th Censuses, 1870, 1880, 1900, *Population Schedules: Wisconsin* (Washington, D.C.: Census Office, 1870, n.d., 1901); U.S. Bureau of the Census, 13th Census, 1910, *Population Schedules: Wisconsin* (Washington: Government Printing Office, 1913) National Archives microfilm publications. Wisconsin Dept. of State, *Wisconsin State Census, 1905: Population Schedules,* Wisconsin Secretary of State's Office (Madison: State Historical Society of Wisconsin, n.d.). Microfilm.

[47] Practitioners were identified from the Milwaukee Health Department licenses, the licenses filed with the state of Wisconsin, advertisements in city directories, and the names of women who signed birth records as the attendant.

[48] The four counties were Milwaukee, Dane, Trempeleau, and Price. Milwaukee, of course, was the urban center for my study. Madison, Wisconsin's capital city, is located in Dane County, which nevertheless was largely rural in the late nineteenth and early twentieth centuries. Trempeleau County was chosen both because it was the site of Merle Curti's famous study (*The Making of an American Community: A Case Study of Democracy in a Frontier County,* with the assistance of Robert Daniel et al. [Stanford, California: Stanford University Press, 1959]) and because, like Dane County, it was a fine example of a place with a settled, rural population. Price County, in northern Wisconsin, was a frontier county at the turn of the twentieth century. The major industry in this county was lumbering, but a few farmers had settled on the clear-cut land.

[49] Within the four counties, I identified 893 different midwives in the fifty-year period of my study (1870–1920), and I was able to link 398 of these practitioners to the census (44.5 percent). I attempted to link practitioners to the census nearest to the date when they first began to practice, because I wanted to capture the full effects of family life on midwifery practice.

Note that the large number of practitioners linked to the 1900 census reflects the absence of the 1890 manuscript census schedules. The unavailability of this schedule (it was burned in a Washington, D.C., fire near the turn of the century), coupled with the high mobility of midwives' artisan families, was the reason more midwives were not found in the census.

To couple the data describing midwives with information about their patterns of practice, I sampled birth certificates in the same four counties every five years between 1870 and 1920.[50] The data sets contain 28,924 cases for all four counties and provided information on the patterns of practice of 893 midwives and 1,149 physicians.[51]

Neal Devitt's assertion that midwives were predominantly elderly women and that "advanced age" played a role in the demise of midwifery fails to acknowledge the fact that midwifery has always been an occupation for mature, married women. As historian Jane Donegan relates, "The acceptable midwife was the respectable older woman who carried herself correctly."[52] In fact, young, single women often were not allowed to become midwives. Dr. Marie Zakrzewska, the famous director of the New England Hospital for Women and Children, trained as a midwife in one of the most prestigious schools in Berlin. Even with influential support, however, her first application for admission was refused specifically because she was not married and because, at age eighteen, she was considered too young to be a midwife.[53]

Wisconsin's midwives, on the whole, were middle-aged but not elderly women. Two data sets provided information on the ages of these practitioners. The mean age of the women who registered with the state after 1909 was 48.1 years. Many women from this group were even younger—the numerous school-educated practitioners were a significantly more youthful group.[54] The census study, which identified practitioners earlier in their careers, confirmed the findings in the state file. In the census study, the 398 midwives had a mean age of 43.9, almost five years younger than the average age of the state licentiates.

The census study provided crucial information on marital status. This

[50] To obtain an estimate of the frequency of practice and to maintain a reasonable sample size, every birth was coded for a given three-month period in Milwaukee County, every birth for six months in a sample year for Trempeleau County and Price County after 1905, and every birth for the entire year for Dane County and Price County between 1880 and 1905.

[51] The birth data are contained in two data sets. Milwaukee, Price, and Trempeleau counties were coded together in one set, with a total of 24,302 cases. Dane County data were coded alone, having a total of 4,622 cases. For each case, the following variables were coded: date of birth, including day, month, and year; place of the birth, including county, city, whether home or at a hospital; parents' home address; the occupation of the father; the birthplace of the mother and the father; the mother's parity; the attendant at the birth (if noted, otherwise a code for parents or other was included); the status of the attendant (midwife, physician, or other). Each attendant was given a unique five-digit number that reflected his or her county and the year he or she first appeared in any record of birth attendants. The data were analyzed using various statistical subroutines of SPSS.

[52] Jane B. Donegan, " 'Safe Delivered,' But by Whom? Midwives and Men-Midwives in Early America," in *Women and Health in America: Historical Readings*, ed. Judith Walzer Leavitt (Madison: University of Wisconsin Press, 1984), p. 304.

[53] Agnes C. Vietor, ed., *A Woman's Quest: The Life of Marie E. Zakrzewska, M.D.* (New York: D. Appleton, 1924), p. 39. Zakrzewska was eighteen years old when she first sought admission to the Midwifery School of Berlin. Finally, by the entreaties of a powerful physician friend to the King of Prussia, she was admitted to the school (pp. 39–42).

[54] The mean age of all of the respondents was 48.05, with a standard deviation of 12.37 (Midwife License File, State Board of Medical Examiners, Archives Division, Wisconsin State Historical Society). When age was cross-tabulated with kind of training, the chi-square was 43.88, highly significant at the .00001 level.

data emphasized the close relationship between the occupation of midwifery and the attendant's family situation. Of the 394 women whose marital status could be determined, 316 (80.2 percent) were married women living with their husbands. Another 59 women (15.0 percent) were widowed, and 11 midwives were divorcées. Thus, only 8 women (2.0 percent) were single. In Wisconsin, at least, midwifery was most definitely practiced by married women.

Like the midwives in Declercq's Lawrence, Massachusetts, study, almost all of the midwives identified in Wisconsin were literate. Of the 359 midwives in the census study for whom literacy could be ascertained, 343 (95.5 percent) were described as being able to read or write (not always in English). Only 16 women in this sample were not literate.[55]

Midwifery, like the prevalent nineteenth-century custom of keeping boarders and lodgers, was ideally suited in many ways as an occupation for mature, married women. Though midwives worked outside of their own homes, they delivered babies in other women's homes, remaining protected from the shop floor and the effects of working there, popularly accepted as deleterious. Like women who kept boarders, midwives could adjust their job schedule to meet the needs of their families, taking fewer cases if they needed the time at home, and taking more cases if they needed the extra money. Indeed, when viewed in the context of home-based wage earning, midwifery may have replaced keeping boarders. Only 31 midwife families in the census study (7.8 percent) also had boarders present. Instead, most midwife families were nuclear—287 midwives (72.1 percent) lived solely with their husbands and children.[56]

Many Wisconsin midwives were responsible for their own minor children even as they were on call to deliver other women's babies. Of the midwives in the census study, 323 (81.2 percent) had at least one child living at home with them, and most women had more. The mean number of children was three. The range was quite narrow, however, as most practitioners (77 percent) had four or fewer children at home.[57] Married,

[55] Although the number of illiterate midwives was small, I tested for significance for birthplace and county. When literacy was cross-tabulated with county, the chi-square was significant (15.82, significant at .001). Half of the sixteen illiterate midwives lived in Milwaukee County, three lived in Dane County, four in Trempeleau, and one in Price. The Dane and Trempeleau figures account for the significance, as fewer illiterate women were expected. Birthplace was also significant, with a chi-square of 22.9, significant at 0.0008. The Scandinavian group had more illiterate women than expected. For thirty-nine midwives, literacy was either not reported or unknown.

[56] The percentage of families with boarders is significantly low when midwife families are compared with families sharing similar ethnic and class backgrounds and age distributions. Modell and Hareven's study of New England families, for example, found that as high a proportion as 40 percent of immigrant, working-class families where the head of the household and his wife were in their forties had boarders. John Modell and Tamara Hareven, "Urbanization and the malleable household: an examination of boarding and lodging in American families," *J. Marriage and the Family*, 1973, *35*: 472. The implications of boarding for women's work in the home are discussed on p. 473.

[57] Information on children was obtained for 387 women. The mean number of children was 2.99, with a standard deviation of 2.41. Number of children ranged from 1 to 13; 77 percent of midwives had 4 or fewer children, 48 percent had 2 or fewer.

divorced, and widowed women did not have significantly different numbers of children.[58]

These children were not preschoolers. Midwives, unlike other mothers working outside of the home, did not work until their children were of school age. Most working-class mothers who earned money outside of the home, on the other hand, *stopped* working when their oldest child reached the age of nine or ten. Thus, the mean age of midwives' oldest children, 16.0, and the mean age of the youngest, 8.1, deviated considerably from Lynn Y. Weiner's findings for children of working mothers.[59]

To measure the effect of the number and the ages of midwives' children on the midwives' patterns of practice, data from the vital statistics were aggregated and merged with data from the census study. In a multivariant analysis of 274 midwives who were found in both the census study and in the birth records, no relationship was found between the number of babies a midwife delivered and the number of children she herself had, the age of her oldest child, or the age of her youngest child. A very weak relationship did exist between the size of a midwife's practice and her age.[60]

Like the age distribution of midwives' children, the distribution of midwives' husbands' occupations did not fit the usual pattern for married women workers. The census sample provided data on 314 husbands of midwives. Unlike the husbands of other married women wage earners, who were employed in very low status jobs, most husbands of midwives were skilled artisans (41 percent, n = 131). The next largest occupational groups were much smaller: 17.2 percent (54) of spouses of midwives worked on farms or in the lumber trade, and 16.9 percent (53) worked as laborers. Most midwives, therefore, were the wives of men who earned a comfortable working-class income.[61]

The data on patterns of practice provide no direct evidence of a mid-

[58] An analysis of variance showed no significant differences between groups. The f-ratio was .99 and the f-probability equalled .37.

[59] Lynn Y. Weiner, *From Working Girl to Working Mother: The Female Labor Force in the United States, 1820–1920* (Chapel Hill: University of North Carolina Press, 1985), p. 85. The median age of oldest children of all midwives was 15, with a modal (most frequent number) age of 9. However, the range of ages was quite broad, from 1 to 54 years.

[60] In order to analyze the number of births attended by each practitioner or group of practitioners, the births file was sorted and aggregated by county, attendant number, and attendant status. The resulting file had information on 274 midwives who had also been found in the census data. A new variable "NBABES" captured the total number of babies delivered by each midwife. Matching the midwives in this file to the census data by linking the attendant numbers, I performed a stepwise multiple regression, with number of babies as the dependent variable. The number of children, age of the oldest, age of the youngest, and the midwife's own age were entered as independent variables. Only age of the midwife had an F-value significant enough to enter into the regression equation (f = 7.101, significance = .0082). The resultant R^2 equalled 0.02194, showing age only very weakly explaining the total number of babies delivered.

[61] To code for husbands' occupation, I used the categories defined by the federal census of 1940. The categories are roughly hierarchical, except for the two categories in the middle defining farming and mining. To approximate "white-collar" versus "blue-collar" occupations, I grouped the professional, managerial, clerical, and sales groups together, and the transport, artisan, laborer, and service categories together. Only 39 midwife husbands fell into the "white-collar" group (12.4 percent), while 201 husbands (64.0 percent) were clearly "blue-collar." The farm husbands, all located in the three rural counties, were hard to categorize, as farm property values varied tremendously.

wife's economic incentive. When data on the size and the pace of practice were linked to the midwives' demographic characteristics found in the census study, tests of significance showed no connection between the total number of births a midwife attended and her husband's occupation. Thus, poorer practitioners were not more active than those who might be better off. Midwives married to day laborers were not significantly more active practitioners than midwives married to men having more secure occupations like carpentry or clerking.

While midwife families may not have been as desperately poor as the families of women working in factories, the status of their homes demonstrates their precarious financial position and provides some evidence of a midwife's economic incentive. Of the 320 midwife families whose housing status could be verified, 47 percent rented their homes (151 families), 29 percent held mortgages (92), and only 24 percent (77) owned their homes outright. As Stephan Thernstrom's work shows, home ownership in the nineteenth century was the result of a very slow and expensive process.[62] About half of the midwife families in my study were involved in the process of home ownership, a figure consistent with Thernstrom's findings for working-class families in Newburyport, Massachusetts.[63]

Although a simple analysis of the size of a midwife's practice showed no statistical connection with the status of her family's house,[64] a more complex examination of home ownership by midwife artisan and laborer families revealed one potential use of a midwife's income. One would perhaps expect artisan families predominantly to own their homes and laborers mostly to rent, but in actuality many artisan families rented and not many of them owned houses outright. On the other hand, fewer laborer families rented and more of them had mortgaged or fully owned homes than would be expected (see table 2). These surprising findings for artisan midwife families indicate that some of them lived near the edge of economic hardship. Thus, for those families who did not own their homes, the practice of midwifery may have been an important secondary source of income or a supplement to be saved for a house.

While the Wisconsin data provided no clear-cut economic rationale for midwifery practice, they demonstrated strongly the relationship between midwives' immigrant status and their chosen occupation. Most of the midwives in the Wisconsin census sample were first-generation immigrants; all but 10 had parents born in Europe. The majority of these women, not sur-

[62] Amortizing mortgages, common to today's home buyers, were rare. Working-class families were required, therefore, to save a large sum of money to pay off the principal at the end of the loan period, even as they paid the interest in semi-annual payments. See Stephan Thernstrom, *Poverty and Progress: Social Mobility in a Nineteenth-Century City* (New York: Atheneum, 1974), pp. 121–22, 129–30.

[63] *Ibid.*, pp. 120–21.

[64] One-way ANOVAs were used to test whether the size of a midwife's practice was related to her socioeconomic status. For husbands' occupation, the f-ratio was 1.3700, with 7 d.f., and an f-probability of .2202. For home status, the f-ratio was 1.7649, with 2 d.f., and an f-probability of .1736.

Table 2. Home Ownership in Artisan and Laborer Midwife Families

Occupation	Renters	Mortgage Holders	Owners
Artisan			
Number	61.0	34.0	10.0
Expected number	47.1	31.8	25.2
Laborer			
Number	13.0	13.0	13.0
Expected number	18.4	12.4	9.8

SOURCES: U.S. Census Office, 9th, 10th and 12th Censuses, 1870, 1880, 1900, *Population Schedules: Wisconsin* (Washington, D.C.: Census Office, 1870, n.d., 1901); U.S. Bureau of the Census, 13th Census, 1910, *Population Schedules: Wisconsin* (Washington: Government Printing Office, 1913) National Archives microfilm publications. Wisconsin Dept. of State, *Wisconsin State Census, 1905: Population Schedules*, Wisconsin Secretary of State's Office (Madison: State Historical Society of Wisconsin, n.d.). Microfilm.

NOTE: A chi-square analysis of midwives' husbands' occupations by home status was significant at the 0.0001 level. The artisan and laborer groups provided most of the deviations from the expected numbers.

prisingly, came from Germany (52 percent). Indeed, the German-speaking midwives, including those born in Germany, the Austro-Hungarian empire, and Switzerland, constituted 72 percent of all of the foreign-born women in the sample. The next largest groups of foreign-born attendants were Poles (13.3 percent) and Scandinavians (12.7 percent). An analysis of the 274 midwives who appeared both in the census study and as birth attendants in the vital record study strengthened the evidence for a link between an immigrant background and the practice of midwifery. The 166 German and Swiss midwives delivered an average of 28 babies each, both the Austrian and the Polish practitioners averaged 22 births, and the American-born midwives attended an average of only 11.8 births. Scandinavian-born birth attendants, living mostly in rural counties, were called to a mean of only 7.6 births, the smallest average number of deliveries of any group (see table 3).

Historians of midwifery in the United States have long noted the large percentage of midwives who were immigrant women. Frances Kobrin describes turn-of-the-century midwives as being employed primarily by the foreign-born and their children.[65] Jane Pacht Brickman writes, "Wherever immigrants settled, the midwife flourished.... The midwife, almost always foreign-born and living in the community, lay in the buffer that immigrant groups maintained against an already overwhelming cultural shock."[66] Almost none of the midwives Eugene Declercq found practicing in early twentieth-century Lawrence, Massachusetts, were born in the United States: "Like the population they served, they came predominantly from Southern and Eastern Europe."[67]

[65] Kobrin, "American midwife controversy," p. 350.

[66] Brickman, "Public Health, Midwives, and Nurses," p. 70.

[67] Declercq, "Nature and style," p. 115. Table 2 (p. 117) indicates that all of the "major midwives" in Lawrence were born in Europe.

Table 3. Distribution of Births by Ethnic Group

Group	Mean Number of Deliveries	95 Percent Confidence Levels
German	28.01	23.11 to 32.91
Swiss	28.01	23.11 to 32.91
Scandinavian	7.61	2.37 to 12.85
Austrian-Hungarian	22.05	8.47 to 35.62
Polish-Russian	22.45	14.45 to 30.45
American	11.83	2.05 to 21.62

SOURCES: Milwaukee, Price, Trempeleau, Dane Counties [Wisconsin], *Births*, Vital Records Office, Wisconsin Dept. of Health and Human Services, Madison. Also U.S. Census Office, 9th, 10th and 12th Censuses, 1870, 1880, 1900, *Population Schedules: Wisconsin* (Washington, D.C.: Census Office, 1870, n.d., 1901); U.S. Bureau of the Census, 13th Census, 1910, *Population Schedules: Wisconsin* (Washington: Government Printing Office, 1913) National Archives microfilm publications. Wisconsin Dept. of State, *Wisconsin State Census, 1905: Population Schedules*, Wisconsin Secretary of State's Office (Madison: State Historical Society of Wisconsin, n.d.). Microfilm.

NOTE: A one-way analysis of variance was very significant when attendants' total number of deliveries was tested with their own birthplace. The f-ratio was 2.86, with a probability of 0.0103. The German and Swiss groups also showed the least variability around the mean as measured by the 95 percent confidence levels about the mean. Other groups had much wider-ranging confidence levels.

Contemporary observers of these birth attendants documented their prevalence and popularity in immigrant communities all over the United States. As Michael Davis noted in his study of immigrant health:

> The immigrant mother has rarely been accustomed to a man doctor at the time of confinement. She and her friends have used the midwife, who, in most European countries, is a woman of some standing, trained, and in many countries, carefully supervised. The midwife is the most important single element in the general question of the care of immigrant mothers.[68]

Grace Abbott's 1917 work on immigrant communities in Chicago also recounted immigrant women's preference for midwives at their lyings-in,[69] and Joseph B. DeLee, a virulent opponent of midwives, noted approvingly in 1915 that midwives seemed to be "slowly disappearing in America," and were only found in "crowded communities of foreigners."[70]

"Communities of foreigners" were not restricted to crowded urban areas, however. A 1924 study of Minnesota midwives found that only 13.7 percent of them were born in America; 86.3 percent (101 of 117 midwives found by the survey) were foreign-born. Twenty-five percent of the practi-

[68] Michael M. Davis, "Birth Rates and Maternity Customs," in his *Immigrant Health and the Community* (New York: Harper and Brothers, 1921), p. 195.

[69] Grace Abbott, *The Immigrant and the Community* (New York: Century, 1917). Abbott declared that "sometimes the traditions and the prejudices of our immigrant population must be consulted in determining an important health policy. An admirable illustration of this necessity may be found in the question of what should be our state and city policy with regard to the practice of midwifery" (pp. 145–46).

[70] Joseph B. DeLee, "Progress towards ideal obstetrics," *Amer. Assoc. Study of the Prevention of Infant Mortality*, 1915, *6*: 119.

tioners in this sample were from Scandinavian countries; 17.1 percent were German.[71] The Children's Bureau study of maternity conditions in Marathon County, Wisconsin, found that the large number of German and Polish settlers clung "to their foreign customs and habits of thought and to a certain extent their languages, making the district as a whole distinctly foreign in its atmosphere."[72] Midwife attendance at childbirth was one of the "foreign customs" these women clung to. Twenty-four midwives, all members of the German or Polish communities, practiced in the county, though only a few were reported to be earning their living as midwives. Most female attendants "went out" as an accommodation to their neighbors. Of all of the midwives in the region, only six practitioners delivered more than ten babies per year, and only two oversaw the births of more than thirty babies.[73]

As the Children's Bureau study suggests, midwives practiced not only in cities but also in rural areas where there were significant communities of immigrants. In my census study of Wisconsin midwives, American-born midwives were a distinct minority in each of the four counties. Tests of significance showed distinct ethnic groupings in each county, with Scandinavians predominantly located in the three rural areas and Polish and Russian midwives exclusively based in Milwaukee. Very few American-born midwives practiced anywhere in Wisconsin. In Milwaukee County, only 7.6 percent of 301 birth attendants were born in America, and the numbers of American-born midwives in the other counties were so small that statistical analysis was not feasible.[74] In Milwaukee, the overwhelming majority of practitioners, 185 (61.5 percent), had emigrated from Germany. In Dane County, German-born midwives formed the largest group (n = 15, 45.5 percent), but Scandinavian attendants were almost as numerous (n = 10, 30.3 percent). In the other two rural counties in the study, Scandinavian women were the dominant ethnic group. In Trempeleau County, 65.4 percent (n = 17) of the attendants were Scandinavian, and in Price County, 34.2 percent (n = 13) were born in Norway, Sweden, or Denmark. Five midwives in Price County were natives of Finland.

Summarizing the practice of midwifery in the early twentieth century on the lower East Side of New York City, historian Elizabeth Ewen described it as an occupation that was "an acceptable means for a married woman to contribute to her family economy."[75] This article has demon-

[71] E. C. Hartley and Ruth E. Boynton, "A survey of the midwife situation in Minnesota," *Minn. Med.*, 1924, 7: 439–46.

[72] Sherbon and Moore, *Maternity and Infant Care*, p. 10.

[73] *Ibid.*, pp. 30–32.

[74] A chi-square analysis of county by birthplace was highly significant; the chi-square = 207.949, significant at 0.00001. The three rural counties had many more Scandinavian midwives than would be expected, and Milwaukee City had many fewer. (Milwaukee had only 4 Scandinavian midwives; 37.1 were expected.) Trempeleau and Price counties also had many fewer German birth attendants than would be expected. (Trempeleau had no German midwives; 13.9 were expected. Price had 8 instead of 20.3.)

[75] Elizabeth Ewen, *Immigrant Women in the Land of Dollars: Life and Culture on the Lower East Side, 1890–1925* (New York: Monthly Review Press, 1985), p. 133.

strated that midwifery certainly was an occupation practiced by mature, married women. But the evidence from the analysis of Wisconsin's midwives also demonstrates that these women were not motivated solely by economic necessity. Though many Wisconsin practitioners undoubtedly contributed significant amounts of money to their family's economy, they practiced midwifery because it was an acceptable occupation in the *immigrant* community. Both historians and contemporary observers of European settlers in the United States found that these settlers had high regard for midwives. Midwives respected birth traditions of immigrant families and, as female birth attendants, they did not offend religious families' sense of moral propriety.[76]

Midwifery, unlike nursing or medicine, was practiced according to established customs. Women could adapt the size of their practice to the needs of their families and themselves. Unlike other wage-earning women, midwives did not have to punch a time clock or listen to a boss. However, despite having some very highly developed skills and the autonomy to set their own hours, midwives had more in common with housewives who kept boarders and lodgers than they did with nurses and female physicians. In these two branches of the healing arts, practitioners worked full time, accepting people from many diverse classes and ethnic groups as patients. Midwives, despite their education, practiced part time, birthing babies in their own immigrant neighborhoods. While the trained midwife may have delivered more babies and attended births on a more regular basis than the neighbor-woman, midwifery for both kinds of female practitioner was an occupation that was fitted in around the needs of the family. Given this priority within the context of women's lives, midwifery could not develop into a profession.

The demise of American midwives can best be understood through an analysis of the practitioners themselves. Many historians have emphasized the physician debates concerning midwives at the beginning of the twentieth century. This work on Wisconsin's midwives adds a new dimension to these authors' analyses, one that is important to complete the story: that dimension is an understanding of the process of change within this quintessentially female occupation. Midwives, I have argued, faced the same problems as nurses and physicians at the turn of the century: practitioners needed to build a professional ethos based on standardized education, licensure, and self-governance. Nursing and medical leaders of the early twentieth century gave direction to and helped shape their professions' goals. But there were no positions as leaders for midwives to assume. The presumptive leaders—trained midwives—relied on others to decide their

[76] Several historians have noted the almost outright ban on male doctors in immigrant women's birthing rooms. See Brickman, "Public Health, Midwives, and Nurses," p. 70; Kobrin, "American midwife controversy," p. 321; Ewen, *Immigrant Women in the Land of Dollars*, pp. 70–71.

qualifications and to license them for practice. In addition, the personal characteristics of the practitioners and the characteristics of their practice prevented the emergence of a united profession.

The demise of midwifery at the beginning of the twentieth century in this country was due to a number of complex and interrelated factors, including, but not limited to, physician opposition to these birth attendants. In an era when Americans increasingly insisted on professionals in many aspects of their lives, the evidence on Wisconsin's midwives shows that midwifery remained a traditional female occupation. Failing to achieve any professional goals or identity, and lacking any outside professional support, midwives faced the inevitable: their replacement in the birthing room by the self-consciously professional obstetrician.

THE IMMIGRANT MIDWIVES OF LAWRENCE: THE CONFLICT BETWEEN LAW AND CULTURE IN EARLY TWENTIETH-CENTURY MASSACHUSETTS*

Eugene Declercq and Richard Lacroix

In 1907 the Massachusetts Supreme Court ruled that the practice of midwifery was in violation of the state's Medical Practice Act of 1894.[1] In the same year, three midwives living in Lawrence, Massachusetts—Virginia Pedrazzini, Angelina DeMarco and Laura Carpenito—attended thirty-nine of the registered births. In 1913, six years after their practice was declared officially illegal, these three women attended 442 births and registered them with the Lawrence City Clerk.[2] Overall, the proportion of all registered births in Lawrence attended by midwives increased after the court's ruling from 38 percent in 1907 to 41 percent in 1913. The campaign to abolish midwifery in Massachusetts, the influence of the practice of midwifery in Lawrence and the factors that limited the initial impact of the campaign against it are the subject of this paper.

The national campaign against midwifery in the United States has been extensively documented elsewhere.[3] The center of opposition was Massachusetts. As a Pennsylvania physician remarked in 1915, "I know of no other section of the country which has been more successful in prohibition (of midwifery) than Massachusetts."[4] The legislative restriction of midwifery in the state was accomplished in several stages during the nineteenth century culminating with the Medical Practice Act of 1894.[5] It required that physicians

* Revised version of a paper presented at the fifty-fifth annual meeting of the American Association for the History of Medicine, Bethesda, Maryland, 30 April 1982. Research was supported by a grant from the Faculty Development Committee of Merrimack College.

[1] Supreme Judicial Court of Massachusetts, *Massachusetts Reports: Cases Decided and Determined by the Supreme Judicial Court of Massachusetts* (Boston: Little, Brown, 1908), 196 (1907): 326–29 (hereafter cited as *Massachusetts Reports*).

[2] Lawrence City Clerk, Birth Registration Sheets, 1907–13, City Hall, Lawrence, Massachusetts.

[3] See Judy Barrett Litoff, *American Midwives, 1860 to the Present* (Westport, Conn.: Greenwood Press, 1978); Jane B. Donegan, *Women and Men Midwives: Medicine, Morality and Misogyny in Early America* (Westport, Conn.: Greenwood Press, 1978); Frances E. Kobrin, "The American midwife controversy: a crisis of professionalization," *Bull. Hist. Med.*, 1966, *40*: 350–63.

[4] J. M. Baldy, "Is the midwife a necessity?" *Amer. J. Obstet. Diseases Women Children*, 1916, *73*: 400.

[5] Secretary of the Commonwealth of Massachusetts, *Acts and Resolves of the Commonwealth of Massachusetts*, 1894, chap. 458, pp. 530–33 (hereafter cited as *Acts and Resolves*).

be licensed and established penalties for those practicing obstetrics without a license.[6]

Was midwifery the same as obstetrics? The physicians who controlled the Massachusetts Board of Registration in Medicine contended that it was, and in the 1907 case of Commonwealth vs. Hanna Porn, the Massachusetts Supreme Court ruled that the midwife Porn's attendance at births was "contrary to the plain intent of the statute." The judge also noted Mrs. Porn's good character and suggested that it was possible for the Massachusetts legislature to "separate by a line of statutory demarcation the work of the midwife from that of the practitioner of medicine."[7] Such a distinction was never made as neither the Board of Registration in Medicine nor the Massachusetts Medical Society showed any interest in creating a system of education and licensing of midwives, such as the one which had been successfully developed in New Jersey.[8] Instead, the Porn decision, combined with an 1897 Birth Registration Act, put midwives in a double legal bind. The registration law required midwives and doctors to report all the births that they attended to the local city clerk, under penalty of a twenty-five dollar fine.[9] Therefore, if a midwife attended a birth and reported it, she was, in fact, admitting that she was practicing obstetrics. If she did not report the birth, she was breaking the Birth Registration Law.

When faced with these legal problems, a midwife could pursue one of four options. The first was to give up her practice, as many midwives in the Boston area apparently did.[10] Second, she could continue to practice surreptitiously, hoping to avoid the attention of legal authorities. Third, she could become, in essence, an obstetrical nurse who worked under the close supervision of a physician.[11] Her final option was to join with other midwives and battle politically the medical associations opposing her. However, the midwives never developed sufficient political self awareness and organization to respond successfully to their male competitors.[12]

It has generally been believed that, given these limited options, midwifery practice in Massachusetts quickly died out.[13] This was not the case. Midwifery practice continued and, as the example of Lawrence will illustrate,

[6] Reginald Fitz, "The Legislative Control of Medical Practice," in *Medical Communications of the Massachusetts Medical Society* (Boston: Massachusetts Medical Society, 1894), pp. 277–360.

[7] *Massachusetts Reports*, 196: 328.

[8] John A. Foote, "Legislative measures against maternal and infant mortality: the midwife practice laws of the states and territories of the United States," *Amer. J. Obstet. Diseases Women Children*, 1919, *80:* 534–51. See also James L. Huntington, "The midwife in Massachusetts: her anomalous position," *Boston Med. Surg. J.*, 1913, *168:* 418–21; Julius Levy, "Maternal mortality and mortality in the first month of life in relation to attendant at birth," *Amer. J. Public Health*, 1923, *13:* 88–95.

[9] *Acts and Resolves*, 1897, chap. 444, pp. 424–29.

[10] J. Whitridge Williams, "Medical education and the midwife problem in the United States," *JAMA*, 1912, *58:* 1–7. See also Baldy, "Midwife a necessity?" p. 401.

[11] Arthur B. Emmons and James L. Huntington, "A review of the midwife situation," *Boston Med. Surg. J.*, 1911, *164:* 261.

[12] Litoff, *American Midwives*, pp. 91–121.

[13] David Stewart, *Five Standards for Safe Childbearing* (Marble Hill, Missouri: NAPSAC, 1981), p. 126.

flourished openly in some areas for almost a decade after the Porn decision. Since changes in the law did not free Massachusetts of midwives, opponents began to focus their efforts on influencing public opinion.

The most vocal opponent of midwifery in Massachusetts, Dr. James L. Huntington, argued in a 1912 article in the *Boston Medical and Surgical Journal:* "But in (trying to stop midwifery), legislation is powerless. The keynote is education. We must educate the general public to realize the necessity for good obstetrics, and place the means of obtaining it at the disposal of all, and then with or without laws against the midwife, she will disappear from our cities."[14] The public education campaign advising women to consult physicians about obstetrics and not midwives, which began in the late nineteenth century, was to continue into the 1930s.[15] One of the factors involved was a redefinition of birth as a disease rather than a natural process.[16] Physicians stressed the pain of childbirth and the value of obstetrical anesthesia such as Twilight Sleep which could be administered only by a doctor in a hospital.[17] Class and racial bias were also invoked against the midwives.[18] Finally, hospital births were portrayed as a matter of status as well as of safety.[19] The cumulative impact of this educational campaign was substantial. A woman having a midwife as a birth attendant would not only be putting herself and her baby at needless risk of death, she would also be experiencing much greater pain than necessary by trusting her fate to a woman who generally served only those too ignorant to know any better. This was the message carried directly to women by physicians in lectures and public forums and also in articles in popular women's magazines.[20] Ironically, two of many women's greatest fears, the episiotomy and the use of forceps, became symbols of the technology medicine could bring to the aid of the parturient woman.[21] Still midwives continued to practice in Massachusetts in part because of the support of the only class of medical practitioner that medical societies viewed as worse than the midwife: the "ignorant and often unscrupulous" doctors who helped them.[22]

At a meeting of the Massachusetts Medical Society on 30 December 1913, Dr. Hugh Cabot complained that some members of the society were assisting

[14] James L. Huntington,"The regulation of midwifery," *Boston Med. Surg. J.*, 1912, *167:* 86.

[15] Litoff, *American Midwives*, pp. 64–90.

[16] Joseph B. DeLee, "Progress toward ideal obstetrics," *Trans. Amer. Assoc. Study Prevention Infant Mortality*, 1915, *6:* 117.

[17] Judith Walzer Leavitt, "Birthing and anesthesia: the debate over Twilight Sleep," *Signs*, 1980, *6:* 147–64.

[18] One article described the immigrant midwife as a woman who "knows nothing of hygiene, asepsis or antisepsis and is often practically responsible for the death and invalidism of mothers as well as the death, blindness and mental and physical impairment of infants. The immigrant midwife is herself a dirty, unkempt person living in a squalid tenement." See Carolyn C. Van Blarcom, "Midwives in America," *Amer. J. Public Health*, 1914, *4:* 198.

[19] Richard Wertz and Dorothy Wertz, *Lying-In: A History of Childbirth in America* (New York: The Free Press, 1977), pp. 146–60.

[20] Litoff, *American Midwives*, pp. 81–82.

[21] Joseph B. DeLee, "The prophylactic forceps operation," *Amer. J. Obstet. Gyn.*, 1920, *1:* 34–44.

[22] Huntington, "Regulation of midwifery," p. 84.

midwives by "... handing in birth returns for cases over which they have had no supervision."[23] Cabot felt that physicians must be enjoined from this practice. One step had been taken in 1912 by a change in the Birth Registration Law which required physicians to indicate on birth returns whether or not they had personally attended. A second, more direct attempt to control physicians' activities was manifested in a 1917 amendment to the Medical Practice Act. By this a physician could lose his license for one year if he "... acted as principal or assistant in carrying on the practice of medicine by an unregistered person."[24]

With this background in mind, the status and practice of midwives in Lawrence, Massachusetts, in the early twentieth century was studied. The primary focus of the investigation was to learn how the midwives and their clients coped with this changing legislation. Did the restrictions cause midwives to stop practicing? If not, how did they avoid prosecution? Did they serve only the immigrant community as some have suggested? What kind of relationships did they have with local doctors? Finally, if legal restrictions did not diminish their practice, what did?

This study of midwives in Lawrence is based on data drawn from several distinct sources. Most important are the birth registration records filed with the Lawrence City Clerk between the years 1896 and 1915. Lawrence was chosen as the focus of the study because it fitted the description of an "immigrant city," and its birth records had been kept intact and were made available by a cooperative city clerk. The choice of the years studied was dictated by the information contained in the records of this period. Prior to 1896 birth attendants often did not sign returns and consistent records begin in Lawrence only in that year.[25] In late September 1914, midwives in Lawrence all stopped signing returns as a result of the prosecution of several of them.[26] Returns through 1915 were analyzed to get a sense of the midwives' practice after they stopped signing returns, but empirical analysis is complete only through 30 September 1914.

The number of births attended by each midwife and each doctor in every year was recorded. As a result of limited resources, complete information from all 30,000 birth returns in this period could not be coded. Only in selected years—1896, 1900, 1905, 1910, 1914—was more information gathered, and when doctors attended, only the birthplace of the parents was noted. However, in the case of midwives attending in those years, the following information was coded: child's sex and name, father's name and mother's maiden name, address of parents, birth place of each parent, oc-

[23] Walter Burrage, *A History of the Massachusetts Medical Society, 1789–1922* (Norwood, Massachusetts: Plimpton Press, 1923), p. 255.

[24] *Ibid.*

[25] Some signed records from 1892–93 were found but these were not consistently recorded.

[26] Starting in October of 1914 and continuing through May of 1915, most of the major Lawrence midwives were prosecuted for "practicing medicine without a license." All pleaded guilty and were fined $100.

cupation of the father and, in 1914, ages of each parent, occupation of mother and the ward of Lawrence in which the birth took place. Complete information was also gathered on each midwife's deliveries in her first year and, when possible, the date of each midwife's first and last recorded delivery was identified. While more complete data is desirable, overall the study is based on the examination of over 30,000 individual records and almost 3,000 coded midwife returns.

In 1913, Lawrence began to undertake a census of births in the city. In the first few weeks of a year, canvassers visited every household in the city and recorded any births in that family in the previous year. The canvasser's cards were then matched to the birth return slips. These records were valuable in two ways. First, they filled in apparent gaps in the birth returns with approximately 100–200 births recorded each year that were not covered by the returns. Second, when midwives stopped signing returns, the canvasser's card provided a clue to their continued practice. For example, a birth return would be signed by a doctor who noted that he had not personally attended the birth, while the corresponding canvasser's card would list the attendant as one of the midwives. This discrepancy was probably the result of parents, perhaps unaware of the legal situation faced by midwives, candidly responding to the canvasser's question of who attended the birth. When the census cards and birth returns were later collated, the midwife's name would frequently be crossed out or erased and the doctor's name inserted. Also, on occasion, the census card would indicate two attendants, such as "Mrs. Riehm and Dr. Birmingham." This information enabled the authors to get a sense of the relationship between various doctors and midwives. The canvasser's cards by 1918 seldom noted midwives' deliveries, although interviews suggest that some women continued to practice for two more decades. The change in 1918 may have indicated a decline in midwife activity or simply a change in data collection procedures by the canvassers. Therefore, while the canvasser's cards will be used in certain cases, primary empirical analysis will focus on the period between 1896 and 1914 when more data is available.

Since all of the midwives who attended births in Lawrence have been dead for at least twenty years, the best source of information concerning the informal side of their practice was interviews with their relatives, friends, acquaintances and clients. These people were traced through an examination of the midwives' death records, obituaries and various marriage and birth records. Those contacted were sent a questionnaire and, when possible, interviewed. The same process was used to trace the relatives of important doctors. Also, the local newspapers from 1890–1915 that were available on microfilm were examined and records of midwives' and doctors' advertising were made.[27] Finally, the data on population and immigration was drawn from United States Census records and the annual reports of state births.[28]

[27] The *Evening Tribune* was the primary newspaper studied for this period.

[28] Secretary of the Commonwealth of Massachusetts, *Annual Reports of Births, Marriages and Deaths in the Commonwealth*, Public Document No. 1 (Boston: State Printers, 1860–1930). These will be cited hereafter

Before examining the impact of the Porn decision on the midwives of Lawrence, it is important to understand why the setting of this study so aptly fitted the description of an immigrant city. Lawrence, Massachusetts, was chartered in 1847 as a six-square-mile planned mill community on the Merrimack River twenty-five miles north of Boston. Founded as a model town by Abbott Lawrence, the community soon flourished, and by 1850 the population was 6,000, most of whom were mill workers drawn from the New England states. They had the services of a library, eleven schools and a dozen doctors, all provided under the paternalistic arrangements of the mill owners.[29]

The character of the city changed as the labor supply was drawn in greater numbers from Europe. The first wave of immigrants consisted of Irish families in the 1850s and 1860s. Changes were particularly rapid in the city from 1890–1910 when Germans, Russians, Italians, Turks and Poles moved there (see Table 1). In 1890, the city population was 44,654; by 1915, it had grown to 90,259 with 45.8 percent of the population being foreign born.[30] While the birth rate changed little in this period, by 1915 the rapidly increasing population resulted in a doubling of the number of births in the city to 2,948.

By the turn of the century the city was hardly a model community as it now had a large, poor, overcrowded, generally illiterate population. Relations between laborers and mill owners were continually strained as the workers sought to organize into unions. The culmination of this tension was a strike in 1912 which drew national attention to Lawrence as a symbol of both the struggle faced by immigrants and the bitterness of union-management disputes nationally.[31] It was in this setting that the midwives worked, in many cases renewing a practice they had established in another country, and which had been interrupted when they joined their husbands who sought a better life in the mills of Lawrence.

Figure 1 illustrates the proportion of recorded births attended by midwives, male doctors and female doctors. It suggests why physicians felt so threatened by midwives that Hanna Porn was taken to court. In 1903, male doctors attended 67.6 percent of all births in Lawrence, but by 1908 the figure had dropped to 56.5 percent. Even more striking is the consistency of the gains made by midwives, who more than doubled the proportion of

---

as *Births, Marriages and Deaths*. The total births for Lawrence listed in these volumes are typically higher than the number of birth registration sheets remaining in the Lawrence Clerk's office. This is partly the result of the state totals being drawn from the summary birth registration books which have a number of entries made after the end of the calendar year. It also probably reflects a loss of some sheets—a problem largely rectified with the initiation of the birth canvass. Also an examination of ethnicity suggests that there was no systematic bias in the data.

[29] Donald Cole, *Immigrant City: Lawrence, Massachusetts, 1845–1921* (Chapel Hill: University of North Carolina Press, 1963), pp. 17–26.

[30] *Ibid.*, p. 209.

[31] *Ibid.*, pp. 1–13; 177–94. See also William Cahn, *Lawrence 1912: The Bread and Roses Strike* (New York: Pilgrim Press, 1977).

Table 1. Immigration Into Lawrence, 1890–1910

Year	Total Population	Nativity							
		Ireland	England	Canada	Germany	Russia	France	Italy	Turkey
1890	44,654	7,697	4,985	4,459	1,830	60	0	46	5
1895	52,164	7,487	5,486	5,665	2,402	426	68	263	213
1900	62,559	7,058	5,131	8,682	2,465	780	147	936	277
1905	70,050	6,557	5,153	7,597	2,388	1,297	435	2,804	1,332
1910	85,892	5,943	5,659	9,498	2,301	4,366	788	6,693	2,077
Net Change	+41,238	−1,754	+674	+5,039	+471	+4,306	+788	+6,647	+2,072

Source: Derived from Donald Cole, *Immigrant City* (Chapel Hill: University of North Carolina Press, 1963), p. 209.

births they attended between 1900 and 1908 from 19.9 percent to 41.3 percent. Female doctors played a small role in this period, except between 1898 and 1901. Their activity in those years was primarily the result of the arrival of one woman—Johanna Koehler. Dr. Koehler personally accounted for 60 percent of the 813 deliveries attended by six different female doctors in those four years. After Dr. Koehler's departure in 1902, female doctors attended, on the average, only about one in thirty births in the city. In 1914 this dropped to a low of attendance at 0.6 percent of all births.

Figure 1 also makes clear the limited immediate effect of the Porn decision of 1907. Seven years after the decision, midwives were still attending 40 percent of the deliveries in the city and *signing the birth returns* in clear violation of the law. In another sense, however, the outcome of the Porn case may well have had an effect. The steady growth in the number of midwife deliveries stopped after 1908, though other factors, discussed below, also influenced that change. Overall, however, figure 1 clearly illustrates that the legal campaign to drive out the midwives was only partially successful at the outset. As contemporary opponents of midwives noted, legislation alone was not sufficient to stop immediately midwives' practice in Lawrence or in similar communities.[32]

Another factor that probably concerned opponents was the growing number of midwives in Lawrence and the changing nature of their practice. If midwifery practice in a community was limited to one or two women, they could be singled out for special sanctions or attention.[33] However, if there were greater numbers of midwives, it was more difficult to control them and there was greater public support for their activities. In 1896, there were 327 midwife-attended deliveries in Lawrence. One woman, Louise Beck, was responsible for all of these. By 1913, there were more than a dozen active midwives attending almost 100 births a month with the busiest woman, Minnie Riehm, accounting for only 18 percent of all midwife deliveries.[34] Therefore, the opponents of midwifery could not succeed by shutting down the practice of even the most prominent midwife in the city. The fact that there were a number of midwives practicing is important to keep in mind as the reasons why the proportion of midwife-attended births remained so high are examined.

Midwifery in Lawrence continued for a number of reasons including the immigrant nature of the population, the lack of a hospital providing obstet-

[32] For example, in a 1909 study of five Massachusetts cities, Huntington found that 15 percent of all births were attended by midwives, and argued for the public education campaign described above. See Huntington, "Regulation of midwifery," p. 86.

[33] Roger W. Cobb and Charles D. Elder, *Participation in American Politics: The Dynamics of Agenda-Building* (Boston: Allyn & Bacon; paperback, Baltimore: The Johns Hopkins University Press, 1972), pp. 124–29.

[34] Lawrence City Clerk, Birth Registration Sheets, 1913. Physician obstetrical practice was also decentralized. In 1913, there were forty-one physicians, as opposed to twelve in 1896, who attended at least ten births. They averaged about 140 deliveries a month with the busiest physician, Dr. Alphonse Riopelle, attending 14 percent of all physician deliveries.

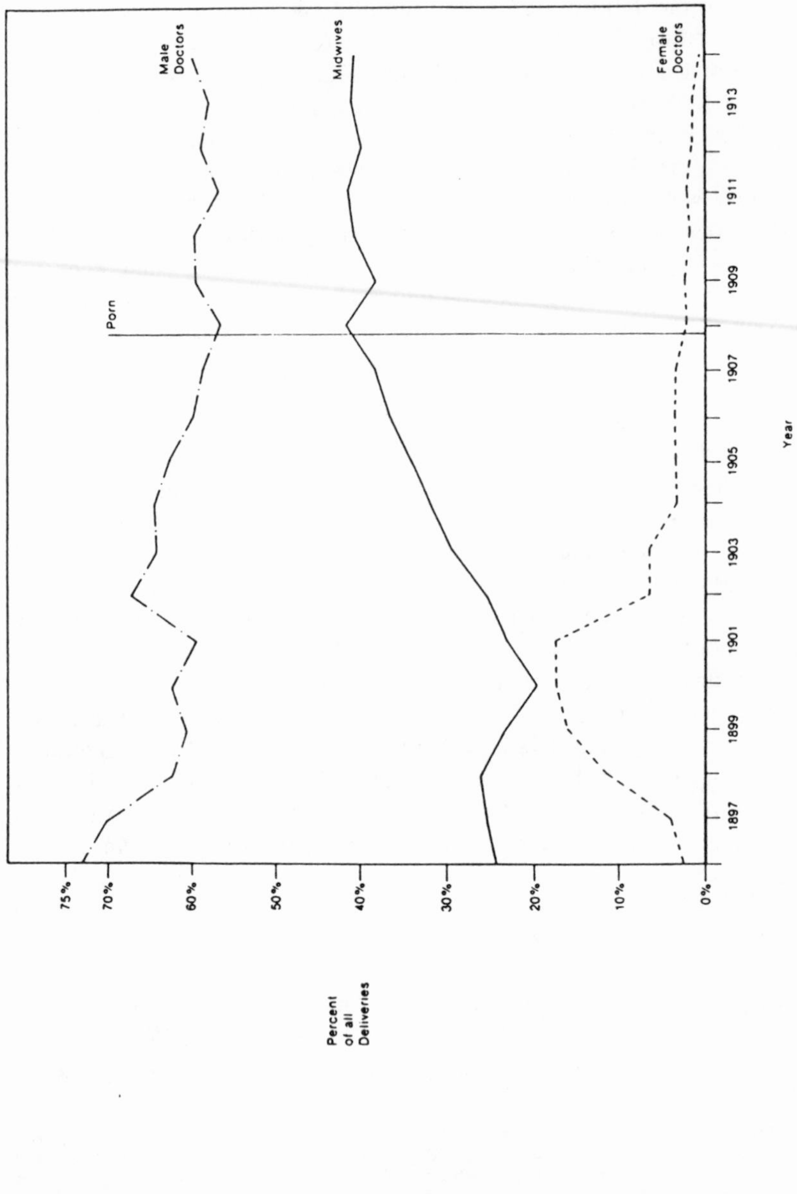

Figure 1. Percent Distribution of Recorded Deliveries in Lawrence, 1896–1914

rical care, the quality of the care provided at limited cost by the midwives, their use of advertising and finally, when the laws were more rigorously enforced, the midwives' ability to circumvent them. Of these factors, the most important was the immigrant nature of Lawrence's population. The relationship between immigration and midwifery was apparent to some midwife critics, such as Dr. Ira Wile, who noted that: "the interrelation of immigration and midwifery may not seem patent at first glance. A moment's thought, however, will show that in the problems relating to the restriction of immigration and the restriction of midwifery there are many points in common."[35] In Lawrence, between 1896 and 1914, the proportion of births to immigrants to total births was never less than 81.8 percent and was as high as 86.2 percent in 1909.[36] More important, as noted earlier (see Table 1), the immigrants in Lawrence at this time were primarily from Southern and Eastern Europe where, as Wile pointed out, "... the midwife exists as an established institution."[37] While English, Irish, and Canadian immigrants accounted for 84 percent of the total foreign-born Lawrence population in 1890, by 1910 they made up only 51 percent of all immigrants in the city.[38] Van Blarcom was prophetic when she stated:

> The desire of foreign women to employ midwives and the determination of midwives to be employed is so strong that even were midwives legislated out of existence they would still be called in by expectant mothers ... (and) among the Italians the husbands of the expectant mothers will seldom permit a physician or "man midwife" to be employed.[39]

Also, among those moving into the city were many women with formal training and established practices overseas who expected to continue their professional activities in their new home. In contrast, the ethnic groups already in the city had generally relied on doctors of their own nationality to deliver their babies. This was particularly true for Canadian mothers.[40]

Of the three active midwives in the city in 1900, two were generalists in terms of the ethnic groups they served while the third—Ann Curran—was a specialist. Almost all of the forty-two births she attended were either to women born in Ireland (thirty-one) or women born in the United States with Irish maiden names. The other two active midwives, Hedwig Himmer and Minnie Riehm, served a heterogeneous mix of women from Italy, Ireland, Poland, Germany and England. Five years later, the situation had changed considerably. In 1900, births to Canadian mothers accounted for

[35] Ira S. Wile, "Immigration and the midwife problem," *Boston Med. Surg. J.*, 1912, *167*: 113.
[36] *Births, Marriages and Deaths*, 1910, vol. 68.
[37] Wile, "Immigration and the midwife problem," p. 113.
[38] Cole, *Immigrant City*, p. 209.
[39] Van Blarcom, "Midwives in America," p. 203.
[40] For example, in 1900, 95 percent of births to Canadian mothers were attended by five doctors, primarily Omer Routhier (160), A. J. Riopelle (126) and Alphonse Petit (52). Likewise all but eighteen of the almost 400 deliveries attended by these five doctors were either to mothers born in Canada or women whose husbands were Canadian. Only nine births to Canadian mothers were attended by midwives.

almost one-third of all births in the city and those to Italian mothers one-twentieth. By 1905, the proportion of births in the city to Canadian mothers dropped to one-fourth while births to Italian mothers now numbered over 200 or one-eighth of all city births. It should also be noted that these Italian mothers relied almost exclusively on midwives, most notably, Angelina Lopiano (120 of 122 deliveries to Italian mothers), Serafina Isabella (36 of 38 deliveries) and the ubiquitous Minnie Riehm (37 of 192 deliveries). There was also a midwife who served only Syrian mothers and yet another whose clients were all Portuguese.

As these new immigrant groups moved into the city, they apparently were not deterred by, and perhaps were even unaware of, the legislative and court battles being waged in Boston. It is also noteworthy that as late as 1907 there were no Italian doctors in Lawrence attending births, and they did not become involved in more than a handful of births until 1914. An examination of city directories reveals that the two doctors who were most active in delivering babies after the midwives stopped signing returns did not appear in any directory until 1905 (Dr. Joseph Sinclitico) and 1909 (Dr. Constant Calitri).[41] Hence, when the first wave of Italian immigrants arrived in Lawrence at the turn of the century they not only brought with them a preference for midwives but they also found few Italian doctors to serve them. Therefore, a cultural preference for midwives in certain key ethnic groups and a dearth of doctors to serve these populations combined to provide a major impetus for continued midwifery practice in the city.[42]

A second factor contributing to the popularity of the midwives was the lack of opportunity to give birth in a hospital. Although much of the effort of the "reformers" of midwifery legislation was aimed at moving birth from the home to the hospital, this was not possible in Lawrence.[43] Lawrence in this period had only one major hospital. The Lawrence General Hospital had opened in 1875, but it was not until 1903 that the city records note a birth there, and it was 1911 before the hospital first recorded one hundred births in a single year. In 1913 there were 145 deliveries at Lawrence General and twenty-three more at two smaller hospitals, accounting for 6 percent of all births in the city. Four midwives and one physician had more home deliveries individually that year. The overwhelming majority of Lawrence physicians attended only home births prior to 1915. By 1920, however, obstetrical practice at the hospital had increased to 354 births or about one out of every eight in the city. The births were attended by a variety of doctors though

[41] *The Lawrence Directory* (Boston: Sampson & Murdock), annual editions from 1900 to 1910.

[42] For a more complete study of the relationship between midwifery practice and ethnicity in Lawrence, see Eugene Declercq and Richard Lacroix, "The Politics and Sociology of Immigrant Midwifery Practice in Lawrence, Massachusetts," a paper presented at the annual meeting of the Organization of American Historians, Cincinnati, Ohio, April 1983.

[43] Wertz and Wertz, *Lying-In*, pp. 132–77.

certain ethnic groups, notably Italians and Canadians, still gave birth almost exclusively at home.[44]

A third factor working on behalf of the midwives was that many of them did not fit the dire images drawn of them by their adversaries. The busiest of the midwives in Lawrence were all literate, had specialized education, and were apparently concerned about hygiene. Many had been trained in Europe where, as an article in the *Journal of the American Medical Association* noted, "They are admitted to lying in asylums for poor women and are taught cleanliness and the physiology of labor theoretically as well as practically."[45] For example, Louise Beck had graduated from the "principal school of obstetrics" in Germany, Hedwig Himmer from the Royal School of Obstetrics in Dresden, Serafina Isabella from the Clinica Obstetrica in Naples and Virginia Pedrazzini from the University of Milan. Some of the midwives received their training in the United States with Minnie Riehm (New York School of Midwifery) the most prominent example.[46] In addition, the midwives were often members of various social and religious groups in the city and, in serveral cases, their spouses were also prominent men in the community.

A fourth incentive for the use of midwives was financial. The standard cost for midwifery services was five dollars for which a mother often received, "... attendance at confinement, simple nursing care, help in the household duties and general nursing and oversight of the home until such time as she is able to resume her work."[47] Daily checkups from midwives for eight to ten days were common, while physicians would usually limit themselves to one or two postpartum visits. Of equal importance to some of those interviewed was the fact that most midwives would accept payment in kind such as food and services, while doctors usually wanted to be paid their ten dollar fee in cash. Flexible payments of this type were important since the average mill worker in Lawrence in 1912 made eight dollars and seventy-six cents for working a fifty-four hour week.[48]

Another reason why midwives retained their prominence was advertising. At almost no time between 1896 and 1911 did a daily edition of the *Lawrence Evening Tribune* appear without an advertisement for at least one midwife in it. Louise Beck, Minnie Riehm, and Hedwig Himmer advertised most frequently. With the exception of Serafina Isabella, the Italian midwives never advertised in the *Tribune*, perhaps either relying on informal communication networks or a more specialized Italian newspaper. The *Tribune*

[44] Thomas Leavitt, *Mill Owners and Missionaries: A History of the Lawrence General Hospital* (Lawrence: Lawrence General Hospital, 1975), pp. 1–28. See also City of Lawrence, Birth Registration Summary Volumes, 1900–1930, City Hall, Lawrence, Massachusetts.

[45] "Supervision of midwives," *JAMA*, 1907, *48*: 712.

[46] This information was drawn from the midwives' obituaries, their advertisements, and interviews with their descendants.

[47] S. Josephine Baker, "The function of the midwife," *Women's Med. J.*, 1913, *23*: 197.

[48] U.S., Congress, Senate, *Report on Strike of Textile Workers in Lawrence, Massachusetts, 1912*. Prepared under Charles P. Neill, Commissioner of Labor. 62nd Cong., 2nd Sess., 1912, Document #870, p. 72.

advertisements were generally straightforward and were quite similar to those used by physicians at the time. In fact, several of the midwives, including Hedwig Himmer, referred to themselves as doctors.[49] Minnie Riehm, for most of her eight years of advertising, also stated that she was a "successor to Louise Beck," which she was, quite literally. The last advertisement for Louise Beck appeared on 8 May 1899 (her last recorded delivery in Lawrence was on 11 May before she moved to Paterson, New Jersey), followed the next day by a "successor" advertisement for Mrs. Riehm who even lived at the same address that Mrs. Beck had. This use of advertising continued after the Porn decision but only by Mrs. Himmer. Minnie Riehm's final advertisement appeared on 7 May 1907 and a month later Serafina Isabella's last advertisement was published. Mrs. Himmer continued running advertisements until 3 June 1911, but under the category of doctors.

Finally, when pressure from some physicians forced the midwives to be less conspicuous, they adopted several tactics which allowed them to continue their practice. As noted earlier, in 1911 and 1912 a change on birth registration forms was gradually implemented. A line was added which read, "I did (doctor's name) personally attend this birth." This was a means used to identify those doctors who might be aiding midwives and, indeed, in some cases most of a doctor's birth returns would involve deliveries he did not attend. The fact that midwives attended the mothers involved in these cases is based on evidence drawn from the canvasser's census cards attached to the birth returns. There were references to virtually all midwives on the canvasser's cards even after they had stopped signing birth returns in 1914. In many cases, a midwife appears in conjunction with a doctor. However, from examining the records, the exact nature of the relationship between the midwives and the physicians is unclear.

There were apparently two strategies used by midwives in these cases. In some instances, midwives working with a cooperative physician would deliver the baby and the doctor would later check on the mother and complete the birth return. According to the descendants of several of the midwives, several doctors agreed to these informal cooperative relationships. In other cases, involving less friendly doctors, the midwives would attend the birth, clean up and leave, and have the parents get in touch with a doctor. According to one physician from that era, it was this practice that led him to contact state authorities in an unsuccessful attempt to have several midwives arrested in the early 1920s.[50]

[49] The first "doctor" advertisement for Hedwig Himmer appeared on 4 June 1902 in the *Evening Tribune*, p. 7 and similar ads ran until 3 June 1911. The text of Himmer's advertisement was as follows: "Dr. H. Himmer, Midwife, 190 Park St., Lawrence. Graduate from the Royal School of Obstetrics at Dresden, Germany. I am prepared to respond to all calls made upon me. Telephone 740-11," *Lawrence Evening Tribune*, 5 January 1905. Serafina Isabella's advertisement referred to her as a "Lady Physician and Midwife" and noted that she was bilingual and responded to calls at all hours while charging five dollars for eight to ten days' attendance, *Lawrence Evening Tribune*, 4 January 1904.

[50] Interviews with Dr. N. F. Decesare and descendants of midwives Angelina Lopiano, Elena Stella, Concetta Leonardi, Virginia Pedrazzani, Angelina DeMarco and Paola Leone.

A careful examination of birth records revealed another strategy adopted by certain midwives: having fathers complete birth returns and sign as attendants. Prior to 30 September 1914, the proportion of birth returns signed by fathers was never greater than 1 percent. In the final three months of 1914, it was almost 5 percent. However, the assumption of midwives' involvement was based on more than inference. Canvasser's cards attached to birth returns signed by fathers frequently listed midwives as the attendant. The midwife most commonly cited in this regard was Minnie Riehm.

It might appear from the above discussion that the midwives were organized in some formal or informal relationship. Interviews with their descendants and analysis of the dates of births attended by midwives indicate that relations between the midwives were cordial at best. There was no formal organization of midwives and no evidence to suggest that, for example, one midwife might cover for another on vacation. It appears that some of the busiest midwives never took a vacation since no clear break in the dates of their deliveries occurs. Even within ethnic groups, the relationships between midwives appear complex with factors such as birthplace within a country, neighborhood and training all playing a role. For example, most of the Italian midwives were divided into Sicilians and Neapolitans, which in turn influenced the area of the city in which they lived and, therefore, practiced.[51]

Eventually, the combination of legal restrictions, physicians' opposition, the greater influx of foreign-born doctors, the rise in hospital births, the passage of restrictions on immigration in the 1920s, and the assimilation of immigrants resulted in the diminution of midwifery practice. By the late 1920s midwives were the attendants at only a small fraction of the births in the city. Interviews with their descendants, however, indicate that several of the midwives continued to practice into the 1930s. Also, in the *Lawrence Directory,* in the listing of nurses, Angelina DeMarco (1933), Concetta Leonardi (1938) and Angelina Lopiano (1938) continued to have the word midwife added after their name well into the 1930s.[52] By this time, their practice was limited primarily to family and friends.

Overall, this research has shown that changes in the legal restrictions regulating midwifery practice had little immediate impact on the activities of midwives in Lawrence, Massachusetts. For at least seven years after supposedly being legislated out of existence, midwives continued to practice openly in the city and, for some time after, practiced surreptitiously. The factors that enabled them to continue practicing despite the legal changes were primarily the immigrant nature of the population, the midwife's own skill and standing in the community, their minimal fees in a working-class community, and their ability to circumvent the statutes, often with the help of cooperative

[51] *Ibid.*

[52] *The Lawrence Directory*, annual editions from 1933 to 1938.

physicians. As conditions changed, most notably with the arrival of doctors who were also immigrants and who did not cooperate with midwives, the law became a larger barrier for the midwives. In essence, the first thrust of the campaign against midwifery—legal changes—served as a necessary, but clearly not a sufficient, condition to eliminate midwifery practice. Eventually, by 1940, as limits on immigration cut down the flow of newcomers who supported the midwives, midwifery practice in Lawrence largely disappeared.

MARY BRECKINRIDGE, THE FRONTIER NURSING SERVICE AND THE INTRODUCTION OF NURSE-MIDWIFERY IN THE UNITED STATES*

Nancy Schrom Dye

Since the early decades of the twentieth century, physicians have been the sole legitimate birth attendants in the United States. In contrast to Great Britain and many nations in continental Europe, where trained midwives attend a significant percentage of births and where home delivery continues to be an accepted alternative to hospital confinement, the United States has little in the way of a tradition of professional midwifery, and most Americans regard the hospital as the only safe place to give birth. Early in the twentieth century, Mary Breckinridge attempted to transplant the European tradition of professional midwifery to the United States. To demonstrate the desirability of trained midwives, she established the Frontier Nursing Service (FNS) in 1925. Since that time, the FNS has provided maternal and infant care to many of the residents of Leslie County, Kentucky, an impoverished rural county in the heart of Appalachia. By establishing the FNS, Breckinridge introduced the profession of nurse-midwifery to the United States, pioneered in establishing an autonomous professional role for nurses, and helped demonstrate that the quality and consistency of care rather than surgical intervention or hospitalization were the critical factors in maintaining low maternal and infant mortality. Despite the success that the FNS has enjoyed and despite the fact that it has served as a model for health care organizations throughout the world, nurse-midwifery is still an anomaly in the American health care system. Nurse-midwives attend only 1 percent of American births.[1] The history of the establishment and early work of the Frontier Nursing Service illustrates some of the difficulties nurse-midwifery has faced in the United States and sheds light on the question of why professional midwifery continues to be a relatively little known system of maternity care.[2]

* This article is a revised and expanded version of a paper originally presented at the "Women and Medicine" conference held at Boston College, November 1980. I wish to thank Janet James, James Klotter, Anne Campbell, Eric Christianson, and Judith Walzer Leavitt for their helpful comments and suggestions at various stages of its evolution. I also wish to thank Ellen Douglas Breckenridge for sharing her work on the Frontier Nursing Service and Manfred Waserman for his help with manuscript collections at the National Library of Medicine.

[1] American College of Nurse-Midwives, *Nurse-Midwifery in the United States, 1976–1977* (Washington, D.C.: American College of Nurse-Midwives, 1978), p. 1.

[2] A second organization, the Maternity Center Association of New York City, has also been closely identified with professional midwifery and has been active since the 1930s in training and employing nurse-midwives

I. During the 1920s, Mary Breckinridge often appealed to potential supporters of her nurse-midwifery organization in the Kentucky mountains by comparing childbirth to war. "The bearing and rearing of children," Breckinridge declared,

> is as much a national service as fighting—and should receive equal recognition. . . . But up until now, although we have hospitalized and pensioned our soldiers . . . we have left our rural women and infants to die unattended or, far too often, to drag out a mutilated existence unrelieved. We have lost more women in childbirth than men on the battlefield, and exact the heaviest toll from our youngest and most defenseless citizens.[3]

In the years immediately following World War I, as the public became increasingly alarmed by the United States' high maternal and infant mortality rates, many found this analogy compelling. Until the mid-1930s, an American woman of childbearing age was more likely to die from complications related to parturition than from any other medical cause except tuberculosis. What was more, although the number of deaths from diseases such as typhoid, diphtheria, and tuberculosis had been steadily declining since the late nineteenth century, maternal and infant mortality remained persistently high. In 1923, when Mary Breckinridge began to make plans for the Frontier Nursing Service, the national maternal death rate stood at 6.7 per 1000 live births, one of the poorest rates in the Western world.[4] The United States' infant mortality record was also poor. Each year in the first decades of the twentieth century, more than 250,000 infants, nearly one out of every ten born, died before they reached their first birthdays. By the twenties, infant mortality had begun to decline steadily, but the United States remained among the countries with the highest death rates.[5]

---

and exploring alternatives to physician-managed, hospitalized childbirth. Although the Maternity Center Association was established in 1918, it did not begin to make use of nurse-midwives as birth attendants until the 1930s. For a general overview of nurse-midwifery in the United States, see Judy Barrett Litoff, *American Midwives, 1860 to the Present* (Westport, Conn.: Greenwood Press, 1978), chapter 7.

[3] Mary Breckinridge, "Outline for a Demonstration of a Children's Public Health Service," copy of proposal submitted to the American Child Health Association, 1923, p. 1. Frontier Nursing Service Papers, Special Collections, University of Kentucky Libraries, Lexington, Kentucky (hereafter cited as FNS Papers).

[4] Carolyn Van Blarcom, "Provisions for maternity care in the United States," *Amer. J. Obstet. Gyn.*, 1925, *9*: 697–703; Grace Meigs, *Maternal Mortality From All Conditions Connected with Childbirth in the United States and Certain Other Countries*, U.S. Department of Labor, Children's Bureau Pub. No. 19 (Washington, D.C.: G.P.O., 1917), pp. 7–13, 51; Robert Morse Woodbury, *Maternal Mortality: The Risk of Death in Childbirth and from All Diseases Caused by Pregnancy and Confinement*, U.S. Department of Labor, Children's Bureau Pub. No. 148 (Washington, D.C.: G.P.O., 1926). For a brief historical overview, see Joyce Antler and Daniel M. Fox, "The movement toward a safe maternity: physician accountability in New York City, 1915–1940," *Bull. Hist. Med.*, 1976, *50*: 569–95, especially pp. 569–77.

[5] In 1915, the infant mortality rate stood at 99.9 per 1000 live births. By 1934 the figure was reduced to 60.1 per 1000 live births. These statistics cited from Robert Woodbury, "Infant mortality in the United States," *Annals Amer. Acad. Polit. Soc. Sci.*, 1936, *188*: 94–106. Also see *Save the Youngest*, U.S. Department of Labor, Children's Bureau Pub. No. 61 (Washington, D.C.: G.P.O., 1920), pp. 2–7. Infant mortality varied widely from community to community and was closely correlated with socio-economic status. For specific community studies during this period, see, e.g., Elizabeth Hughes, *Infant Mortality: Results of a Field Study in Gary, Indiana, Based on Births in One Year*, U.S. Department of Labor, Children's Bureau Pub. No. 112 (Washington, D.C.: G.P.O., 1923) and Anna Rochester, *Infant Mortality: Results of a Field Study in Baltimore, Maryland, Based on Births in One Year*, U.S. Department of Labor, Children's Bureau Pub. No. 119 (Washington, D.C.: G.P.O., 1923).

Mary Breckinridge founded the Frontier Nursing Service at a time when physicians, public health officials, and social reformers were engaged in a well-publicized debate over the causes of and remedies for America's high maternal and infant death rates. The twenties also witnessed the first federal legislation, the Sheppard-Towner Maternity and Infancy Act, enacted specifically to provide public funds for maternal and child health programs. Although physicians and public health experts recognized that many factors, including poverty, lack of prenatal care, unsanitary living conditions, and contaminated milk and water, contributed to maternal and infant deaths, most focused their concern upon the competence of American birth attendants.[6] Who, they asked, was qualified to manage childbirth?

By the 1920s, three types of practitioners served as birth attendants. General practitioners cared for a significant majority of American women. Midwives still attended about 20 percent of American births, considerably less than the 50 percent of births they had managed in 1910. Although their numbers had declined appreciably by the 1920s, they continued to be important as birth attendants in cities with large foreign-born populations and in the rural south and west.[7] Obstetrical specialists attended a small but increasing number of urban women.

Obstetricians, striving to win full professional recognition for their new specialty and anxious to impress Americans with the necessity of skilled medical help at all births, argued that ideally every woman should be attended by a specialist. They were particularly vocal in identifying poor obstetrical technique and inadequate clinical training as the major reasons for the high American death rates. At least half of all maternal deaths, obstetricians pointed out, were preventable. Nearly 40 percent were due to puerperal sepsis, but the incidence of infection could be drastically reduced by observing aseptic techniques. Other potentially fatal diseases and complications of pregnancy and labor—toxemia, pelvic disproportion, malpresentations, and hemorrhage, for example—could often be prevented or remedied by conscientious prenatal care and skilled practitioners. According to obstetricians, however, the average American birth attendant's skills left much to be desired. Many general practitioners were apparently ignorant of the necessity of prenatal care, careless about asepsis, and ill-prepared to handle any complication of pregnancy or labor.[8] Obstetricians argued that to remedy

[6] Antler and Fox, "Movement toward a safe maternity," pp. 569–77. See, e.g., "Minimum standards for the public protection of the health of children and mothers," *Standards of Child Welfare: A Report of the Children's Bureau Conferences, May and June, 1919,* U.S. Department of Labor, Children's Bureau Pub. No. 60 (Washington, D.C.: G.P.O., 1919); Julius Levy, "Maternal mortality in the first month of life in relation to attendant at birth," *Amer. J. Public Health,* 1923, *13:* 88–95; Joseph B. DeLee, "Progress toward ideal obstetrics," *Trans. Amer. Assoc. Study Prevention Infant Mortality,* 1919, *6:* 114–23.

[7] Litoff, *American Midwives,* p. 114, for percentage of births attended by midwives. For good discussions of lay midwives' declining role as birth attendants, see Litoff, especially chapter 3, and Frances Kobrin, "The American midwife controversy: a crisis in professionalization," *Bull. Hist. Med.,* 1966, *40:* 350–63.

[8] Antler and Fox, "Movement toward a safe maternity," pp. 569–95; Kobrin, "American midwife controversy," pp. 352–53; Meigs, *Maternal Mortality;* New York Academy of Medicine Committee on Public Health Relations, *Maternal Mortality in New York City: A Study of All Puerperal Deaths, 1930–1932* (New York:

this situation the public must be educated to realize that obstetrics was a highly skilled branch of surgery, requiring "a surgical operating room, all the applicances [*sic*] of modern surgery, all the precautions of modern surgery."[9] Once that state of affairs was reached, they maintained, maternal and neonatal mortality would decline dramatically.

Physicians also blamed midwives for America's high death rates, although by the 1920s the midwifery controversy was considerably less heated than it had been in the previous decade. Obstetricians and general practitioners alike regarded midwives as ignorant, dirty anachronisms, incapable of appreciating the need for cleanliness or of understanding the basic anatomical and physiological principles of the birth process. Most advocated doing away with midwifery altogether, either by outlawing its practice or by licensing midwives and enacting requirements so demanding that few would be able to meet them.[10]

Many public health officials, particularly those in rural regions, disagreed. Pointing out that midwives still attended a significant percentage of American births, they argued that abolishing midwifery altogether was impractical, if not impossible.[11] Although urban midwifery patients might be accommodated by expanding public maternity clinics and hospitals, there were no alternatives to the midwife in rural areas, where doctors were scarce and hospitals virtually non-existent. Kentucky, for instance, counted 2500 practicing midwives in 1922. Although they attended only 18 percent of births in the state as a whole, they were the primary birth attendants in the thirty-six eastern mountain counties. Kentucky board of health figures for three rural Appalachian counties illustrate midwives' importance. In Leslie County, for instance, forty-three midwives attended 265 births while the one physician in the county attended forty. In Knott County, fifty-five midwives attended 396 births; five doctors attended thirty-seven. In Owsley County, fifty-three midwives attended 163 births; three physicians attended sixty-six.[12]

Public health officials also pointed out that despite midwives' lack of systematic training, many appeared competent. Early twentieth-century comparative investigations of physician- and midwife-managed births in Provi-

---

Commonwealth Fund, 1933), pp. 32–50, 183–212. For an early influential critique of American obstetrical education, see J. Whitridge Williams, "Medical education and the midwife problem in the United States," *JAMA*, 1912, *58:* 1–7.

[9] Comment by Dr. Mary Sherwood in response to Charles V. Chapin, "The control of midwifery," *Standards of Child Welfare; A Report of the Children's Bureau*, U.S. Department of Labor, Children's Bureau Pub. No. 60 (Washington, D.C.: G.P.O., 1919), p. 166.

[10] See, for example, Chapin, "Control of midwifery," pp. 157–63. For a detailed and illuminating discussion of physicians' attitudes toward midwifery, see Litoff, *American Midwives*, especially chapters 5 and 6, and Kobrin, "American midwife controversy."

[11] Kobrin, "American midwife controversy"; Annie Veech, "For blue grass and hill babies," *Survey*, 1924, *52:* 625–26.

[12] Annie Veech, "A Practical Solution of Kentucky's Midwife Problem," typescript, William Hutchins Papers, Berea College, Berea, Kentucky (hereafter cited as William Hutchins Papers); Mary Breckinridge, *Midwifery in the Kentucky Mountains: An Investigation* (privately printed, 1923), Mary Breckinridge Papers, University of Kentucky Special Collections, Lexington, Kentucky (hereafter cited as Mary Breckinridge Papers).

dence, Rhode Island, Newark, New Jersey, and New York City found that midwives had fewer maternal deaths than general practitioners, reported births more promptly, and had fewer cases of infant ophthalmia, presumably because state health departments required them to use silver nitrate.[13] Educate and supervise rather than eliminate the midwife, the public health argument ran; upgrade her status and competence with required instruction and certification. In this way, intelligent young women, especially in rural communities where physician shortages made the need for skilled birth attendants most pressing, could be encouraged to take up midwifery as an occupation.[14]

Mary Breckinridge introduced a new alternative to this debate over who should manage childbirth. She argued that the major difference between countries such as the United States, with one of the highest maternal and infant death rates, and countries such as the Netherlands and Sweden, with the lowest, was the existence of skilled professional midwives.[15] If the United States would rely upon nurses with intensive postgraduate training in midwifery (i.e., nurse-midwives) to provide maternal and infant care, particularly in rural areas, the nation could significantly improve the health and survival rates of mothers and children. Nurse-midwives could effectively perform several essential functions. They could provide primary health and prenatal care in communities that lacked physicians and they could handle the great majority of normal, uncomplicated labors. At the same time, they could work as health educators, instructing families in the basic principles of sanitation, nutrition, and child care. It was to demonstrate the efficacy of nurse-midwifery as a system of maternity and infant care that Breckinridge established the Frontier Nursing Service in 1925.

II. Mary Breckinridge was herself a nurse and a trained midwife. Born in 1881 into one of the most politically and socially prominent of Southern families, she made the decision to become a nurse shortly after the untimely death of her first husband. She entered St. Luke's Hospital in New York for training in 1907 at the age of twenty-five. Upon her graduation in 1910 she did not pursue her new occupation, but returned to her parents' home in Fort Smith, Arkansas, to care for her invalid mother. There she met and

[13] Kobrin, "American midwife controversy." The best known of these comparative studies is Julius Levy, "The maternal and infant mortality in midwifery practice in Newark, New Jersey," *Amer. J. Obstet.*, 1918, *77*: 41–53 and Julius Levy, "Maternal mortality and mortality in the first month of life in relation to attendant at birth," *Amer. J. Public Health*, 1923, *13*: 88–95. For a critique of Levy's statistical methodology, see M. Pierce Rucker, "The relation of the midwife to obstetric mortality with especial reference to New Jersey," *Amer. J. Public Health*, 1923, *13*: 816–22. See also Chapin, "Control of midwifery," and comments, pp. 157–66, and S. Josephine Baker, "Schools for midwives," *Amer. J. Obstet.*, 1912, *65*: 256–70.

[14] Veech, "Practical Solution of Kentucky's Midwifery Problem"; comments of S. Josephine Baker, *Standards of Child Welfare*, pp. 165–66. See Kobrin, "American midwife controversy," for a clear statement of the public health argument.

[15] See, for example, Mary Breckinridge, "A Frontier Nursing Service," *Amer. J. Obstet. Gyn.*, 1928, *15*: 667–72.

married Richard Thompson, president of Crescent College in Eureka Springs, Arkansas. As Mrs. Thompson, Breckinridge postponed a nursing career indefinitely. In 1914, she bore a son, Clifton. Two years later, she bore a premature daughter who died shortly after birth.[16]

Although Breckinridge took great delight in her little boy, life in Eureka Springs proved unfulfilling and her marriage unhappy. To stave off discontent, she served as secretary for the local women's suffrage organization and as board member for several state public health committees. She also developed a course on child care for the students at Crescent College. In the midst of Breckinridge's involvement with these activities, her four-year-old son died. The tragedy changed the course of her life immediately and irrevocably. Determined to end her unhappy marriage and to break all ties with the past, she separated from Thompson and, after her divorce several years later, resumed her maiden name.[17]

When she left Eureka Springs in 1918, she had few concrete plans for the future. Her son's death, however, had made her fiercely determined to devote her life to child welfare. She worked briefly as a researcher for the federal Children's Bureau, took up nursing again during the influenza epidemic, and, shortly after the Armistice, left for France as a volunteer for the American Committee for Devastated France (C.A.R.D.), a private relief organization.

C.A.R.D. assigned Breckinridge to Vic-sur-Aisne, a town in a war-devastated area of northern France where she coordinated food and medical relief for some seventy villages. During her two years there, Breckinridge discovered and developed her considerable administrative abilities. Her primary achievement was organizing a visiting nurse service that offered both general and maternity nursing. This organization was the prototype of the FNS.[18]

In France, Breckinridge became well acquainted with the European tradition of professional midwifery. Impressed by the work of both French midwives and British nurse-midwives, she wrote in her autobiography that "After I met British nurse-midwives, first in France and then on my visits to London, it grew upon me that nurse-midwifery was the logical response to

[16] Mary Breckinridge's autobiography, *Wide Neighborhoods: A Story of the Frontier Nursing Service* (New York: Harper and Bros., 1952; reissued by the University Press of Kentucky, Lexington, Kentucky, 1981) remains the standard source for biographical information. She destroyed most of her personal papers, but a small collection of letters and memorabilia is preserved in the Mary Breckinridge Papers, Special Collections, University of Kentucky Libraries, Lexington, Kentucky. The Frontier Nursing Service Papers and the Frontier Nursing Service Oral History Project interviews, both held by University of Kentucky Special Collections, also contain much biographical material. For published studies, see Drew Gilpin Faust, "Mary Breckinridge," *Notable American Women: The Modern Period* (Cambridge, Mass.: Harvard University Press, 1980), pp. 103–5, and Carol Crowe-Carraco, "Mary Breckinridge and the Frontier Nursing Service," *Register of the Kentucky Historical Society*, 1978, *76*: 179–91.

[17] Faust, "Mary Breckinridge."

[18] Breckinridge, *Wide Neighborhoods*, pp. 83–110; Faust, "Mary Breckinridge"; Anne Campbell, "Mary Breckinridge and the American Committee for Devastated France: Foundations of the Frontier Nursing Service," unpublished seminar paper, University of Kentucky, 1981. This period of Breckinridge's life is extensively documented in the correspondence contained in the Mary Breckinridge Papers.

the needs of the young child in rural America."[19] A major obstacle existed to introducing nurse-midwifery to the United States, however, for midwifery went unrecognized as a profession, and neither health care professionals nor the general public drew a distinction between professional and lay midwives.

When Breckinridge returned to the United States late in 1921, she began a systematic program of preparation for introducing nurse-midwifery to this country. Her schedule included graduate study in public health nursing at Columbia Teachers' College, midwifery training at the British Hospital for Mothers and Babies and at the York Road General Lying-In Hospital in London, and an extended visit to Scotland to observe the operation of the Highlands and Islands Medical and Nursing Service, a decentralized health care organization staffed by nurse-midwives that provided skilled care to an impoverished rural population.[20]

By 1923, Breckinridge had decided upon the eastern Kentucky mountains as the site for her nurse-midwifery demonstration. She chose this location for several reasons. Like many health professionals concerned with maternal and infant health, Breckinridge believed that rural women and children were at higher risk than urban ones. Although vital statistics consistently indicated that both infant and maternal death rates were lower in rural regions than in cities, Breckinridge argued that the rural figures were unreliable.[21] In keeping with the emphasis on the attendant as the critical variable in explaining maternal and neonatal mortality, she insisted that more rural mothers died because they often went without skilled care.[22]

Other factors influenced Breckinridge's choice. As a Breckinridge, she had political and social connections in Kentucky that would automatically insure interest and support. Personal influence, she wrote in 1923, "counts for so much in the South. I would be too doubtful of success to push this project in Vermont or Michigan, but I *know* I can put it across in Kentucky—and probably anywhere in the South."[23] Then, too, Breckinridge emphasized the geographical isolation, rugged terrain, and poverty of eastern Kentucky. If nurse-midwives could succeed in improving maternal and infant health under such adverse conditions, she argued, their success would be all the more impressive.

Breckinridge spent the summer of 1923 in eastern Kentucky, surveying

[19] Breckinridge, *Wide Neighborhoods*, p. 111.

[20] Faust, "Mary Breckinridge"; Breckinridge, *Wide Neighborhoods*, pp. 122–56.

[21] See, e.g., Van Blarcom, "Provisions for maternity care," pp. 689–99; Woodbury, "Infant mortality in the United States," p. 97; Frederick Adair, "Maternal, fetal, and neonatal morbidity and mortality," *Amer. J. Obstet. Gyn.*, 1935, *29:* 384–94. In 1923, the urban maternal mortality rate was 7.7 per thousand live births; the rural rate was 5.9. Van Blarcom, "Provisions for maternity care," p. 699.

[22] Mary Breckinridge, "Memorandum concerning a suggested demonstration for the reduction of the infant and maternal death rate in a rural area of the South," typescript, 1923, FNS Papers; Van Blarcom, "Provisions for maternity care," pp. 698–99; Elizabeth G. Fox, "Rural problems," *Standards of Child Welfare,* pp. 186–93.

[23] Mary Breckinridge to Ellen Phillips Crandall, 20 October 1923, Mary Breckinridge Papers.

the health needs of women and children. In Leslie County, where the FNS was eventually located, ten thousand people lived in scattered hollows, almost completely isolated from modern transportation and communication networks. "There was no motor road within sixty miles in any direction," Breckinridge recalled. "Horseback and mule team were the only mode of travel. Brought-on supplies came from distant railroad points and took from two to five days to haul in."[24] Hyden, the county seat, with 313 people, was the largest settlement. Without a railroad, Leslie County was not yet visibly penetrated by the coal industry that had already transformed the traditional culture and the economic patterns of many Kentucky mountain counties. But by 1925, most of Leslie County's land was already owned by northern investors and industrialists, most notably Henry Ford, whose Fordson Coal and Timber Company made him the largest landowner in the area.[25]

The county ranked as one of the most impoverished in Appalachia as well as one of the poorest in the United States. Most families eked out a precarious living from the land, cultivating small crops of corn on tenant farms. In 1929, the average per capita income was $102 a year, compared to the southern Appalachian average of $433 and the national average of $776.[26]

Missionaries and settlement school teachers in the region described Appalachian families as patriarchal and invariably commented upon the hard life of mountain women. In this preindustrial mountain society, a woman's life was indeed hard. In addition to the daily domestic round of cooking (generally over an open fire), spinning, weaving, knitting, making quilts, hauling water, and chopping wood, women performed much agricultural labor, helping to plant corn on the steep hillsides in the spring, hoeing throughout the summer, and helping to harvest the crop in the fall. Leslie County women married young and bore an average of nine children. Frequent childbearing, then, represented a central and virtually inevitable part of mountain womanhood.[27]

Because the great majority of women relied upon lay, or "granny," midwives as birth attendants, Breckinridge concentrated upon investigating these folk practitioners' practices. Not surprisingly, the fifty-three midwives whom Breckinridge interviewed in three mountain counties appear to have regarded her with considerable suspicion. None allowed her to be present at a birth or confided any details about deaths or problems. Nevertheless, Breckinridge's report on mountain midwifery is valuable for providing information on the identity, economic status, and practices of early twentieth-century Appalachian midwives.[28]

[24] Breckinridge, *Wide Neighborhoods*, pp. 169–74.

[25] Harry Caudill, *Night Comes to the Cumberlands* (Boston: Little, Brown, 1962), pp. 254–55.

[26] Mary Willeford, "Income and Health in a Remotely Rural Area" (Ph.D. dissertation, Columbia University, 1927), pp. 37, 55.

[27] For a typical description of Appalachian culture and gender roles, see John Campbell, *The Southern Highlander and His Homeland* (New York: Russell Sage Foundation, 1921), pp. 126–42.

[28] Breckinridge, *Midwifery in the Kentucky Mountains*.

Almost all of the mountain midwives with whom Breckinridge talked were elderly; their mean age was sixty. They delivered an average of nineteen babies a year in a small radius of territory around their homes. All but two had borne children themselves; most had eight or nine. Like the population they served, midwives generally made their living off the land. Those who were married had farmer husbands; eighteen widows cultivated land themselves. Their midwifery practices provided small supplementary incomes. Two of the four midwives who lived in the region's small towns kept boarding houses; one served as county jailer.[29]

Unlike midwives in other parts of the American South, Appalachian midwives did not have a prescribed training system, although nine women in Breckinridge's sample were daughters of midwives and had worked as their mothers' assistants and almost all had observed other midwives at work before taking on cases of their own.[30] The one black midwife in Breckinridge's sample had been taught the rudiments of midwifery by her mistress when she had been a Virginia slave. A few had assisted physicians. Breckinridge made her survey just when the state had begun to regulate midwifery practice. The Kentucky legislature mandated midwifery registration in 1920, and in 1922, with Sheppard-Towner funds, the new Bureau of Maternal and Child Health instituted a series of one-day classes for mountain midwives. Ten of the women in Breckinridge's sample had attended these classes, where they had listened to lectures on how to equip and care for a regulation midwifery bag, maintain personal cleanliness, and administer silver nitrate. For the most part, though, these women were self-taught, and had taken up their occupation by accident. As they explained to Breckinridge, neighbors had turned to them for help in childbirth and they had discovered an aptitude for the work. Twelve of the midwives could read and write. Breckinridge stressed, however, that literacy was not necessarily a correlate of competence in a region in which the literacy rate, particularly among the older population, was low.[31]

One aspect of the existing situation that most distressed Breckinridge was the lack of prenatal attention. Rarely did a mountain mother visit her midwife before delivery. A family did not summon the midwife until active labor began. Because travel was difficult, and often involved fording mountain streams and climbing steep slopes, a midwife could take hours to reach

[29] *Ibid.*, pp. 5–7.

[30] *Ibid.*, p. 9. Also see K. Osgood, D. L. Hochstrasser, and K. W. Deuschle, "Lay midwifery in southern Appalachia: the case of a mountain county in eastern Kentucky," *Arch. Envir. Health*, 1966, *12*: 759–70. Black midwives in the Deep South often had to complete a prescribed apprenticeship and give evidence of a special calling in order to take on the role of midwife. See Beatrice Mongeau, Henry Smith, and Ann Maney, "The 'granny' midwife: changing roles and functions of a folk practitioner," *Amer. J. Sociol.*, 1961, *66*: 497–505; Molly Dougherty, "Southern Lay Midwives as Ritual Specialists," in Judith Hoch-Smith, ed., *Women in Ritual and Symbolic Roles* (New York: Plenum Press, 1978), pp. 151–64; Linda Holmes, "Granny midwives in Alabama," paper delivered at the American College of Nurse-Midwives' convention, Lexington, Kentucky, March 1982.

[31] Breckinridge, *Midwifery in the Kentucky Mountains*, pp. 7–8.

her patient. In spring, when floods made many creeks impassable, transportation was even more difficult.

Breckinridge found that eastern Kentucky midwives, like traditional folk practitioners elsewhere, relied upon empirical knowledge, local customs, and herbal remedies. Contrary to the notion of midwives as non-interventionist practitioners who "let nature take its course," however, Breckinridge found that most midwives actively managed their patients' labors.[32] Although few went as far as the one young midwife who employed pituitrin extract, all performed vaginal examinations to see if the baby was "coming straight" and if the "womb was open." Most examined women frequently, often trying to stretch the birth canal to speed labor. In the event of an abnormal presentation, midwives usually attempted to turn the baby, either by external or, more commonly, internal version. Usually midwives delivered women in a sitting position, either on the knees of their husbands or on a rude birthing chair. One sometimes delivered patients standing. Breckinridge believed that such birth postures caused perineal lacerations and prolapsed uteri.[33] Once a mother was "fixed up" and the baby dressed, the midwife collected her small fee, usually paid in kind, and took her leave. Unless a family called her back to deal with a postpartum complication, she did not return.

Midwives reported a variety of empirical and folk practices to deal with complications. Hemorrhage, the most dreaded, generated the greatest variety of responses. Some midwives relied upon a practice called "cording the leg" which involved binding the leg and hand on one side of a woman's body. Others employed ergot or paregoric or applied cold cloths to the woman's perineum. Teas of yellowroot, rattlewood, service bark, comfrey, pepper, assafetida, nutmeg, pine needles, and coal soot were also named as remedies to stop hemorrhage. An ax under the bed blade up, a practice found not only in Appalachia but throughout the rural United States, was believed to stop bleeding and alleviate labor pains. One midwife reported that she laid her hands upon a hemorrhaging patient and repeated Ezekiel 16:6: "And when I passed by thee, I saw thee polluted in thine own blood. I said unto thee when thou wast polluted in thy own blood, live."[34] For other complications of pregnancy and labor, such as the convulsions of eclampsia, the "granny" midwives had no remedies to suggest. Eastern Kentucky midwives only rarely sent for a physician. Because the region's few doctors lived in towns or coal camps, they were usually too far away to offer much help.[35]

Breckinridge was able to glean only an impressionistic sense of mortality and morbidity in this first survey. Kentucky had entered the federal govern-

[32] *Ibid.*, pp. 17–20. For discussion of midwives as non-interventionist practitioners, see Dorothy Wertz and Richard Wertz, *Lying-In; A History of Childbirth in America* (Glencoe, Ill.: Free Press, 1977), chapter 1.

[33] Breckinridge, *Midwifery in the Kentucky Mountains*, p. 20.

[34] *Ibid.*, p. 22. The reading of Ezekiel 16:6 is also reported as a practice of black midwives in the Deep South. See Dougherty, "Southern Lay Midwives as Ritual Specialists," p. 161.

[35] Breckinridge, *Midwifery in the Kentucky Mountains*, pp. 20–25.

ment's birth registration area in 1915 and had been collecting its own vital statistics since 1912, but Breckinridge had no doubt that births and deaths in rural mountain counties were at best haphazardly recorded. She insisted that maternal mortality was exceedingly high, but had no statistics to document this assertion. The midwives she interviewed were reluctant to discuss deaths. As one stated, "Everybody lost women."[36]

As a professional nurse, Breckinridge felt little sense of sisterhood with these traditional female practitioners. The report that resulted from her survey was highly critical. She conceded that some midwives were clearly competent and knowledgeable, and acknowledged that they performed necessary services. They charged much less than the region's doctors ($2.00 to $5.00 for a delivery as compared to $30.00 to $35.00) and were willing to accept in-kind payments in lieu of money, essential in a cash-poor region. Nevertheless, she judged most midwives and their houses dirty and assumed that because they lacked formal training their practices were harmful. "It is unthinkable," she concluded, "that childbirth should be without skilled care."[37]

In the fall of 1923, just before she sailed for England to take her own midwifery training, Breckinridge sent a copy of her report, accompanied by a detailed proposal for a maternal and child health project, to the American Child Health Association. Her proposal outlined a five-year demonstration project in eastern Kentucky to be called the Children's Public Health Service. The program would provide "each child from antenatal life through the school age free health facilities similar to the opportunities now provided for a free public school education" and would be staffed by public health nurses with specialized midwifery training. Breckinridge assured the Child Health Association that the region's few physicians had already expressed their support for such a program so that "such obstetrical cases as fall to the share of the Children's Nurses can be met legally and without friction." To fund the project, she estimated an annual budget of $16,300 to pay the salaries of six nurses and purchase supplies.[38]

Breckinridge stressed that the rural communities in need of such a program could never bear the greater part of its cost. But, she argued, "the bearing and rearing of children is not a local, or even a state, but a national function. Nobody," she concluded, "expects a poor city district to finance its health centers. Why should one expect it of a poor rural county which has already contributed to the wealth of the city?"[39] Although Breckinridge envisioned local taxes and charity eventually covering some of the expense, she stressed that the greater part of the cost should be assumed by public

[36] *Ibid.*, pp. 19–20.

[37] *Ibid.*, p. 27.

[38] Breckinridge, "Outline for a demonstration of a children's public health service," pp. 3, 7, FNS Papers; Mary Breckinridge to Ellen Phillips Crandall, 20 October 1923; 14 November 1923; Ellen Crandall to Mary Breckinridge, 12 February 1924, Mary Breckinridge Papers.

[39] Breckinridge, "Outline for a demonstration," p. 2. This early emphasis on the necessity for public funding

funds such as those provided through the Sheppard-Towner Maternity and Infancy Act.

The Child Health Association responded warmly to Breckinridge's proposal and indicated serious interest in sponsoring the Children's Health Service. Soon, however, the proposal ran into difficulties. The Commonwealth Fund, which underwrote Child Health Association projects, required the approval of the appropriate state health officials before financing a project. In this instance, Annie Veech, a physician who headed Kentucky's Bureau of Maternal and Child Health, was the individual consulted to evaluate Breckinridge's plan.

Veech was strongly and unalterably opposed to the Children's Health Service proposal, largely because it advocated the employment of nurse-midwives. Veech disagreed with Breckinridge's assessment of Kentucky's lay midwives and accused her of exploiting and misrepresenting mountain people. Many lay midwives, she maintained, were competent, as evidenced by the fact that the mountain counties had among the lowest maternal mortality and puerperal sepsis rates in the state. Rather than hire professional midwives, the state should recruit young mountain women for state-sponsored midwifery training programs. Such local midwives, Veech argued, would be far less expensive than trained nurses and would be content to reside permanently in the region. Implicit in Veech's argument, too, was the idea that lay midwives would be more amenable to state control and supervision than professional nurse-midwives would be.[40]

Professional rivalry and the customary opprobrium physicians displayed toward midwives also colored Veech's objections. "If high-type young women want to be of real value in isolated areas," she asked, "why do they not prepare themselves by taking a medical degree? They could then practice obstetrics instead of midwifery and meet the other medical needs of the people they wish to help. After all," Veech concluded, "a nurse-midwife is only a midwife. There appears to be a tendency among certain groups of nurses towards practicing medicine for which they are in no way prepared without graduating in medicine."[41]

Annie Veech's opposition proved to be an important and longstanding obstacle to Breckinridge's plans. Without Veech's support, Breckinridge had no choice but to withdraw her proposal from the Child Health Association. Nor could she look to the state for financial assistance. Thus, from the beginning, funding agencies, state officials, and health professionals viewed nurse-midwifery as an impractical, professionally questionable anomaly.

is very different from Breckinridge's later insistence that the FNS always remain privately funded. The FNS did not accept federal funds until 1965, the year of Breckinridge's death.

[40] Annie Veech to Mary Breckinridge, 31 October 1923, FNS Papers; Annie Veech to William J. Hutchins, 6 May 1927, William Hutchins Papers; Veech, "Practical Solution of Kentucky's Midwife Problem."

[41] Annie Veech to William Hutchins, 6 May 1927, William Hutchins Papers.

III. Breckinridge, a singularly strongminded individual, remained undaunted by Veech's hostility. When she returned to the United States in 1925, certified as a nurse-midwife by England's Central Midwives' Board, she was still determined to work in eastern Kentucky and chose Leslie County as the site for her project. Working with her wide network of personal friends and family connections, Breckinridge bypassed the state bureaucracy and established the Kentucky Committee for Mothers and Babies (renamed the Frontier Nursing Service in 1928) as a private philanthropic organization. She quickly won the support of some sixty of the state's most influential citizens and social reformers, including Frank McVey, president of the University of Kentucky, and three of the most prominent individuals in mountain welfare work: William Hutchins, president of Berea College; Katharine Pettit, founder of the Pine Mountain Settlement School; and Linda Neville, well known for her efforts to eradicate trachoma, a disease then endemic to mountain communities.

The first problem for the fledgling Committee for Mothers and Babies was funding. Breckinridge remained hopeful about receiving state or foundation support in the future, after her nurse-midwifery demonstration had proved itself, but she realized that in the short run she would have to depend upon individual donations. She soon proved to be a talented fund raiser, persuading Appalachian coal operators and railroad officials to donate money and services and Lexington and Louisville physicians to promise free treatment. As it became apparent that large-scale support was not going to be forthcoming, Breckinridge established FNS women's committees in cities throughout the nation. The money they raised soon provided the basis of the FNS budget. Some of them, most notably those in Detroit and New York, served as kinds of women's auxiliaries to industrialists with investments in the region. One of the largest contributors, for example, was Henry Ford's wife, Clara, who was active in the Detroit committee.

Breckinridge's success as a fund raiser was due partly to the fact that the establishment of the FNS coincided in time with growing interest in the people and culture of the southern highlands. As industrialists penetrated the region, attracted by its vast timber and mineral wealth, Americans created a romantic image of mountain people. Appalachia, according to this conception, was populated by rugged, self-reliant, but essentially childlike people, with quaint folkways, dialects, and rustic crafts.[42] Breckinridge capitalized upon this image to raise funds and attract attention to the FNS's work. In the strongly nativist decade of the 1920s, she also carefully presented a picture of mountain people as "the finest American stock" living on the "last frontier." As one FNS spokeswoman made this appeal, "They are the nursery of the race for children of the old American stock. Could money be better spent

[42] For a discussion of Appalachia's image, see Henry Shapiro, *Appalachia on Our Mind* (Chapel Hill: University of North Carolina Press, 1978).

than in protecting the motherhood and infancy of such a race?"[43] In characterizing mountain culture in this fashion, the FNS ignored the cataclysmic social changes that were reshaping mountain society in the early twentieth century, and remained aloof from the labor strife and economic exploitation of the region.

In the summer of 1925, two nurse-midwives, Edna Rochstroh and Freda Caffin, settled in Hyden. Both had worked as nurses with the Maternity Center Association of New York City, and taken their midwifery training in England. In Hyden, they confronted a second major problem: gaining community acceptance for a new and unfamiliar system of health care. Leslie County families were initially suspicious of the new nurse-midwives and the region's "granny" midwives understandably felt threatened. As Betty Lester, an English nurse-midwife who joined the FNS staff in 1927, recalled, "At first they ... questioned us. Did we know what we were talking about? And then, of course, the midwives who were here before us didn't like us very much because we were taking their work away from them."[44]

The first staff members worked gradually to win community acceptance. Their initial project was a house-to-house survey of maternal and infant mortality. The Kentucky Committee for Mothers and Babies was determined to compile accurate mortality statistics to serve as a baseline against which to evaluate future work.[45] At the same time, the nurse-midwives, under the auspices of the state department of health, initiated a general public health program, inoculating residents against typhoid, diphtheria, and smallpox, and treating the parasitical infections common in mountain communities.

From 1925 through 1927, the staff attended few deliveries. Although pregnant women began to register with the FNS for prenatal care, about one-third of them called a local midwife during labor.[46] Betty Lester recalled,

> They wanted their friends ... the midwife they'd had for all their other babies. ... So it was all right by us. And if a Midwife went to a call we never went. I mean, if we had done all the prenatals and ... when the time came for the delivery, they called the local midwife, we didn't go; we just left it. But if a midwife got into trouble, if she needed us, or if the husband wanted us to go, then we would go and help out. ... If they wanted them, it was all right ... whoever they wanted they had.[47]

The FNS aimed to replace the granny midwives gradually. While significant numbers of them continued to practice in Leslie County, Breckinridge in-

[43] Ella Woodyard, "Statement of costs of the Frontier Nursing Service during the fiscal year May 1, 1926–27," *Quart. Bull. FNS,* 1928, *4:* 3–8. Also see Edna Rochstroh, "Enter—the nurse-midwife," *Amer. J. Nurs.,* 1927, *27:* 149–64.

[44] Interview with Betty Lester, Frontier Nursing Service Oral History Project, FNS Papers.

[45] Breckinridge, *Wide Neighborhoods,* pp. 162–63; W. Bertram Ireland to Kentucky Committee for Mothers and Babies, 1 September 1925, FNS Papers.

[46] *Quart. Bull. FNS,* 1927, *3:* 3.

[47] Betty Lester Interview, FNS Oral History Project, FNS Papers.

sisted that they be treated courteously and with respect. FNS facilities and services were always available to them.[48]

Women's reluctance to trust the FNS diminished rapidly. During 1928–1929, only fifteen of a total of 106 prenatal patients turned to lay midwives for delivery. By 1931, the FNS nurse-midwives averaged "a baby a day," and were attending the majority of reported births in Leslie County. The opposition of local midwives also seems to have declined, perhaps because few mountain midwives depended upon midwifery for their livelihoods, or because folk traditions associated with midwifery were not as strong in eastern Kentucky as they were in other parts of the South. By 1928, no new midwives were starting practice in the FNS area.[49]

The FNS expanded rapidly throughout the rest of the 1920s. By 1930, Breckinridge had a staff of more than thirty nurses. Because nurse-midwives were not trained in the United States, she recruited most of her staff from English lying-in hospitals, and required prospective staff members to be certified by the English Central Midwives' Board. The American nurses who worked for the FNS also took their training and certification in England.

By 1930, too, the FNS covered more than 700 square miles of territory in Leslie County and in neighboring Perry and Clay counties, had raised more than $40,000 toward a permanent endowment, and had arranged for Louis Dublin, a vice president at Metropolitan Life Insurance Company and a statistician well-known for his knowledge of and concern for maternal and infant welfare, to tabulate the FNS medical statistics. Breckinridge, upon Dublin's advice, hoped to cover 1000 square miles by 1934. The expanse of territory, she maintained, "will give us the statistical data . . . necessary in order to demonstrate adequately that the work of nurse-midwives in remotely rural areas does lower the death rate of women and young children, and raise the general level of health of the population."[50]

The FNS's decentralized organization followed that of the Highlands and Islands Medical and Nursing Service that Breckinridge had studied in Scotland. FNS operations were based in Hyden, where a nursing center and, by 1928, a twelve-bed hospital were located. Most services were provided by teams of two nurse-midwives who lived and worked in outpost nursing centers. Five such centers opened between late 1925 and 1929. Each nurse covered approximately seventy-five square miles of territory in the district surrounding her center. To visit their patients, nurses rode on horseback through the mountain creekbeds and trails. Although maternal and infant care were the FNS's primary concerns, the district nurses also provided general health care, occasionally referring family members to periodic FNS

[48] Mary Breckinridge to Louis Dublin, 14 April 1928, Louis Dublin Papers, National Library of Medicine, Department of the History of Medicine, Bethesda, Maryland (hereafter cited as Dublin Papers).

[49] *Quart. Bull. FNS*, 1929, 5: 2; *Quart. Bull. FNS*, 1931, 7: 3; Mary Breckinridge to Louis Dublin, 14 April 1928, Dublin Papers.

[50] *Quart. Bull. FNS*, 1929, 5: 1.

specialty clinics in Hyden or to Lexington and Louisville physicians and hospitals.

Breckinridge was proud of the FNS's cost efficiency. For $1.00 a year, a family could register for general health care. The midwifery service cost $5.00. Both fees could be paid either in cash or, as was more common, in kind. As Breckinridge had predicted, these fees covered only a small portion—less than 10 percent—of the FNS's costs. However, despite the distances FNS nurses travelled to visit their patients, the costs of the FNS compared favorably to other health projects. The average cost of an FNS visit in 1926–1927 was $1.45, considerably lower than the costs of several other rural health projects. Breckinridge was convinced that decentralization and reliance upon nurse-midwives rather than physicians were important innovations in the economics of health care.[51]

Breckinridge was also proud of the maternity care her organization provided. The FNS nurse-midwives' management of pregnancy, labor, and the puerperium compared very favorably to and differed significantly from the care American women customarily received. The first FNS departure was in prenatal care. At a time when few physicians routinely monitored their patients' pregnancies and only an estimated 20 percent of American women received prenatal care of any kind,[52] the FNS urged women to register for care as soon as they believed themselves pregnant, and followed patients closely from the time their pregnancies were diagnosed. The nurse-midwives visited their patients biweekly until the seventh month, and weekly thereafter, concentrating upon remedying dietary deficiencies and parasitic infections. Once summoned during labor, the nurse-midwife stayed continuously with her patient. "We think the support and help given through the long hours of the first stage [of labor] has a bearing on the outcome," Mary Breckinridge stated, explaining this hallmark of a nurse-midwife's care.[53] Although more than 90 percent of FNS-managed deliveries took place in patients' homes, Breckinridge's staff was rigorous in maintaining aseptic conditions. A nurse-midwife carried everything she needed to conduct an aseptic delivery in her horse's saddlebags.

FNS nurse-midwives' management of childbirth differed from contemporary practices in other ways as well. At a time when medical attendance at birth was becoming nearly universal, the FNS staff handled nearly 95 percent of its deliveries without medical assistance.[54] Initially, medical supervision consisted only of signed protocols drawn up by Lexington physi-

[51] Woodyard, "Statement of costs of the FNS," p. 4. The best discussion of FNS costs within the context of rural health care generally is Ellen Douglas Breckenridge, "The Frontier Nursing Service: A Demonstration Project in Rural Health Care" (B.A. thesis, Harvard University, Cambridge, Massachusetts, 1979), pp. 40–43. This thesis is also on file in the Appalachian Collection, University of Kentucky Libraries, Lexington, Kentucky.

[52] J. Stanley Lemons, *The Woman Citizen* (Urbana, Illinois: University of Illinois Press, 1973), p. 154.

[53] Mary Breckinridge, "The nurse-midwife—a pioneer," *Amer. J. Public Health*, 1927, *27*: 1149.

[54] "Report on the third thousand confinements of the Frontier Nursing Service," compiled by Elizabeth Steele for the Metropolitan Life Insurance Company, 1937, p. 8, FNS Papers.

cians. Regional doctors handled the small number of major complications. In the early 1930s, the FNS hired its first resident physician who served as medical director. Given the distances involved, the physician's presence could not be counted upon in the event of an emergency. In practice, then, FNS nurse-midwives were essentially autonomous practitioners, forced to handle most complications on their own.

At a time when surgical intervention increasingly characterized American obstetrics, and when approximately 25 percent of American hospital deliveries could be characterized as involving operations, FNS nurse-midwives did little to interfere with the birth process.[55] In its first 3000 deliveries, conducted in the years 1925 to 1937, intervention was minimal. Caesarean sections, which, of course, had to be performed by physicians, were very rare; only six were done. Other operations were also uncommon. Forceps, which only physicians could employ, were used very infrequently. Only fourteen of the first 3000 deliveries involved forceps. Episiotomies, which were not general in obstetrical practice at the time, were extremely rare. In the first 3000 deliveries, only two were performed, although lacerations of the perineum were listed among the most frequent complications.[56] One nurse-midwife observed after a forceps delivery in 1938 that "this is my second forceps case in seven years. I know doctors and midwives will think this almost incredible. I do myself when I am in the cities and around hospitals. Of course in the mountains I only average around thirty deliveries in a year, but even so I think two forceps deliveries an unusual record for more than two hundred cases."[57]

How successful was the FNS in its original goal of reducing maternal and infant mortality? No reliable figures exist for 1925. The mortality survey that the Kentucky Committee for Mothers and Babies conducted in that year has been lost.[58] It is clear, however, that the FNS maintained excellent maternal mortality and morbidity records. In its first two series of 1000 deliveries, spanning the years 1925 through 1934, the FNS did not have a maternal death; in its third series, covering the years 1935 through 1937, it had two. Its maternal mortality rate for these twelve years stood at 0.68 per thousand live births.[59] For white women in rural Kentucky, the closest comparable population for which figures are available, the maternal death rate during these years averaged from 4.4 to 5.3 per thousand live births. The maternal death rate for white women in the city of Lexington, where physicians at-

[55] *Ibid.*, pp. 8–9; New York Academy of Medicine Committee on Public Health Relations, *Maternal Mortality in New York City*, pp. 116–23.

[56] "Report on the third thousand confinements of the Frontier Nursing Service," pp. 8–9, FNS Papers; "Analysis of the records of 1004 women during pregnancy, labor, and postpartum" (First Series), FNS Papers.

[57] Nora Kelly, "Put out the fire and all the lamps," *Quart. Bull. FNS*, 1938, *14*: 12.

[58] Although there are numerous references in early FNS publications and correspondence to this survey, its results were never reported upon in any detail and all copies of the completed survey apparently have been lost. Although Breckinridge turned a copy of the compiled statistics over to the state health department, no state agency has a record of the transaction or a copy of the survey. There is no copy in the FNS records.

[59] "Report on the third thousand confinements," pp. 1, 8.

tended almost all deliveries and where, by the 1930s, hospitalization was common, stood between 8 and 9. In the United States as a whole, the maternal death rate during these years ranged from 5.6 to 6.8 per thousand live births.[60]

The percentage of FNS patients who developed complications during pregnancy was also significantly lower than that of the general population. Toxic symptoms were the most common complication. In the first series of 1000 deliveries, for instance, 172 women exhibited symptoms of toxemia, but only two developed eclampsia. This low incidence of eclampsia was one of the best indicators of the efficacy of FNS prenatal care.[61]

FNS patients also developed fewer complications during labor and the puerperium than the general childbearing population. The incidence of puerperal sepsis, the major cause of maternal death in the United States, was very low. In the first thousand FNS deliveries, for example, five women were judged to be suffering from a septic condition. By the third series, for reasons that are unclear, the incidence of sepsis had increased to twenty-five cases.[62] Finally, FNS end results were extremely favorable. In each of the first three series of one thousand deliveries, more than 95 percent of the patients were in satisfactory or better condition at the end of the puerperium.[63]

The FNS's strikingly low maternal death rate and its unusually good morbidity statistics must be set within the context of general health conditions in Leslie County. As John Kooser, the FNS's resident physician during the 1930s, summarized, "Our major problems are related directly or indirectly to poor diet, intestinal parasites, chronic infection, and to a high birth rate."[64] Many FNS mothers were high risk maternity patients. The patient population included significantly greater numbers of very young and very old mothers than the national norm, and the number of women who were bearing their fourth or subsequent child was also greater than the national average. Then, too, the FNS came to believe that their patients' inadequate diets may have made them particularly susceptible to toxemia. Kooser reported that "there appears to be a direct relationship between the seasonal incidence of toxemia and the seasonal scarcity of food."[65]

During its first years, the FNS achieved an impressive neonatal mortality record also, although again, in the absence of reliable figures it is difficult

[60] *Ibid.*, pp. 8–9, compares FNS mortality to that of rural white Kentucky women generally and to the United States as a whole. Lexington statistics are compiled from State Board of Health of Kentucky, *Preliminary Vital Statistics Reports*, 1925 through 1938.

[61] Louis Dublin to Mary Breckinridge, 4 May 1932, Dublin Papers; Frontier Nursing Service, "Analysis of the records of 1004 women during pregnancy, labor, and postpartum," (First series of 1000 deliveries), FNS Papers; "Report on the third thousand confinements," pp. 3–9.

[62] "Report on the third thousand confinements," pp. 7, 9.

[63] *Ibid.*, p. 9.

[64] John Kooser, "Rural obstetrics: a report of the work of the Frontier Nursing Service," *South. Med. J.*, February 1942, *35*: 123–31, quotation from p. 130.

[65] John Kooser, "Observations on the possible relationship of diet to the late toxemia of pregnancy," *Amer. J. Obstet. Gyn.*, 1941, *41*: 288–94; quotation from p. 294.

to generalize. In the first thousand deliveries, fifteen infants were stillborn and twenty-five babies died within a month of life; in the second thousand, eighteen infants were stillborn and twenty-six died. In both these series, the incidence of stillbirths and neonatal deaths was one-third less than that of the surrounding population.[66] In the third series, covering the years 1935 through 1937, the neonatal death rate rose to 39.8 compared to the national figure of 33 per 1000 live births. Two-thirds of these infants were premature.[67] This rise corresponds in time to the depths of the Great Depression. The FNS's greater difficulty in reducing neonatal mortality may well be related to the closer and more direct connection between infant mortality and poverty.[68]

In its first years, then, when Mary Breckinridge was most determined to use the Frontier Nursing Service as an example of how efficacious nurse-midwives could be as providers of maternal and infant care, the FNS was clearly a successful organization. Its maternal mortality and morbidity rates were considerably lower than those found in virtually any other patient population. Breckinridge attributed this success to the quality and consistency of care throughout pregnancy, labor, and the puerperium, and maintained that the FNS record proved that nurse-midwives were highly competent professionals, able to handle the vast majority of maternity cases without medical assistance.

The record of the FNS was well known and publicized. Several prominent American obstetricians, including George Kosmak, editor of the *American Journal of Obstetrics and Gynecology,* praised the FNS's work. The statistician and maternal welfare expert Louis Dublin was the most steadfast champion of the FNS and of nurse-midwifery generally. After tabulating the statistics of the first FNS series of deliveries, he stated that "... the first need today is to train a large body of nurse-midwives." The Frontier Nursing Service, he announced, demonstrated that "a decentralized service of nurses who are properly qualified as midwives, under the general direction of a competent obstetrically qualified doctor at a central point ... does effectually reduce our high national maternal death rate to a negligible level."[69]

The FNS came in for a good deal of popular journalistic coverage as well. Articles in the *Survey,* the *Nation,* and *Harper's,* among many others, helped give favorable publicity to its work. Ernest Poole's "Nurses on Horseback," serialized in *Good Housekeeping* in the early 1930s, was widely read.[70]

[66] "Report on the third thousand confinements," p. 2; Louis Dublin to Mary Breckinridge, 4 May 1932, Dublin Papers.

[67] "Report on the third thousand confinements," pp. 7–8.

[68] The early twentieth-century infant mortality surveys made by the Children's Bureau are valuable for documenting the relationship between infant mortality and social class. See especially Rochester, *Infant Mortality: Results of a Field Study in Baltimore, Maryland,* pp. 105–6.

[69] Louis Dublin to Mary Breckinridge, 4 May 1932, Dublin Papers; *Quart. Bull. FNS,* 1932, *7:* 3; George Kosmak, "The trained nurse and the midwife," *Amer. J. Nurs.,* 1934, *34:* 425.

[70] See, for example, Mary Sumner Boyd, "Why mothers die," *Nation,* 18 March 1931, *132:* 293–95; Dorothy Bromley, "What risk motherhood?" *Harper's,* June 1929, *149:* 11–22; Mary Breckinridge, "An adventure in

Although the FNS was well known and regarded as a highly successful health care organization, nurse-midwifery was not adopted in the United States to any significant extent. Why not? Part of the answer to this question lies in the history of the FNS itself. Developments and trends over which the FNS had no control, however, were far more important in explaining why nurse-midwifery has not flourished. Specifically, funding problems, the attitudes of leading nursing professionals toward nurse-midwifery, physician opposition, and general trends in American obstetrics during the 1920s and 1930s help explain why professional midwifery had such difficulty taking root.

Mary Breckinridge remained unable to find any large-scale backing for the FNS's work. In 1927 she appealed directly to the Commonwealth Fund. She bolstered her proposal with testimonials from several leading public health experts and statisticians who stressed their support for both the concept of nurse-midwifery and the work of the FNS. Louis Dublin, for instance, wrote that "I frankly do not know of anything that has come to my attention in the long list of health projects in the last year which has stirred me so much and has made its appeal with such intensity."[71] The new type of midwifery that Breckinridge was developing, Dublin stressed, had "national implications with reference to the health and welfare of mothers and children in our frontiers."[72] The Commonwealth Fund, however, rejected her proposal, arguing that her work was too expensive, and ill-suited for private financial backing.[73] In 1929, Congress refused to extend the Sheppard-Towner Act, thus ending all federal spending for maternal and child welfare.

By the early 1930s, Breckinridge despaired of receiving public funds and of convincing foundations to support nurse-midwifery. Accordingly, she redoubled her efforts to find private philanthropic backing for her work. Breckinridge's organization remained dependent upon piecemeal private largesse. As Ellen Breckenridge has argued in her work on the FNS as a rural health care demonstration, the FNS's permanent dependence upon private charity affected its image as a health care organization during a period in which large scale foundation and government funding were rapidly becoming the customary means of financing major health projects. Since the 1930s, the American public has viewed the FNS primarily as a romantic group of heroic "nurses on horseback," rather than as a demonstration of a viable alternative in childbirth management.

midwifery," *Survey*, October 1926, *57*: 25–27; Ernest Poole, "Nurses on horseback," *Good Housekeeping*, June 1932, *94*: 38–39, 203–10. Poole's work was published later in 1941 as a book by Macmillan.

[71] Louis Dublin to Barry Smith, 10 February 1928, Dublin Papers.

[72] Louis Dublin to Mrs. John Sloane, 29 November 1930, Dublin Papers.

[73] Barry Smith to Louis Dublin, 8 November 1928; Barbara Quin to Louis Dublin, 11 February 1928, Dublin Papers. I am indebted to Ellen Douglas Breckenridge for her discussion of the FNS's difficulties with the Commonwealth Fund during 1927 and 1928 and for her perceptive discussion of the changing patterns of foundation funding and private philanthropy as they related to rural health care generally. See Breckenridge, "Frontier Nursing Service: a demonstration project in rural health care," pp. 34–46.

The Great Depression also had a profound effect upon the organization's history. The economic collapse hit the southern Appalachian region hard, and exacerbated the FNS's already chronic financial difficulties. "Everything has been harder this past year than before," Mary Breckinridge wrote in the summer of 1930. "Thousands of mountaineers who sought better opportunities for their families in the towns along the railroads have lost their jobs and poured back into the mountains, without cows, chickens, pigs, or gardens—seeking subsistence in a county where the margin of living is never far above the hunger level."[74] A severe drought in 1930–1931 added to the already desperate economic situation. By the winter of 1931, famine existed throughout Leslie County. The FNS strained its existing budget to help families meet their immediate needs. During the same years, however, outside contributions began to diminish. By 1932, despite salary cuts and staff layoffs, the FNS was operating with a large deficit. Before the Depression, Breckinridge had planned to expand FNS services further into surrounding eastern Kentucky counties. In addition, in 1930, the FNS surveyed health conditions in the Ozark region of Arkansas and southern Missouri with the idea of establishing a second nurse-midwifery demonstration there. The Depression ended these plans.[75]

Even had the FNS expanded, however, it is clear that neither Breckinridge nor her supporters advocated the FNS model for cities where physicians were plentiful or for middle-class women who could afford a doctor's care. In other words, she did not envision an organization that directly challenged the medical system. The Frontier Nursing Service, Breckinridge repeatedly stressed, was designed for impoverished "remotely rural areas" without physicians. Indeed, Breckinridge may well have realized that given many physicians' hostility toward nurse-midwifery and their fears of nurse-midwives as potential competitors, an impoverished rural area offered the only location in which she could carry out her project independently.[76] Breckinridge, then, may have had no choice but to advocate a two-tiered system of maternity care: private obstetrics for affluent women, nurse-midwifery for poor women. To the limited extent that nurse-midwifery has been adopted in the United States, it has been largely along such lines. The two other major nurse-midwifery organizations, the Maternity Center Association of New York and the Catholic Maternity Institute in Santa Fe, New Mexico, have existed primarily for women who could not afford a private physician or hospitalization.

Nurse-midwifery's difficulty in making headway was also due in part to the American nursing profession's ambivalence about this new specialty. Even nurse-midwifery supporters disagreed over how nurse-midwifery

[74] "Fifth annual report of the Frontier Nursing Service," *Quart. Bull. FNS*, 1930, *6*: 1–2.
[75] *Quart. Bull. FNS*, 1932, *8*: 2–3.
[76] On physicians' attitudes toward nurse-midwifery, see Seth Goldsmith, John W. C. Johnson, Monroe Lerner, "Obstetricians' attitudes toward nurse-midwives," *Amer. J. Obstet. Gyn.*, 1971, *111*: 111–18.

should be defined. Breckinridge and her predominantly British staff pushed for full recognition of nurse-midwifery as a profession. In 1928, the FNS staff founded the Kentucky State Association of Midwives (later renamed the American Association of Nurse-Midwives) to work for professional recognition and certification standards along the lines of those required by the English Central Midwives' Board. Breckinridge, however, stood largely alone in this early determination to professionalize midwifery, and in her insistence that nurse-midwives work as autonomous practitioners. Indeed, during these same years the Maternity Center Association took the opposite tack, stressing that nurse-midwives were "at best only careful conscientious routine assistants to the physician. None would attempt or pretend to be more. . . . Several have tucked away the title of midwife as, in some situations, it would be far more of a handicap than a help because of the prejudices it arouses."[77]

This ambivalence helps explain the slowness with which American nurse-midwifery training programs were established. In 1923, the Maternity Center Association contemplated starting a midwifery training program, but found that the obstetrical community and many leading public health nurses opposed the idea.[78] The Maternity Center Association reversed itself in 1932, when it established the first nurse-midwifery training program in the United States, although, in keeping with the Association's ambivalence about professionalization, its graduates were not granted any sort of professional certification. During World War II, when the FNS lost the better part of its staff as its English nurse-midwives returned home, it, too, established a midwifery training program for registered nurses.

Finally, the FNS innovation of using nurse-midwives to offer complete care to women with normal pregnancies and labors differed too radically from the dominant trends being established in American obstetrics by the 1920s to make much headway. Certainly influential medical advocates of professional midwifery existed. The physician Dorothy Mendenhall, for instance, did much to publicize the desirability of professional midwives in her research and writing for the federal Children's Bureau.[79] Obstetricians such as Frederick Adair and Ralph Lobenstine, both of whom were concerned primarily with maternal welfare programs, were longtime proponents of nurse-midwifery. In its report on maternal mortality published in 1933, the New York Academy of Medicine recommended the training and employment of nurse-midwives.[80]

[77] For information on the Kentucky State Association of Midwives, see "Kentucky State Association of Midwives, Inc.," *Quart. Bull. FNS,* 1939, *15:* 19–22; Hattie Hemschmeyer, "Midwifery in the United States," *Amer. J. Nurs.,* 1939, *39:* 1183; Interview with Hazel Corbin, FNS Oral History Project, FNS Papers.

[78] *Maternity Center Association, 1918–1943* (New York: Maternity Center Association, 1943), p. 25; Hemschmeyer, "Midwifery in the United States," p. 1186.

[79] Dorothy Reed Mendenhall, *Midwifery in Denmark,* U.S. Department of Labor, Children's Bureau (Washington, D.C.: G.P.O., 1929).

[80] New York Academy of Medicine Committee on Public Health Relations, *Maternal Mortality,* pp. 207–12, 221–22; Litoff, *American Midwives,* pp. 125–26.

In practice, however, the possibilities for developing nurse-midwifery as a viable alternative in maternity care have been undermined by the dominant medical conceptions of childbirth and its management. In the 1920s, medical attendance, preferably by an obstetrical specialist, was rapidly becoming the sine qua non of safe, competent care. A physician's attendance was mandatory at birth because birth had come to be defined as a medical event, always potentially pathological. As a result, many health care professionals came to believe that they "must eliminate midwifery entirely from our social scheme."[81] Medical practitioners who argued for nurse-midwifery were increasingly overshadowed by physicians such as Joseph DeLee who argued that ideally birth should be managed as a surgical procedure. "All my medical life," DeLee wrote in 1934, "I have striven to eradicate the low opinion of obstetrics. . . . It is therefore with great pain and some alarm that I notice a trend in Britain and in parts of our Eastern seaboard, a reactionary trend, toward the state of midwifery."[82] Thus, many obstetricians still refused to make much distinction between professional and lay midwifery and would always regard a midwife's care as inherently inferior to that provided by a physician.

The work of the Frontier Nursing Service became increasingly anomalous in another way as well. The FNS attended the great majority of its patients at home at a time when doctors began to insist that only in hospitals could birth be safe. Although only 36.9 percent of American births took place in hospitals in 1935, by 1950, fully 88 percent did.[83] The widespread construction of rural community hospitals in the late 1920s and 1930s also undercut this aspect of the FNS's care. Significantly, at the same time the Commonwealth Fund refused to consider a grant to the FNS, it underwrote a major rural hospital construction program.[84]

Thus, greater reliance upon operative intervention, hospitalization, and universal medical management became the hallmarks of American birth practices during the 1920s and 1930s. In such a system that emphasized the pathology of birth, nurse-midwifery had little role to play. Most health care professionals have seen the FNS model as a successful but marginal, even anachronistic system, useful for poor, geographically isolated Appalachian women, but with little relevance for the nation as a whole.

[81] *Maternity Center Association, 1918–1943* (New York: Maternity Center Association, 1943), p. 25. The classic articulation of the idea that childbirth should always be regarded as a medical crisis is Joseph B. DeLee, "The prophylactic forceps operation," *Amer. J. Obstet. Gyn.*, 1920, *1*: 34–44.

[82] Joseph B. DeLee, "Obstetrics versus midwifery," *JAMA*, 1934, *103*: 308.

[83] For discussion and statistics concerning the increase in hospital births, see Neil Devitt, "The transition from home to hospital birth in the United States, 1930–1960," *Birth Family J.*, 1977, *4*: 47–58.

[84] "Rural community hospitals established by the Commonwealth Fund," *Amer. J. Public Health*, 1926, *16*: 315; W. S. Rankin, "Rural medical and hospital services," *Amer. J. Public Health*, 1927, *17*: 15–19; "Federal hospital construction act of 1940," *Amer. J. Public Health*, 1940, *30*: 719.

PREGNANCY AND PARTURITION FROM A WOMAN'S POINT OF VIEW*.

MARGARET E. COLBY, M. D., Clear Lake, Iowa.

Pregnancy and parturition from a woman's point of view may seem a superfluous venture on my part, yet there is a side to this subject that lies in shadow on which I hope future investigation will throw more light.

As physicians, we are supposed to be skilled in the art of healing, to understand the causes of disease and abnormal function. Yet with all our boasted skill we have much to learn.

Doctors we are and should be in the best significance of the term, and because of this we should find nothing trivial or beneath our careful investigation and serious thought, which may conduce to the health and well being of our patients.

Natural laws are powerful and all pervading. Helpful or hurtful as we are in harmony or otherwise.

Birth and death are alike cruel and alike blessed; they are the beginning and the end of the phenomenon we call life.

There is a plane from which those most interested view the subject, to which the physician and obstetrician seldom descend and that is the woman's. It is hard to put yourself in another's place. To woman is given the charge of the beginning of life. Death shows no favor on account of sex, men and women share alike at his hands. On the other hand the merest chance has placed woman, who has all the natural instincts, love of life, fear of death and dread of pain, in a class the exercise of whose functions under present conditions incur the danger of pain and death to a marked degree. There is so much flippant criticism of women, especially American women, because they are supposed to be averse to child bearing. Child bearing while safe as regards life in a large percentage of cases is like an ocean voyage or a journey by rail, it may end in appalling disaster.

Disease and death are inevitable. Yet death as the result of pregnancy may be placed, as a rule, in the category of preventable causes. A woman takes the chances willingly or unwillingly as the case may be , too often ignorantly.

For defense and perpetuity, all healthy individuals owe a duty to the state, yet the state has no right to enforce the duty of defence unless her existence is emperilled. A natural law provides for the perpetuity of the race so the perpetuity of the state is assured.

In every case a man claims for himself the right to be a father or not. Is there any justice in his insisting on the female of the race fulfilling her functions whether she chooses or not? How often is this condition forced upon a woman whose health from disease due to too frequent child bearing, makes the pregnant condition, instead of a welcome one a horror, against which her whole being rebells and protests. We cannot blame her

*Read before the Austin Flint-Cedar Valley Medical Society, Waverly, Iowa, Oct. 14, 1902.

and we must pity her, but that does not help her.

Men and women are alike ignorant of the ethics of marriage, and a woman realizes this often when it is too late. Ninety-five out of every hundred women are ignorant of the physiological conditions of pregnancy and parturition and are possessed with exaggerated ideas of the possible pathological condition incident to the condition. Ignorant and apprehensive she shrinks from both.

There is another factor in the count, and that is the hyperesthetic condition of these functional organs under the stimulus of the new force and their physical effects.

Never is the interdependent physical and mental condition more apparent than in the pregnant state. Nothing is more worthy our closest observation and careful study not only as regards the mother but her offspring. As physicians, are we not too prone to regard this condition of pregnancy and parturition as a natural every day affair, where there are some risks, considerable discomfort—but on the whole the majority get along.

To our credit be it said, great advance has been made in the knowledge and skill of the obstetrician, on the surgical side, in aseptics and antiseptics. But along the lines of prophylactics there is much that is lacking.

The following statement made by Prof. King in commending Turks suggestion in regard to preparation on the part of the obstetrician against possible post-partum hemorrhage, is significant: "It has long been my opinion that some such preparation on the part of the physician should be required by legal enactment." When called to a case of labor the physician is usually fortunate if he or she has obtained some knowledge of the patient during the last few weeks of pregnancy. Without such knowledge the physician can have little positive knowledge of what he may have to face. Labor may be ideally normal or it may be any one or more of the complications liable to occur in pregnancy. Some of these complications are appalling to the most nervy practitioner and require courage, coolness, and skill. He must be master of the situation or the worst will happen.

Pregnancy and parturition ought to be in the natural ordering of things, a simple physiological process. But is it? Turn to the literature of Obstetrics and Gynecology—the latter so often the result of bad management of the first—and we are forced almost to the conclusion that it is altogether a pathological process that is imposed upon woman. This is not right. Observation and experience teach that there is attainable either by a natural gift or careful management for the pregnant and parturient woman a reasonably healthy condition and one of comparative safety. But in order to have this we must begin at the period of woman's functional development. We must teach and insist on the importance of the woman having a chance to develop physiologically the functions of her sex. Much of the suffering and disease of women is due to ignorance and misapprehension of the importance and value of these functions to the individual. These are an integral part of her being. She is not created a mere machine for the propagation of the race.

Aside from her specialized function, she is an individual, she has a right to the control of her own body, to enjoy the privilege of wifehood, the sacred and exalted state of motherhood voluntarily, of her own free will and accord. She has a right to look to her physician for instruction in regard to the best care of herself during pregnancy, and the best preparation possible for the process of parturition.

As physicians we should give more careful study to the concomitant disturbances of pregnancy and their cause. The so often distressing and depressing condition of nausea is an example. There is nothing like it but mal de mere, and often the woman like the voyager, fears she will die from the loathsomness, the physical and mental depression, and if it continues, a more or less constant symptom for weeks, perhaps months, she like him, wishes she could die. Unless her life is in danger she is told, "it is due to her condition, and she must bear it as best she can, she will get over it by and by." Yes thanks to natural recuperative powers in some individuals, time brings a measure of relief, and she does get over it.

Turn to any volume on obstetrics and you find a chapter on diseases of pregnancy; one says, "The diseases of pregnancy are numerous and varied." Disease is a derangment of vital function in which natural action is interrupted or disturbed. Why should the fulfilling of a natural law, essential to life bring "many and varied diseases?" Something is wrong and there is a fault somewhere.

We should expect in the very nature of things to find an exaltation of nutritive activity. This must demand a corresponding activity in elimination, in order to keep the balance. In a healthy individual this balance is very nearly if not quite perfect, and in a pregnant woman with this balance, preserved, we find exalted function, but no disease. This is so true that a woman totally incapable of bearing a living child owing to structural incapacity, may come to the period of parturition without herself or friends having the least suspicion that things are not, in all respects, perfectly normal.

No woman should be allowed to marry without

examination by a competent physician, so that she, and her husband that is to be, shall know if she is incapacitated for child bearing, by structural inadequacy or disease. Then if they choose to face the danger it is their right.

Many a woman would rather face a caesarean section than live a loveless, childless life.

The choice of methods of relief should be left with those where the danger to the mother is of small moment when compared with the destruction of their child.

Heaven save to humanity, the sentiment, the romance, the glow, and glory of a mutual love that finds its highest fruition in fatherhood and motherhood, but give to the individual and the state a little practical common sense in regard to the duties and responsibilities of marriage and parentage.

Take a classification of diseases of pregnancy and we can charge most of them to faulty elimination, the nutritive functions are exalted, the nervous system, in consequence, more sensitive, the sympathetic especially so, and with the cumulative effect of retained waste material poisons the organism. We might expect "numerous and varied disturbances."

Faulty elimination may be due to a vicious organism inherited or acquired, it is often due to the ignorance of the patient in regard to the kind and amount of nutriment required, and the necessity and importance of elimination.

Mechanical disturbances if not entirely relieved may be modified. On the whole, much that is distressing and painful in the pregnant state may be eliminated and prove a prophylactic in the parturient condition.

Structural inadaquecy is a serious difficulty, yet in this condition, if in a minor degree, a measure of relief may be obtained by previous knowledge and care. Over-feeding with increase of fat and muscle does not conduce to conservation of pelvic space. Over work and exhaustion does not husband the muscular and nerve force for the supreme effort.

Nature is better to us than we are to ourselves. She gives to the woman in the pregnant condition added forces for the benefit of the new life developing, and if the woman is not robbed of these by mismanagement, by demands outside of her duty to her offspring and to herself, (which is the supreme one, if she is to become a mother), she is herself enriched.

While parturition is a natural process it is in a large percentage of cases an extremely painful one. The future physiologist and physician should discover why this is so, and be able to find a remedy. At present while physiological, it approaches too near a pathological, and not infrequently a surgical condition. Outside of malformations of the pelvis, or malproportion between fetus and pelvis, preventable conditions arising during pregnancy have much to do with the gravity of the case and the chief of which is deficient elimination. Barring emergency cases, what surgeon will not choose to give his patient the advantage of the best possible physical condition, attainable under existing circumstances, as a preparation to assure successful operation. Yet the woman comes to parturition with but little, if any preparation whatever, placing both herself and her physician at a perhaps dangerous disadvantage. How helpful in many instances would be the ascertained history of previous labors, as to the occurrence of post-partum hemorrhage, uterine inertia, prolonged, or difficult labors from any cause, or the knowledge of organic disease, indeed anything that increases the hazard in labor, should be known of each case by the physician in charge. This lack of preparation and previous knowledge places the majority of labors in the category of emergency cases.

- The physicians responsibility should not end with the termination of labor. The case should be under the charge of the doctor till the end of the period of involution. Watchful care should be given to prevent, if possible, any outward conditions arising as a sequel of incomplete physiological change. Much of the unhealthy condition following labor are due to the want of physical and mental rest at the lying-in period. It is a noticeable circumstance, that the females of animals in the wild state, seek isolation from their kind for a time, instinctively yielding obedience to a natural law that demands rest and quiet. Domestic animals will do the same if given a chance. Reason should teach the human being as much as instinct teaches the animal. Lack of proper nutrition, deficient elimination, over-work before the completion of the process of involution all tend to hinder the natural physiological change. It would be of great benefit if the laity could be made to understand that while the puerperal period is of four weeks duration there is still other four weeks that a woman requires to complete the organic changes that takes place as a result of parturition.

It is a common statement of patients to the Gynecologist, "I have not been well since my last baby was born, or since some previous child was born. I had a nurse, had good care for four weeks and was all right, but the nurse went away, or the girl, and I had company, or was obilged to do a large washing, or there was house cleaning," some one or other, or all of the causes for over-work on the part of most mothers and house-wives, and all avoidable if she and those about her realized the need of further care. As a result of this she is doomed to drag out a miserable existence for months, sometimes years, the victim of ignorance and mismanagement. And this is the price paid for wifehood and motherhood, and in view of the misery it entails, it is small wonder if she considers it too great. We, as physicians, must come to the rescue of this portion of our clientage with a better understanding of her disability on account of ignorance in regard to the functions of her sex, and a keener sense of our own responsibility for her wellfare.

As a body of scientists, doctors, we are teachers no less than practitioners, and it is of the utmost importance that the mother of this nation should have, as far as lies in our power, not only scientific treatment when her need demands it, but instruc tion in regard to the vital importance of health, normal development, and hygiene, if she is to assume the office of wife and mother.

That the conditions of pregnancy are of importance in relation to their effect on the parturient; that both need, in view of the possibilities of danger to life and health, the best advice and scientific care. I for one, protest against licensed midwives, every woman who has come to maturity should be instructed in a few things she can do for the parturient woman in emergency, but no one should be licensed to assume the responsibility of caring for the parturient woman but a physician well equipped. I would urge a higher appreciation of the importance of a healthy motherhood. The need of a better understanding of the causes of the pathological conditions accompanying so large a proportion of pregnant cases. The duty we owe women to encourage in them a better understanding of the inestimable value of a perfect physical development, a higher ideal of their privilege as a factor in the perpetuity of the nation, and hope that the skill and knowledge of the physician will lessen the intolerable discomfort and pain of prospective motherhood.

Soc. Sci. Med. Vol. 19, No. 11, pp. 1201–1206, 1984
Printed in Great Britain. All rights reserved

0277-9536/84 $3.00 + 0.00

PREGNANCY, LABOR AND BODY IMAGE IN THE UNITED STATES

EMILY MARTIN
Department of Anthropology, The Johns Hopkins University, Baltimore, MD 21218, U.S.A.

Abstract—Although substantial research has been done showing how alienation (which in Ollman's terms means that "an essential tie has been cut in the middle") manifests itself in the domain of human *labor*, little has been done to discover how similar processes might operate in other domains. I examine women's images of their bodies while they are pregnant, using Lakoff and Johnson's method of looking for metaphors that are presupposed in our ordinary language. I uncover several central images, all of which display a marked sense of separation of self from the parts of the body, and a passive stance in which events are described as happening to rather than being brought about by the speaker.

Turning to written texts, I find that the literature on childbirth (from opposite ends of the spectrum—popular literature advocating prepared childbirth on one end and obstetrical texts for medical students on the other) holds an assumption in common: that the uterus is an involuntary muscle. This is so despite evidence to the contrary cited in the texts themselves. I explore the implications of this imagery for obstetrical treatment of 'uterine inertia' and show the similarity between this imagery and ideas about women's physiology that were current in the 19th century.

Paul Schilder, an early gestalt psychologist, argues that in some phase of its development disturbed body images are part of almost every neurosis [1]. He presents one example in detail: "A case of loss of unity in the body-image". The patient reports:

> When I get this anxiety state I cannot walk further. I run into myself. It breaks me into pieces. I am like a spray. I lose my centre of gravity. I have no weight. I am quite mechanical. I have gone to pieces. I am like a marionette. I lack something to hold me together [2].

In other cases, "The patient sees his face in the mirror changed, rigid, and distorted. His own voice seems strange and unfamiliar to him, and he shudders at the sound of it as if it were not himself speaking. Gottfried feels that his movements are interrupted. His body feels as if it were dead and he has the sensation that a dynamo is hissing in his head. The body feels too light, just as if it could fly. A patient of Loewy's says, 'I feel the body not for me but for itself'. The patients look for their limbs in the bed. A patient of Pick's complains that his eyes are like two holes through which he looks" [3].

Although these compelling experiences of bodily fragmentation frequently accompany neurosis, Schilder contends that they are never far from us, even in the everyday life of a normal person.

> The important conclusion we may draw is that feeling our body intact is not a matter of course. It is the effect of self-love. When destructive tendencies go on, the body is spread over the world.... One would like to say that we lose the unity of our body only under special pathological conditions; but we also have to remember how much the feeling of our body varies under normal conditions. When we touch an object with a stick we feel with the end of the stick. We feel that clothes eventually become a part of ourselves. *We build the picture of our body again and again* [4].

Schilder believes loss of unity in the body image comes from the individual's experience in the family during the course of psychic development: "There are forces of hatred scattering the picture of our own body and forces of love putting it together" [5]. In this essay, I will describe how the social and cultural organization of experience can also affect body image, and can lead to fragmentations similar in form to those Schilder describes.

Originally, my interest in this topic came from Marx's concept of alienation, which Ollman has interpreted as entailing a certain distortion of human nature:

> An essential tie has been cut in the middle. Man is spoken of as being separated from his work (he plays no part in deciding what to do or how to do it)—a break between the individual and his life activity. Man is said to be separated from his own products (he has no control over what he makes or what becomes of it afterwards)—a break between the individual and the material world. He is also said to be separated from his fellow men (competition and class hostility has rendered most forms of cooperation impossible)—a break between man and man.... The whole has broken up into numerous parts whose interrelation in whole can no longer be ascertained. This is the essence of alienation, whether the part under examination is man, his activity, his product or his ideas. The same separation and distortion is evident in each [6].

It struck me that although considerable research has been done showing the details of how human *labor* becomes alienated in a capitalist economy [7] much remained to be done to find out whether similar processes might operate in other areas, such as the human body itself. On the one hand I was inspired by historical treatments of developments in the economy and society of seventeenth, eighteenth and nineteenth century Western Europe and the United States and how these developments affect images of the body and how it should be treated in health and illness [8]; on the other I was inspired by Michael Taussig's paper 'Reification and the Consciousness of the Patient' which spelled out the details of how a patient's view of her body as a 'broken up whole' affected her concept of her disease and its treatment [9].

I designed a research project intended to elicit rich verbal material from people who are in some situation—such as being pregnant—which could be expected to make them particularly aware of their physical states, but who are not normally institutionalized for any length of time [10]. (My hope was to avoid, in so far as it is possible, the particular effects that being institutionalized itself would inevitably have.)

My goal was to analyze ordinary language in order to get at images people have of their bodies and physical states. I was convinced this was a fruitful approach by reading George Lakoff and Mark Johnson's *Metaphors We Live By*: by examining our everyday language they identify certain metaphors that structure the way we think, talk, and act. For example, the conceptual metaphor 'argument is war' is reflected in ordinary language in many ways:

Your claims are *indefensible*.
He *attacked every weak point* in my argument.
His criticisms were *right on target*.
I *demolished* his argument [11].

And so is the metaphorical concept 'time is money':

You're *wasting* my time.
This gadget will *save* you hours.
I've *invested* a lot of time in her.
You need to *budget* your time.
He's living on *borrowed* time.
You don't use your time *profitably* [12].

Eventually I hope to show whether metaphors about the body discovered in everyday language vary by class or by the kind of work people experience. At this point in the research, perhaps because—despite my efforts—my sample is too homogeneous or for other reasons, I have not discovered these kinds of differences [13]. Consequently, in what follows I treat the interviews and birth reports I have as a single body of data.

PREGNANCY IN THE UNITED STATES

Before looking at metaphors and images in everyday language about pregnancy, I must point out that many women in the United States are subjected to treatments and procedures that seem powerfully disposed toward affecting—one way or another—images of the body. Examples are almost endless, but here are three key ones, with an indication of how women respond.

Use of doptone to amplify baby's heartbeat

(Secondary school teacher) When I hear the heart beat they have to peel me off the wall [14].

(Pentacostal minister) But the first time I heard the *heart*beat of the baby, I mean I just, it sound like somebody building a house—boom, boom, boom...'cause they got this li'l thing they use and it amplifies the sound, you know, and it sound, it just remind you of a carpenter up on a house bammin' on the house and goin' boom, boom, boom....

(Proofreader) The first time we heard the heartbeat I wasn't as excited as my husband... and I couldn't figure out myself why wasn't I so excited, and then I finally realized that the reason I wasn't is because my doctor gave me the heartbeat. It's like he took it away from me because he said here's the heartbeat. I mean he is the one who arranged that I could hear it and I sort of felt like, well this is my baby's heartbeat but I can't hear it unless he does it for me and I don't know, maybe I'm a real independent person or something, you know, but I sort of felt funny about that, that we had to rely on him. I wanted to do it myself, and I since I really like the doctor I was surprised at myself that I wouldn't like that because he's so supportive...

Moving woman from labor room to delivery room during last stage of labor

(Former Special Education Teacher) Our teacher last time gave us the analogy of if you're having sex with your husband and you're all ready to climax and they put you on this table and said now hang on, wait there, we'll take you to this other room and now you can climax, she says it's almost the same thing—you're getting all ready to have this baby and then they switch you to this table and rush you down and say O.K. now you can do it. It's just so intrusive and just interrupts the whole thing and it's hard to get back where you were.

Use of epidural (*spinal anesthesia* which usually necessitates forceps delivery)

(Birth report) At this time I asked for an epidural, not knowing that I was actually in transition and nearly fully dilated. At 6:00 I was ready to push, but with the epidural I couldn't feel the urge, we had to watch the monitor to know when to push.

(Pentacostal minister)...oftentimes [...] when you're havin' a baby the doctors act like they're watchin' T.V. 'cause your legs are up and you're all draped over and all them are just goin' like this [gestures in the air with a poking motion], you know, and you kind a wonder what do they have, a portable T.V. down there watchin' or are they really workin' on me, you know. *What* is goin' on, 'cause half of you is numb down there so you can't feel what they doin', you know. And everybody just [mumbles] and you want to know, my god, what is goin' on, you know, then they start to talk in terms that you don't know what they talkin' about and you must know automatically, I've got cancer, I'm bleeding to death, you know. Some *drastic* something, you know, and whew! It's just too much sometimes, almost you need them few days a rest after the labor just to get your mind together.

THE LANGUAGE OF PREGNANCY, LABOR AND DELIVERY

Applying Lakoff and Johnson's method to interviews and birth reports, the following images emerge.

Your body sends you signals

(Teacher of exceptional children) So, um, I just think it's amazing your body tells you these things anyway. You know, like I just started getting really tired—it's like my body was saying, all right you've had enough, now you stop [...] It's really—I don't know any other word to use—amazing. Your body tells you these things. Like, with not being able to sleep, well, I think well maybe the baby's going to be getting up at this time anyway during the night, you know, to get fed or whatever. So in a way it's preparing me for what's coming. Just like when you needed more sleep in the beginning. It's, you know, when the baby's developing the most anyway. So, it sends out signals you know.

(Interviewer) Sure.

Sometimes you pick them up, sometimes you don't.

Labor, delivery and its component stages are states one goes through

(Nurse) Number one, I didn't think I could go through the process of delivery—that has always terrified me.

I just can't see myself going through labor and delivery without some kind of anesthesia.

(Insurance firm clerk) I mean you gotta go through the labor before you go through the delivery.... I mean you've gone through all that, why not go through another hour or two?

(Former Special Education Teacher) Women have been going through this for millions of years, even if it is terribly painful for 5 hours or whatever I can go through it like everybody else did. Look at all the babies around...

Contractions are outside the self: they are "*the* contractions" (not *my* contractions); they "come on"; we "get them".

(Nurse) In the beginning I didn't even know (if) it was true or false labor, because there were continuous like contractions. The contractions came from the back to the front like the, I guess the overall term is like they were like a girdle from around front, but they were not regular like everyone else... when a contraction would come on I would do my breathing...

(Birth report) While we were relaxing eating our dinner, the contractions were 5 minutes apart but very mild... while in the store the contractions were still strong and 5–6 minutes apart.... every time I would get a contraction Tom would massage my lower back.

(Birth report) At 4:00 a.m. I had my first contraction. About 7 minutes later I had another. They they began to get stronger and came about every 3 minutes.

These are the dominant ways women talk about pregnancy and labor in my interviews and birth reports. Far less common, but occasionally present, are expressions—such as "*my* contractions", "*my* labor"—that depict the woman as the active experiencer of events. One woman objected to the prevailing attitude that pregnancy is something that she was afflicted by: "It's like all of a sudden you've got, well, like some people would say you've got this condition. What condition? I'm pregnant, I don't have this condition". And one woman developed an image for thinking about her approaching labor that made it an act she would perform:

I'm hoping it'll be a beautiful experience, like, almost like a blossom opening or something, it'll be beautiful. I'm, this is my fruit, you know [...] it'll be a wonderful experience because of the joy in it, the joy will take away from the pain and I'm willing to face the pain.

But this image stands alone among the many others in which the woman is passively being *done to*: "like being spread from wall to wall wide open", "it felt like I imagine it would feel if someone was trying to pull your tongue out with a pliers".

PREPARED CHILDBIRTH TEXTS AND CLASSES

The dominant verbal images in the interviews and birth reports describe a separation of the self and the body (your body sends you signals, labor is a state one goes through) which entails a separation between the self and physical sensations ("*the* contractions *come* on" etc.) When I realized the extent of these locutions in the interviews I had done, my first reaction was that these women had probably not thoroughly learned the lessons of prepared childbirth texts and classes. Surely, I thought, in Lamaze texts and prepared childbirth classes, birth would be described as an act a woman *did* and physical sensations would be described as being intrinsically a part of her. When I looked at these texts, I did indeed find many statements that affirm childbirth is an act a woman does and that convey a sense of unity about the person and her physical experiences.

Birth "is not something that you simply let happen to you... it is something that you *do*" [15].

Labor is called labor because it is the hardest work most people will ever do [16].

However, I found alongside statements such as these, statements of very different import. It is the uterine contractions that do the birthing: "Labor is the time when your uterus contracts to bring about the birth of your baby" [17]. And more emphatically, uterine contractions are *involuntary*. "The uterus is an involuntary muscle, like the heart, and it is impossible to stop it from contracting" [18]. Lamaze himself, who I had always thought stressed the ability of women to control the process of birthing, clearly separates areas of labor and delivery that a woman can control from those she cannot. On the one hand a woman is the "force which directs, controls, and regulates her labor" [19]; on the other her uterine contractions are 'involuntary' [20]. In labor the voluntary and involuntary actions come together: "pushed by the coupling of these combined effects—the one involuntary (uterine contraction), the other voluntary (mother's muscular efforts)—the child will make headway through the pelvic cavity" [21,22].

OBSTETRICAL TEXTS

Like Lamaze texts, obstetrics texts assert that the uterus is an involuntary muscle. An early (1950) edition of *Williams Obstetrics* says: "Uterine contractions are *involuntary*. Their action is not only independent of the mother's will, but of extra-uterine nervous control" [23]. The latest edition, a standard text used at Johns Hopkins Medical School, adds only that anesthesia sometimes interferes with contractions: "Uterine contractions are involuntary and, for the most part, independent of extrauterine control. Neural blockage from caudal or epidural anesthesia, if initiated quite early in labor, sometimes reduces the frequency and intensity of uterine contractions but not after labor is well established" [24].

Despite assertions of how involuntary uterine contractions are, at least some obstetrics texts themselves contain direct evidence that this is not entirely so. As the 1950 *Williams* puts it, in a particularly contorted fashion, "... clinical observation indicates clearly that extraneous causes may interfere reflexly with the activity of the uterus. It is a matter of common experience that the entrance of the obstetrician into the lying-in chamber is frequently followed by a temporary cessation of the labor pains. Extreme nervousness, profound mental emotions, or excruciating pain may have a similar effect" [25]. This clinician's observation that labor contractions are affected by a woman's environment and emotional state has been omitted from the 1980 edition of *Williams Obstetrics*.

However, a similar statement can be found in another standard text at Johns Hopkins: "Of utmost importance is that the physician take a reassuring, kindly attitude toward the patient. The degree of rapport that can be established with her will frequently determine the degree of comfort with which labor is experienced and perhaps also the length of labor" [26]. And specific studies have shown that aspects of the environment such as light, noise and being moved from one place to another affect the length of labor in animals and humans [27]. However, this knowledge is not brought to bear on obstetrical treatment. For example, if a woman's labor slows down because her contractions are not sufficiently strong (hypotonic uterine dysfunction) most obstetrics texts suggest these causes: the pelvis is too small; the fetus is not positioned properly; or the uterus is overly distended [28]. Nowhere is it suggested that the woman's general state of mind (fear, anxiety) might have led her to stop her labor, *even though "in many instances—perhaps one-half—the cause of uterine dysfunction is unknown"* [29].

Seeing uterine contractions as 'involuntary' has implications for what treatment is recommended, of course [30]. The only measures obstetrics texts suggest are external manipulations of the woman's body: rupturing the membranes, an oxytocin drip, which acts chemically on the uterus, or a Caesarean section, which makes further contractions unnecessary [31].

The point I am moving toward is not that a uterine contraction is actually an act a woman does just as she might make a fist, but that along a range from an act a woman does to an event that happens inside her but independently of her, it is probably far closer to an act she does than we have imagined. Another organ, the heart, ought to be—by our physiologic lights—an even more 'involuntary' muscle than the uterus, because it is controlled deep in the brain stem, whereas the uterus is controlled by the hypothalamus [32]. Yet it is now common knowledge that the heart and its activity are profoundly affected by a variety of emotional states [33]. Where are comparable insights about the uterus? Why does the uterus lag behind the heart? A simple—but not false—answer is that men have most of the heart attacks, but women have uteruses.

In the nineteenth century, women were seen as controlled in mind, body and soul by the ovaries and the 'paramount power' of the uterus [34]; the female was pictured as "driven by the tidal currents of her cyclical reproductive system" [35]. Special care had to be taken when the uterus was active (during puberty, menstruation and pregnancy) not to overtax it by mental or physical exertion. "Because reproduction was woman's grand purpose in life, doctors agreed that women had to concentrate all their energy downward toward the womb" [36].

Perhaps we no longer see women as so enslaved by their reproductive organs. But we are still a long way from seeing quintessentially female functions as acts women *do* with body, mind and emotional states working together or at least affecting one another. Perhaps there will come a day when instead of speaking of 'hypotonic uterine dysfunction' or 'non-progressive labor' which has to be treated by chemical or operative intervention, we will say "the woman has stopped contracting her uterus" and focus on alleviating the fear and anxiety which have probably, in the majority of cases, led her to do so.

I have by no means done a systematic study of whether in some cases women might be said to stop their uteruses from contracting [37]. One woman I interviewed, however, suggests they might. This woman labored for about 12 hours at home and found it a "*really* nice experience". When she and her husband arrived at the hospital:

> I wasn't upset or anything and we got there and the obstetrician wasn't there and this young female resident examined me and just got freaked out, she got really nervous. She observed some blood and she felt that perhaps the placenta was [...] down over the cervix and she just didn't know what was going on. She just scared [my husband] and I to death and she wouldn't let me walk and rushed us into this room next to the O.R. and it was freezing cold and they wouldn't give me a blanket. It was just *very* upsetting.

Later a sonogram showed the placenta was positioned normally: the resident's exam had produced the blood.

> It was a horrible experience, I mean, we were just both so scared and from then on things just didn't progress terribly well. I just didn't dilate over 4 cm after a few hours of patocin so they decided to do a C-section. So I was *real* disappointed.

Aside from the fear, there were certainly other factors that bore on the difficulty of this woman's làbor after she got to the hospital: she was made to wear an oxygen mask, she was not allowed to stand or walk around, and she felt both her obstetrician and the labor room nurses could have been more supportive and sympathetic. But in her own perception of events, the fact that she was frightened to death was central:

> Immediately after the birth I thought that I just got so scared that, and so uptight, that my contractions then sort of stopped for a while and then eventually started again and I thought maybe that, you know, it's been something voluntary, you know, for whatever reason.

But her perception that her fear may have led her to stop her labor is then immediately denied:

> But I read in the paper that there was a woman who was in a coma from a car accident and when it was time for the child to be born so the child was born and she still is in a coma, so I guess it really is involuntary [...] whether you're really upset just doesn't make any difference.

In her final statement, her inkling that having her baby might have been something she could *do* is paradoxicallly reasserted, then denied again. Finally, the fact that she did not accomplish the birth herself is put down to her own failure.

> I felt like my body had failed me. I really felt that, you know, having a baby is something you should be able to just *do*. And it's so much out of your control, I mean you can't really try harder [...] I felt sort of angry at myself.

The convolutions in this woman's interview are the result of a very real dilemma: if she takes seriously the possibility that childbirth is an act she as a whole person might have done, then the fact that she did not do this must be laid at the feet of the hospital, of herself, or both. Any of these possibilities would probably be more threatening than seeing childbirth as something that 'happens to you' and labor as something that 'really is involuntary'.

In an insight that anticipates the core finding of this paper, Adrienne Rich observes that:

> I think now that my three unconscious deliveries...(approved and implemented by my physician) and my friends' exhilaration at having experienced and surmounted pain [through some version of natural childbirth] (approved and implemented by their physicians) had a common source: we were trying in our several ways to contain the expected female fate of passive suffering. None of us, I think, had much sense of being in any real command of the experience. Ignorant of our bodies, we were essentially nineteenth-century women as far as childbirth (and much else) was concerned [38].

It would be beyond the scope of this paper to show what forces, operating in the social realm in which pregnancy and birth are culturally managed, scatter our picture of our bodies. Almost certainly these forces will not turn out to be simply the same ones Schilder identified within the self and the family. Quite probably they will include ideological features of the medical system such as the dichotomy between mind and body, emotional and physical states. But I suspect in addition (though the answer must await future research) that, as we "build the picture of our body again and again", we do it just as surely in response to something akin to 'forces of hatred' in the social fabric outside the medical system—forces that deny the potential wholeness and autonomy of people—as the individual does in response to destructive tendencies within the self.

REFERENCES

1. Schilder P. *The Image and Appearance of the Human Body*, Psyche Monographs No. 4, p. 139. Kegan Paul, Trench, Trubner, London, 1935. I thank Sidney Mintz for referring me to Schilder's work.
2. *Ibid.*, p. 159.
3. *Ibid.*, p. 139.
4. *Ibid.*, p. 166, emphasis added.
5. *Ibid.*, p. 166.
6. Ollman B. *Alienation: Marx's Conception of Man in Capitalist Society*, pp. 133–35. Cambridge University Press, Cambridge, 1976.
7. Braverman H. *Labor and Monopoly Capital.* Monthly Review Press, New York, 1974; Aronowitz S. *False Promises: The Shaping of American Working Class Consciousness.* McGraw–Hill, New York, 1973.
8. Reiser S. J. *Medicine and the Reign of Technology.* Cambridge University Press, Cambridge, 1978; Illich I. *Medical Nemesis: The Expropriation of Health.* Bantam, New York, 1976; Brown E. R. *Rockefeller Medicine Men: Medicine and Capitalism in America.* University of California Press, Berkeley, 1979; Berliner H. A larger perspective on the Flexner reforms. *Int. J. Hlth Serv.* **5,** 73–92, 1975; Foucault M. *The Birth of the Clinic: An Archeology of Medical Perception.* Pantheon, New York, 1973; Wertz R. W. and Wertz D. L. *Lying-in: A History of Childbirth in America.* Schocken, New York, 1977.
9. Taussig M. T. Reification and the consciousness of the patient. *Soc. Sci. Med.* **14,** 3–13, 1980.
10. The complete project will involve intensive interviews with girls in the first years after the menarche, women in their childbearing years, and women during or after menopause. This paper is based on the 24 interviews completed so far, 20 of which were with women in their childbearing years, and on 14 'birth reports', written by parents at the request of a childbirth education instructor soon after their baby's birth. Volunteers were found at the Early Pregnancy Series and various Prepared Childbirth classes run by the Childbirth Education Association of Baltimore. I thank Beverly Daniels, Marsha Hettinger, C-Anne Fischer, Bonnie Cooper, Susan Doering and Jean Hahn for permission to explain the project and request volunteers during class time. All interviews were conducted in the woman's home or office by me or Suzanne Siskel, a graduate student in my department, whose help was invaluable. Financial support was provided by Johns Hopkins' Biological Research Support Grant funded by NIH.
11. Lakoff G. and Johnson M. *Metaphors We Live By*, p. 4. University of Chicago Press, Chicago, 1980.
12. *Ibid.*, pp. 7–8.
13. I include a wide cross-section of class and occupation, by soliciting volunteers from classes taught in downtown hospitals with clinic patients as well as suburban hospitals with private patients. I am beginning to suspect that the uniformity I find in the interviews reflects aspects of experience women *share* in this society, regardless of class or occupation.
14. In quotes from interviews, "..." indicates part of a sentence has been omitted; "[...]" means part of a sentence is unintelligible.
15. Karmel M. *Thank You, Dr Lamaze: A Mother's Experience in Painless Childbirth*, p. 42. Lippincott, Philadelphia, 1959.
16. Walton V. E. *Have it Your Way*, p. 143. Bantam, New York, 1976.
17. Todd L. *Labor and Birth: A Guide for You*, p. 3. International Childbirth Education Association, Minneapolis, 1981.
18. Walton V. E., *op cit.*, p. 132.
19. Lamaze F. *Painless Childbirth: Psychoprophylactic Method*, p. 16. Pocket Books, New York, 1965.
20. *Ibid.*, p. 131.
21. *Ibid.*, p. 137.
22. Suzanne Arms has pointed out that the Lamaze method gives a woman control over her body, but that she remains separate from the "sensations, smells and sights of her body giving birth" *Immaculate Deception: A New Look at Women and Childbirth in America*, p. 146. Houghton Mifflin, Boston, 1975.
23. Eastman N. J. *Williams Obstetrics*, 10th Edition, p. 326. Appleton–Century–Crofts, New York, 1950.
24. Pritchard J. A. and MacDonald P. C. *Williams Obstetrics*, 16th Edition, p. 376. Appleton–Century–Crofts, New York, 1980.
25. Eastman N. J. *op. cit.*
26. Niswander K. R. *Obstetrics: Essentials of Clinical Practice*, 2nd Edition, p. 184. Little, Brown, Boston, 1981.
27. Haire D. *The Cultural Warping of Childbirth*, p. 13. International Childbirth Association, Minneapolis, 1972; Newton N. The effects of disturbance on labor. *Am. J. Obstet. Gynec.* **101,** 1096–1102, 1968.
28. Pritchard J. A. and MacDonald P. C. *op. cit.*, p. 789.
29. *Ibid.*
30. There are complex issues involved in deciding whether the dichotomy (or range) 'voluntary–involuntary' adequately captures the variety and sense of human actions. I do not take up these issues here, but see Anscombe G. E. M. *Intention*, pp. 89–90. Cornell University Press, Ithaca, NY, 1963, for relevant discussion.
31. Benson R. C. *Current Obstetric and Gynecologic Diagnosis and Treatment*, pp. 905–907. Lange Medical Publishers, Los Altos, 1982; Niswander K. R. *op. cit.*, p. 207; Pritchard J. A. and MacDonald P. C. *op. cit.*, p. 789.
32. Benson R. C. *op. cit.*, p. 63.

33. Lynch J. J. *The Broken Heart: The Medical Consequences of Loneliness*. Basic Books, New York, 1977.
34. Wood A. D. The fashionable diseases: women's complaints and their treatment in nineteenth-century America. *J. Interdisc. Hist.* **4**, 25–52, 1973.
35. Smith-Rosenberg C. Puberty to menopause: the cycle of femininity in nineteenth-century America. In *Clio's Consciousness Raised* (Edited by Hartman M. and Banner L. W.), pp. 23–37. Harper, New York, 1974.
36. Ehrenreich B. and English D. *For Her Own Good*, p. 127. Doubleday, New York, 1978.
37. Alice Rossi finds it surprising that "so little research has been done on the impact of pregnancy upon women from the point of view of the impact of medical management of pregnancy and birth". (A biosocial perspective on parenting. *Daedalus* **106**, 19, 1977.) I find it surprising, given that the (medical) cause of uterine dysfunction is as often as not unknown, that a sytematic study of the relationship between fear and anxiety, say, and uterine dysfunction has not been done.
38. Rich A. *Of Woman Born*, pp. 171–172. Bantam Books, New York, 1976.

Forgotten Women: American Midwives at the Turn of the Twentieth Century

By

JUDY BARRETT LITOFF*

At the turn of the twentieth century, the total number of births in the United States was attended about equally by midwives and by physicians. Indeed, although statistics relating to the number of early-twentieth-century midwives are not always reliable, conservative estimates indicate that as late as 1910 at least 50 percent of all births were attended by midwives.[1] Southern black families relied heavily on midwives. In fact, in Mississippi in 1918 midwives attended 87.9 percent of all Negro births. Immigrant women, especially those newly arrived from southern and eastern Europe, also used midwives. A 1908 study revealed that 86 percent of all Italian-American births in Chicago were reported by midwives. In isolated rural areas of the United States, a friend or relative often acted as a midwife in the absence of the doctor. Moreover, there was no clear demarcation line between the woman who acted in the official capacity of a midwife once or twice a year and the neighbor who occasionally came to the aid of a friend in need.[2]

Until recently, historians have devoted little attention to this topic. The dearth of scholarly research on the turn-of-the-century

*The author is Assistant Professor of History at Bryant College. An earlier version of this article was presented at the Missouri Valley History Conference in 1976.

[1]The Bureau of the Census did not begin publishing statistical information with regard to the attendant at birth until 1937. Annual Volumes, *Vital Statistics of the United States,* 1937 - 1970 (Washington, D.C.). Most turn-of-the-century observers estimated that at least 50 percent of all births were attended by midwives. See, for example, Thomas Darlington, "The Present Status of the Midwife," *American Journal of Obstetrics and the Diseases of Women and Children* 63(1911): 870; Grace Abbott, "The Midwife in Chicago," *American Journal of Sociology* 20(1915): 684.

[2]Helen M. Dart, *Maternity and Child Care in Selected Rural Areas of Mississippi,* U.S. Department of Labor, Children's Bureau Publication No. 88 (Washington, D.C., 1921), 27; "The Midwives of Chicago," *Journal of the American Medical Association* 50(1908): 1346; Viola I. Paradise, *Maternity Care and the Welfare of Young Children in a Homesteading County in Montana,* U.S. Department of Labor, Children's Bureau Publication No. 34 (Washington, D.C., 1919), 30 - 32.

midwife is particularly striking in light of her ubiquity.[3] Certainly, an event as universal as childbirth deserves careful study by the historian. Moreover, a complete understanding of the circumstances of childbirth requires knowledge of the customs, practices, and education of those persons who have served as birth attendants.

Although the early-twentieth-century midwife was numerically significant, middle- and upper-class white Americans had little first-hand contact with her. The woman of wealthy or moderate means usually enlisted both a physician and a monthly nurse for the confinement period.[4] A growing number of well-to-do women were even choosing to have their babies in hospitals in order to employ obstetrical specialists. Hospitalization for childbirth was the exception, however, and in 1900 the vast majority of all births occurred in the home.[5]

Only the very poor, who accepted public or private charities, and the well-to-do, who could afford competent obstetricians and private rooms, went to the hospital. Families with moderate incomes could not pay the high price of hospital deliveries. Still,

[3] Recent studies of the American midwife include the following: Frances E. Kobrin, "The American Midwife Controversy: A Crisis of Professionalization," *Bulletin of the History of Medicine* 40(1966): 350 - 63; Judy Barrett Litoff, "Midwives in America, 1870 - 1930" (Ph.D. diss., University of Maine, 1975); Richard Wertz and Dorothy Wertz, "Lying-in, A History of Childbirth in America; Its Technologies and Social Relations" (unpublished manuscript, Boston, 1974). For an examination of the status of the American midwife prior to 1860, see Janet B. Donegan, "Midwifery in America, 1760-1860: A Study in Medicine and Morality" (Ph.D. diss., Syracuse University, 1972).

[4] The monthly nurse was a woman who was hired to perform medical and housewifely duties. Her employment began at the onset of labor and lasted for about one month. She did not deliver the baby, however, unless the physician was late in arriving at the home of the parturient woman.

[5] Statistics on hospital and home deliveries are sketchy for the years prior to 1940. One of the first attempts to collect information of this type occurred in 1916 when the Department of Health of New York City reported that 19.1 percent of all births in the borough of Manhattan took place in hospitals. City of New York, Department of Health, "The Supervision of Midwives in New York City," by Lee W. Thomas, *Monthly Bulletin* 9(1919): 117. A report issued by the United States Children's Bureau for the year 1921 pointed out that hospital confinements occurred more frequently in cities than in rural areas. For example, "of the births in Baltimore, 18.7 percent occurred in hospitals, as compared with only 2.6 percent of the births in Maryland outside of Baltimore." Robert M. Woodbury, *Maternal Mortality: The Risk of Death in Childbirth and From All Diseases Caused by Pregnancy and Confinement,* U.S. Department of Labor, Children's Bureau Publication No. 158 (Washington, D.C., 1926), 86. The first national statistics on hospital deliveries appeared in 1935. In that year, a study conducted by the United States Children's Bureau reported that 36.9 percent of all births occurred in hospitals. Elizabeth Tandy, *Infant and Maternal Mortality Among Negroes,* U.S. Department of Labor, Children's Bureau Publication No. 243 (Washington, D.C., 1937), 7.

prohibitive fees were not the only factor limiting the number of hospital births. Just as important was the widespread belief that the hospital was a disease-ridden institution where people went to die. Surgical operations and other types of obstetrical interference were also closely associated with hospital delivery. Thus, poor women often preferred to pay the small fee of the midwife rather than enter a charity hospital where they would allegedly be subjected to experimentation and obstetrical interference.[6]

Clearly, many immigrant and black women were attracted to midwives because of the small fees which they charged. The average fee of midwives in Waterbury, Connecticut, in 1913 was eight dollars while physicians charged between fifteen and twenty-five dollars. In rural Mississippi in 1916, the midwife's fee ranged between five and ten dollars, depending on the difficulty of the case, the distance the midwife had to travel, and the ability of the family to pay. Often, informal arrangements were made for payment in chickens, pigs, or grain or on a neighborly give-and-take basis. One black Mississippi midwife even admitted that she charged more for boys than girls because "boys are harder to handle and mothers want them more." In contrast, Mississippi physicians charged between ten and fifteen dollars for normal deliveries, plus additional fees for prenatal and postnatal visits.[7]

Midwives not only charged substantially less than did physicians but they also provided a variety of services which general practitioners did not offer. The midwife served as the birth attendant, nurse, and housekeeper. In addition to caring for the mother and newborn infant, she often cleaned the house, prepared the meals, and looked after other children in the family for three to ten days after the birth of the baby. Women who used midwives were well aware that they were receiving services not offered by physicians. A 1924 Texas midwifery survey revealed that many women preferred midwives because they were "really worth more." One woman pointed out that the midwife "did my washing and charged only $5.00." Another stated that she paid $7.50 for three and one-half days of service. Thus, the midwife combined the

[6] George W. Kosmak, "Maternity Hospital Care for the Woman of Moderate Means," *Transactions of the American Association for the Study and Prevention of Infant Mortality* 4(1913): 208; Edward P. Davis, "The Need of Hospitals for Maternity Surgical Cases," *Transactions of the American Association for the Study and Prevention of Infant Mortality* 5(1914): 196 - 97.

[7] Estelle E. Hunter, *Infant Mortality: Results of a Field Study in Waterbury, Connecticut. Based on Births in One Year,* U.S. Department of Labor, Children's Bureau Publication No. 29 (Washington, D.C., 1918), 45; Dart, *Maternity and Child Care,* 27 - 32.

duties of the physician and monthly nurse for less than one-half the fee usually charged by the physician.[8]

The preference of immigrants for midwives was reinforced by a long-established European tradition of having them serve as attendants at birth. Hence, most midwives in Europe held high positions in the community and were both well trained and well supervised. When immigrant women arrived in America, they continued to employ midwives even though very few provisions for their training and regulation existed in the United States at the turn of the twentieth century. That the midwife spoke the language of the parturient woman was also frequently important. Physicians, in contrast, were often unable to communicate in any language other than English. Most important, immigrant families employed midwives because there was a "strong sense of shame [associated] with permitting a man to attend a woman in confinement." Almost all contemporary observers agreed that the immigrants' dislike for men in the lying-in chamber had a great deal to do with the popularity of midwives.[9]

Black women also expressed a dislike for male birth attendants. The 1924 survey of the midwife situation in Texas disclosed that many black women were "ashamed to have a man at that time" and that they preferred the midwife because she "is a woman and will be more help than [the] doctor." One woman stated that "I am used to women and not strange men like doctors." The Texas women also indicated that it was traditional to use midwives. One woman remarked that she "never knew anything else but a midwife." A frequent explanation given for employing midwives was that "everybody else had them around here." Another woman commented that the midwife was "just handed down from slavery times." Black women also preferred midwives because of their easy accessibility. Southern physicians usually lived in the towns and cities and often had to travel long distances in order to reach the parturient woman. In contrast, midwives were scattered throughout the rural south. Accordingly, Texas women chose midwives because "way out in the country doctors charge too much."[10]

[8] S. Josephine Baker, "Discussion," *Transactions of the American Association for the Study and Prevention of Infant Mortality* 3(1912): 254; "Report of the Midwife Survey in Texas," mimeographed (State Board of Health of Texas, 1924), 3.

[9] Hunter, *Infant Mortality*, 45; Jessamine S. Whitney, *Infant Mortality: Results of a Field Study in New Bedford, Massachusetts. Based on Births in One Year*, U.S. Department of Labor, Children's Bureau Publication No. 68 (Washington, D.C., 1920), 31; F. Elisabeth Crowell, "The Midwives of New York," *Charities and the Commons* 17(1907): 668.

[10] "Report of the Midwife Survey in Texas," 1 - 3.

Any attempt to present a composite picture of turn-of-the-century midwives is complicated by the fact that no diaries, letters, autobiographies, or other manuscript materials have been uncovered. Moreover, midwifery legislation was almost nonexistent prior to 1910, and the few regulations that were enacted were usually not enforced.[11] Thus, little information with regard to the practices of midwives can be garnered from legal sources. In addition, much of the existing information about them was written by hostile observers. For example, both Northern and Southern writers frequently portrayed midwives as "dirty," "ignorant," "evil," and "inveterate quacks."[12]

Northern observers were especially concerned about the growing number of midwives who were believed to be performing illegal abortions. One early-twentieth-century study of midwives in New York City reported that 176 of 500 midwives under investigation had participated in such "criminal work." A similar study of Chicago midwives classified one-third of them as "criminal." The Chicago study also disclosed that physicians often supported the "criminal midwives" by falsifying death certificates for them.[13]

Southern reports dwelt on the superstitions and folklore of the "granny" midwives which they alleged were "so gross in many cases as to rival voodooism." Black midwives in Maryland were reported to treat hemorrhaging by placing either ice or hot

[11] As late as 1930, the health departments of ten states reported that their midwives were neither controlled nor licensed. Significantly, Massachusetts was the only state which outlawed midwives during the 1900 - 1930 period. The 1901 Medical Practice Act of Massachusetts listed midwifery as a major branch of medicine, but it did not exempt midwives from the provisions of the Act. Thus, only qualified physicians could legally serve as birth attendants. Despite the 1901 Act, many Massachusetts women continued to serve as midwives. At least one woman, Hanna Porn, a trained and practicing midwife, was convicted of violating the 1901 Act. White House Conference on Child Health and Protection, *Obstetric Education* (New York, 1932), 178 - 92, 195; Louis S. Reed, *Midwives, Chiropodists, and Optometrists: Their Place in Medical Care* (Chicago, 1932), 15 - 16; *Revised Laws of the Commonwealth of Massachusetts, 1901*, Chapter 76, Section 8; Commonwealth v. Hanna Porn, *Massachusetts Reports, 1907*, 195, 196.

[12] See, for example, J. Clifton Edgar, "The Remedy for the Midwife Problem," *American Journal of Obstetrics and the Diseases of Women and Children* 63(1916): 881; Arthur B. Emmons and James L. Huntington, "Has the Trained and Supervised Midwife Made Good?" *Transactions of the American Association for the Study and Prevention of Infant Mortality* 2(1911): 207 - 8; H. J. Garrigues, "Midwives," *Medical News* 72(1898): 235; W. C. Gewin, "Careless and Unscientific Midwifery with Special References to Some Features of the Work of Midwives," *Alabama Medical Journal* 18(1905 - 6): 629 - 35; W. A. Plecker, "The Midwife Problem in Virginia," *Virginia Medical Semi-Monthly* 18(1914): 475; W. S. Smith, "Careless and Unscientific Midwifery with Special References to Some Features of the Work of Midwives," *Maryland Medical Journal* 33(1895 - 96): 146 - 49.

[13] Crowell, "Midwives of New York," 673; "Midwives of Chicago," 1348-49.

potatoes in the hand of the parturient woman. Many "grannies" were said to believe that a difficult labor might be made easier by throwing hot coals on hen feathers and placing the ashes under the prospective mother's bed. The Southern black midwife was also characterized as an old woman who was qualified for nothing except "catching babies." A favorite quotation included in studies about Southern midwives came from an eighty-year-old "granny" who supposedly stated, "I am too old to clean, too weak to wash; too blind to sew; but, thank God! I can still put my neighbors to bed."[14]

Even studies conducted by the United States Children's Bureau, which was generally sympathetic to midwives, reported considerable ignorance and superstition among "grannies." An investigation of maternity care in rural Mississippi for the years 1916-1918 reported that two-thirds of the black midwives were illiterate. Although the white midwives were "superior" to their black counterparts, they "did not differ either in training or practice from the Negro midwives." Nine-tenths of the rural Mississippi midwives, most of whom were black, used no antiseptics. Only the "more intelligent ones" realized that puerperal fever was caused by uncleanliness, and many midwives continued to adhere to the old custom of not changing the bed coverings for at least three days after the birth. A similar study of maternity care in the north Georgia mountains for the years 1916-1918 also disclosed that midwives were "steeped in the superstitions and practices of a bygone generation." Significantly, most of the Georgia mountain midwives were white. Although the majority of the accounts of Southern midwives focused on the "ignorance" and "superstition" of the black "grannies," contemporary reports also indicated that white midwives were sorely lacking in the basic knowledge required of competent attendants at birth.[15]

Very few satisfactory training programs for midwives existed at the turn of the twentieth century. This helps to explain why contemporary observers frequently characterized them as "dirty," "ignorant," and "evil." In most instances, those midwives who wanted to learn more about pregnancy and parturition could not attain adequate training. Between 1916 and 1918, for example,

[14] Guy Steele, "The Midwife Problem and Its Legal Control," *Maryland Medical Journal* 48(1905): 2, 4; Helmina Jeidell and Willa M. Fricke, "The Midwives of Anne Arundel Country, Maryland," *Johns Hopkins Hospital Bulletin* 23(1912): 281; Marie Campbell, *Folks Do Get Born* (New York: Rinehart, 1946), 33; Carolyn C. Van Blarcom, "Midwives in America," *American Journal of Public Health* 4(1914): 198.

[15] Dart, *Maternity and Child Care*, 22 - 23; Glenn Steele, *Maternity and Infant Care in a Mountain County in Georgia*, U.S. Department of Labor, Children's Bureau Publication No. 120 (Washington, D.C., 1923), 17.

Mississippi midwives expressed an interest in having public health nurses advise them, but there were no nurses available for that purpose. Some midwives did serve apprenticeships with physicians. Usually, however, they were taught by their mothers or older midwives. Most of their training consisted of first-hand experience and observation. An important part of the educational process for many midwives was the birth of their own children. Most midwives were, in fact, married women who had borne several children.[16]

Many black women felt that proper midwifery training included a "calling by the Lord." In a recent interview, Elizabeth Singleton, a 104-year-old black midwife from Wayne County, Georgia, underscored the importance of the "calling." In recounting her experiences as a midwife, Singleton remarked,

> The good Lord taught me how to catch babies. . . . I don't know how many babies I caught; must have been more than 1000, both black and white babies. . . . When I was catching babies, I would pray for you. I would ask the Lord to help you and to help me to take care of you. I've prayed so hard, and my soul would be so full of joy. . . .[17]

A few schools for midwives existed in the northeastern and midwestern areas of the United States at the turn of the twentieth century, but the course of instruction provided at most of them was totally inadequate. Early twentieth-century midwife studies concurred in the opinion that the schools were "veritable diploma mills run for revenue only, usually by some conscienceless physician who receives lucrative returns for a minimum outlay of time and trouble." The fees charged by the physicians were, in fact, exorbitant. The average cost of a six-month course of study in Chicago in 1907 was $100-$175. This fee usually included nothing more than lectures for three to four hours each week in the physician's back office. Often, the instructor required no textbooks and provided no clinical instruction for his students. Schools for midwives in New York City provided the same type of low-quality education. A 1907 study of New York City midwives disclosed that some women who could neither read nor write held midwifery diplomas from New York schools. The quality of the education in the New York schools was so deplorable that the author of the 1907 study concluded that the "eighty-eight midwives who had no diplomas were quite as efficient and capable as

[16] Dart, *Maternity and Child Care,* 22; Glenn Steele, *Maternity and Infant Care,* 16; Hunter, *Infant Mortality,* 45 - 47; Paradise, *Maternity Care,* 31 - 33.

[17] "104 Years Old: God, Herbs Get Credit for Long Life," *Atlanta Constitution,* 16 September 1974, p. 6-a, cols. 1 - 4.

the 209 who held those worthless New York diplomas."[18]

It would be a mistake, however, to leave the impression that all turn-of-the-century midwives were poorly trained. Newly-arrived immigrant midwives were often graduates of excellent European midwifery schools. Of course, most Americans were not familiar enough with the European situation to know whether or not the foreign diplomas displayed by the midwives were from reputable schools. Moreover, the European-trained midwife often became careless in her practice once she arrived in America where few rules and regulations existed. It was also economically impossible for her to compete with the American idea of the midwife and still maintain the high standards she had been taught in Europe.[19]

There were a few American schools which provided adequate training programs for midwives. In 1883, the New York State Supreme Court granted the College of Midwifery of New York City the right to confer the diploma of Graduate in Midwifery. The College advertised that it was "the only regular institution in America incorporated for the purpose of conferring on women the diploma of Graduate in Midwifery." The faculty consisted of seven physicians. The lone woman, Sarah E. Post, M.D., served as Professor of Physiology. She was also the attending physician to the Out-door Department of the New York Infirmary for Women and Children.[20]

The course of instruction at the College included lectures, recitations, and demonstrations in English, German, French, and Spanish for a three-month period. This was followed by three months of practical work at the Metropolitan Dispensary for Women and the Woman's Infirmary and Maternity Home. The cost of the six-month course was $105. If a candidate failed to pass the final examination, she could enroll in the College a second time by paying a fee of $30. Fourteen students attended the first session, and ten satisfactorily completed the course. The faculty believed that the "interests of the institution and the community would be better served by teaching a limited number thoroughly and well, rather than send out a large class unqualified."[21]

[18] "Midwives of Chicago," 1348; Abbott, "Midwife in Chicago," 690; Crowell, "Midwives of New York," 671.

[19] Taliaferro Clark, "Training of Midwives," *Chicago Medical Recorder* 46(1924): 297 - 304; S. Josephine Baker, "Schools for Midwives," *American Journal of Obstetrics and the Diseases of Woman and Children* 54(1912): 258 - 62.

[20] *Woman's Work in the Field of Medicine* (New York: College of Midwifery, 1883), 19 - 20, 27.

[21] *Ibid.*, 7, 21 - 24.

In the original bulletin announcing the opening of the College, the Board of Trustees explained why such a school was necessary. Predictably, they pointed to the lack of training programs for midwives in the United States. They included extensive quotations from the congratulatory and commendatory notices that had appeared in some of the leading medical journals of America. These notices reiterated that the United States had "thus far . . . been destitute of any adequate provision" for the education of midwives. The Trustees defended the concept of formal training for midwives on the grounds "that it teaches, what, by experience, had been found to cover the sphere of woman's usefulness in medicine." They argued that women were especially suited for training in midwifery but that they were not, in most instances, capable of undergoing the "whole curriculum of surgery, pharmacy, materia medica, jurisprudence, etc." The failure of female medical schools to attract large numbers of students was used as proof that women did not have the ability to become regular practitioners of medicine. Moreover, the Trustees also believed that it was "not necessary to lay open the entire subject of obstetrics to the midwife. On the contrary, her practice should be strictly limited; the knowledge imparted serving mainly to indicate to her when the accoucheur is to be called in and that the office of midwife is to attend women in natural labor."[22]

The Playfair School of Midwifery in Chicago, Illinois, was another institution that provided quality education for midwives. It was established in 1896 for the purpose of interesting "such a class of women in this work that by proper opportunities we will remove the present prevailing prejudice against the midwife and finally restore her to her proper reputable position in the profession, not as a competitor to the physician but [as one] able to conduct normal labor with safety, and . . . to recognize early the nature of pathological cases and send for the timely help of the physician." The Playfair School offered diplomas in midwifery as well as certificates in obstetrical nursing to those women who did not wish to qualify fully as midwives.[23]

The 1889 annual announcement of the Playfair School listed eleven faculty members, ten of whom held the M.D. degree and six of whom were women. There were two major admission requirements. First, the prospective midwife had to present a certificate of good moral character, signed by either a physician or a minister. Second, she had to show evidence of an ability to

[22] *Ibid.*, 6 - 9.

[23] *The Playfair School of Midwifery: Annual Announcement, 1899* (Chicago: n.p., 1899), 6 - 7.

read and write in English or German. Instruction was given in both languages. The course consisted of two terms, each five months in length. Lectures, demonstrations, and recitations were scheduled daily from 9:00 A.M. to noon. Clinical work was conducted at the Playfair Lying-in Hospital Dispensary. The cost of the complete ten-month course was $110. To graduate, the student had to attend at least 80 percent of all the lectures, pass all examinations given by the various branches of the school, and show satisfactory evidence of having attended twelve cases of labor. Because of the "high grade requirements" of the Playfair School, the Illinois State Board of Health exempted its graduates from any additional state examinations required for midwives.[24]

A third school for midwives whose diplomas were of real value was the St. Louis College of Midwifery. One of its founders, Dr. A. A. Henske, was so concerned about improving the quality of midwife education that he also established a monthly journal, the *American Midwife,* which appeared from November 1895 through October 1896. The College offered a five-month course of instruction in English and German. Six of the eight faculty members were men with the M.D. and/or Ph.D. degrees. The two women on the faculty, Katherina Zratz and Annie J. Byrns, were trained midwives.[25]

Unfortunately, information about these three schools is incomplete and of uneven quality. The pioneering studies of the midwife situation in New York City and Chicago, published in 1907 and 1908 respectively, made no specific mention of either the College of Midwifery of New York City or the Playfair School of Midwifery. Possibly, the two schools were no longer in existence or else were judged to be "diploma mills." Moreover, the number of students enrolled in the three schools represented only a tiny minority of the women who acted as midwives at the turn of the century.[26]

A fourth, but more well known and longer lived, school for midwives was the municipally sponsored Bellevue School for Midwives in New York City. It was established in 1911 as a branch of the Bellevue Hospital. The original course of instruction was six months in length, but it was soon expanded to cover eight months. Both practical and theoretical instructions were given by physicians and nurses. The training program included instruction in the proper care of cases of normal labor, the necessity

[24] *Ibid.*, 7, 13 - 16.

[25] See advertisements for the St. Louis College of Midwifery appearing in each issue of the *American Midwife*, 1 - 2(1895 - 96).

[26] Crowell, "Midwives of New York," 667 - 77; "Midwives of Chicago," 1346 - 60.

of breast-feeding, the manner and methods of artificial feeding, and the essential housewifely duties to be performed by the midwife. Most of the students were of Italian, German, Polish, or Hungarian descent. There were no special admission requirements other than that the prospective midwife be able to read and write and be of good moral character. There was no charge for the course. Most reports issued by physicians and public health officials during the second decade of the twentieth century agreed that the Bellevue School for Midwives was the *only* institution in the United States that provided an adequate education for midwives. One early report, published by the New York City Department of Health, described the students as "intelligent, earnest women with sufficient education to grasp the principles of normal obstetrics, asepsis, and to recognize abnormalities. . . ." Interestingly, nurses were not allowed admission to the Bellevue School for Midwives because it did "not have accommodations suitable" for them. The School remained in existence until 1935, when a lack of students forced it to close its doors.[27]

Very few printed materials were published expressly for American midwives during the late nineteenth and early twentieth centuries. A variety of maternity and mothers' manuals appeared during this period, but they were written for the educated, middle- and upper-class woman. Significantly, most books of this type omitted any mention of the role of the midwife in the birth process. They assumed that both a physician and monthly nurse would be in attendance.[28] The growing number of obstetrical textbooks for students and practitioners that were published at this time were also unsuitable for midwives. These works were often quite lengthy and written in highly technical language that was difficult for people untrained in the medical sciences to under-

[27] Thomas, "Supervision of Midwives," 118. Information on the origins and history of the Bellevue School for Midwives is also included in the following works: Baker, *Child Hygiene* (New York, 1925), 118 - 19; Baker, "Schools for Midwives," 263 - 66; City of New York, Department of Health, "Supervision of Midwives in New York City," *Weekly Bulletin* 10(1921): 154; Hazel Wedgewood, "Midwifery in Massachusetts," *Commonhealth* 8(1921): 81 - 83.

[28] See, for example, Pye Henry Chavasse, *Woman as a Wife and Mother* (Philadelphia, 1870); John H. Dye, *Painless Childbirth: or Healthy Mothers and Healthy Children* (New York, 1884); Anna M. Fullerton, *A Handbook of Obstetrical Nursing for Nurses, Students and Mothers* (Philadelphia, 1890); Frederick Hollick, *The Origin of Life and Process of Reproduction in Plants and Animals, with the Anatomy and Physiology of the Human Generative System Male and Female, and the Causes, Prevention and Cure of the Special Diseases to Which it is Liable* (Philadelphia, 1878); George H. Napheys, *The Physical Life of Woman: Advice to the Maiden, Wife, and Mother* (Philadelphia, 1890); Prudence B. Saur, *Maternity: A Book for Every Wife and Mother* (Chicago, 1890); Alice B. Stockham, *Tokology: A Book for Every Woman* (Chicago, 1885); Tullio Suzzara Verdi, *Maternity: A Popular Treatise for Young Wives and Mothers* (Philadelphia, 1885).

stand.[29] A number of British manuals for midwives were also accessible to American women. However, midwives in the United States were discouraged from reading the British works because they were considered to be too difficult. The British manuals were deemed inappropriate for study by the "typical midwife" in the United States who was "not to be trusted with a . . . bag equipped with needles, hypodermic syringes, [and] ergot."[30]

The aforementioned *American Midwife* was probably the only turn-of-the-century journal published expressly for American midwives.[31] It was published monthly, in both English and German, from November 1895 through October 1896. Its founder, Dr. A. A. Henske, and many of the contributing authors were members of the faculty of the St. Louis College of Midwifery. Judging from the notices that appeared within its pages, the *American Midwife* was a much-needed and greatly appreciated periodical. Congratulatory letters from midwives from such diverse locations as St. Louis, Missouri, Bensenville, Illinois, and San Francisco, California, were published. Edna L. Bartells of Bensenville wrote that "the midwife ought to have something to read just as well as the doctor, so as to remind her of her duties to her calling. As a rule she does not read any journal or book but even forgets what she has learned." One early subscriber, Emma E. Peters, urged other midwives to send in their subscription fees of one dollar since "midwives are never through learning." Medical practitioners also congratulated the editors for their endeavors. In assessing the success of the periodical, the editors reported in January 1896 that "our list of subscribers increases daily; in fact, far beyond our expectations. There is no city in the country which has not added a number of names to it." They concluded that the *American Midwife* "would be the journal that will furnish the midwife with well written articles, treating on different subjects of her science, and will keep her well advised of all the

[29] See, for example, Rodney Glisan, *Textbook of Modern Midwifery* (Philadelphia, 1881); Charles Jewett, *Outlines of Obstetrics: A Syllabus of Lectures Delivered at the Long Island College Hospital* (Philadelphia, 1894); J. Whitridge Williams, *Obstetrics: A Textbook for the Use of Students and Practitioners* (New York, 1903).

[30] See, for example, Comyns Berkeley, *A Handbook for Midwives and Maternity Nurses* (New York, 1909); A. B. Calder, *Lectures on Midwifery for Midwives* (New York, 1905); Henry Fly Smith, *The Handbook for Midwives* (London, 1873); "Handbook for Midwives," *American Journal of Obstetrics and the Diseases of Women and Children* 6(1873 - 74): 700; "Reviews and Book Notes," *Public Health Nurse* 17 (1925): 220 - 21.

[31] Only three libraries in the United States possess copies of the *American Midwife*. The two most complete files of the journal are located at the Library of the College of Physicians of Philadelphia and the St. Louis Society for Medical and Scientific Education.

latest advances and of all important discoveries in the different branches of obstetrics."[32]

Most of the articles published in the *American Midwife* were written by physicians. Occasionally, however, midwives wrote about their own experiences. Lillian V. Young, a student at the St. Louis College of Midwifery, discussed the importance of clinical training. She stated that her first childbirth case was normal but that she considered it an "ordeal" because of her lack of clinical experience. Young did not hold the St. Louis College of Midwifery responsible for her predicament. She maintained that this was a problem common to midwifery colleges throughout the United States.[33]

Annie J. Byrns, a trained midwife and a faculty member of the St. Louis College of Midwifery, wrote an article in which she was highly critical of those physicians who called for the elimination of midwives on account of their alleged ignorance relating to matters of childbirth. Byrns argued that "the United States can boast of midwives who not only wash their hands with soap, but scrub them to their elbows, clean their finger nails, and then use antiseptic solution." She pointed out that thousands of concerned midwives read the *American Midwife* and that St. Louis had a Scientific Association of Midwives.[34]

The last issue of the *American Midwife* appeared in October 1896. The editors provided no explanation for its cessation. Certainly, the claim that its subscribers came from every city in the United States was an exaggeration. Nevertheless, the significance of the *American Midwife* must not be underestimated. The fact that Henske and the other faculty members at the St. Louis College of Midwifery believed that American midwives would support the journal is important. Apparently, there were conscientious midwives in St. Louis and other areas of the United States at the turn of the century who were very concerned about improving their knowledge of pregnancy and parturition and keeping abreast of developments in midwifery. For such women, the only publication to turn to was the *American Midwife*. For a brief period, American midwives had a journal in which they could relate their experiences and even reply to the criticisms of those physicians who maintained that they were "dirty and ignorant." When it ceased publication in October 1896, over two decades

[32] "Letters," *American Midwife* 1(December 1895): 6; "Editorial," *ibid.* 1(January 1896): 1.

[33] "Letters," *American Midwife* 1(February 1896): 15.

[34] Annie J. Byrns, "Midwives," *American Midwife* 1(December 1895): 4-5.

elapsed before another journal for American midwives was established.[35]

Although it is not possible to create an exact prototype of the American midwife at the turn of the twentieth century, it is possible, through the examination of a variety of sources, to envisage what she may have been like. She would, most likely, have been a married woman who had borne several children and who had received her first training from her mother or some other older woman. Whether she was an immigrant midwife living in the urban Northeast or a black "granny" living in the rual South, she would have had little, if any, formal midwifery training. Only a very small percentage of the total number of midwives were trained at quality schools such as the Bellevue School for Midwives in New York City. The smattering of education that she received focused on how to manage normal confinements and which symptoms merited calling in a physician. Her lack of formal training and her unfamiliarity with obstetric techniques meant that she usually let "nature work its course." Her practice was frequently limited to friends and relatives. Not surprisingly, therefore, the income she derived from her work was quite small. The average fee charged by the midwife was less than one-half that of the general practitioner. Moreover, arrangements were often made for payment "in kind." In addition, she also performed a variety of housewifely functions which were not required of the doctor. She had only limited contact with other midwives and, for the most part, went about her work unobtrusively. Indeed, in 1900 the American midwife appeared to be a forgotten person.

By 1910, however, physicians, public health advocates, and to a lesser extent the general public were embroiled in fierce debate over the midwife question.[36] Two unforeseen developments had brought an end to the apathy of 1900. First, physicians had become increasingly concerned about medical education reform and the related issue of "overcrowding" in the medical profession during the early years of the twentieth century.[37] Doctors maintained that "overcrowding" had resulted in a loss of their status

[35] During 1918, Ferdinand Herb, a Chicago physician, published a monthly bulletin, *Topics of Interest to Midwives.*

[36] For additional information on the early-twentieth-century midwife debate, see Kobrin, "American Midwife Controversy," 350 - 63; Litoff, "Midwives in America," chapters 3 - 5.

[37] The most thorough examination of physicians' attitudes about "overcrowding" during the 1890 - 1910 period can be found in Gerald E. Markowitz and David Karl Rosner, "Doctors in Crisis: A Study of the Use of Medical Education Reform to Establish Modern Professional Elitism in Medicine," *American Quarterly* 25(1973): 87 - 97.

and a diminution of their income. Many physicians believed that eliminating the midwife, or at least substantially reducing her numbers, would help alleviate this problem.[38]

The second and equally significant development which was responsible for the upsurge of interest in the midwife around 1910 was the revelation that the maternal and infant mortality rates in the United States were alarmingly high. Prior to 1900, birth and death statistics were too sketchy to be meaningful. During the early years of the twentieth century, however, the United States Bureau of the Census, the United States Children's Bureau, and state and local health boards began to collect and publish birth and death statistics.[39] For example, a study conducted by the United States Children's Bureau revealed that the 1910 infant mortality rate in the United States was 124 deaths per 1,000 live births. The Bureau also reported that "childbirth caused more deaths among women fifteen to forty-four years old than any disease except tuberculosis" and that in only two of fifteen countries it had investigated had maternal mortality rates been higher than in the United States for the 1900-1910 period.[40] The publication of these birth and death statistics caused physicians and public health officials to take a closer look at the midwife's role in the birth process and to try to determine why so many American mothers and infants were dying.

The question of the midwife's status in American society was heatedly debated between 1910 and 1930. The United States Children's Bureau, the various state and city Bureaus of Child Hygiene, the American Association for the Study and Prevention of Infant Mortality, the New York Committee for the Prevention of Blindness, the Babies Milk Fund Association, and numerous physicians and public health officials examined the midwife question in detail at this time. In addition, scores of articles on the "midwife problem" appeared in most of the major American medical journals. To a lesser extent, articles on the midwife

[38] A classic example of this viewpoint can be found in J. Whitridge Williams, "The Midwife Problem and Medical Education in the United States," *Transactions of the American Association for the Study and Prevention of Infant Mortality* 2(1911): 184 - 93.

[39] U.S. Bureau of the Census, *Historical Statistics of the United States, Colonial Times to 1957* (Washington, D.C., 1960), 18.

[40] Dorothy E. Bradbury, *Five Decades of Action for Children: A History of the Children's Bureau* (Washington, D.C., 1962), 6; Grace L. Meigs, *Maternal Mortality From All Conditions Connected with Childbirth in the United States and Certain Other Countries*, U.S. Department of Labor, Children's Bureau Publication No. 19(Washington, D.C., 1917), 7, 17.

were also published in many of the popular magazines of the period.[41]

The midwife debate was lengthy and acrimonious. At one end of the spectrum were the physicians, many of whom were obstetric specialists, who were vehemently opposed to the midwife. They believed that the obstetrician would never receive due recognition in the "overcrowded" medical profession as long as women untrained in the medical sciences continued to deliver one-half of all the births in the United States.[42] At the other end of the spectrum were the public health officials who argued that the utilization of properly trained and regulated midwives would help significantly to reduce maternal and infant mortality rates in the United States.[43] There were, of course, a wide array of views that fell within these two extremes. For example, many midwife critics supported training and regulatory programs as temporary measures to be discontinued as soon as an adequate number of physicians could be properly trained to take over the clientele being served by midwives.[44] Public health officials in several states supported the regulation of the midwife with the hope that ultimately she would be regulated out of existence.[45] Finally, some anti-midwife physicians were optimistic that nurses, who had received special training in obstetrics, could be substituted for midwives.[46]

[41] Litoff, "Midwives in America," chapters 3 - 5.

[42] A leading figure in the movement to eradicate midwives was Dr. Joseph B. DeLee, founder of the Chicago Lying-In Hospital (1895). See, for example, Joseph B. DeLee, "Report of Sub-Committee for Illinois," *Transactions of the American Association for the Study and Prevention of Infant Mortality* 5(1914): 231; DeLee, "Progress Toward Ideal Obstetrics," *ibid.* 6(1915): 117.

[43] Dr. S. Josephine Baker, Director of the Bureau of Child Hygiene of the Board of Health of the City of New York for over twenty-five years, was a leading figure in the movement to establish programs for the training and regulation of midwives. See, for example, Baker, "Schools for Midwives," *Transactions of the American Association for the Study and Prevention of Infant Mortality* 2(1911): 232 - 43; Baker, "The Function of the Midwife," *Woman's Medical Journal* 23(1913): 196 - 97.

[44] A leading spokesman for this viewpoint was Dr. J. Clifton Edgar, a New York City physician. See, for example, Edgar, "The Remedy for the Midwife Problem," 881; Edgar, "Why the Midwife?" *American Journal of Obstetrics and the Diseases of Women and Children* 77(1918): 243.

[45] See, for example, S. W. Newmayer, "The Status of Midwifery in Pennsylvania and a Study of the Midwives in Philadelphia," *Monthly Cyclopedia and Medical Bulletin* 4(1911): 713, 718; Jessie L. Marriner, *Midwifery in Alabama* (Alabama State Board of Health, [1925]), 1 - 2, 10; *Report,* Board of Health of Mississippi (1919 - 1921), 198.

[46] Edgar, "The Education, Licensing and Supervision of the Midwife," *American Journal of Obstetrics and the Diseases of Women and Children* 73(1916): 391; George W. Kosmak, "Certain Aspects of the Midwife Problem in Relation to the Medical Profession and the Community," *Medical Record* 85(1914): 1015.

The midwife debate raged for two decades. When it finally subsided around 1930, the ranks of the midwives had been dissipated and their historical role appeared to be relegated to near-total obscurity.[47] The revival of the feminist movement and the increased interest in women's history over the last ten years, however, have caused historians to take a fresh look at the role of the midwife in American society.[48]

[47] Between 1900 and 1930, the percentage of midwife-attended births declined from 50 to 15 percent. White House Conference on Child Health and Protection, *Obstetric Education*, 169.

[48] In addition to the previously cited works by Donegan, Kobrin, Litoff, and the Wertzes, the question of childbirth in America has also been examined at several recent historical conferences including the 1974 Berkshire Conference on the History of Women, the 1976 Missouri Valley History Conference, and the 1976 Annual Meeting of the Organization of American Historians. In addition, feminists and women's health reformers, who are not necessarily trained historians, have also contributed to the literature on this topic. See, for example, Suzanne Arms, *Immaculate Deception: A New Look at Women and Childbirth in America* (Boston, 1975).

SOCIAL PROBLEMS, Vol. 30, No. 3, February 1983

WORKING-CLASS WOMEN, MIDDLE-CLASS WOMEN, AND MODELS OF CHILDBIRTH*

MARGARET K. NELSON
Middlebury College

This paper suggests that middle-class feminists who have urged a new vision of childbirth are out of touch with the needs of working-class women. This conclusion is derived from a review of the childbirth literature and an analysis of data collected from 322 women who gave birth in a northern New England teaching hospital. The literature on childbirth—whether written from a feminist or non-feminist perspective—ignores the variable of social class. The data demonstrate, however, that working-class and middle-class women have different attitudes towards childbirth during pregnancy, different experiences during childbirth, and different post-partum evaluations of their childbirth experiences. A single set of prescriptions for childbirth may not, therefore, be appropriate for all women.

This paper began as an attempt to describe the class differences in childbirth procedures in a teaching hospital in New England. I assumed that I could show, as have others, that middle-class women received better medical services than working-class women, and that although the hospital had a long way to go before it provided all women with an optimum birth experience, one's chances of reaching that goal were greatly enhanced if one entered as a middle-class client. This paper, then, began as a critique of birth practices in a modern hospital and the class biases implicit among medical personnel.

Responses to my questionnaires and interviews suggested that women could be divided along certain significant social dimensions which corresponded to kinds of birth experiences. One group of women experienced birth in a relatively active and involved way. Another group had more passive births, involving more monitoring, medication, transfers to the delivery room, and use of forceps.

I then went one step further: were the women within each of these groups having the kind of birth experience for which they had indicated preferences during the prenatal period? I hypothesized that middle-class women would be more likely to have their choices respected in the hospital than working-class women. I suspected that doctors would use low social status as a justification for disregarding client preferences, and select the procedures *they* deemed necessary for working-class women.

I was wrong on two counts. First, neither of the two groups of women that I studied received the precise treatment they wanted. Second, there were few differences between the two groups in the extent to which the women got what they wanted. What I found instead was that the middle-class women generally wanted active, involved births free from medical interventions; some of their requests were respected in the hospital. The working-class women wanted more passive birth experiences with more medical intervention; some of their requests also were met within the hospital.

The data thus suggested that within the hospital at least three different models of an appropriate childbirth were operating: two different client models and a medical model. My initial approach to this topic had overlooked the existence of more than one client model distinct from that of the doctors' model.

The literature on women's control over childbirth has highlighted the importance of this life

* An earlier version of this paper was presented at the annual meetings of the Society for the Study of Social Problems, Toronto, August 22–24, 1981. The author thanks the Department of Obstetrics and Gynecology of the Medical Center Hospital of Vermont for supporting the collection of data, Middlebury College for leave, and Helen McGough for her criticism and encouragement. Correspondence to: Department of Sociology-Anthropology, Middlebury College, Middlebury, VT 05753.

event for women. It has also used a set of implicit assumptions: (1) women share a set of common desires and make choices about childbirth independent of their social backgrounds; and (2) women's desires are different from those of doctors. In short, the literature assumes that women all want (or will come to want) conscious control over a basically "natural" childbirth experience and that doctors have resisted these demands. However, my data suggest that these assumptions are not entirely accurate. There is clearly more than just one client model of childbirth: not all women want the same kind of birth experience. And doctors, in fact, appear to resist and reject aspects of *each* client model in favor of their own approach.

SOCIAL CLASS IN CHILDBIRTH RESEARCH

Three traditions of research and theory on childbirth have, each for its own reason, ignored social class differences. These are: (1) the feminist literature; (2) the general medical sociology treatment; and (3) studies which focus on the effects of preparation for childbirth.

Middle-class feminists, eager to take childbirth out of the hands of male obstetricians, seem more intent on describing how childbirth has become distorted over the years than in examining whether the "warping" (Haire, 1978) has affected all groups in the same way. Numerous historical studies document how male physicians wrested control of childbirth from female midwives, and class biases in the way new technologies were distributed (Ehrenreich and English, 1973; Kobrin, 1966; Wertz and Wertz, 1977). They do not tell us whether women of different social classes felt the same way about the changes—or whether there are class differences in attitudes towards childbirth today. (For an exception see Hubert [1974] who explicitly notes the lack of homogeneity in attitudes about childbirth.)

Shaw (1974) found class differences in treatment during the prenatal and hospital periods: the medical staff were aware of the class origins of their clients and treated them according to preconceived notions of what was most appropriate for each group of women. However, the solutions Shaw offers at the end of her book suffer from the opposite type of discrimination and totally ignore social class differences. She assumed her solutions would be equally acceptable to all clients, but never asked how they—as individuals or as representatives of social categories—would like childbirth to proceed.

The heavy emphasis on personal experience in much of the childbirth literature has resulted in the emergence of a single critique which presumes to speak for all women (e.g., Comaroff, 1977; Hart, 1977. Oakley, [1979:627] makes a similar point.) Most of those who write about childbirth are middle-class women. They are motivated by a feminist consciousness and possess the verbal ability which is part of class privilege. Thus, those who are most interested in women defining for themselves the nature and meaning of childbirth are, perhaps, guilty of prescribing a perfect birth for all women, regardless of individual needs or motivations.

The non-feminist literature has also ignored social class, albeit for different reasons. First, the ahistorical bias in much social science has inhibited attention to the social history of childbirth movements, a history which is important for understanding different class attitudes towards childbirth. Second, the use of some of the major sociological concepts might also inhibit a consideration of social class. For instance, although Stewart and Erickson (1977) attack mainstream sociology for failing to seriously consider childbirth, their own emphasis on "roles" as the best possible approach to understanding pregnancy and childbirth leads them to ignore social class. They note, for example, that "the manner in which women describe their labor and delivery varies by many factors including the setting, the difficulty of the labor, their definitions of themselves as sick or healthy, their expectations and how well these were met" (1977:41).

The view of childbirth expounded by middle-class feminists has been the best articulated and, therefore, frequently adopted by academic writers as the only view. Thus, as feminist critiques of contemporary obstetric care became more frequent and solidified, they were adopted by academic writers as representative of a single model or paradigm of childbirth which conflicted with that offered by the medical establishment. For example, Nash and Nash (1979:493) argue that "In American society at the present time there exist two primary interpretations of the meaning and practice of childbirth: the medical and the 'natural' view."

Oakley (1979) offers by far the most sophisticated exploration of competing childbirth paradigms. She notes the paucity of client-oriented studies and emphasizes the different models at work in medical science, clinical psychiatry and psychology, and academic psychology and sociology. Until recently the sociologist's contribution had not been "to investigate the women's experience but to extend the limits of the medical model and propose a more elastic conception of the variables which can be seen to influence the biological outcome of maternity" (1979:624). However, Oakley emphasizes the "natural" model of childbirth as the principal one in conflict with the various medical models. Danziger (1978), in a paper on prenatal encounters between physicians and clients, also assumes that there is only one *client* paradigm.[1]

Some research on childbirth has ignored social class by matching or controlling independent variables too rigorously. Studies on preparation for birth clearly fall within this category. Most research shows that preparation for childbirth has both physiological and psychological effects on the birth process (Doering *et al.*, 1980; Norr *et al.*, 1977), though some studies indicate little or no effect (Zax *et al.*, 1975). Motivation to learn about birth does not appear to be the key factor at work here: studies which compared women who wanted to take classes in childbirth preparation but could not, with women who did take classes (or were randomly assigned to classes) found that preparation was the critical variable (Enkin *et al.*, 1972; Huttel *et al.*, 1972). To test whether preparation is a significant factor in the birth process, researchers have been careful to control for socio-economic status and education in experimental and control groups. For example, Doering and Entwisle's (1975) work on the effects of preparation on the ability to cope with labor and delivery dismisses social class after noting that there were no significant social differences between trained and untrained subjects, even though only 18 percent of the total number of subjects were working-class women. Gaziano's (1979) work only looked at working-class women who used a clinic, and therefore could not compare clients at different levels of income or education. Other researchers who compared the effects of different kinds of preparation for childbirth also controlled for the class origins of the client groups (e.g., Zimmerman–Tansella *et al.*, 1979). Researchers have given only cursory attention to the issue of who chooses to attend childbirth classes and why.

Much of this research assumes that the outcomes of preparation—knowledge, control, cooperation, and an avoidance of medication—are definite, clearcut, and desirable. That is, the studies assume that everyone wants an identical birth. Yet not all women, and particularly working-class women, may perceive the outcome of preparation as a "benefit." In fact, the data I collected for this paper suggest otherwise, and I am no longer willing to assume that all women want the same kind of birth.

METHOD

Background

In response to the declining U.S. birth rate, the movement for home birth, and the criticisms of women committed to hospital birth without extensive intervention, the Department of

1. Comaroff (1977) offers an alternative approach. She notes that within the medical world, the midwives and the physiotherapists have different paradigms of pregnancy, and that clients adopt one or the other of them (though sometimes sequentially). The clients are assumed not to bring with them their own paradigms of pregnancy but to be *tabula rasa* upon which competing health personnel write a script. Comaroff does not see social background determining which script is adopted, although she does acknowledge the influence of psychological factors.

Gynecology and Obstetrics of the Medical Center Hospital of Vermont (MCHV) in Burlington, Vermont, decided in 1979 to revamp its maternity services. It hired certified nurse-midwives to work with clients (both alone, for low-risk clients, and in conjunction with obstetricians for all other clients), opened a labor lounge, altered labor rooms to do double duty as delivery rooms, and stated that clients could choose their own birth style. It immediately hired two social scientists to evaluate whether these changes were satisfying the demands of knowledgeable, consumer-conscious, low-risk clients. I was one of those hired.[2] We thus collected the data on which this paper is based within the context of a specific mandate. The sponsors of the project were not interested in the particular issues underlying my present concern.

The Data

We collected data in three stages from all clients who were served by a private group of obstetricians in MCHV; this group accounted for 80 percent of all births in the hospital during a six-month period in the winter of 1979–80. During the ninth month of their pregnancy we gave the women questionnaires about their previous birth experiences, their feelings about pregnancy and childbirth, and their choices for childbirth procedures. Three or four days after the women gave birth we interviewed them in the hospital, asking them about the birth itself. When the women returned to the doctor for a post-partum check-up (generally six weeks after birth) we gave them a second questionnaire, which asked for their feelings about the birth. A total of 322 women completed the first questionnaire (94 percent of those asked to participate in the evaluation study); 273 were interviewed in the hospital; and 226 completed the second questionnaire. The attrition rate of 30 percent was the result of a number of factors, including early hospital discharges and client failure to keep scheduled appointments. Equal numbers of working-class and middle-class women were lost in this process.

Independent Variables

There are three independent variables in this study: social class, preparation, and parity. The first of these is the most important.

1) *Social Class*: Since I suspected that social class affected choices about childbirth, I sought a way to distinguish class position among the clients. I chose education as the best indicator: women with no more than a high school diploma were categorized as working class; those with at least four years of college were categorized as middle class. Women with some college education or vocational training beyond high school were assigned to a category on the basis of the type of job they held at the time of participation in the evaluation study, or their prior work experience.[3] These procedures resulted in a total of 127 working-class women and 124 middle-class women included in the analysis.

Information on client income was not available. In any case, the fact that my sample includes a small proportion of "voluntary poor"—highly educated women who chose subsistence farming or craft work as a way of life—made income a poor indicator. Nor could occupation alone be used. I found that differences in education made a significant difference in the kind of occupations the women held, either at the time of the study or prior to it, and the pattern of their work involvement. However, not all the women were working or reported ever holding jobs. Occupa-

2. The other was Helen McGough, who became director of the Mary Johnson Day Car Center in Middlebury, Vermont. Our evaluation of the maternity services (McGough and Nelson, 1981) showed that although most clients were pleased with the treatment they received, those clients who did not have the treatment they anticipated were less satisfied than those who got what they wanted. We also found that the medical staff was not giving full support to the innovations and that there remained a high level of medical intervention in childbirth.

3. I excluded all nurses and allied health workers from the study on the grounds that their perspective on birth would reflect their training and level of knowledge, rather than their class position. In fact, most of the women with some college education or vocational training were nurses.

tions of the husbands was also inappropriate, because I wanted a variable which would reflect the background experiences of the women themselves. Furthermore, whether a woman had a college education made a difference along a range of other experiences, such as whether she was a native of Vermont, her religion, and her age at the birth of her first child. I wanted to investigate whether education also made a difference in women's attitudes towards, and experiences during, childbirth. If it did, I could at least argue that childbirth, for these women, was experienced through a set of social mechanisms. I use the notions of middle class and working class to denote the two groups in my sample, although I am aware of the weakness of my indicator for examining social class.

2) Preparation for childbirth: There is a strong relationship between social class and formal preparation for childbirth. Seventy-nine percent of the middle-class women in my sample took childbirth classes, compared with 50 percent of the working-class women. The middle-class women also read an average of three books about pregnancy and childbirth, compared with an average of one for working-class women. These two kinds of preparation were themselves related: women who took childbirth classes were more likely to read about pregnancy and childbirth than women who didn't. This naturally raises the question of whether I'm not, in fact, examining differences in knowledge rather than social class differences. My data do not support this conclusion. Among middle-class women, preparation for childbirth made little difference in the kinds of attitudes I am studying in this paper. Among working-class women, however, preparation for childbirth was extremely important: the attitudes of working-class women who were prepared for childbirth—whether through classes, reading, or a combination of both—were closer to those I defined as "middle-class" than the attitudes of their unprepared peers. Had I examined the unprepared women in each social class I would have found even greater differences between the two groups than I did when I combined the prepared and unprepared women. Therefore, preparation is not relevant to the issue of attitude, except insofar as the findings about class differences in preparation reinforce my fundamental conclusions. When preparation is relevant to my analysis, I mention it below.[4]

3) Parity: One would expect women who have given birth before (multiparas) to have different attitudes about childbirth, compared to women who have never given birth (primiparas). Surprisingly, this was not the case. There were relatively few instances where multiparas and primiparas had different attitudes during the prenatal period about what they wanted, although parity was an important determinant of what actually happened once a woman entered the hospital: multiparous women were less likely to have births marked by extensive medical intervention. In those cases where parity was important, it had the same effects among middle-class women as it did among working-class women. Therefore, since parity—like preparation—is not crucial to most of the issues under consideration, I will discuss it only when it can help clarify my findings.

ATTITUDES TOWARD CHILDBIRTH

The middle-class and working-class women I studied had different attitudes towards childbirth during pregnancy, different experiences during childbirth, and different post-partum evaluations of their experiences.

Attitudes During Pregnancy

Working-class women were more likely than middle-class women to have negative feelings about pregnancy. They were less likely to say they felt good about the way they looked or felt

4. When I did include childbirth preparation as a variable, I measured it by attendance at a series of childbirth classes for the latest, or an earlier, pregnancy. For a more complete discussion of these issues, see Nelson (1982).

and they did not feel they received sufficient consideration from others. Pregnancy was not an unambivalently positive state for working-class women. There were obvious material causes for this attitude: working-class women were more likely to become pregnant by accident (39 percent) than middle-class women (22 percent) and less likely to have the resources with which to find space to rest and relax. Ideological issues may also have been at the root of these differences. Middle-class women felt that pregnancy, labor, delivery and the post-partum presence of a baby were interrelated pleasures, and this was true regardless of either their preparation for childbirth or parity. Working-class women made a greater distinction between the stages that led to birth and the presence of the baby itself: the former was not necessarily desirable, though the latter was. As with middle-class women, these attitudes were unaffected by parity and preparation.

Furthermore, during pregnancy working-class women were more apprehensive than middle-class women about labor and delivery. They worried about their own knowledge and competence; they worried that they wouldn't know when they were actually in labor; they worried that they didn't know what would happen in the hospital. Clearly, the context in which birth occurred was somewhat threatening. In addition, working-class women were more worried than middle-class women about the discomfort of labor and delivery, and whether or not their personal physician would be present for the birth.[5] On the other hand, working-class and middle-class women had almost identical attitudes toward one birth issue and one issue pertaining to the baby itself: they all expressed strong concern about whether the birth would proceed according to personal desire, and they all felt certain that they would love the baby once it was born.

Among both groups, parity was relevant for all issues except concern about whether the baby would be healthy, and whether the mother would love it; those who had given birth before were less anxious. Yet among the primiparas within each social class the differences between the two social groups remain. Moreover, preparation for childbirth was related to concern about two issues: bearing the discomfort and having the birth as planned. However, in both groups those with more preparation were *more* concerned about these issues than were those with less preparation. Different levels of preparation do not account for the differences between the two groups.

Two additional questions in the questionnaire I gave to women in their ninth month of pregnancy were designed to elicit their *general* attitudes towards childbirth. The first of these asked clients whether they agreed with the statement: "I feel that the birth experience can affect the quality of the parent's relationship with the baby." Seventy-eight percent of middle-class women agreed with this statement, compared with 48 percent of working-class women. This indicates that middle-class women place more emphasis on the birth experience itself as a critical stage in becoming a parent. In response to a second question, 69 percent of middle-class women and 54 percent of working-class women agreed with the statement: "I feel that a natural childbirth will be best for my baby." This suggests that, not only do middle-class women think that the birth experience is significant in and of itself, but they have a *specific* idea about what kind of experience will be most appropriate.

Parity was not relevant to either of these attitudes. Among working-class women, preparation altered both attitudes; among middle-class women, only the feeling about a natural childbirth was affected by preparation. The class differences remained when I compared prepared women in each group (Nelson, 1982).

Similar differences between the middle-class and working-class women emerged when the multiparas within each group commented during pregnancy about how they would like the impending birth to differ from past ones. Middle-class women and working-class women shared

5. This is not related to the fact that more working-class women than middle-class women see doctors in a clinic rather than in a private office, since clinic patients, like private patients, are assigned a regular doctor whom they see at each prenatal visit. In fact, no client in the group practice was promised that her personal doctor would be present at the birth.

TABLE 1

Client Attitudes Towards Pregnancy, Labor, and Delivery During the Ninth Month of Pregnancy

Statements	Percentage of Working Class (N = 127)	Percentage of Middle Class (N = 124)
A. Attitudes Towards Pregnancy:[a]		
When I'm pregnant people don't take my feelings seriously.	25	14
I like the way I look when I'm pregnant.	27	40
When I'm pregnant I feel well most of the time.	67	77
When I'm pregnant I feel depressed a lot of the time.	35	17
B. Concerns About Labor and Delivery:[b]		
I worry that I won't know when I'm in real labor	50	30
I worry that I won't be able to bear the discomfort.	43	33
I worry that I don't know what is going to happen to me in the hospital.	36	25
I worry that I don't know enough about the process of childbirth.	32	17
I worry that my doctor won't be there for the birth of my baby.	49	31
I worry that the birth of my baby won't be the way I want it to be.	52	58
I worry that there will be something wrong with my baby.	49	31
I worry that I won't love the baby.	7	6
C. General Attitudes Towards Childbirth:[a]		
I feel that the birth experience can affect the quality of the parent's relationship with the baby.	48	78
I feel that a natural childbirth will be best for my baby.	54	69

Notes:

a. Five response options were offered for these statements: "Strongly Agree," "Agree," "Neither Agree Nor Disagree," "Disagree," and "Strongly Disagree." Percentages include those who gave either the "Agree" or "Strongly Agree" responses.

b. Four response options to indicate the importance of each of these concerns were offered for these statements: "Very Important," "Pretty Important," "Not Too Important," and "Not Important At All." Percentages include those who gave either the "Very Important" or the "Pretty Important" responses.

a desire to have a spouse or partner present during the birth if he was not present before. Women in both groups complained about the treatment they received, but the content of their complaints was not the same. Middle-class women complained in detail about personality conflicts with their doctors, mismanagement (by the staff) of various aspects of labor, and the fact that medication was offered too frequently. Working-class women complained mostly about the lack of information offered to them during labor and delivery.

There were other differences as well. Working-class women often mentioned medical complications and the discomfort of labor and delivery. They expressed a wish that the next birth be faster and easier—if necessary, through the use of more intensive medication. Two comments were typical: "I would have liked to have been put to sleep—it was a long and painful labor"; and "Next time I want a quicker labor—my first was only eight hours but the pains were hard and came every two minutes." Middle-class women rarely mentioned either the intensity of pain or the length of labor. They stressed instead obstacles to a pleasurable experience: "I did not see my baby being born"; and "I would have wanted a more creative delivery."

Both working-class and middle-class clients felt they had a right to evaluate the past performance of medical personnel: if any of the women felt constrained by the professionalism of the staff, the constraint did not totally inhibit subsequent evaluation. Women in both groups felt the medical structure could be modified. However, there were significant differences between the groups. Working-class women felt that the responsibility for providing information rested with the doctor. In addition, they wanted to reduce the length of labor and avoid medical complications. They wanted a birth marked by less pain. In contrast, middle-class women were looking for a pleasurable, and often "natural," experience, and a more cooperative—rather than instructional—relationship with the doctor.

Two issues must be considered in evaluating the women's responses to concrete questions about what they had wanted to happen when they entered the hospital to give birth: (1) whether the women even thought about the birth experience in detail during their ninth month of pregnancy; and (2) if so, what they wanted to have happen to them.

1) With respect to the first issue, we can clearly divide the women into two groups: those who knew about and had already considered the different aspects of the birth process and those who had not. In fact, more middle-class women than working-class women considered the procedures surrounding birth: with the exception of the delivery site, less than 90 percent of the working-class women considered any of the procedures; over 95 percent of the middle-class women considered each of the procedures except artificial rupture of membranes, fetal monitoring, and position at birth. However, these differences between the groups were not found for those procedures directly related to family formation: having the husband present, watching the birth, and holding the baby after it is born. Over 90 percent of the women in *both* groups considered each of these procedures. Middle-class women as a group thus considered each step of the birth process; a substantial minority of working-class women focused only on the final stage—the creation of a new family.

Parity was not relevant here, with the exception that those who had given birth before (in both groups) were less likely to say that they wanted medication during labor than those who had never before given birth. Preparation for childbirth was important here but, as indicated above, preparation changed the attitudes of working-class women but not the attitudes of middle-class women. Most of the unprepared working-class women had not thought about the issues; those who had, wanted medical intervention (Nelson, 1982).

2) Among the women who actually thought about what they wanted to happen, there were differences in the content of their choices—differences which again corresponded to class. Working-class women selected medication during labor and delivery, artificial rupture of membranes, delivery room births, and fetal monitoring more often than middle-class women. The differences between the two groups on these issues ranged from 35 percent for delivery room births and fetal monitoring to 55 percent for the artificial rupture of membranes. Differences between the two groups with respect to the "social" aspects of birth were never more than 7 percent. Over 90 percent of both the middle-class and the working-class women chose to have the birth become an event in which they could participate by having a partner present during labor and delivery, watching the birth, and holding the baby as soon as it was born.

The working-class women seemed to be striving for speed (enema, episiotomy, artificial rupture of membranes), less pain (medication) and technological safety (delivery room birth, monitoring). They favored intervention because they thought it could bring the product easily, quickly, and safely. The middle-class women favored a process which entailed safety (as they defined it) and personal participation, but excluded medical intervention in a "natural" process.

In sum, the content of middle-class choices was "non-intervention." When middle-class women did not make choices, it was often because they wanted to leave control in the hands of the physicians or were unwilling to commit themselves in advance—not because they had not thought about the issues. "I'll let the doctor decide" was often how they responded to questions about their attitudes to specific procedures. Middle-class women selected this option more fre-

TABLE 2

Prenatal Planning for Childbirth: Choices about Procedures and Hospital Events

	Percentage of Clients Who Had Thought About the Procedures		Percentage Who Wanted Each Procedure[a]		Hospital Events: Percentages of Clients Who Had Each Procedure		
Procedures	Working Class (N = 127)	Middle Class (N = 124)	Working Class	Middle Class	All Clients (N = 203)	Working Class (N = 105)	Middle Class (N = 98)
Shave	85	97	20 (47)	20 (67)	65	61	67
Enema	85	98	42 (55)	46 (68)	49	54	47
Labor Medication	81	98	57 (51)	11 (55)	44	50	35
Delivery Medication	84	96	58 (55)	17 (64)	41	52	30
Artificial Rupture of Membranes	44	68	59 (17)	4 (25)	70	69	75
Episiotomy	64	95	64 (14)	62 (37)	85	83	88
Fetal Monitoring	53	86	90 (31)	55 (44)	85	87	80
Lithotomy Position	64	77	— —	— —	75	84	67
Delivery Room Birth	90	95	80 (89)	45 (76)	87	96	82
IV Attached During Labor	—[b]	—	— —	— —	85	91	74
Forceps or Vacuum Suction	—	—	— —	— —	20	24	13
Support Person Present During Labor	96	100	88 (116)	98 (107)	94	96	98
Support Person Present During Delivery	93	100	83 (112)	96 (108)	87	83	94
Watch the Birth	91	92	89 (112)	93 (99)	45	44	46
Hold Baby at Birth	94	99	92 (106)	97 (111)	85	81	88

Notes:

a. Based on the number of clients, in parentheses, who made a choice – positive or negative – with respect to the procedure.

b. Items for which there is no data, as indicated by a dash, were not asked on the pregnancy questionnaire or, in the case of the lithotomy position, did not have simple response options.

quently than working-class women. In contrast, when working-class women made choices they favored intervention; when they didn't make choices it was more often because they had not thought about the issues.

Among working-class women there was a tension between not thinking about the impending event and preparing for it by making decisions, between avoidance and self-determination. For the middle-class women the tension was between self-determination and reliance on professional expertise. Moreover, the goals of self-determination were also different for the two groups. Among middle-class women the goal was a definition of childbirth free from the prevailing medical and technological model embodied in the authority of the male physician. Working-class women sought freedom from the birth process itself through the use of strategies which would reduce pain and effort. Working-class women were not trying to give the experience a unique definition. They were trying to survive it with a minimum of embarrassment, discomfort, and isolation. There were exceptions to all of these generalizations. We can begin to speak about models only in the most general terms.

Experiences During Childbirth

Not only did working-class and middle-class women have different ideas about what they wanted to happen during childbirth, they also had different experiences during the actual birth. Working-class clients had births marked by more medical intervention and less client participation than middle-class women. To a certain extent these features are interrelated: the use of medication during labor and delivery makes it more likely that the delivery will take place in the delivery room and that forceps (or vacuum suction) will be required; the use of forceps ensures that an episiotomy will be necessary and that the client will have to be in a prone rather than

TABLE 3

Discrepancy Between Choice and Event: Percentage of Clients Who Got What They Wanted

Procedure	Working Class		Middle Class		Percentage Difference: Working Class v. Middle Class
	Percentage	Number	Percentage	Number	
Shave	65	(34)	56	(52)	9
Enema	50	(38)	65	(51)	-15
Labor Medication	59	(37)	59	(37)	0
Delivery Medication	53	(51)	78	(60)	-15
Artificial Rupture of Membranes	50	(20)	38	(13)	12
Episiotomy	67	(9)	61	(28)	6
Fetal Monitoring	87	(23)	58	(31)	29
Delivery Site	86	(66)	70	(60)	16
Support Person Present During Labor	86	(85)	98	(87)	-12
Support Person Present During Delivery	79	(84)	95	(87)	-16
Watch the Birth	48	(84)	44	(73)	4
Hold the Baby at Birth	83	(76)	84	(81)	-1

sitting or semi-sitting position. And the more intervention, the less likely the woman is to be able to respond to the baby immediately.

Many middle-class women also had a great deal of intervention during labor and delivery: over 70 percent of them experienced an artificial rupture of membranes, an episiotomy, fetal monitoring, a delivery room birth, and forceps or vacuum suction. The proportion of middle-class women who had such intervention was less than for working-class women by more than 10 percent for the issues of labor and delivery medication, lithotomy position, delivery room birth, IV attached during labor, and forceps or vacuum suction. The middle-class women were more able to be active participants in the birth: they controlled the contractions with breathing techniques, pushed the baby out themselves, and held the baby as soon as it was born. Ninety-eight percent of them had a partner during labor and 94 percent had a partner present at the birth.

Parity was relevant to the birth experience in the same way within each group: primipara women had births which involved more extensive medical intervention. Preparation for childbirth influenced only the use of medication and the presence of a support person—and these only among working-class women. Therefore, preparation cannot explain the class differences (Nelson, 1982).

The difference between the births of middle-class and working-class clients corresponded roughly to what the women actually selected during pregnancy—although not, perhaps, to the motives behind the choices, in the case of working-class women. Working-class women wanted speedy access to the product. They didn't necessarily get speed, nor did they get the baby immediately. The middle-class concern with the entire process might have ensured a speedier labor and delivery and more immediate access to the baby (Doering *et al.*, 1980; Huttel *et al.* 1972; Zax *et al.*, 1975).

Neither working-class nor middle-class women had all their choices met during labor and delivery. The extent to which client choices about shaves, labor medication, episiotomies, watching the birth, and holding the baby were respected was about the same for both groups. But the two groups differed in the extent to which other choices were respected. More middle-class women than working-class women had their wishes met with respect to enemas, delivery medications, and the presence of a support person during labor and delivery. More working-class women had their way with respect to fetal monitoring, artificial rupture of membranes, and the delivery site. The data do not indicate that doctors impose their will on working-class clients more frequently than they do on middle-class clients; nor do the data indicate that working-class clients are less effective in stating what they want than are middle-class clients. Both working-class and middle-class models of birth conflict with hospital protocol. Women in neither group were entirely successful in getting their way.

Post-Partum Evaluations

In the second questionnaire after birth, I asked a final question about their impressions of the experience: "If you were to tell a woman who had never given birth what it was like, what would you say?" Both working-class and middle-class women indicated that birth was worthwhile. But working-class women said that it was worthwhile in spite of the pain because you have a baby at the end: "I think it's worth it, a day of pain for all this." At the same time, many working-class women said that they forgot about the pain quickly. They said they wouldn't frighten other women, the way they had been frightened, with predictions of pain, and that women shouldn't believe all the terrible things they heard about giving birth.

Middle-class women had a very different set of ideas about what it was appropriate to tell a woman who had never before given birth. First, they would focus on the process as an experience itself: "a high and painful experience"; "a fantastic experience." Second, they would talk about the work involved: "It's hard work; preparation is important"; and "Now I know why it is called labor. You really have to work." Third, they would be more likely to give details. Indirectly, they suggested that they had been told positive things about birth by their peers.: "It's as good an experience as I had been told it was"; "It was even better than my friends said it would be."

These responses reveal a further basic difference between the two groups of women. The working-class women did not value the process of birth itself. They almost never used the word "experience." They focused, instead, on the product—the baby. Labor and delivery were something to be endured to get the product. Middle-class women valued the process as well as the product. They felt fortunate to be able to enjoy the experience of birth en route to motherhood.

DISCUSSION

There are at least two possible explanations for the different attitudes of middle-class and working-class women towards birth, and for their different experiences during the birth process.

One possibility is that working-class women are simply more inhibited by the context in which birth occurs. They don't become interested in the birth process because they don't think that they can determine what is going to happen to them. Caught between medical experts and nature, they see little room for individual initiative. This is clearly an element of working-class disaffection with birth. But working-class feelings of impotence cannot entirely explain the particular nature of the choices that some working-class women make, particularly since these choices are *not* congruent with those of the professionals who manage their care.

A second possibility is that the greater interest of middle-class women in childbirth issues derives from greater knowledge. Not only are middle-class women more likely to prepare themselves for childbirth through reading and attending classes, but they draw on different kinds of information.[6] But these facts cannot entirely explain different attitudes either: the middle-class women hold their attitudes regardless of whether they have educated themselves about child-

6. Working-class women were far more likely than middle-class women to say that they relied on mothers and other relatives to provide them with information about childbirth. The middle-class rejection of their mothers as a source of information about childbirth suggests that they were aware that the ideology about childbirth has changed since they themselves were born and that their mothers' retelling of a painful—but ultimately anesthetized—birth offered nothing to them. Working-class women felt that more information could be obtained from their mothers, indicating that they had not recognized the changes that have occurred since the time when they were born.

birth; the same is not true among working-class women: those who are prepared frequently hold different attitudes from those who are not. In any case, the search for information is probably the result—not the cause—of interest in childbirth.

Why, then, do working-class women remain uninterested in learning about childbirth? Why do they reject the middle-class ideology which not only stresses the value of information but holds as its goal a client-structured, "natural" childbirth? The answer lies in the contexts in which each group of women gives birth and in the fact that the movements that have created the middle-class model of the birth experience do not speak to both contexts equally.

The working-class women in our study started their families at a younger age, and at every subsequent age level had more children living in the home than did the middle-class women. The working-class women were more likely to have accidental pregnancies. Furthermore, when they chose to become pregnant, a small number of factors were taken into account, and these factors concerned only the family structure. Middle-class women were more likely to plan their pregnancies, and the total number of children they want to have. As with working-class women, the ideal family structure was central to their thinking, but this factor was diluted by self concerns (work and career, child care facilities) and world concerns (overpopulation). Working-class women, then, had their children earlier, had more of them, and frequently had them without planning. They also had fewer material resources with which to raise these children.

The movements that created the "middle-class" model of birth experience do not clearly address the working-class context of childbirth. In the early 1960s, middle-class women were not very interested in childbirth; in the early 1970s only elite portions of the middle-class took childbirth classes or attempted to define birth independently of the prevailing, high-technology, medical model. The change came with the convergence of (and occasional tension among) four social movements: (1) the natural childbirth movement; (2) feminism; (3) consumerism; and (4) "back to nature" romanticism.

Advocates of natural childbirth initially glorified it as a step toward motherhood. The movement appealed largely to middle-class women who were eager to be more active participants in this important event but who continued to accept professional (as distinct from technological) control over it. The feminist movement, on the other hand, told many of these same women that it was time to reject the authority of men in specifically female experiences, in order to gain personal control of their lives and their bodies. Feminists also rejected the notion that childbirth was primarily important as a step towards motherhood. Consumerism advocated questioning attitudes towards any and all prevalent medical practices. And the "back to nature" movement advocated rejecting modern technology and returning to great-grandmother's way. None of these movements speaks to working-class women.

Natural childbirth preparation requires time, money, and a willing alliance with professionals. Within the feminist movement, the focus on middle-class concerns of access to professional jobs and consciousness-raising has alienated many working-class women who face employment problems of a very different cast and who are less able to live independent of their husbands' paycheck. Moreover, an "educated" contempt for professionals is easier for those who live among them than it is for those who may have to submit to experts in a wider range of life experiences. The failure of middle-class feminists to make contact with working-class concerns has been noted frequently: this analysis merely points to an additional consequence of this failure.

Consumerism depends on a steady income, mobility, and time. Middle-class women can afford to shop around for goods and services. But working-class women pay more out of necessity, not out of choice (Caplovitz, 1967); clinic clients see the doctor assigned to them. And a rejection of technology is the luxury of those who have already benefitted from it. That class of women who have always had access to the most sophisticated medical technology may make the decision to reject some aspects of that class privilege; those who have not yet consistently received these benefits may not be ready to abandon them.

In sum, I argue that the middle-class model of childbirth has its roots in social movements which do not have immediate relevance for working-class women. The model is also predicated of the idea of choice, the idea that one can take control of one's life and one's body (e.g., Boston Women's Health Book Collective, 1976). Working-class women have fewer opportunities for making choices; even pregnancy often appears to be outside their control.

The emerging model of hospital birth is, in fact, closer to the middle-class model than it is to the working-class model. Doctors have been greatly influenced by the criticisms of their most vocal clients and, in "progressive" hospitals, are making a conscious effort to "humanize" birth. Some of these changes may converge with working-class goals. But reducing the frequency with which medication is administered is foreign to working-class concerns and may seem like a threat to a woman (of any social class) who is unprepared to do without medication. In fact, doctors may use the middle-class model to force working-class compliance: one woman we interviewed said she was told by a doctor that if she "didn't stop yelling he would make [her] go natural."

Each of the two models of childbirth makes sense within the context of the lives its adopters lead. Each model confers benefits on the women who adhere to it. Each model may also have drawbacks. The middle-class model mystifies childbirth. Accepting it produces a sense of guilt and personal failure in women whose births fail to conform to its high standards. The working-class model engenders a dependence on potentially harmful medication and creates an anxiety which can prolong and complicate labor.

This kind of evaluation brings me back onto thin ice. My reading and research suggest that a single model of childbirth has too often been held up for all women. Initially, doctors defined the experience for all women. Then one group of women began speaking for all women. But women are not a single, undifferentiated category. Childbirth is a biological experience mediated by class position. We have to learn more about what women at different locations in the social structure want for themselves rather than pass judgment on what they do. If changes are to come in either working-class or middle-class birth styles, they must come from the women themselves—not from one group of women speaking for another, or from doctors dictating to all of them.

REFERENCES

Boston Women's Health Book Collective
1976 Our Bodies, Ourselves. New York: Simon and Schuster.

Caplovitz, David
1967 The Poor Pay More. New York: The Free Press.

Comaroff, Jean
1977 "Conflicting paradigms of pregnancy: Managing ambiguity in antenatal encounters." Pp. 115-134 in Alan Davis and Gordon Horobin (eds.), Medical Encounters: The Experience of Illness and Treatment. London: Croom, Helm Ltd.

Danziger, Sandra Klein
1978 "The uses of expertise in doctor-patient encounters during pregnancy." Social Science and Medicine 12(5A):359-367.

Doering, Susan G., and Doris R. Entwisle
1975 "Preparation during pregnancy and ability to cope with labor and delivery." American Journal of Orthopsychiatry 45(5):825-837.

Doering, Susan G., Doris R. Entwisle, and Daniel Quinlan
1980 "Modeling the quality of women's birth experience." Journal of Health and Social Behavior 21(1):12-21.

Ehrenreich, Barbara, and Deidre English
1973 Witches, Midwives and Nurses. Old Westbury, N.Y.: Feminist Press.

Enkin, N. W., S. L. Smith, S. W. Dermer, and J. D. Emmett
1972 "An adequately controlled study of the effectiveness of PPM training." Pp. 62-67 in Norman Morris (ed.)., Psychosomatic Medicine in Obstetrics and Gynecology. New York: S. Karger.

Gaziano, Emanuel P., Marlene Garvis, and Elaine Levine
1979 "An evaluation of childbirth education for the clinic patient." Birth and the Family Journal 6 (Summer):89-94.

Haire, Doris
1978 "The cultural warping of childbirth." Pp. 185-201 in John Ehrenreich (ed.), The Cultural Crisis of Modern Medicine. New York: The Monthly Review Press.

Hart, Nicky
1977 "Parenthood and patienthood." Pp. 98-114 in Alan Davis and Gordon Horobin (eds.), Medical Encounters: The Experience of Illness and Treatment. London: Croom, Helm Ltd.

Hubert, Jane
1974 "Beliefs and reality: Social factors in pregnancy and childbirth." Pp. 37-51 in Martin Richards (ed.), The Integration of a Child Into a Social World. New York: Cambridge University Press.

Huttel, F. A., I. Mitchell, W. M. Fischer, and A. E. Meyer
1972 "A quantitative evaluation of psychoprophylaxis in childbirth." Journal of Psychosomatic Research 16(2):81-92.

Ilsley, Raymond
1967 "The sociological study of reproduction and its outcome." Pp. 45-141 in Stephen A. Richardson and Alan F. Guttmacher (eds.), Childbearing—Its Social and Psychological Aspects. Baltimore: Williams and Wilkins.

Kobrin, Frances E.
1966 "The American midwife controversy: A crisis of professionalization." Bulletin of the History of Medicine 40 (July-August):350-363.

McGough, Helen, and Margaret K. Nelson
1981 "An evaluation of obstetric services at Medical Center Hospital of Vermont." Unpublished report, Department of Obstetrics and Gynecology, Medical Center Hospital of Vermont, Burlington.

Nash, Anedith, and Jeffrey E. Nash
1979 "Conflicting interpretations of childbirth: The medical and natural perspectives." Urban Life 7(4):493-511.

Nelson, Margaret K.
1982 "The impact of childbirth classes on women of different social classes." Journal of Health and Social Behavior 23(4):339-352.

Norr, Kathleen L., Carolyn R. Block, Allan Charles, Suzanne Meyering, and Ellen Meyers
1977 "Explaining pain and enjoyment in childbirth." Journal of Health and Social Behavior 18 (September):260-275.

Oakley, Ann
1979 "A case of maternity: Paradigms of women as maternity cases." Signs: A Journal of Women in Culture and Society 4(4):607-631.

Shaw, Nancy Stoller
1974 Forced Labor. New York: Permagon Press.

Stewart, Mary, and Pat Erickson
1977 "The sociology of birth: A critical assessment of theory and research." Social Sciences Journal 14 (April):33-47.

Wertz, Richard W., and Dorothy C. Wertz
1977 Lying-In: A History of Childbirth in America. New York: Schocken Books.

Zax, Melvin, Arnold J. Sameroff, and Janet E. Farnum
1975 "Childbirth education, maternal attitudes, and delivery." American Journal of Obstetrics and Gynecology 123(2):185-190.

Zimmerman-Tansella, Ch., G. Dolcetta, V. Assini, G. Zacche, P. Bertagni, R. Siani, and M. Tansella
1979 "Preparation courses for childbirth in Primipara: A comparison." Journal of Psychosomatic Research 23(4):227-233.

Acknowledgments

Ballantyne, J.W. "The 'Byrth of Mankynde.'" *The Journal of Obstetrics and Gynæcology of the British Empire* Part I, Vol.10, No.4 (1906): 297–325; Part II, Vol.12, No.3 (1907): 175–94; Part III, Vol.12, No.4 (1907): 255–74.

Atkinson, Colin B., and William P. Stoneman. "'These Griping Greefes and Pinching Pangs': Attitudes to Childbirth in Thomas Bentley's *The Monument of Matrones* (1582)." *Sixteenth Century Journal* 21, No.2 (1990): 193–203. Reprinted with the permission of the *Sixteenth Century Journal*.

Harley, David. "Historians as Demonologists: The Myth of the Midwife-Witch." *Social History of Medicine* 3 (April 1990): 1–26. Reprinted with the permission of Oxford University Press.

Gardiner, Anne Barbeau. "Elizabeth Cellier in 1688 on Envious Doctors and Heroic Midwives Ancient and Modern." *Eighteenth-Century Life* 14 (February 1990): 24–34. Reprinted with the permission of the Johns Hopkins University Press.

Wilson, Adrian. "The Perils of Early Modern Procreation: Childbirth With or Without Fear?" *British Journal for Eighteenth-Century Studies* 16 (Spring 1993): 1–19. Reprinted with the permission of the Voltaire Foundation, Taylor Institution.

Schnorrenberg, Barbara Brandon. "Is Childbirth Any Place for a Woman? The Decline of Midwifery in Eighteenth-Century England." *Studies in Eighteenth-Century Culture* 10 (1981): 393–408. Reprinted with the permission of the American Society for Eighteenth Century Studies.

McGregor, Deborah Kuhn. "'Childbirth-Travells' and 'Spiritual Estates': Anne Hutchinson and Colonial Boston, 1634–1638." *Caduceus* 5 (1989): 1–32. Reprinted with the permission of Southern Illinois University School of Medicine.

Ulrich, Laurel Thatcher. "'The Living Mother of a Living Child': Midwifery and Mortality in Post-Revolutionary New England." *William and Mary Quarterly* 66, No.1 (1989): 27–48.

Leavitt, Judith Walzer. "'Science' Enters the Birthing Room: Obstetrics in America since the Eighteenth Century." *Journal of American History* 70, No.2 (1983): 281–304. Reprinted with the permission of the *Journal of American History*.

Kass, Amalie M. "'Called to Her at Three O'Clock AM': Obstetrical Practice in Physician Case Notes." *Journal of the History of Medicine and Allied Sciences* 50 (1995): 194–229. Reprinted with the permission of the *Journal of the History of Medicine and Allied Sciences*.

Gunn, John C. "Midwifery" and "Directions for Midwives." In *Gunn's New Domestic Physician: or, Home Book of Health* (Cincinnati: Moore, Wilstach, Keys & Co., 1862): 414–27.

Stowe, Steven M. "Obstetrics and the Work of Doctoring in the Mid-Nineteenth-Century American South." *Bulletin of the History of Medicine* 64 (Winter 1990): 540–66. Reprinted with the permission of Johns Hopkins University Press.

King, Charles R. "The Woman's Experience of Childbirth on the Western Frontier." *Journal of the West* 29, No.1 (1990): 76–84. Copyright 1990 by the Journal of the West, Inc. Reprinted with the permission of the *Journal of the West*, P.O. Box 1009, Manhattan, KS 66505–1009, USA.

Siddall, A. Clair. "Bloodletting in American Obstetric Practice, 1800–1945." *Bulletin of the History of Medicine* 54 (1980): 101–10. Reprinted with the permission of Johns Hopkins University Press.

Borst, Charlotte G. "The Training and Practice of Midwives: A Wisconsin Study." *Bulletin of the History of Medicine* 62 (1988): 606–27. Reprinted with the permission of Johns Hopkins University Press.

Declercq, Eugene, and Richard Lacroix. "The Immigrant Midwives of Lawrence: The Conflict between Law and Culture in Early Twentieth-Century Massachusetts" *Bulletin of the History of Medicine* 59 (1985): 232–46. Reprinted with the permission of Johns Hopkins University Press.

Dye, Nancy Schrom. "Mary Breckinridge, the Frontier Nursing Service and the Introduction of Nurse-Midwifery in the United States." *Bulletin of the History of Medicine* 57 (1983): 485–507. Reprinted with the permission of Johns Hopkins University Press.

Colby, Margaret E. "Pregnancy and Parturition from a Woman's

Point of View." *Woman's Medical Journal* 4 (December 1902): 272–75.

Martin, Emily. "Pregnancy, Labor and Body Image in the United States." *Social Science Medicine* 19 (1984): 1201–06. Reprinted with the permission of Pergamon Press, Ltd.

Litoff, Judy Barrett. "Forgotten Women: American Midwives at the Turn of the Twentieth Century." *Historian* 40 (February 1987): 235–51. Reprinted with the permission of the International Honor Society in History.

Nelson, Margaret K. "Working-Class Women, Middle-Class Women, and Models of Childbirth." *Social Problems* 30 (1983): 284–97. Reprinted with the permission of the University of California Press. Copyright (1983) by the Society for the Study of Social Problems.

EDITORS

Series Editor

Philip K. Wilson, MA, Ph.D., is an assistant professor of the history of science at Truman State University (formerly Northeast Missouri State University) in Kirksville, Missouri. After receiving his undergraduate degree in human biology from the University of Kansas, he pursued work towards an MA in medical history at the William H. Welch Institute for the History of Medicine at The Johns Hopkins School of Medicine and received his Ph.D. in the history of medicine from the University of London. He has held postdoctoral positions at the University of Hawaii-Manoa and Yale University School of Medicine before settling in Missouri.

Wilson has received scholarly support including a Logan Clendening Summer Fellowship, an Owsei Temkin Scholarship, a Folger Shakespeare Library Fellowship, a Wellcome Trust Research Scholarship, and grants from the Hawaii and Missouri Committees for the Humanities for medical and science history projects. He was a founding member of the Hawaii Society for the History of Medicine and Public Health. Wilson has contributed chapters to volumes including *The Popularization of Medicine 1650–1850* (Routledge), *Medicine in the Enlightenment* (Rodopi), and *The Secret Malady: Venereal Disease in Eighteenth-Century Britain and France* (University Press of Kentucky), articles in the *Annals of Science,* the *London Journal,* and the *Journal of the Royal Society of Medicine,* and is a regular contributor of medical and science history entries to many dictionaries and encyclopedias. Currently, Wilson is pursuing research on women's diseases, osteopathy, and eugenics in Kirksville, Missouri, where he lives with his wife, Janice, and son, James.

Assistant Editors

Ann Dally, MA, MD, received her Master's degree from Oxford University, having been an exhibitioner in modern history at Somerville College. She then studied medicine at St. Thomas' Hospital, London, qualifying in 1953. After some years of general medical practice, she specialized in psychiatry, a specialty she practiced until her retirement in 1994. Meanwhile she pursued her interests in the history of medicine, receiving her doctorate in that subject in 1993. The book based on her doctoral thesis, *Fantasy Surgery, 1880–1930,* will shortly be published as part of the Wellcome Institute for the History of Medicine (London) series. Her most recent book, *Women Under the Knife. A History of Surgery* (Routledge), follows a long publishing history of books including *The Morbid Streak, Why Women Fail, Mothers: Their Power and Influence, Inventing Motherhood: The Consequences of an Ideal,* and a book of memoirs, *A Doctor's Story*. Currently a Research Fellow at the Wellcome Institute for the History of Medicine (London), she lives with her husband Philip Egerton in West Sussex, England and has four children and seven grandchildren.

Charles R. King, MD, MA, is a professor of obstetrics and Gynecology at the Medical College of Ohio. He received his BA from Kansas State University, an MD from the University of Kansas, and has completed post graduate medical training at the University of Kansas and the University of Oregon. He has since received an MA in medical history from the University of Kansas. King has been the recipient of Rockefeller Foundation, National Endowment for the Humanities, American College of Obstetricians and Gynecologists-Ortho, and Newberry Library Fellowships for projects in medical history. He is the author of numerous publications regarding women's health, including articles in the *Bulletin of the History of Medicine, Kansas History,* and the *Great Plains Quarterly,* and has recently completed *Child Health in America* (Twain). He currently lives with his wife, Lynn, in Temperance, Michigan.